Mixing Process Technology

Industrial mixing processes often present multiple optimization challenges to producing desirable products. The resulting processes must be cost effective, "first-time right," and frequently, the designated most-effective technology for the global manufacture of specific products. *Mixing Process Technology: A Guide to Industrial Applications* shares the authors' extensive knowledge of mixing research and industrial practice. It features 20 industrial mixing chapters that are purposely light on mixing fundamentals, while heavy on practical mixing applications for practical process design and manufacturing. This text serves as an applied guide to industrial mixing for practitioners who want brief explanations of mixing concepts with real-life examples and software to help perform associated design calculations. This book also:

- Offers side-by-side discussion of mixing systems including impellers, static mixers and rotor-stators, as offered by several major manufacturers
- Describes the authors' innovative mixer designs to meet manufacturing needs
- Includes a chapter by a mixer manufacturing representative describing design, sizing, and expensing of industrial mixers
- Presents a chapter by a mixing equipment manufacturing leader that explains mechanical design considerations in clear terms
- Contains a chapter on emerging mixing technologies, including mixing via resonant acoustics and controlled cavitation
- Discusses computational fluid dynamics in mixing with multiple practical examples by a contributing author from a leading pharmaceutical company
- Includes Excel-based mixing worksheets throughout book examples and Excel-based input/output (mixit-io) interface hosted on the publisher's website

This book is aimed at chemical and process engineers as well as students seeking to understand industrial mixing technology.

Kishore Kar dedicated 36 years of service to The Dow Chemical Company and retired as an R&D Fellow. Currently he is serving as a chemical process technology consultant. Kishore holds 18 US patents and has authored more than 35 publications, including two chapters in the *Encyclopedia of Chemical Processing* published by Taylor & Francis Group. He is the recipient of 2025 NAMF Award for excellence and sustained contributions to mixing research and practice. During his Dow career, Kishore led the Global Fluid Mechanics and Mixing Discipline. Kishore received his MS and PhD in mechanical engineering (fluid, thermal, and aerospace sciences) from Case Western Reserve University, Cleveland, Ohio, USA.

Richard Cope received his PhD in chemical engineering from Brigham Young University, Provo, Utah, USA, and spent more than 19 years in chemicals R&D (Dow) and more than 7 years in biopharma R&D (Eli Lilly and Alkermes). In 2017, he conceived/co-designed a mixing experiment for the International Space Station. In 2020, he joined SPX FLOW, the world's largest mixing solutions provider. He has co-authored 19 papers/encyclopedia chapters and 7 US or world patents for mixing technologies. Richard was treasurer of the North American Mixing Forum (NAMF) from 2012 to 2024 and recipient of the NAMF Mixing Award in 2020.

Juergen Lueske enjoyed 35 years with The Dow Chemical Company. His focus in the Process Engineering group was project planning and process development, with an emphasis on modeling and simulation for large projects. In the Fluid Mechanics and Mixing group, his focus was calculations and modeling of gas–liquid contactors, especially fermenters and bioreactors. He has authored several papers and co-authored two encyclopedia chapters. Juergen received his MS degree in mechanical and process engineering from Leibniz University of Hannover, Germany.

Mixing Process Technology

A Guide to Industrial Applications

Kishore Kar, Richard Cope, and Juergen Lueske

CRC Press
Taylor & Francis Group
Boca Raton London

CRC Press is an imprint of the
Taylor & Francis Group, an **informa** business

Designed cover image: Shutterstock; the cover photograph is courtesy of Dr. Luciano Piras, Milan, Italy.

First edition published 2025
by CRC Press
2385 NW Executive Center Drive, Suite 320, Boca Raton FL 33431

and by CRC Press
4 Park Square, Milton Park, Abingdon, Oxon, OX14 4RN

CRC Press is an imprint of Taylor & Francis Group, LLC

ISBN: 978-1-032-99861-9 (hbk)
ISBN: 978-1-032-99863-3 (pbk)
ISBN: 978-1-003-60639-0 (ebk)

DOI: 10.1201/9781003606390

For Product Safety Concerns and Information please contact our EU representative: GPSR@taylorandfrancis.com
Taylor & Francis Verlag GmbH, Kaufingerstraße 24, 80331 München, Germany

Disclaimer

The contents of this book, including the user-interfaced worksheets (mixit-io), have been prepared in good faith and carefully reviewed and edited. The authors and publisher, however, cannot be held liable for errors of any sort in these chapters. Furthermore, because the authors have no means of checking the reliability of all data presented in published literature, but can only examine it for suitability for the intended purpose herein, this information cannot be warranted. Also, because the authors cannot vouch for the experience or technical capability of those who use the information, or the suitability of the information for each user's specific purpose, use of the contents must be at the user's discretion.

The authors are not biased towards any commercial entity or its products as discussed in the book. The recommendations and constructive suggestions, if given, are solely based on author experience with the particular commercial product in their specific process and must not be construed as discriminatory. Excerpts in the book are diligently referenced to their origins. Any omissions are unintentional on the author's part.

Contents

Foreword

Mixing technology underpins countless critical process operations across a wide range of sectors, including chemical, petrochemical, agriculture, food, pharmaceutical, personal care, water treatment, and mining. From ensuring the homogeneity of life-saving pharmaceuticals to optimizing yields in large-scale chemical reactors, effective mixing is not merely a step in the process – it often is a critical factor in dictating product quality, operational safety, and ultimately, economic viability. For this reason, I am honored to provide the Foreword for this book, Mixing Process Technology, which I believe provides an excellent foundation, and more importantly a practical guide, in the design, operation, and optimization of mixing processes across diverse industries and applications.

The book – Mixing Process Technology – benefits from decades spent in industrial practice, a collective journey of the authors of learning, problem-solving, and delivery of industrially relevant solutions. The insights shared within these pages are rooted in the extensive careers of the authors, Dr. Kishore Kar, a former R&D Fellow at DOW, Dr. Richard Cope, with a career in chemicals (DOW) and biopharma (Eli Lilly and Alkermes), and Juergen Lueske, a former R&D and process engineering expert at DOW, who, combined, have over 97 years of industrial experience in developing and commercializing mixing solutions. Through the incorporation of additional input from several expert contributing authors across technology, equipment design/manufacturing, and sales, topics including specialty mixing technologies, such as cavitation and resonant sonics, along with equipment design and cost/economics considerations are enhanced. This reenforces the critical goal of delivering mixing technology solutions that are practical, scalable, and cost effective. The combined experience of the authors teaches that while the theoretical underpinnings of fluid dynamics and transport phenomena are essential, and are inherent in the book, the true mastery of process mixing lies in the practical application of this knowledge, developed through experience and collaboration with fellow experts and equipment manufacturers.

The book highlights one of the key challenges in today's environment, that simply achieving the desired mixing outcome is no longer sufficient. Practicing engineers are tasked with doing so in the most cost-effective manner, minimizing energy consumption, reducing waste, and maximizing production efficiency. This book champions that philosophy, providing insights into designing simple state-of-the-art mixing systems that meet process requirements without unnecessary complexity or expense. The authors emphasize the need for critical thinking, guiding engineers to troubleshoot effectively and determine if mixing is indeed the limiting factor, and they address the detrimental effects of overmixing, a common pitfall that can lead to wasted energy and compromised product integrity. In addition, the authors introduce mixit-io, a user-friendly, MS Excel®-based software tool that complements the principles discussed throughout these pages. This knowledge-based expert system is designed to empower practicing engineers to design new mixing systems and optimize existing ones.

I believe this book will be an invaluable resource for students, practicing process engineers, R&D scientists, and technical professionals who can leverage the comprehensive information and enabling tools provided to enable and accelerate their efforts.

BOB MAUGHON

Bob Maughon is currently president, Circular Logic Executive Advisors, LLC. He retired in 2024 as the executive vice president, chief technology & sustainability officer at SABIC, and prior to that served as an R&D vice president at DOW, with over 27 years of experience in the Petrochemicals sector. He received his PhD in Chemistry from the California Institute of Technology in 1998.

Preface

In this book, the authors disseminate their extensive knowledge of mixing research and industrial practice, without divulging any proprietary information. This knowledge was accumulated from hands-on mixing experience, which includes successful design and construction of pilot and production plants across the world. They also share mixing principles assimilated from books, technical articles, consortiums, conferences, and extensive communications with fellow researchers and engineers, including many employed by commercial mixing system manufacturers.

The authors succinctly present 20 industrial mixing chapters that are purposely light on mixing fundamentals and heavy on practical mixing applications for practicing process design and manufacturing engineers. This work is also a suitable textbook for students seeking to understand industrial mixing technology. It contains numerous citations to and excerpts from mixing literature. Excel-based input/output (mixit-io) support material is available on the Routledge website (www.routledge.com/9781032998619). Readers are advised to pursue what is most relevant to them, and thus, further their practical and applied mixing knowledge in the manner most meaningful to their interests and needs.

Generally, industrial mixing processes present multiple optimization challenges while producing desirable products. The resulting processes must be cost effective, "first-time right," and frequently, the designated most-effective technology (MET) for the global manufacture of specific products. Although the energy-and-maintenance-related running costs of a production process are typically important, safety (operational and environmental) and capital cost are likely more dominant in accepting or rejecting a given process design.

Admittedly, the book contains a few references to specific mixer manufacturers and consortiums, but this is only because such have long been familiar to the authors and not because only cited mixer sources can be trusted. Readers should not construe such citations as authors bias. If any appropriate references are missing, such is unintentional and should be excused awaiting correction in future book editions.

Finally, the authors express their gratitude to colleagues at The Dow Chemical Company (Dow Inc.) and to the global mixing community. Additionally, the principal authors wish to thank the book's contributing authors for their invaluable contributions. In deference to stronger feelings of kinship and love, however, this book is dedicated to the authors' beloved parents, partners, and family members.

With compliments and best wishes,

Kishore, Richard, and Juergen

Introduction

Process mixing is vital to many process industries, including chemical, petrochemical, agriculture, food, pharmaceutical, personal care, water treatment, and mining. It is frequently coupled with chemical reaction, heat transfer, mass transfer, and other phenomena that are critical to required product properties (e.g., homogeneity, selectivity, yield, quality) and operational safety. Mixing is particularly critical to processes involving fast reactions that are otherwise limited by micromixing, diffusion, mass transfer, and/or heat transfer. Mixing systems operate in batch, semi-batch, and continuous modes, as illustrated below.

Batch:	Stirred Tank Reactors (STR)
	Impeller Agitated Systems
	Jet Mixed Systems
	Stand-Alone Gas Sparged Systems
Semi-batch:	Continuous Stirred Tank Reactors (CSTR) with Cascading Inlet(s) and Outlet(s)
Continuous:	Inline Mixing (ILM)
	Inline (Dynamic) Agitation Systems
	Tee Mixers
	Motionless or Static Mixers
	Rotor-Stator Dynamic Mixers

The many mixing technologies presented in this book are an ensemble of those utilized and developed by expert industrial practitioners whose knowledge was obtained through decades of personal experience, collaboration with mixing equipment manufacturers, and an ever-ongoing review of relevant books and publications in the mixing arena. A chief focus is designing simple state-of-the-art mixing systems that meet a process requirement in the most cost-effective manner (often called "most-effective technology," or MET). Before delving into such an expensive endeavor as the redesign of the mixing system, it is essential for practicing engineers to critically analyze and troubleshoot the process issues that can establish mixing as the probable process bottleneck. Overmixing wastes energy and often produces other detriments like cell rupture in bioreactors, fragmentation in crystallizers, and unwanted byproducts in chemical reactors.

The mixit-io software complementing this book is a knowledge-based expert system for mixing applications. This software helps design new mixing systems and improve production efficiencies (e.g., desirable product yield) or reduce byproducts and/or energy consumption in existing systems. Mixit-io is an intentionally uncomplicated MS Excel® spreadsheet-based, user-friendly mixing software for practicing process engineers. The pseudo-dialogue that it establishes with the user leads to an appropriate mixing system design for the given process. The design bases are emphasized in the book, and thus, clearly accessible to the users. Mention of commercial products or tradenames is not an endorsement, nor is the absence of such a bias against those not so mentioned.

In summary, the *Mixing Process Technology* book and its supplemental mixit-io software address some of the most effective mixing technologies for cost-effective production in a broad spectrum of process industries.

Author

Kishore Kar dedicated 36 years of service to The Dow Chemical Company and retired as an R&D Fellow. Currently, he is serving as a chemical process technology consultant. He holds 18 US patents and has authored more than 35 publications including two chapters in the *Encyclopedia of Chemical Processing* published by Taylor & Francis Group. He is the recipient of 2025 NAMF Award for excellence and sustained contributions to mixing research and practice. During his Dow career Kishore led the Global Fluid Mechanics and Mixing Discipline. Kishore received his MS and PhD in mechanical and aerospace engineering (fluid, thermal, and aerospace sciences) from Case Western Reserve University, Cleveland, USA.

Richard F. Cope received his PhD in chemical engineering from Brigham Young University and spent 19+ years in chemicals R&D (Dow) and 7+ years in biopharma R&D (Eli Lilly, and Alkermes). In 2017 he conceived/co-designed a mixing experiment for the International Space Station. In 2020 he joined SPX FLOW, the world's largest mixing solutions provider. He has co-authored 19 papers/encyclopedia chapters and seven US or world patents for mixing technologies. He was treasurer of North American Mixing Forum (NAMF) from 2012 to 2024 and recipient of the NAMF Mixing Award in 2020.

Juergen Lueske enjoyed 35 years with The Dow Chemical Company. His focus in the Process Engineering group was project planning and process development, with emphasis on modeling and simulation for large projects. In the Fluid Mechanics and Mixing group, his focus was calculations and modeling of gas-liquid contactors, especially fermenters and bioreactors. He has authored several papers and co-authored two encyclopedia chapters. Juergen received his MS degree in mechanical and process engineering from Leibniz University of Hannover, Germany.

Contributors

Zhao Yu received his PhD in Chemical Engineering from the Ohio State University, where his research was focused on multiphase flows and computational fluid dynamics. He worked in the Fluid Mechanics and Mixing group in Dow Chemical's Core R&D organization for three years before joining Eli Lilly, where he is currently the director of Engineering. Zhao has over 20 publications in peer-reviewed journals and conference proceedings. He is also a co-author of the book *Dynamics of Multiphase Flows* published by Cambridge University Press.

Douglas Mancosky has worked in cavitation technology for over 20 years and currently serves as chief science officer for Hydro Dynamics, Inc. of Rome, GA. He holds six US and international patents, has authored dozens of articles, and spoken globally about cavitation technology. At Hydro Dynamics Doug leads the company's R&D efforts and launched the company's highly successful brewery initiative. Doug received his PhD from the Institute of Paper Science and Technology at Georgia Tech in Atlanta, GA, USA.

Lawrence Farrar is the founder and owner of Resodyn Acoustic Mixers, a company which he established in 2005. Mr. Farrar holds 15 US and European patents. He has authored numerous publications over a broad range of technical subjects, which include ResonantAcoustic® mixing processes and systems. Lawrence received his BS and MS degrees in mechanical engineering from the University of California, Santa Barbara, CA.

Andrew Creathorn is vice president of Global Sales at SPX Flow Mixing Solutions. He has over 25 years in mixer manufacturing, including nearly three years as vice president/general manager of a manufacturing plant in Rochester, NY where he oversaw the mechanical design of numerous "engineered-to-order" solutions across a broad range of complex applications. He also led design and development of "standard" products currently delivering mixing solutions in multiple end markets. Andrew is a BS mechanical engineer (Teesside University, Middlesborough, England), with two mixing equipment patents.

Timothy Geiger is the proprietor and president of Mattoon & Lee Equipment, a chemical process equipment manufacturers' sales representative firm. He has over 30 years of experience in the design, sales, marketing, and troubleshooting of fluid mixing and separations equipment used in the food, pharmaceutical, automotive coatings, and industrial chemical markets. Tim holds BS and MS degrees in civil engineering from Michigan Technological University, Houghton, MI.

1 Mixing Processes and Working Media

1.1 INTRODUCTION

In nature, substances typically exist in the three distinct phases of solid, liquid and gas. The phenomenon of mixing often brings these different phases into intimate contact with each other as a working media involving separate components or a combination of components/products. Characteristics of the working media depend on the varying degrees of component miscibility, which can be differentiated as follows: (1a) a combination of miscible liquids is called a blend, while (1b) that of immiscible liquids is known as a liquid–liquid dispersion; (2a) a liquid containing dissolved solids is considered a solution, whereas (2b) a liquid with undissolved solids is known as a slurry (solids dispersed within the liquid); and (3) a gas distributed as bubbles within a liquid is considered a gas–liquid dispersion. These mixing processes are also identified by such terms as blending, dispersion, dissolution and contacting.

1.2 PROCESS IDENTIFICATION

The mixing unit operation depends on the process to be completed. Simply, the process could be categorized as shown in the following diagram.

Mixing Process

- Single phase
- Two-phase (Solid-Liquid/Gas-Liquid Dispersion)
- Multi-phase (Gas-Liquid-Solid Mixing)

Examples of common industrial applications pertinent to the above processes are briefly outlined below.

Single Phase Applications

- Simple blending
  Formulations:
  home care, health care and food products
- Blending with reaction
  Catalytic reaction products:
  polymers (polyolefins and polyols)

Two-Phase Applications

- Gas-Liquid Mixing
  Oxidation, Hydrogenation:
  (a) Fermentation, Wastewater treatment (b) H_2O_2, Vegetable oil
- Liquid-Liquid Dispersion
  Emulsion and Suspension Polymerizations:
  Latex and Epoxy products
- Solid-Liquid Dispersion
  Product formation:
  Polyvinylidene chloride (PVDC), High density polyethylene (HDPE)

Multi-Phase Applications

Oxdation, Chlorination, Hydrogenation, Stripping
Products: Crude Terephthalic Acid (CTA), Chlorinated Polyethylene (CPE)

DOI: 10.1201/9781003606390-1

Mixing applications are too numerous and diverse for all to be given detailed descriptions within this book. Thus, the aim is to familiarize process/manufacturing engineers with some common mixing processes used in chemical and allied industries, and to explain how they fit within at least one of the above categories.

1.3 PROPERTIES OF WORKING MEDIA

Practitioners must know some physical and thermal properties of the working media (i.e., viscosity, density, surface tension, interfacial tension, specific heat, thermal conductivity, etc.) to understand its flow, and its mass and heat transfer characteristics. The determination of these properties for multi-phase working media can be very complex. Since multiple references and physical property data banks address this complexity with experimental correlations, it is anticipated that the reader can access these references as needed. Some homogeneous and heterogeneous working media are briefly discussed in the next section.

TABLE 1.1
Classification of Complex Mixtures from the Viewpoint of Their Flow Behavior[1]

<table>
<tr><td rowspan="2">Single Phase</td><td colspan="5">Multi-Phase (Gas-Liquid, Liquid-Liquid, Gas-Solid, Liquid-Solid)</td></tr>
<tr><td colspan="2">Fine Dispersion</td><td>Coarse Dispersion</td><td>Macro-Mixed</td><td>Stratified</td></tr>
<tr><td rowspan="2">True Homogeneous</td><td colspan="2">Pseudo-homogeneous</td><td colspan="3">Heterogeneous</td></tr>
<tr><td>Laminar
Turbulent</td><td>Turbulent only</td><td colspan="3"></td></tr>
<tr><td colspan="3">← Flow Behavior as Single Phase →</td><td colspan="3">← Flow Behavior as Multi-Phase →</td></tr>
</table>

1.4 PHYSICAL PROPERTIES

1.4.1 Viscosity and Fluid Rheology

Time-independent viscous fluids could be categorized as simply Newtonian or non-Newtonian. Newtonian fluids follow Newton's hypothesis represented by the following rheological equation:

$$\tau = \mu . \dot{\gamma} \tag{1.1}$$

where, τ = shear stress [Pa], μ = dynamic viscosity [Pa-s], $\dot{\gamma}$ = shear rate [s^{-1}]

This constitutive relationship means shear stress is directly proportional to the shear rate and the constant of proportionality is known as viscosity. In other words, viscosity is independent of shear rate. Some examples of Newtonian fluids are air, water, glycerin, glycol and mineral oil, etc.

Fluids that exhibit Newtonian behavior after exceeding certain yield stress (τ_y) are called Bingham fluids. The constitutive equation for a Bingham fluid is

$$\tau - \tau_y = \mu . \dot{\gamma} \tag{1.2}$$

The rheology of many industrial fluids differs significantly from this simple Newtonian behavior. Such are identified as non-Newtonian fluids because their viscosity is a function of the shear rate. A fluid whose viscosity decreases with an increase in shear rate is called a pseudo-plastic or shear thinning fluid. One whose viscosity increases with increasing shear rate is known as a dilatant or shear thickening fluid. Examples of pseudo-plastic fluids are high molecular weight polymeric materials and their solutions, whereas dilatant fluids are mostly slurries with a high (typically, $\gtrsim$60%) solids content. Several constitutive equations are used to capture the rheological behavior of non-Newtonian fluids. One such is the Power Law model:

$$\tau = K . \dot{\gamma}^n \tag{1.3}$$

where, K = consistency index and n = flow behavior index.

$$\tau = \left(K . \dot{\gamma}^{n-1}\right)\dot{\gamma} = \mu_a . \dot{\gamma} \tag{1.4}$$

$$\mu_a = K . \dot{\gamma}^{n'} \tag{1.5}$$

where, μ_a = apparent viscosity [Pa-s].

Apparent viscosity is a function of shear rate. The value of n' (i.e., $n-1$) and K are determined from the slope and the anti-intercept of shear stress versus shear rate plotted on logarithmic coordinates, respectively.

$n < 1$ ⇨ Pseudoplastic fluid (apparent viscosity decreases with increase in shear rate)
$n = 1$ ⇨ Newtonian fluid (apparent viscosity remains unaltered with increase in shear rate)
$n > 1$ ⇨ Dilatant fluid (apparent viscosity increases with increase in shear rate)

Fluids that exhibit non-Newtonian shear-thinning behavior beyond certain yield stress (τ_y) are called yield-pseudoplastic. Gels such as Carbopol® solutions and clay-water suspensions are typical examples of yield-pseudoplastic fluids. Such fluids are represented by a modified power law equation as shown here:

$$\tau - \tau_y = K . \dot{\gamma}^n \tag{1.6}$$

The rheology of time-dependent viscous fluids can be categorized as thixotropic or rheopectic in nature. Fluids experiencing structural decay over time, under constant sustained shear, are thixotropic. Such fluid behavior is often referred to as shear degradation, typical of high-molecular-weight polymers. A rather uncommon form of time-dependent fluid behavior, known as rheopexy, causes structural reformation and increased viscosity under sustained shear. Such fluids, called rheopectic, include gypsum paste and printer ink. Figure 1.1 summarizes these fluid behaviors by plotting shear stress as a function of shear rate.

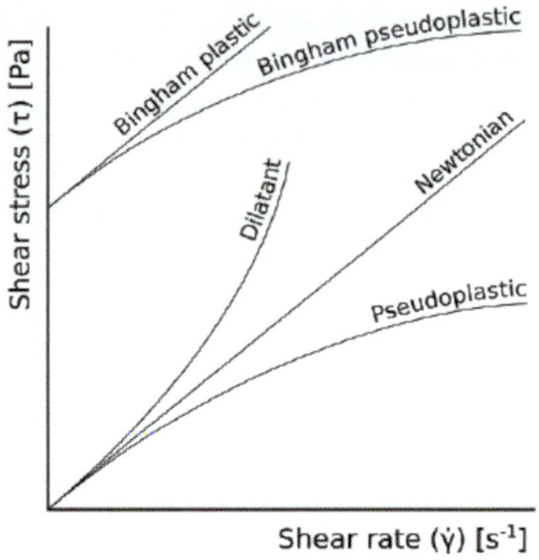

FIGURE 1.1 Shear stress and shear rate behaviors of fluids.[2]

Fluids that exhibit partial elastic recovery along with its aforementioned viscous behavior are called viscoelastic. The positive normal stress, a consequence of elastic recovery, initiates flow normal to the major shear stress, known as Weissenberg effect. In a steady rotational flow field, this fluid tends to rise up the rotating rod. This "rod climbing" effect has important repercussions for polymer processing. In the case of unsteady flow, the importance of elastic effects is determined from a parameter called the Deborah number, De, defined as:

$$\mathrm{De} = \frac{\text{Fluid Relaxation Time}}{\text{Process Time Scale or Residence Time of Fluid}}$$

for De << 1 ⇨ process fluid behaves like a viscous fluid,

De ~ 1 ⇨ process fluid behaves like an elastic solid.

1.4.2 Viscosity-Temperature Relationship

The viscosities of most liquids decrease with increase in temperature. Some polymers such as hydrophobically modified associating polyacrylamide (HAPAM) and its partially hydrolyzed analogues are exceptions to this behavior. These polymers are actually a micro-heterogeneous system where micelles behave as microdomains with a high local hydrophobia concentration. HAPAM is used in enhanced oil recovery applications as a thickener. In the case of gases, note that viscosity increases with temperature, because of increased transfer of momentum between the stationary and colliding molecules.

1.4.3 Viscosity-Pressure Relationship

Fluid viscosity increases with an increase in pressure. The piezo-viscous (viscosity-pressure) behavior is represented by the following equation and typically measured by a falling needle or falling ball viscometer.

$$\mu_p = \mu_{p_0} \cdot e^{\alpha(p-p_0)} \tag{1.7}$$

where, μ_p = viscosity (Pa-s) at absolute pressure p [Pa],

μ_{p_0} = viscosity (Pa-s) at atmospheric pressure p_o [Pa],

α = pressure-viscosity coefficient [1/Pa].

Typically, pressures encountered in chemical reactors are low enough not to measurably increase the viscosity of the working media; thus, the pressure effect is often ignored.

1.4.4 Rheology Instruments

Several viscometry devices are used to understand the complex rheological behavior of single-phase and multi-phase fluids. Rheologists understand the advantages and disadvantages of these instruments. Interpreting viscometry data is extremely critical to accurately assessing the rheological behavior of a fluid. A brief list of simple viscometers that are widely accepted in industry is given as follows. For further understanding of their principles of operation and specific applications, readers can pursue the referenced source.[3]

A Cannon-Fenske tube is used to measure the kinematic viscosity ($\nu = \mu / \rho$) of Newtonian fluids at different temperatures. Gravity causes the fluid within a reservoir bulb to drain through a capillary tube. The time required for the liquid level to change between two predetermined markers is used to determine the kinematic viscosity from a calibration chart provided by the manufacturer. The Cannon-Fenske viscometer is widely used to determine the Viscosity Index[4] (VI—viscosity-temperature behavior) of lubricants and engine oils.

A High-Shear Capillary Viscometer is a simple and inexpensive device consisting of a capillary tube, pressure measuring devices, a flow meter and a pump or pressurized chamber. It is used to determine both Newtonian and non-Newtonian rheology of fluids at reasonably high shear rates. The flow and corresponding pressure drop data across the capillary tube are used to determine the rheological behavior of the fluid (see equations 1.3–1.6). Flow in the capillary tube must be fully developed and laminar. Mixit-io SS1.1 shows how to determine simple Newtonian and non-Newtonian rheology models for a fluid from pressure drop (Δp) vs. volumetric flowrate ($\dot{Q}$) data.

Haake and Brookfield make rotating viscometers that typically rotate a cylinder, four-blade paddle, or cone inside a chamber containing the fluid. The torque experienced by the rotor is translated into the fluid viscosity. Such instruments measure viscosity of Newtonian and non-Newtonian fluids at low shear rates.

A Tapered Bearing Simulator is a special rheometer used to determine the viscosity of fluids at very high shear rates, i.e., up to approximately $1.5 \times 10^6 \text{sec}^{-1}$ and high temperatures.[5]

The Rheometric (cone and plate configuration) is a rotating device capable of measuring normal and shear stresses to describe the rheology of viscoelastic fluids.

A Falling Needle Viscometer determines viscosity from the terminal velocity of a free-falling needle (or ball) in a pressurized tube filled with the liquid whose pressure-viscosity coefficient is to be determined.

1.4.5 Viscosity of Complex Mixtures

In many industrial applications, the fluid of interest is a complex mixture, i.e., a dilute or a concentrated slurry. For a dilute slurry of spherical particles that behave like a pseudo-homogeneous media, viscosity can be determined using the following equations[1] proposed by Einstein[6] and Guth et al.[7]

$$\mu_m = \mu\left(1 + 2.5\, c_v\right),\ c_v < 2\% \tag{1.8}$$

$$\mu_m = \mu\left(1 + 2.5\, c_v + 14.1\, c_v^{\,2}\right),\ c_v < 20\% \tag{1.9}$$

where, μ_m = viscosity of the mixture,

c_v = solid concentration by volume (volume fraction)

The following equation may be used for concentrated slurry[8] with certain precautions:

$$\mu_m = \mu \left[1 + 2.5\ c_v + 10.05\ c_v{}^2 + 0.00273 \times \exp(16.6\ c_v)\right],\ c_v < 90\% \tag{1.10}$$

This correlation for concentrated suspensions does not account for the effects of (a) non-spherical particles and (b) particle-particle interactions due to the hydrophilic-lipophilic balance (HLB ζ-potential), electrostatic energy, and van der Waals energies, which are often media specific. The literature includes multiple widely used correlations for concentrated suspensions (slurries and colloids).[9,10] All heterogeneous slurries with increasing dispersed-phase volume fraction gradually become non-Newtonian.

The following example uses equation 1.10 to determine the viscosity of a 30vol.% purified terephthalic acid (PTA)/water slurry at room temperature (20°C). Let 90% of measured PTA particle diameters, i.e., d_{90}, exceed 150μm. The particles are assumed to be spherical, and their energy interactions are ignored. The viscosity of water, μ, at room temperature is 1cP, and the solid concentration by volume, c_v, is 0.3.

Slurry mixture viscosity,

$$\mu_m = 1\left[1 + 2.5 \times 0.3 + 10.05 \times (0.3)^2 + 0.00273 \times \exp(16.6 \times 0.3)\right] = 3.05\,cP.$$

The mixture viscosity, μ_m, of light emulsions containing a suspension of spherical liquid droplets is given in the following classic theoretical formula derived by Taylor:[11]

$$\mu_m = \mu\left[1 + 2.5c_v\left(\frac{\mu_p + \frac{2}{5}\mu}{\mu_p + \mu}\right)\right] \tag{1.11}$$

where, μ and μ_p are the viscosities of the continuous and dispersed phases.

This equation is useful for obtaining "order of magnitude" viscosities for emulsions containing less than 2–3 volume percent ($c_v < 0.2 - 0.3$) dispersed phase.[1]

Gambill (1959) proposed the following equation for estimating the kinematic viscosity, ν_m, of a two-liquid mixture of two different kinematic viscosities, i.e., ν_a and ν_b. χ_a and χ_b are mass fractions of the respective A and B components.[12,13]

$$\nu_m = \left[\chi_a(\nu_a)^{1/3} + \chi_b(\nu_b)^{1/3}\right]^3 \tag{1.12}$$

Later, Refutas (2000) proposed a method to determine the kinematic viscosity, ν_m, of a mixture of two or more liquids. In this method, a Viscosity Blending Number (VBN) of each component is first calculated, then used to determine the VBN of the liquid mixture shown as follows.[12,13] Also, see mixit-io SS12.1.

$$VBN_i = 14.534\ x\ ln\left[ln(\nu_i + 0.8)\right] + 10.975 \tag{1.13}$$

The VBN of the liquid mixture:

$$VBN_m = \sum_{i=0}^{n} \chi_i \times VBN_i \tag{1.14}$$

The kinematic viscosity of the mixture can be estimated using the following equation:

$$\nu_m = exp\left[exp\left(\frac{VBN_m - 10.975}{14.534}\right)\right] - 0.8 \quad (1.15)$$

1.4.6 Density

Depending on particle morphology, the density (mass divided by the volume) of a solid particle is generally differentiated in two distinct ways, i.e., as true (ρ_s) or apparent (ρ_a) densities. The former density is based on the volume of solid mass only, representing a particle without any internal voids. The latter describes the density of a particle inclusive of any void volume. An example is presented to illustrate the calculation of true and apparent densities. A 300micron size spherical resin particle has 10% void volume. The mass of the particle is 0.01 mg. The true and apparent density of the particle are 1100 and 991 kg/m^3, respectively. See spreadsheet mixit-io SS1.2 for calculation of these densities.

The bulk density, ρ_b, of solid particles within a given container is another commonly determined property. It is defined as the mass per occupied unit volume in the container, including the interstitial gaseous volume. For example, if the mass of a 20liter (~5 US gallon) bag fully filled with polyethylene (PE) pellets is 16kg (~35lb), the bulk density of the PE in the bag would be 800kg/m^3 (7lb/gal). Bulk density is often used to determine a final hopper or silo load, i.e., the weight of containers for shipping and transportation purposes.

The density of a pseudo-homogeneous multi-phase material is calculated from the volume fraction or weight fraction of its components in a simple additive manner. For example:

To compute the density of a mixture of n pure components, from their respective weight fractions, c_{wi}, use

$$\rho_m = \frac{1}{\sum\left(\frac{c_{wi}}{\rho_i}\right)_n} \quad (1.16)$$

To compute mixture density using respective individual component volume fractions, c_{vi}, use

$$\rho_m = \sum_n \left(c_{vi}\,\rho_i\right) \quad (1.17)$$

While mixing a two-phase solid–liquid slurry, densities of the solid (ρ_s) and liquid (ρ_l) phases are used to assess solid suspension characteristics. If the slurry solids weight fraction is χ, the mixture density, ρ_m, is

$$\rho_m = \frac{\rho_s \rho_l}{\rho_s \chi + \rho_l (1-\chi)} \quad (1.18)$$

For example, in a slurry dispersing a χ = 0.3 weight fraction of PET particles ($\rho_s = 1.4g/cc$) in water ($\rho_l = 1.0g/cc$), the calculated mixture density, ρ_m, is

$$\rho_m = \frac{1.4 \times 1.0}{1.4 \times 0.3 + 1.0(1-0.3)} = 1.25g/cc.$$

1.4.7 Surface Tension and Interfacial Tension

Surface tension occurs at gas–liquid surfaces because the two phases are exerting unequal forces on the surface molecules. Interfacial tension exists at interfaces between two or more immiscible or partially miscible liquid phases. Interfacial tension creates the tendency of an interface to become spherical to make its surface energy as low as possible. Though numerically different, interfacial tension (cohesive and adhesive forces) is analogous to surface tension between a liquid and a vapor.

Surface tension (liquid-vapor) and interfacial tension (immiscible liquid–liquid) are measured using a DuNoűy Ring or a Krűss Pendant Drop Tensiometer. The pendant drop is a drop suspended from a needle in a bulk liquid or gaseous phase. The drop results from the relationship between the surface tension or interfacial tension and gravity. Interfacial tension is determined from the shape (curvature) of the drop.[14,15] Although the DuNoűy ring method is traditionally practiced, the pendant drop method is the state-of-the-art technique and considered more accurate.

Interfacial tension plays an important role in preparing emulsions, heterogeneous systems consisting of droplets of at least one immiscible liquid intimately dispersed within another phase. Both surface tension and interfacial tension are a strong function of temperature and decrease with temperature increases. The ability of existing tensiometers to accurately measure interfacial tensions below 2dynes/cm is questionable.

1.5 THERMAL PROPERTIES

In some mixing operations, heat transfer is the controlling mechanism. Appropriate mixing enhances process-side heat transfer and produces a desirable yield while reducing batch or residence time. Heat can be added or removed from the working media depending on the required process temperature. Failure to control temperature often leads to byproduct formation, fouling, and unsafe operation due to run-away reactions. Knowing working media thermal conductivity and specific heat, whether single-phase or multi-phase, is essential. Gas, liquid and solid thermal properties can be obtained from reference handbooks and thermo-physical property data-banks.[16]

1.5.1 Thermal Conductivity

This correlation by Hamilton and Crosser[17] can help determine the effective thermal conductivity, k_{eff}, of such complex mixtures as a heterogeneous two-component slurry of sub-micron size particles $(k_p / k_o > 100)$.

$$\frac{k_{eff}}{k_o} = \frac{\left[k_p + (n-1)k_o - (n-1)\alpha(k_o - k_p)\right]}{k_p + (n-1)k_o + \alpha(k_o - k_p)} \tag{1.19}$$

where, k_p is the thermal conductivity of the particle

k_o is the thermal conductivity of the reference fluid

α is the particle volume fraction.

The empirical shape factor, n, is expressed in terms of sphericity, ψ, as follows:

$$n = {}^{3}\!/_{\psi} \tag{1.20}$$

where, ψ references a sphere of equal volume to the particle of interest. Then, ψ is the surface area of this equal-volume sphere, S_n, divided by the surface area of the actual particle, S.[18]

The true sphericity index, ψ, ranges from 1 (perfect sphere) to 0 (elongated shape). The model shows that non-spherical shapes (all other parameters being the same) increase the conductivity above that of spheres.

$$\psi = \frac{S_n}{S} = \frac{\sqrt[3]{36\pi V^2}}{S} \tag{1.21}$$

Mixit-io SS1.3 shows how to determine approximate mixture thermal conductivities for suspensions/slurries based on equation 1.19. The procedure is illustrated in the following example that determines the effective thermal conductivity of a 40vol.% cobalt-water colloidal suspension at room temperature (20°C). The thermal conductivities of cobalt particles, k_p, and water, k_o, are 100 and 0.6 $Wm^{-1}K^{-1}$, respectively. Cobalt particles have rod-like morphology, 10nm in diameter (d_p) and 50nm in length (l_p).

Particle volume, $V = \frac{\pi}{4} d_p^{\,2} l_p = \frac{\pi}{4}\left(10 \times 10^{-9}\right)^2 \times 50 \times 10^{-9} = 3.925 \times 10^{-24}$ [m³]

Nominal surface area, $S_n = \sqrt[3]{36\pi V^2} = \sqrt[3]{36\pi\left(3.925 \times 10^{-24}\right)^2} = 12 \times 10^{-16}$ [m²]

Particle surface area, $S = \pi d_p l_p + \frac{\pi}{2} d_p^{\,2} = \pi\left(10 \times 10^{-9}\right) \times \left(50 \times 10^{-9}\right) + \frac{\pi}{2}\left(10 \times 10^{-9}\right)^2$

$= 15.7 \times 10^{-16} + 1.57 \times 10^{-16} = 17.27 \times 10^{-16}$ [m²]

Sphericity, $\psi = \frac{S_n}{S} = \frac{12 \times 10^{-16}}{17.27 \times 10^{-16}} = 0.695$

Shape factor, $n = 3/0.695 = 4.31$ (for perfect spherical particles, n = 3)

Since it was initially stated that particles occupy 40 vol.%, the particle volume fraction, α, is 0.4.

From equation 1.19, $k_{eff} = \left\{ \frac{\left[k_p + (n-1)k_o - (n-1)\alpha\left(k_o - k_p\right)\right]}{\left[k_p + (n-1)k_o + \alpha\left(k_o - k_p\right)\right]} \right\} k_o$

$$\left\{ \frac{\left[100 + (4.31-1) \times 0.6 - (4.31-1) \times 0.4 \times (0.6 - 100)\right]}{\left[100 + (4.31-1) \times 0.6 + 0.4 \times (0.6 - 100)\right]} \right\} 0.6$$

$= 2.25 Wm^{-1}K^{-1}$

1.5.2 Specific Heat

Specific heat of a mixture at constant pressure, c_{p-eff}, is calculated from the mass fraction and intrinsic specific heats of components in the working media. For example, let A and B be constituents with specific heats c_{p-A} and c_{p-B}, mass fractions of ζ and $(1-\zeta)$, respectively. Then the effective specific heat is

$$c_{p-eff} = \zeta c_{p-A} + (1-\zeta) c_{p-B} \tag{1.22}$$

1.6 DROPLET AND PARTICLE SIZE DISTRIBUTION

As pointed out earlier, the working media may contain dispersed droplets and/or particles in liquid. The type of dispersion, i.e., liquid/liquid or solid/liquid, and the diameter of the dispersed phase in two-phase systems dictate whether the working media is to be treated as a micro-emulsion, an emulsion, a colloidal suspension or a slurry (see Figure 1.2).

In many instances in the chemical and allied industries, the products of two-phase reactions are solid particles. Examples include products from:

1. suspension polymerization—polyvinyl chloride (PVC) and polyvinylidene chloride (PVDC) resins
2. oxidation of paraxylene—crude terephthalic acid (CTA)
3. ethylene polymerization within a slurry process—high-density polyethylene (HDPE)

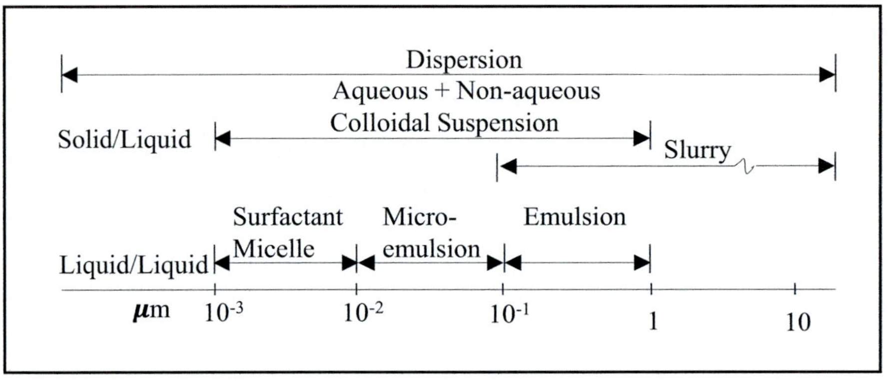

FIGURE 1.2 Classification of a two-phase system.[19]

1.6.1 Particle Size Distribution

The customer often specifies the size distribution (PSD) of solid particles to ensure safe and smooth operation of the downstream processes—most likely involving storage in silos, feed through the hoppers, and extrusion. To satisfy the customer's application, plant production specification (spec) target requires volume median particle size with constraints on fines and coarse particles. For example, a PVC customer may seek a psd with average (volume median) particle size of 150 microns, but no more than 3% fines (say < 10 microns) and no more than 1% of coarser particles (say > 250 microns). Fines are notorious for building up and creating dust and static charge in silos and hoppers. They may lead to occupational safety and health hazard (OSHA) issues. The quality of extruded films and slabs are often susceptible to variations in particle size and extruder operational parameters.

By far the most important physical property of particulate samples is particle size. Measurement of particle size distributions is routinely conducted across a wide range of industries and is often a critical manufacturing parameter. Measuring particle size distributions and understanding how they affect products and processes can be critical to successful manufacturing. For particle size analyses and psd measurements involving nanometer, micrometer or millimeter sizes, various commercial instruments are available. These are based on light scattering or laser diffraction technologies. Beckman Coulter LS (range:17nm to 2mm) and Malvern's Mastersizer® (range:10nm to 3.5mm) and Horiba LA Instruments[20] are widely used for particle size measurements. Sieve analysis is a cost-effective method of analyzing particle size distributions, by mass, for sizes ranging from 0.2 microns to tens of millimeters. Different mesh-size screens and fine filters are used for such measurements.

The filtration process may need liquid flushing and/or vibration to avoid filter plugging and erroneous psd measurement. Mettler-Toledo's Focused Beam Reflectance Measurement (FBRM) technology[21] is capable of inline/in-process and offline measurement of the raw chord-length distribution and particle size and shape analyses. FBRM is used for dsd or psd measurements to understand the equilibrium state in two-phase systems. Details are in Mettler-Toledo's FBRM-PVM-Particle-System-Characterization brochure.[21] PSD measurement techniques are described in detail in Chapter 18, "Experimental Techniques in Mixing."

Here, three different measures of PSD are briefly addressed. There are many correlations relating Sauter mean diameter or surface mean diameter, d_{32}, to agitation parameters (e.g., impeller/rotor tip speed or local energy dissipation rate) and physical properties (e.g., interfacial tension and/or viscosity). By definition, $d_{32} \approx d_v^3 / d_s^2$ is the ratio of volume mean diameter cubed to surface mean diameter squared, or proportional to the ratio of particle volume to surface area. If particle volume, V_p, and surface area, A_p, are known, then

$$d_{32} = 6V_p / A_p \tag{1.23}$$

Although d_{32} is commonly used in the research community, production sites are more familiar with the volume median diameter, d_{43}. To avoid fines and coarse particles, understanding the PSD width or band is also critical to most products. PSD specification often uses d_{10} and d_{90} (respective particle sizes below which 10 vol.% and 90 vol% of all particles reside), as well as the volume median d_{50}, or surface mean d_{32}.

The Sauter mean diameter, d_{32}, is probably the most commonly used mean as it characterizes multiple important processes (e.g., mass transfer and reaction) and their efficiency studies. Volume median diameter d_{43} characterizes evaporation and molecular diffusion phenomena. An example size analysis follows, based on the particle size distribution plotted in Figure 1.3 and its mean diameter characterization.

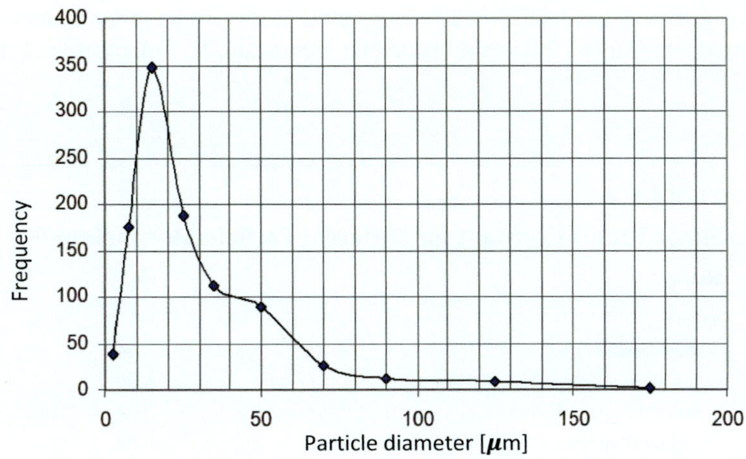

FIGURE 1.3 Particle size distribution.

1. Arithmetic mean or Number mean diameter:

$$\bar{d}_p = \frac{\sum_{i=1}^{N} n_i d_i}{\sum_{i=1}^{N} n_i}$$

2. Sauter mean or Surface mean diameter:

$$\left(d_p\right)_{32} = \frac{\sum_{i=1}^{N} n_i d_i^3}{\sum_{i=1}^{N} n_i d_i^2}$$

3. Mass or Volume mean diameter:

$$\left(d_p\right)_{43} = \frac{\sum_{i=1}^{N} n_i d_i^4}{\sum_{i=1}^{N} n_i d_i^3}$$

where, n_i = number of particles in the i th size class,
= mass of particles in the i th size class / mass of particles of diameter d_i.

In this example (Figure 1.3), the number mean, Sauter mean and volume median diameters are 24, 66 and 94 microns, respectively.

1.6.2 Particle Morphology

In addition to particle size distribution, understanding of the shape and surface features of the particles are also desired. Particle morphology studied under a microscope reveals growth mechanisms and crystal breakage due to impacts and/or attrition, if present.

Sauter mean diameter is inversely proportional to the surface density (surface area per unit volume), s_v, and can be expressed as $d_{32} = k_{sv}/s_v$ by enlisting the use of a characteristic dimensionless shape factor, k_{sv}. Shape factors are commonly used in microscopy and image analysis and can apply to both two-dimensional and three-dimensional objects. They describe a particle shape independent of its actual size and are determined from particle features like area, diameter, chord length, perimeter, centroid, etc. Table 1.2 lists the shape factor constant, k_{sv}, (departure of the particle from

TABLE 1.2
Shape Factor Constant for Different Particle Morphology[22]

Shape	k_{sv}
Sphere	6
Tetrahedron	8.94
Octahedron	8.06
Cube	7.44
Dodecahedron (pentagonal faces)	6.59
Icosahedron (triangular faces)	6.39

its equivalent spherical form), for a few platonic solids (regular polyhedral).[22] Shape factors were determined by dividing "the perimeter around the particle contour area of interest" by the square root of that area. Computational prediction of sphericity and roundness of particles is addressed in reference 23.

REFERENCES

1. Govier, G. W., Aziz, K., *The Flow of Complex Mixtures in Pipes*, p. 2, Van Nostrand Reinhold, New York, 1972.
2. Shear Thinning, https://en.wikipedia.org/wiki/Shear thinning www.rheosense.com/applications/viscosity/newtonian-non-newtonian
3. Mezger, T. G., *The Rheology Handbook: For Users of Rotational and Oscillatory Rheometers*, 2nd ed., Vincentz Network, Hannover, 2006.
4. Wright, W. A., Talbot, A. F., Manning, R. E., *Viscosity Index Tables for Celsius Temperatures*, ASTM, DS 39B, American Society for Testing of Materials, 1975.
5. Selby, T. W., Tolton, T., Very high shear rate, high temperature viscosity using the automated tapered bearing simulator-viscometer, technische akademie esslingen, *7th International Colloquium*, Automotive Lubrication, 8–1, p. 13, January 16–18, 1990.
6. Einstein, A., A new determination of molecular dimensions, *Annals of Physics*, 19(2), pp. 289–306, 1906.
7. Guth, E., Simha, R., Untersuchungen über die Viskositat von Suspensionen und Losüngen, *3. Uber die Visksitat von Kugelsuspensionen, Kolloid-Zeitschrift*, 74, pp. 266–275, 1936.
8. Thomas, D. G., Transport characteristics of suspension: VIII. A note on the viscosity of Newtonian suspensions of uniform spherical particles, *Journal of Colloid Science*, 20, pp. 267–277, 1965.
9. Mooney, M., The viscosity of a concentrated suspension of spherical particles, *Journal of Colloid, Science*, 6(2), pp. 162–170, April 1951.
10. Mukherjee, A., Pisupati, S. V., Interparticle interactions in highly concentrated coal-water slurries and their effect on slurry viscosity, *Energy Fuels*, 29(6), pp. 3675–3683, May 2015.
11. Taylor, G. I., The viscosity of a fluid containing small drops of another fluid, *Proceedings of the Royal Society of London. Series A: Mathematical, Physical and Engineering Sciences*, 138(834), pp. 41–48, 1932.
12. https://neutrium.net/fluid_flow/estimating-the-viscosity-of-mixtures/
13. Gambill, W. R., How to estimate mixtures viscosities, *Chemical Engineering*, 66, p. 151, 1959.
14. Krűss Scientific, *Pendant drop method for interfacial tension measurement*. www.kruss-scientific.com/services/education-theory/glossary/pendant-drop
15. Woodward, R. P., *Surface tension measurements using the drop shape method*. http://firsttenangstroms.com/pdfdocs/STPaper.pdf
16. Thermophysical Properties of Fluid Systems, *NIST chemistry webbook*, SRD69. https://webbook.nist.gov/chemistry/fluid/
17. Hamilton, R. L., Crosser, O. K., Thermal conductivity of heterogeneous two-component systems, *Industrial & Engineering Chemistry Fundamentals*, 1(3), August 1962.
18. Wadell, H., Volume, shape, and roundness of rock particles, *The Journal of Geology*, 40, pp. 443–451, 1932.
19. Takeo, M., *Disperse Systems*, 1st ed., Wiley-VCH, March 1999.
20. Horiba Scientific. *A guidebook to particle size analysis*, Horiba Instruments Inc., 2016. www.horiba.com/us/particle
21. Worlitschek, J., de Buhr, J., *Application note—crystallization study with focused beam reflectance measurement and MultiMax*, Mettler-Toledo, AN-SA-001/2005 V1.1, May 2005. www.mtautochem.com
22. DeCarlo, P. F., Slowik, J. G., Worsnop, D. R., Davidovits, P., Jimenez J. L., Particle morphology and density characterization by combined mobility and aerodynamic diameter measurements. Part 1: Theory, *Aerosol Science and Technology*, 38(12), p. 1185, 2004.
23. Cruz-Matías, I., Ayala, D., Hiller, D., Gutsch, S., Zacharias, M., Estradé, S., Peiró, F., Sphericity and roundness computation for particles using the extreme vertices model, *Journal of Computational Science*, 30, pp. 28–40, 2019.

2 Mixing Terminology

2.1 INTRODUCTION

Mixing technology uses various dimensional and dimensionless (also called non-dimensional) parameters related to fluid flow, mass transfer and heat transfer. The more prominent of these parameters are briefly described as follows to familiarize practicing engineers with their physical meaning and characterization. Dimensionless groups are developed based on Rayleigh's Principle of Similitude (1915) or Buckingham's π Theorem (1914) for expressing the behavior of a physical system in terms of the minimum number of independent variables. If there are n variables in a problem and these variables contain m primary dimensions such as M, L, T (i.e., mass, length and time), then the equation relating all of these variables will have $(n-m)$ dimensionless groups. These dimensionless parameters are used in correlations to scale processes up or down while maintaining dynamic and geometric similarity. Furthermore, apart from process scaling in the chemical industry, characteristic dimensionless numbers play an important role in analyzing transport phenomena such as fluid flow and heat or mass transfer. Since general mixing flow phenomena are often chaotic,[1] most correlations are developed from experiments conducted at different scales. For details, readers are referred to the articles[2,3] "Dimensional Analysis—How to Scale Up Scientifically" and/or "Dimensional Analysis for Fluid Agitation Systems."

2.2 DIMENSIONLESS NUMBERS

2.2.1 Fluid Flow Parameters

Reynolds number is defined as the ratio of the inertia force to the viscous force. Viscous dominance creates laminar flow, whereas inertial dominance creates turbulent flow. Between these flow regimes lies the transitional flow regime.

For flow through a circular pipe, the Reynolds number is defined as

$$N_{Re} = vd / \nu = \dot{m}d / \mu$$

where, v = average velocity [m/s]

$\dot{m}$ = mass flux or mass flowrate divided by the cross-sectional area [kg/m^2-s]

d = inner diameter of pipe [m]

ν = kinematic viscosity [m^2/s]

μ = dynamic viscosity [Pa-s or kg/m-s]

For flow through a non-circular pipe, the familiar Reynolds number expression is

$$N_{Re} = vd_o / \nu = \dot{m}d_o / \mu$$

where, $d_0 = 4 \times$ hydraulic diameter, d_h [m]

d_h = cross-sectional area of the pipe/wetted perimeter [m], and

$d_h = d_0 / 4$ for the fully filled pipe (circular) flow

In stirred tanks, impeller diameter is often the characteristic length used to calculate an impeller Reynolds number. The velocity parameter is related to impeller tip velocity, $v_{tip} = \pi DN$, but does not include the π term. Thus

$$N_{Re} = \rho D^2 N / \mu$$

DOI: 10.1201/9781003606390-2

where, ρ = density of working media [kg/m^3]
D = impeller diameter [m]
N = impeller rotations per second [1/s]

The Reynolds number is useful for designing, optimizing and trouble-shooting chemical processes that involve fluid transport, mixing, heat transfer and mass transfer.

Froude number, N_{Fr}, is the ratio of applied inertia force to gravitational force defined as,

$$N_{Fr} = N^2 D / g$$

where, g = acceleration due to gravity [m/s^2]

The Froude number is useful in studying the vortex motion commonly encountered when mechanically stirring low viscosity fluids in an unbaffled tank. N_{Fr} is often essential to scale-up.

2.2.2 Mechanically Stirred Vessel Parameters

Flow number (also known as Pumping number), N_{Fl}, relates discharge flow from an impeller to its rotational speed, N, and diameter, D. As might be expected, flow generated by an impeller is strongly dependent on the impeller blade geometry.

$$N_{Fl} = \dot{Q}_f / \left(ND^3\right)$$

where, $\dot{Q}_f$ = fluid circulation rate [m^3/s]

On occasion this book may follow a common practice of expressing the Reynolds number, N_{Re}, Froude number, N_{Fr}, and Flow number, N_{Fl}, respectively, as *Re*, *Fr* and *Fl* for purposes of brevity. These dimensionless numbers are occasionally used for specific scale-up requiring dynamic similarity.

Power number, N_p (also called Newton number, N_e), is defined as

$$N_p = P / \left(\rho N^3 D^5\right)$$

where, P = shaft power [N-m/s or W]

Power number relates the hydraulic power consumed by the agitation to flow from the impeller and the ratio of tangential fluid velocity to impeller peripheral speed.[4]

k–factor is defined as the ratio of hydraulic (shaft) power delivered to the working media under a gassed condition (P_g), to that delivered to the same system under an un-gassed condition (P_{ug}).

$$k\text{–}factor = P_g / P_{ug}$$

k–factor is a function of impeller Reynolds number, N_{Re}, and Aeration number, N_{Ae}, defined as

$$N_{Ae} = \dot{Q}_g / \left(ND^3\right)$$

Aeration number has important implications in the design of gas–liquid mixing systems, which are described later in Chapter 9. $\dot{Q}_g$ [m^3/s] is the actual volumetric gas flowrate. N_{Ae}, the ratio of the gas flowrate to impeller discharge rate, characterizes the prominence of the former with respect to the latter and dictates power consumption under the gassed condition.

Dimensional analysis shows that N_p is a function of Reynolds number and Froude number.[5]

Weber number, *We*, is defined as the ratio of inertia force to surface tension force and is an essential parameter in controlling the size of dispersed-phase droplets in an agitated working media.

$$We = \left(\rho N^2 D^3\right) / \sigma$$

where, σ = surface tension or interfacial tension [N/m]

Stirred mixing systems, as schematically shown in Figure 2.1, are characterized by a few well-recognized dimensionless parameters that are listed as follows and based on either impeller or tank diameter as the characteristic length. These parameters are used in mixing correlations and are critical for scaling up or down on the basis of geometric similarity (see Chapter 6).

Stirred Tank Parameters for Scaling Up or Down using Geometric Similarity						
D/T	Z/T	S/D	Z_s/D	C/D	B_w/T	B_c/T

where, Z = Liquid height [m]
S = Impeller spacing [m]
Z_s = Liquid submergence of impeller [m]
C = Impeller off-bottom clearance [m]
B_w = Baffle width [m]
B_c = Baffle clearance from tank wall [m]

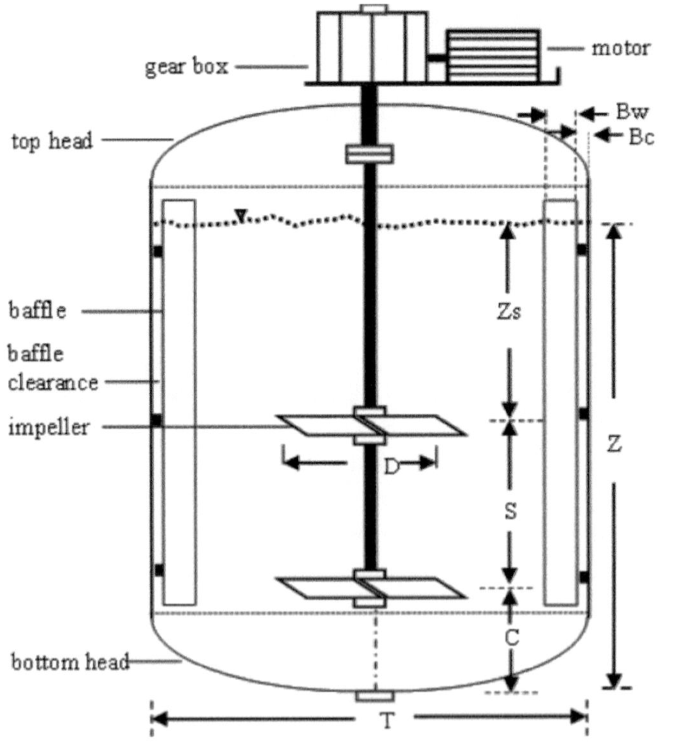

FIGURE 2.1 Schematic of a mechanically agitated tank.

The dimensionless blending/mixing time characteristic, τ_m, of a stirred tank is defined as the product of mixing time, θ [sec], and the shaft rotational speed, N [rps], represented by mixing number

$$\tau_m = N \cdot \theta$$

Other mixing parameters of interest are:

(a) Power per unit volume [W/m³], $\varepsilon_v = P / V$
(b) Power per unit mass (energy dissipation density) [W/kg], $\varepsilon_m = P / \rho V = P / m$
In this book, for all practical purposes, ε_m may be interchangeably expressed as ε.
(c) Superficial velocity of gas [m/s], $v_s = \dot{Q}_g / A_{cs}$
(d) Bulk fluid velocity [m/s], $v_b = \dot{Q} / A_{cs}$
where, $A_{cs} = (\pi / 4)T^2$ [m²] is the cross-sectional area of the stirred tank.
(e) *vvm* [min^{-1}] is the ratio of volumetric gas flowrate, $\dot{Q}_g$, to the volume, *V*, of the working media

2.2.3 Heat Transfer Parameters

Nusselt Number, *Nu*, is the ratio of convective heat transfer to the conductive heat transfer. When applied in a stirred tank of diameter *T*, it is defined as

$$Nu = hT / k$$

where, h = film heat transfer coefficient [W/m²-K]
k = thermal conductivity [W/m-K]
T = tank diameter [m]

Prandtl number, *Pr*, is defined as the ratio of momentum diffusivity ($\nu = \mu/\rho$) and thermal diffusivity $(\alpha = k / \rho c_p)$.

$$Pr = \mu c_p / k$$

where, c_p = specific heat of the liquid at constant pressure [J/kg-K]

Nusselt number, a function of Reynolds number and Prandtl number, assists in the design of reactors, heat exchangers, etc., where estimation of heat transfer is indispensable.

2.2.4 Mass Transfer Parameters

Sherwood number (Mass transfer equivalent of Nusselt number), *Sh*, is defined as the ratio of convective to diffusive mass transport.

$$Sh = k_L T / \vartheta$$

where, k_L = liquid film mass transfer coefficient [m/s]
ϑ = diffusion coefficient [m²/s]

Schmidt number, *Sc*, is defined as the ratio of momentum diffusivity and mass diffusivity as shown:

$$Sc = \upsilon / \vartheta$$

Sherwood number (mass transfer) is analogous to the Nusselt number (heat transfer), and typically a function of Reynolds number and Schmidt number. In chemical processes, it is used to assess mass flux from the dispersed phase to the continuous phase or vice versa considering both diffusive and convective mass transfer. For completeness, additional dimensionless groups and their applications to this effect are noted in Chapter 8 (Table 8.10).

2.3 DIMENSIONAL PARAMETERS

Droplet or bubble surface area per unit volume, a, is calculated as

$$a = A / V \ [\mathrm{m}^{-1}]$$

where, A = total bubble surface area [m^2]

V = gas volume [m^3]

2.4 MATHEMATICAL RELATIONSHIPS

The relationship between droplet/particle surface area, a, and volume fraction of the dispersion, ϕ, is

$$a = 6\phi / d_{32} \ [\mathrm{m}^2]$$

where, the Sauter-mean droplet diameter, d_{32}, is defined as

$$d_{32} = \frac{\sum_{i=1}^{N} n_i d_i^{\,3}}{\sum_{i=1}^{N} n_i d_i^{\,2}} \ [\mathrm{m}]$$

The volume to surface mean diameter, d_{32} is calculated from the drop size distribution spectrum, as shown previously in Figure 1.3.

REFERENCES

1. Ottino, J. M., *The kinetics of mixing: Stretching, chaos and transport*, Cambridge University Press, 1989.
2. Levin, M., How to scaleup scientifically, *Pharmaceutical Technology, Scaling up Manufacturing*, S4, 2005. www.pharmtech.com
3. Dickey, D., Fenic, J. G., Dimensional analysis for fluid agitation systems, *Chemical Engineering*, p. 139, January 1976.
4. Uhl, V. W., Gray, J. B., eds., *Mixing Theory and Practice*, Vol. 1, p. 183, Academic Press, New York, 1966.
5. Holland, F. A., Chapman, F. S., *Liquid Mixing and Processing in Stirred Tanks*, p. 4, Reinhold Publishing Corporation, 1966.

3 Industrial Mixing Systems

3.1 INTRODUCTION

Mixing technology encompasses a range of mixer designs whose capabilities range from simply blending miscible liquids to the complex mixing of multi-phase reacting flows. This chapter briefly introduces typical mixers before later chapters focus on their mixing applications. Stirred tanks, as in Figure 3.1, are a particular focus in this book because they are widely used in mixing operations such as liquid blending, immiscible liquid–liquid dispersion/emulsification, solid–liquid suspension (e.g., slurries) and gas–liquid contacting.

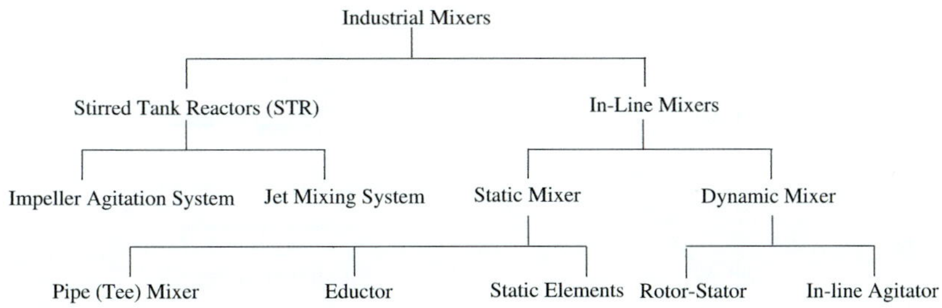

3.2 STIRRED TANK REACTORS

3.2.1 Impeller Agitation System

The operating simplicity of mechanically stirred tanks has promoted their widespread use in the chemical and allied industries for both batch and continuous mixing processes. Mechanical energy is imparted into a fluid phase, and possibly others, by a motor-driven shaft rotating one or more mixing impellers, blades, or ribbons (Figure 3.1). Although the motors are typically electric, pneumatic or hydraulic motors are sometimes used in special circumstances, such as explosive atmospheres or settings without available electricity.

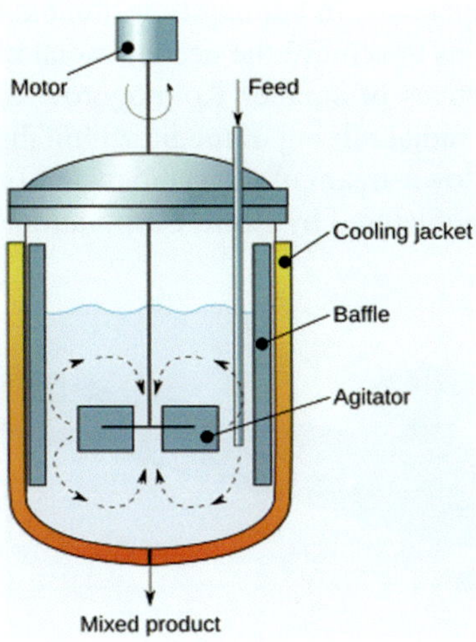

FIGURE 3.1 Schematics of stirred tanks mixed by a mechanical agitator.[1]

DOI: 10.1201/9781003606390-3

3.2.2 Jet Mixer

A mixing alternative to mechanical agitation is high-velocity fluid injection by at least one jet, as shown in Figure 3.2. In this case, the momentum (mass flowrate times velocity, $\rho_f . A_n . v_n^2$) of the fluid imparts mechanical energy into the fluid contents in the tank. The working media itself is often recirculated through a pump to create the required jet. This technique is typical of blending, neutralization and temperature homogenization in large storage tanks. For lighter applications, such as mixing in railcars, gas is injected through jets and spargers.

Whether mechanical agitators or jets deliver the mechanical energy to the fluid, power intensity is altered by equipping the drive motor with a variable frequency drive (VFD) to control agitator or pump speed.

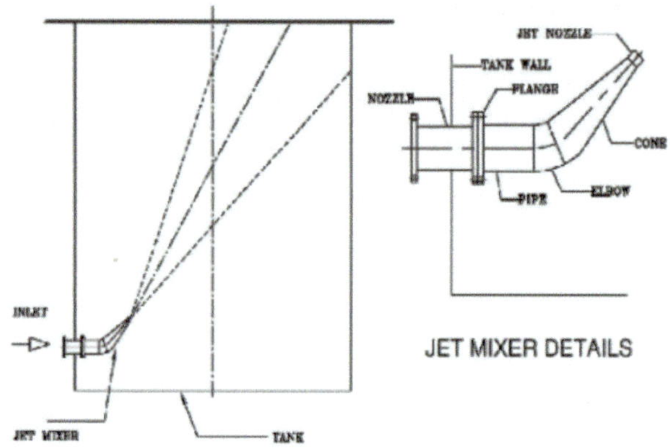

FIGURE 3.2 Schematics of stirred tanks mixed by a jet mixer.[2]

3.3 INLINE MIXERS

3.3.1 Static Mixers

These are motionless mixers, consisting of tubes packed with static mixing elements of various types to accommodate laminar or turbulent mixing of the working media. Typically, the tubes have injection ports to add media components. In laminar flow, fluid lamellae or elements undergo multiple splits, folds and recombinations to achieve the needed radial mixing. In turbulent flow, the static elements produce additional vortices of smaller Kolmogorov length-scales to intensify turbulent fluctuations and enhance chaotic radial mixing of media within the tube. In both cases, the inertia of mixing persists for some length downstream of the static mixer. Figure 3.3 is a marketing image for one commercial static mixer manufactured by Koflo Corporation.[3]

FIGURE 3.3 Marketing image of a Koflo® Static Mixer.[3]

3.3.2 Eductor or Ejector

The typical eductor shown in Figure 3.4 mixes dual components using a venturi as per the associated Bernoulli principle. Pumping the primary stream or motive fluid (component A) through the eductor's venturi creates a negative pressure near the venturi throat.[5] This draws mixing component B, or the suction fluid, into the venturi throat, where it is entrained within the accelerating component A. Mixing continues as the fluids expand in the diverging aft section of the educator. Eductor[4] nozzles are often placed in mixing tanks to prevent the settling of solids or denser fluids.

FIGURE 3.4 Schematic showing the principle of operation of an eductor-nozzle.[4]

3.3.3 Pipe (Tee) Mixer

The tee mixer has no moving parts but is a tee fitting with a reducing branch or a pipe assembly with a transverse/radial port. A smaller injection pipe is inserted through the reducing branch or radial port. The injection pipe discharge location relative to the centerline and wall of the main pipe (i.e., the tee's run) is critical to mixer efficacy. In laminar or turbulent flow systems, inserting static mixer elements (e.g., twisted ribbons) into the pipe, downstream of the injection pipe, enhances mixing efficiency by splitting and combining the fluid components or by enhancing turbulent fluctuations as the fluid proceeds through the pipe. Figure 3.5 shows the design of a simple tee mixer manufactured by Statiflo.[6]

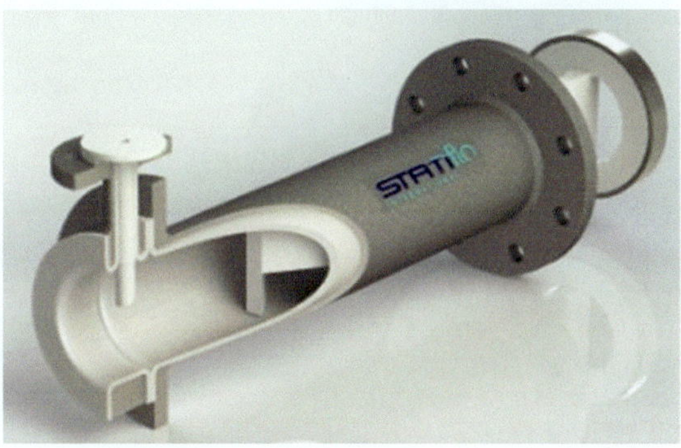

FIGURE 3.5 Marketing image of a Statiflo Tee Mixer.[6]

3.4 DYNAMIC MIXERS

3.4.1 Rotor-Stator

Inline, high-shear rotor-stator mixers are widely used to reduce dispersed-phase droplet diameters to the submicron range after passing through the system only once. Although they come in a variety of different geometries, each consists of one or more rotor-stator pairs (i.e., stages), as shown in Figure 3.6.[7] A central shaft rotates each rotor while each stator remains stationary. A typical annular gap between a paired rotor and stator is on the order of microns. Rotor speeds are usually on the order of thousands of rpm. Fluid passing through the shear gap is subjected to extreme shear rates on the order of 10^5 to 10^6 sec^{-1}. Rotor-stators produce emulsions and dispersions for the home and healthcare, pharma, food, and agricultural markets. Figure 3.7 pictorially illustrates the functioning of a Silverson® high shear in-line mixer widely used as an emulsifier.[8] An emulsifier can be configured to include a second feed stream between the first and second rotor-stator stages.

FIGURE 3.6 Photographic images of IKA multistage rotor-stator elements.[7]

FIGURE 3.7 Partially cutaway image of a Silverson® high shear in-line mixer[8] (also see the movie in Reference 9)

3.4.2 In-Line Dynamic Mixer

The novelty of this technology is continuous mixing using stirrers but no tank. The SPX Flow-Lightnin-Line Blender™ fits inside a pipe and is designed with a stirrer in both the inlet and the outlet zones, commonly flat-blade turbines. The stirrers are mounted on a short shaft and driven by a motor, as shown in Figure 3.8.[10] The pictured separator plate divides the two zones of agitation, while directing flow through the vessel and preventing short-circuiting and back-mixing. The result is a far more intense blending and dispersing action than occurs in a single-stage unit. In some applications, e.g., wastewater, the baffle cage can be removed to reduce pressure drop through the mixer. The benefits of an in-line blender are lower initial investment, space savings, lower power consumption, and simpler implementation.[10]

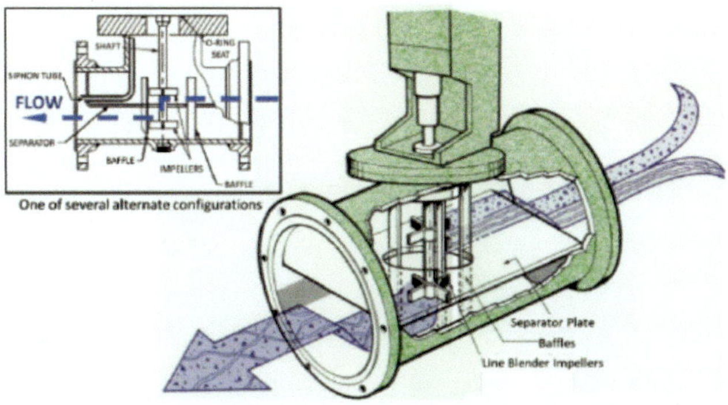

FIGURE 3.8 Cutaway schematic showing working principle of the Lightnin In-Line Blender.[10]

Figure 3.9 is the schematic of the InstoMix™ In-Line Flash Mixer manufactured by Walker Process Equipment Company. InstoMix provides high-energy flash mixing to instantaneously (within milliseconds) disperse coagulant and other flocculant solutions into raw water flow.[11]

Countless mixing systems are available from various vendors. This brief chapter presents a few from various websites to help visualize and classify them by type. Later chapters discuss these systems in detail.

FIGURE 3.9 Cutaway of the Walker Process Equipment InstoMix™ In-Line Flash Mixer.[11]

REFERENCES

1. https://en.wikipedia.org/wiki/Continuous_stirred-tank_reactor
2. www.indiamart.com/mk-design/process-equipments.html#jet-mixer
3. www.koflo.com/
4. www.spray-nozzle.co.uk/spray-nozzles/mixing-eductors
5. www.youtube.com/watch?v=H3TcLoapJBo
6. www.statiflo.com/products/pipe-static-mixers/
7. www.ika.com.my/inlineMixer01.html
8. https://www.silverson.com/images/uploads/documents/Silverson_Brochure_US.pdf
9. https://www.youtube.com/watch?v=mhjZC7B0jEQ
10. *Lightnin Line Blender In-line Mixer*, Brochure B-528C-9M-02-94. https://pdf.directindustry.com/pdf/lightnin/line-blender-in-line-mixers/24564-434217-_2.html
11. www.walker-process.com/pdf/prod_water_InstoMix.pdf

4.2.2 Vessel Material of Construction (MOC)

The material of construction (MOC) of a given vessel depends on the reactants, chemical reaction products, process temperature, pressure, and the desired heat transfer that will be present. Serious consideration must be given to selecting an appropriate MOC. A few MOC examples are outlined in Table 4.5.

TABLE 4.5
Typical Vessel Materials of Construction (MOC) Based on Process Application

Processes	Vessel MOC	Reason(s)
Latex	Glass-lined or Hastelloy	Monomer corrosion, stickiness
Terephthalic acid	Titanium cladded carbon steel	Corrosion w/acetic acid, fire hazard
Pharmaceutical products	Glass-lined carbon steel	Contamination and cleanliness
Agricultural products	Nickel	Acid corrosion

Sometimes instead of the standard surface finish, electro-polished or glass finished surfaces are used to minimize adhesion of sticky materials to the vessel wall. Typical examples are higher-molecular-weight liquids and pastes such as low-density polyethylene (LDPE), polyvinyl chloride (PVC), latex, clay gums (e.g., Kelzan™), and polyurethane (PU) based sealants, etc.

Practicing engineers must remember that the cost of a vessel depends on all of its features including size, material of construction, and surface finish. The latter could significantly increase capital cost. Additionally, it should be remembered that glass vessel linings impose other limitations, including a lower overall heat transfer coefficient, due to the low thermal conductivity of the glass lining, and a higher cost if the vessel size is not one of the standardized glass-lined vessel sizes (i.e., working media volumes). Since non-standard glassed vessels can be much costlier, designers should refer to the De Dietrich and Pfaudler process equipment manuals[4,5] sooner rather than later.

Unless using an existing tank or addressing an extenuating circumstance, one should generally avoid hemispherical and conical bottom tanks. Square tanks can be cost effective as large capacity mixer/settler units. Cylindrical tanks are sometimes mounted horizontally, particularly for batch handling of discontinuous fluids such as viscous pastes and doughs being processed by ribbon and Z-blade mixers, kneaders, etc. One such vessel used in producing cellulosic materials is called the "Littleford Reactor."

In the United States, cylindrical tank diameters are commonly kept below 13 feet due to restrictions when transporting such on the public highways. Tanks larger than 13 feet in diameter may need to be field erected and will be costlier to fabricate on site. Reactor pressure, including both head space pressure and hydrostatic head pressure, dictates vessel wall thickness, which obviously impacts other components of vessel design. Furthermore, in case of bioreactors/fermenters, effects of pressure due to clean-in-place (CIP) with high-pressure steam for sterilization/sanitization must also be considered.

The hoop stress, σ_{hoop}, of thin-walled (radius > 5 times the wall thickness) vessels must not exceed the yield stress of the tank MOC or that of the weld-joint seam at the working media temp, i.e.,

$$\sigma_{hoop} = \frac{pT}{2t} \leq \sigma_{yield} / 2.5 \tag{4.1}$$

where, P = tank internal pressure [psi]
T = tank internal diameter [inch]
t = tank wall thickness [inch]

In equation 4.1, a conservative safety factor of 2.5 is included because of unsteady hydrodynamic loading. For carbon steel, the yield stress, σ_{yield}, is 53,700psi (370MPa), while values for 304 and 316 stainless-steel are 31,200psi (215MPa) and 42,100psi (290MPa), respectively. Since stirred tanks will likely have multiple seam welds and nozzles, practicing engineers are advised to consult with their company's pressure vessel designer and/or vessel manufacturer in designing manufacturing vessels.

4.3 IMPELLERS

At least one shaft equipped with one or more impellers is used in a stirred tank to generate motion in the working media. Impellers that discharge fluid parallel to the rotating shaft are classified as "axial flow" impellers, while those whose discharge flow is perpendicular to the shaft are "radial flow" impellers (see Figure 4.2). A third classification, "mixed flow," describes impellers whose blades discharge a combination of both axial and radial flow. The widely used standard 45° pitch-blade turbine (PBT) is a mixed flow impeller. The flow pattern produced by discharge from an impeller rotating inside a vessel dictates impeller mixing performance. Typical applications of some common commercial impellers are given in Table 4.6.

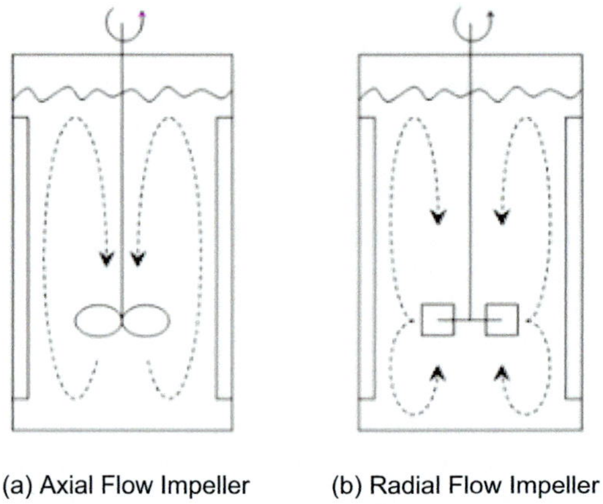

FIGURE 4.2 General flow patterns from an (a) axial flow and (b) radial flow impeller.

TABLE 4.6
Common Impellers and Their Applications

Type	Generic Name	Viscosity Range (cP)	Common Applications
Radial	Disk Turbines: Rushton, Smith, SCABA, BT-6 Flat-Blade Turbine (Paddles)	0.1–2,000	Gas–liquid–solid contacting Excellent gas-dispersion efficiency
	Anchor agitator	5,000–50,000	Blending viscous and sticky materials (process side heat transfer improvement)
Axial	Propeller	0.1–10,000	Blending, Gas–Liquid mixing
	Hydrofoils		Blending, Crystallization
	Pitch-Blade Turbine		Blending, Gas–liquid mixing, Liquid–Liquid dispersion
	Helical Ribbon	25,000–500,000	High viscosity blending, Solid dispersion in high viscosity media
	Kneaders and Z-Blade Mixer	500,000–8,000,000	Blending pastes and Doughs, Dispersing fine particles in high-viscosity liquids

Agitators are also designated as either "open" or "proximity" impellers depending on the typical ratio of their diameter (D) to the diameter of the vessel in which they operate (T). Open impellers such as PBTs and hydrofoils typically have a D/T ratio <0.7, while proximity impellers such as the Paravisc® and anchors have diameters so close to the vessel diameter (D/T ratio ≥0.7) that baffles are not required.

Worldwide, impellers are produced by many agitation systems manufacturers. In Figure 4.3–Figure 4.7, the authors show a few impellers they commonly used during their R&D and manufacturing activities. Here, the intention is to acquaint practicing engineers with their types and trade names with no bias to any specific manufacturer. The cited photographs are directly copied from the manufacturers' (SPX FLOW, NOV, Ekato, De Dietrich, and Pfaudler) websites.[6–10] Design features and details of their applications are addressed in their respective web bulletins. Also, countless special impellers are produced to solve unmet mixing needs in manufacturing processes. Those impellers will be addressed later.

4.3.1 SPX Flow (Lightnin)[6]

Axial Flow Impellers

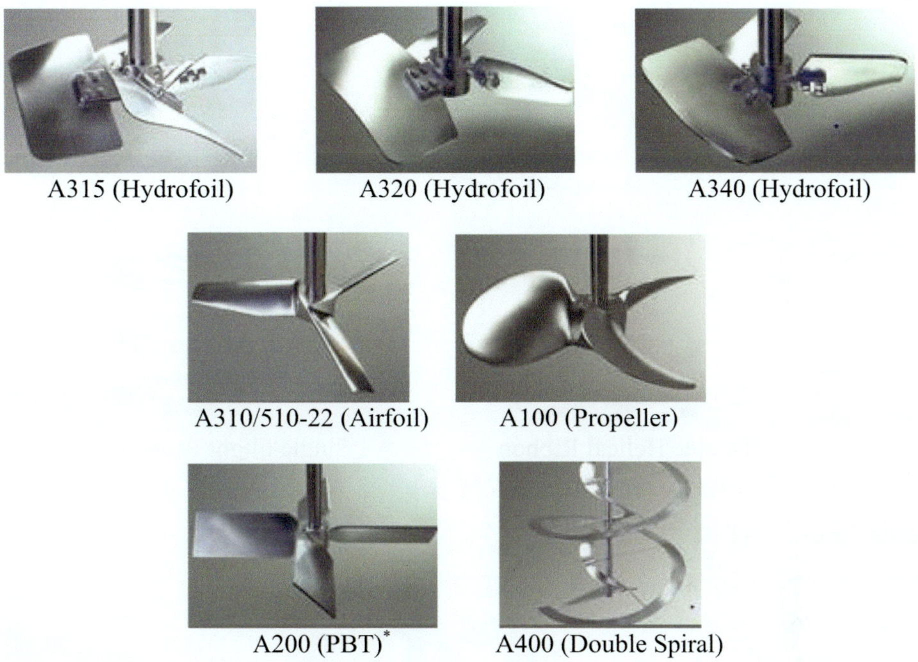

Radial Flow Impellers

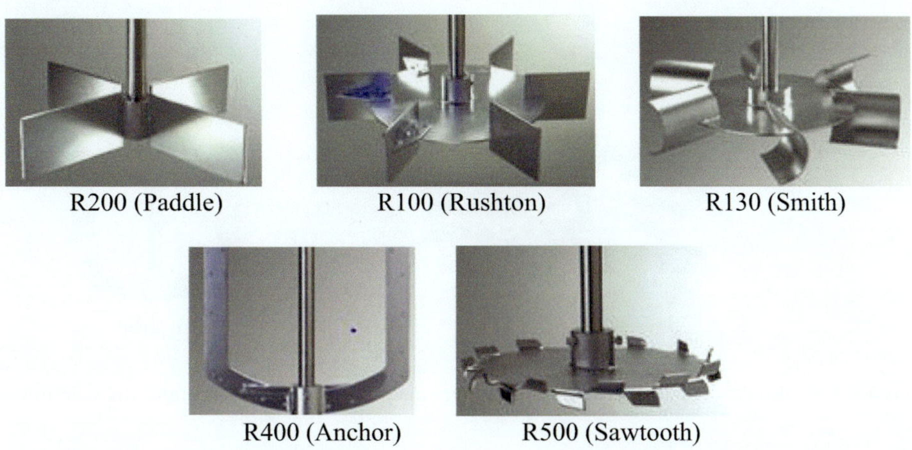

FIGURE 4.3 Images of axial and radial impellers manufactured by SPX Flow under the Lightnin[6] name.

4.3.2 National Oilwell Varco (NOV—Chemineer)[7]

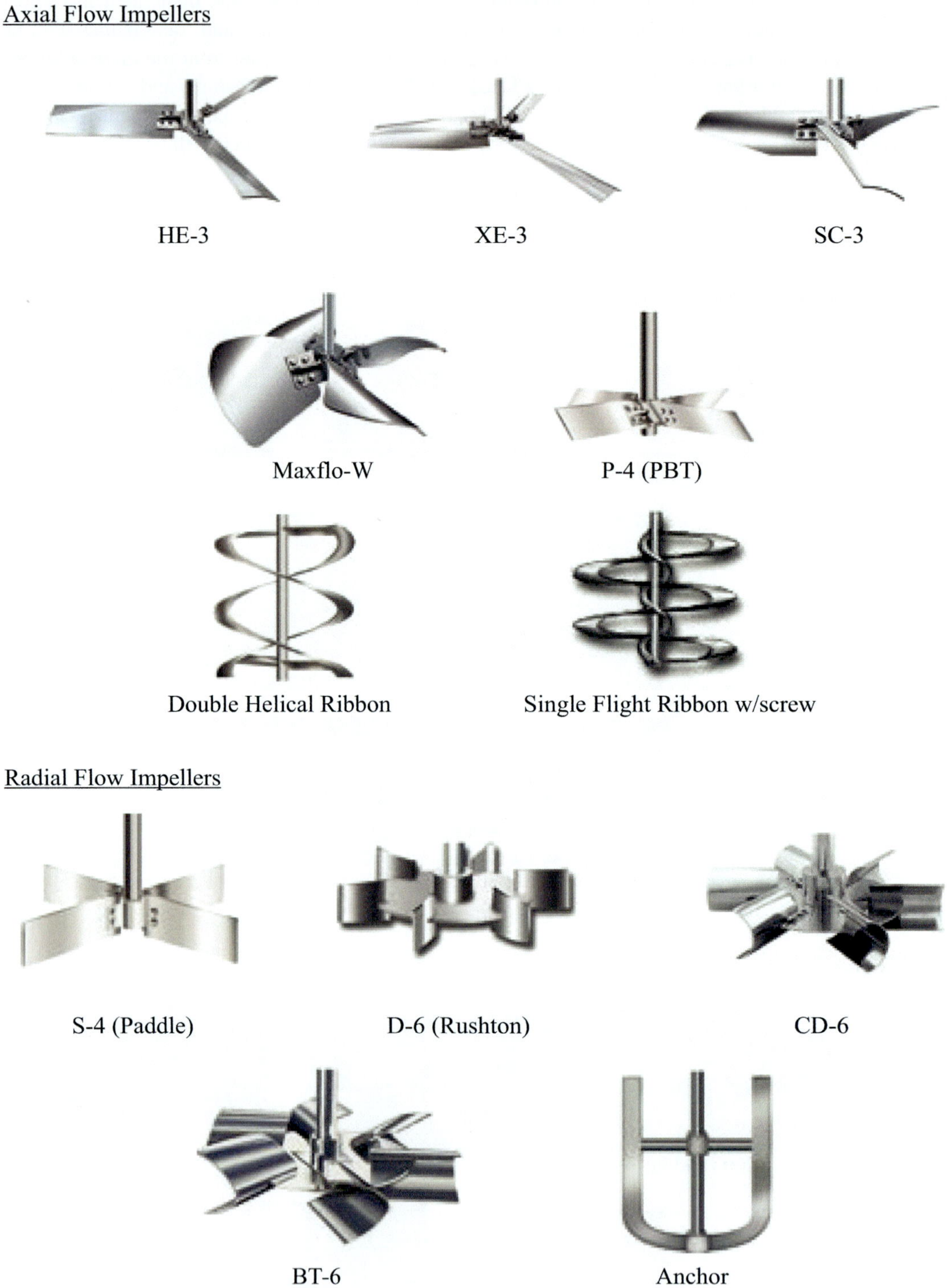

FIGURE 4.4 Images of axial and radial impellers manufactured by NOV under the Chemineer[7] name.

4.3.3 Ekato Systems GmbH (Ekato)[8]

Axial Flow Impellers

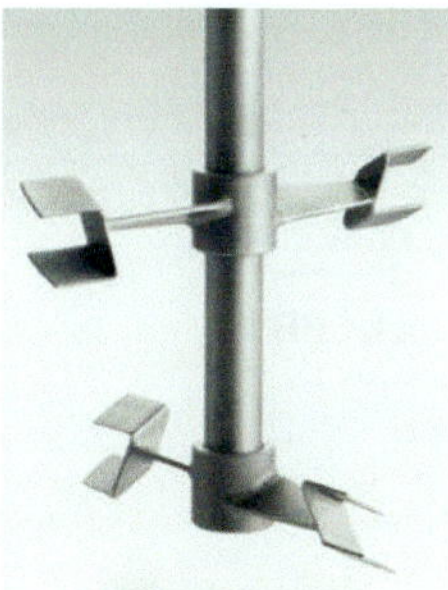

Intermig*

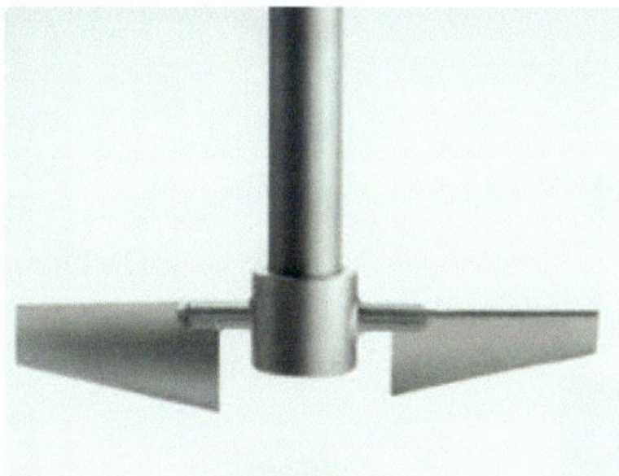

Viscoprop*

Epal

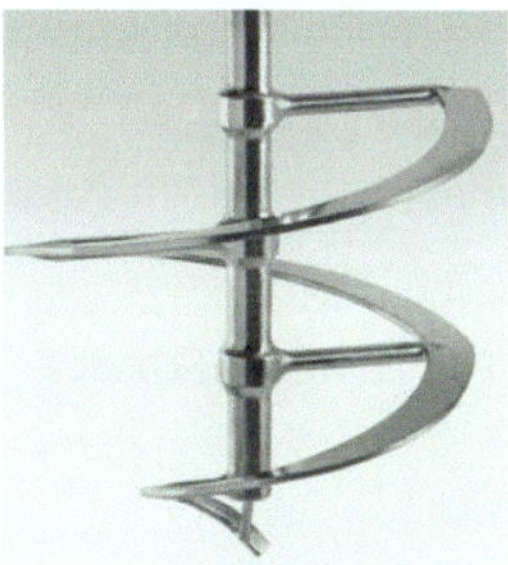

Solids-Paravisc

Paravisc

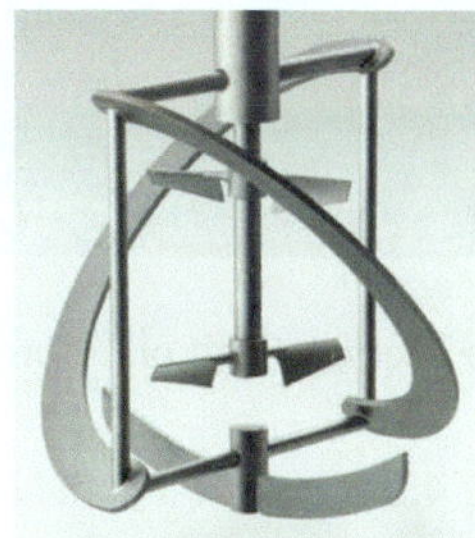

Paravisc Koaxial*

Radial Flow Impellers

Dissolver

FIGURE 4.5 Images of axial and radial impellers manufactured by Ekato.[8]

Glass-lined vessels and impellers are widely used to overcome the processing issues with corrosive, sticky, or high-purity working media, and where product sanitation is particularly important. De Dietrich[4] and Pfaudler[5] are renowned manufacturers of such vessels, impellers, and the tank internal accessories such as baffles and dip/feed pipes, etc. Images of a few of their axial and radial GlasLock® and Cryo-Lock® impellers[9,10] have been retrieved from their respective websites and presented in Figure 4.6 and Figure 4.7.

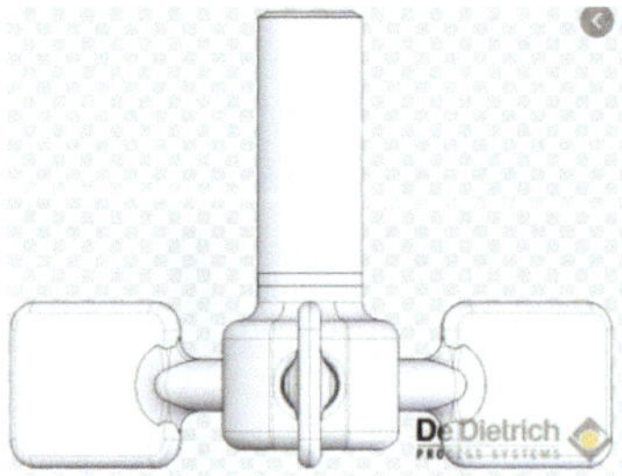

De Dietrich GlasLock® FBT

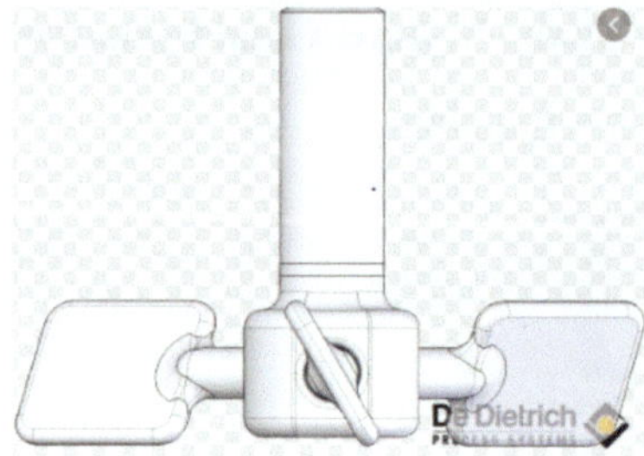

De Dietrich GlasLock® PBT

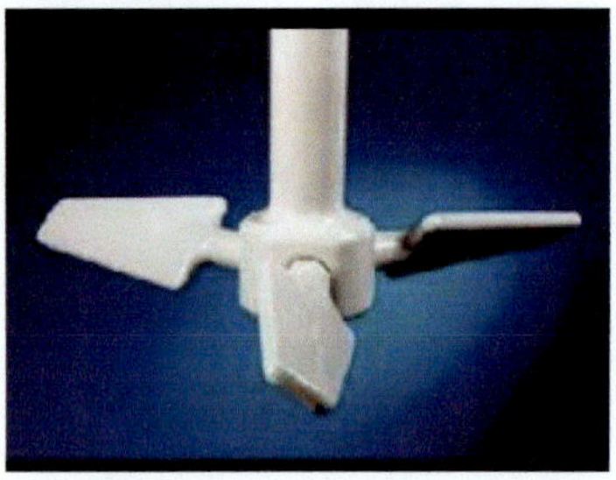

De Dietrich OptiFoil® Aerofoil Blades

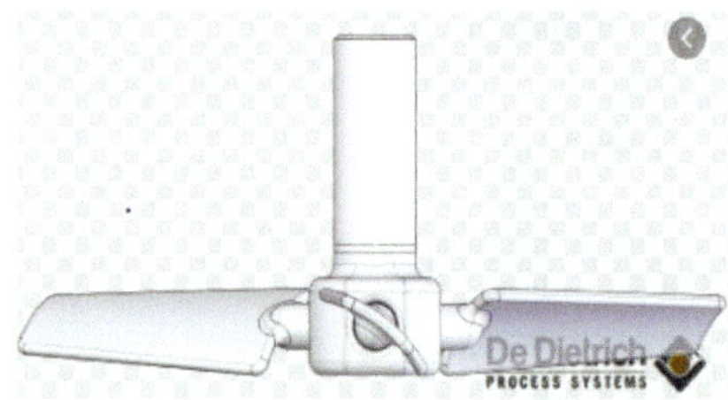

De Dietrich OptiFoil® Hydrofoil Blades

FIGURE 4.6 Images of axial and radial flow impellers manufactured by De Dietrich.[9]

Pfaudler Flat Blade Turbine

Pfaudler Turbofoil®

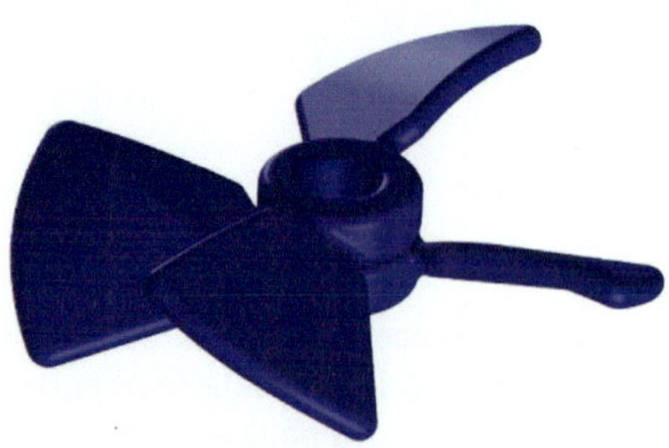

Pfaudler Maxflo® Turbine

Pfaudler Gas Dispersion Turbine

FIGURE 4.7 Images of axial and radial impellers manufactured by Pfaudler.[10]

4.4 PERIPHERALS AND TANK INTERNALS

4.4.1 Ticklers

A traditional tickler has typically been a small PBT (down-pumping or up-pumping configurations) or FBT impeller positioned near the tank bottom to maintain solids suspension in the bottom head. The KT-3 tickler with its curved down-pumping blades was specifically developed to minimize radial flow while inducing downward flow and keeping slurry solids suspended. It was conceptualized and developed to address unmet needs (e.g., minimizing solids heels) in manufacturing facilities of the Dow Chemical Company. It was subsequently licensed to SPX Flow for global sales to either non-Dow or Dow processes. Performance of the KT-3 tickler[11,12] (see Figure 4.8) is superior to the aforementioned traditional ticklers and is especially recommended for operations that drain slurry tanks and reactors through a centerline bottom nozzle. It minimizes solids heels, facilitates faster and more uniform drainage, and also minimizes pump starvation.

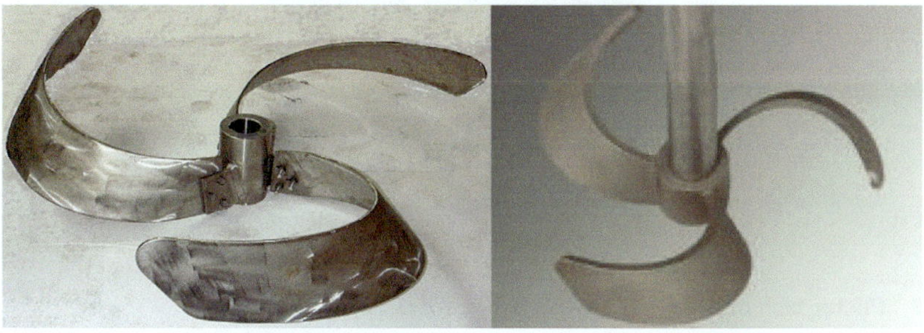

(a) (b)

FIGURE 4.8 (a) First-ever commercial-scale model (72 in)[11] and (b) general CAD image[6] of KT-3 Tickler.

4.4.2 Slingers

Slingers are small disc turbines typically positioned within the headspace of a vessel to uniformly disperse reflux condensate, fresh feeds, or recycled media (liquid) inside the tank. In a boiling reactor with high superficial vapor velocity, however, an additional purpose of the slinger is to also prevent entrainment of fine particulates in vapors entering the plug-susceptible overhead distillation columns and condensers. The Dow-designed KS-8 slinger,[13] seen in Figure 4.9, features curved vanes atop the disk. It creates a much more uniform 360-degree liquid curtain (umbrella) than do traditional straight-vane slingers. It has been implemented in multi-phase boiling reactors to overcome liquid distribution limitations, and thus, to improve manufacturing asset reliability and utilization.

FIGURE 4.9 Photograph of a Titanium Dow KS-8 Slinger.[13]

4.4.3 Foam Breakers

Surface foams are often a process nuisance that decrease vessel capacity and possibly create undesirable loss of containment (LOC) and contamination of tank top internals and overhead process lines. Theoretically, foam breakers have high surface area to contact and destroy the foam without having too much mass or drawing too much power. Traditional foam breakers have used things like multiple vertical loops (constructed from small cylindrical rods) that rotate about the central axis to destroy or minimize foams floating atop the working media. In highly surfactant-laden systems, such as hygroscopic polyols and low-surface-tension acid solutions (e.g., 10wt% glacial acetic acid in water, $\sigma \approx 19$ dynes/cm), mechanical agitation creates undesirable foam that must be controlled. The Dow-patented KBB foam breaker,[14] shown in Figure 4.10, is an alternative to the traditional looped rod style of foam breaker.

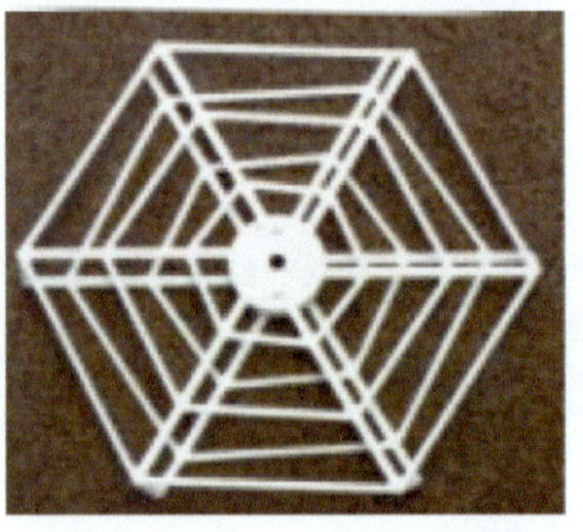

FIGURE 4.10 Photograph of a Teflon® Dow KBB Foam Breaker.[14]

Stirred tank accessories discussed next include baffles and feed/dip pipes, as shown in Figure 4.11.

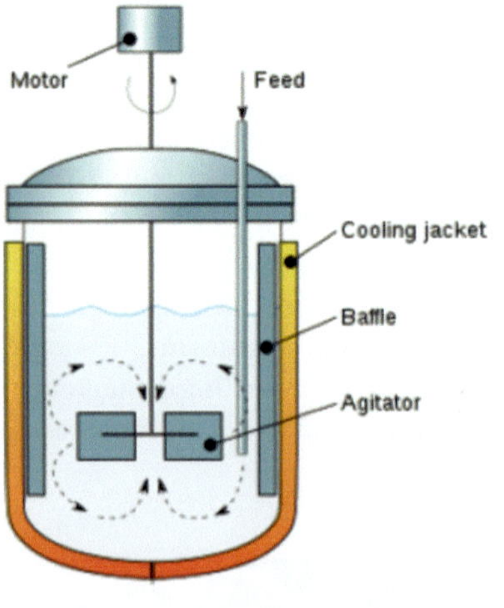

FIGURE 4.11 Typical stirred tank with flat baffles and a dip/feed pipe.[15]

4.4.4 Baffles

Vertical cylindrical tanks with centrally mounted impellers typically use baffles of various shapes and sizes to convert some of the impeller-induced swirling tangential flow into axial flow with top-to-bottom fluid exchange. As noted exceptions, baffles are not typically used in systems with laminar ($Re < 10$) or low transitional flows ($10 \le Re \le 100$). In low viscosity systems, a single baffle typically produces only partial baffling, while three to four baffles circumferentially placed at 120/90 degrees apart are typically required to obtain full baffling (i.e. no visible swirling behavior). Normally, the width of flat baffles, B_w, used in standard stirred tanks is T/12–T/10, where T is tank diameter. Baffles are rigidly fastened directly to the tank wall, or to nozzles in the top and/or bottom heads. A small clearance, B_c, (T/64-T/40) is typically left between the baffle and the wall for flow through the gap. As a rule for open tanks, baffle thickness should be comparable to that of the tank shell, but no less than ¼ inch.[16] Maximum hydraulic force, in pounds, on each baffle is calculated using equation 4.2, assuming that the power delivered to the working media is sustained by the baffles.

$$Force = \frac{5250 \times P \times DF}{n_b \times R_b \times N} \tag{4.2}$$

where, P = shaft power (power delivered to the fluid, sustained by the baffles) [HP]
n_b = number of baffles
R_b = radial distance of outer edge of the baffle from the shaft centerline [ft]
N = impeller speed [rpm]
DF = dynamic load factor (usually 2.0)[16]

Figure 4.12 shows different types of glassed baffle types offered by Pfaudler and De Dietrich. For further information about these individual baffles, the reader is referred to the respective company websites.

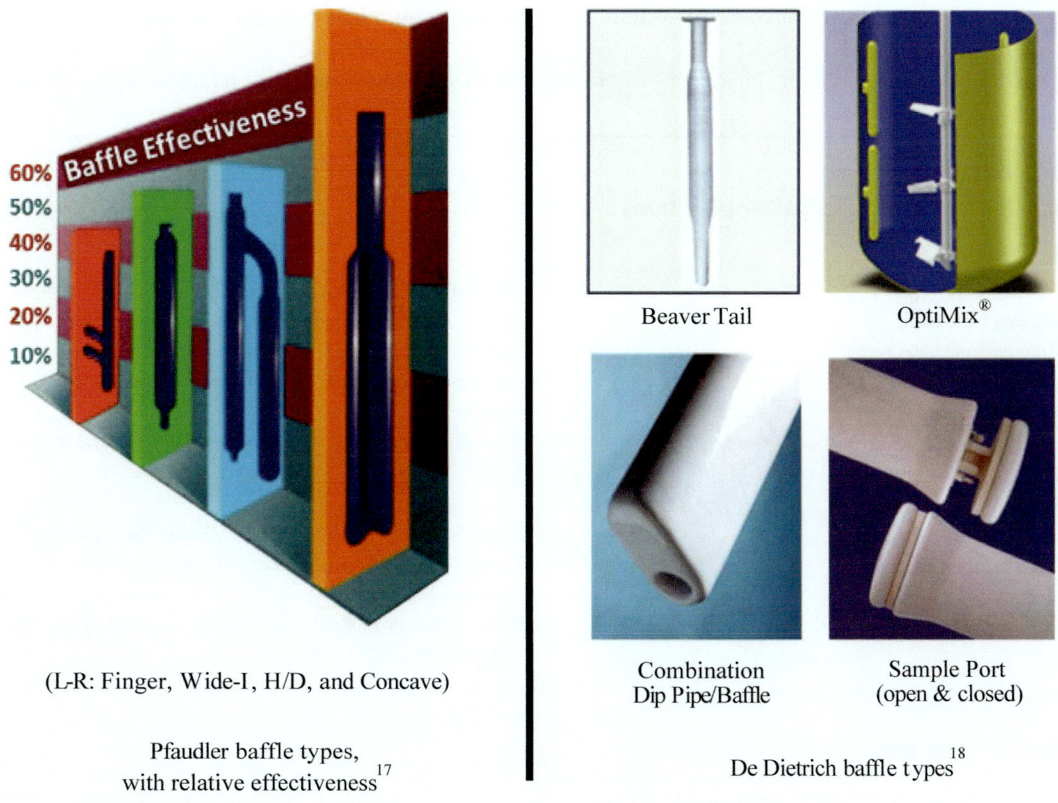

FIGURE 4.12 Different types of glassed baffles offered by Pfaudler[17] and De Dietrich.[18]

Figure 4.13 compares two different heat transfer coil systems that are commonly used in stirred tanks. The left half of the figure shows vertical harp coils that simultaneously serve as both heat transfer surfaces and vertical mixing baffles. On the right in the figure is a traditional helical heat transfer coil. Because it provides no baffling, it is mounted inside of traditional flat plate vertical baffles.

Table 4.7 lists typical applications of various different types of mixing baffles that have been mentioned to this point.

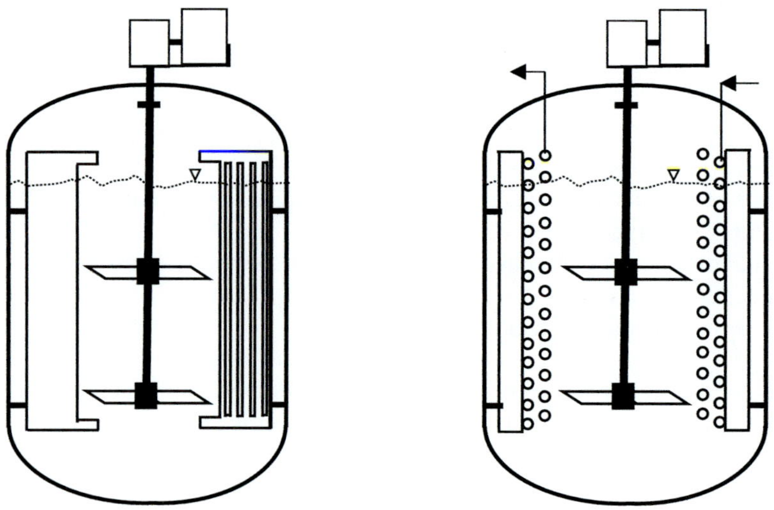

Vertical heat transfer coil system   Horizontal heat transfer coil system

FIGURE 4.13 Heat transfer coil systems commonly used in stirred tanks and their relative baffling effects.

TABLE 4.7
Baffles and Their Typical Applications

Type of Baffle	Applications
Flat baffle	Commonly used in the standard tanks.
Cooling plate baffle	Useful for baffling as well as for heat transfer purpose.
Aerodynamic plate baffle	Slurry or fibrous viscous working media.
Pipe baffle	Commonly used in glass-lined De Dietrich[4] and Pfaudler[5] vessels. Pipe baffles are also useful for the circulation of heat transfer fluid.
Finger baffle	
h or D-baffle	
Beavertail baffle	
Wide-I baffle	
Horizontal baffle	Minimize headspace gas entrainment, inert gas blanketing. Create flow compartmentalization in agitated column for back mixing.
Profiled baffle	Prevent breakage of solids (ex: crystals, mycelia) susceptible to mechanical degradation.

4.4.5 Dip Pipes

Dip pipes are commonly used inside stirred tanks to feed reactants (liquid or gas) or steam, or to house probes for measuring temperature, pH, conductivity, etc. The portion of the dip pipe that is submerged below the liquid surface is often subjected to violent unsteady hydrodynamic forces from the agitated working media. Designing the dip pipe (feed pipe) with sufficient mechanical strength is of paramount importance for safe operation of mechanically agitated vessels. Also, in the case of over-hung dip pipes, special attention must be paid to the weld strength, if it is welded to the flange. The mechanical design of overhung dip pipes is addressed in mixit-io SS4.2 (see dip/feed pipe images in Figure 4.1 and Figure 4.11).

4.4.6 Vortex Breakers

When discharging working media through a centrally located bottom nozzle (discharge hole), a vortex may form as the fluid level drops below the lowest impeller. The root causes are typically persistence of tangential swirling flow (due to fluid inertia) and/or excessive speed of the discharge pump. Such a vortex commonly produces unsteady discharge flow and pump starvation, accompanied by intermittent gurgling, pump noise, and possibly perceived pump vibration. Installing a vortex breaker that enters the drain hole and protrudes into the bottom head (see Figure 4.14) often rectifies vortex formation near the nozzle.[19] While the cross type is common, other vortex breaker types (e.g., angle-mounted disk) also exist. In the absence of a vortex breaker, a simplified approach may be used to calculate minimum liquid submergence of the outlet, S to avoid vortex formation in the tank. S depends on the outlet diameter D and the Froude number Fr of the flow through the nozzle.[20] See mixit-io SS4.3 for calculation of S.

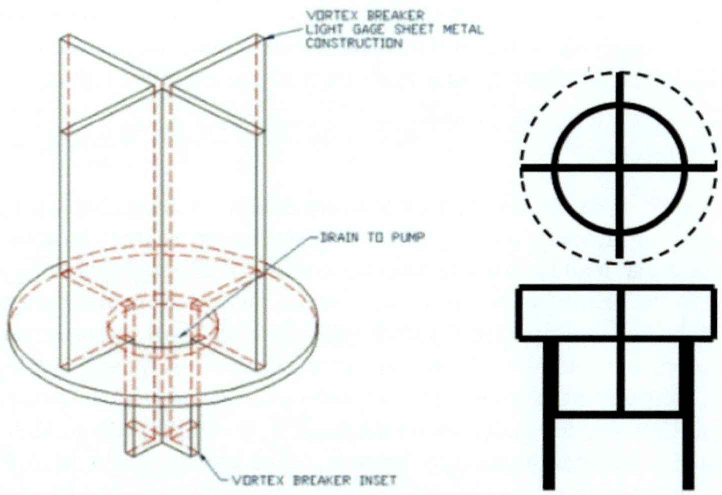

FIGURE 4.14 Schematic of a cross type of vortex breaker.[19]

4.4.7 Draft Tubes

A draft-tube is a cylindrical pipe concentrically located inside a mixing vessel. It is typically anchored to the wall by multiple supporting braces. A draft-tube-impeller (DTI) agitation system incorporates one or more axial impellers inside the draft tube. These are mounted on the main agitation shaft, as depicted in Figure 4.15. The main purpose of the depicted down-pumping DTI agitation system is to create forced axial flow in the mixing vessel and easily release gaseous substances at its top. Design of a typical DTI system for the high volumetric flow of gas to the volume of liquid (vvm) gas–liquid mixing is described in a WPO publication and corresponding paper.[21,22]

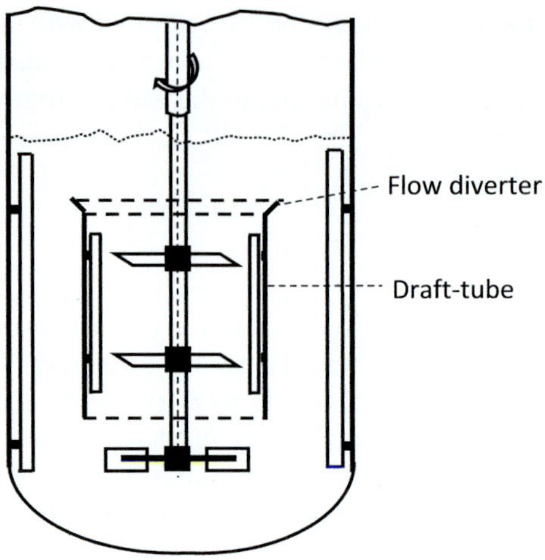

FIGURE 4.15 Schematic of a draft-tube inside a stirred tank for gas–liquid contact.[22]

4.4.8 Ladders

Tall stirred tanks such as fermenters and storage tanks usually have a ladder for mounting, assembling, disassembling, and mechanical maintenance of the vessel internals. The top of the ladder is typically located adjacent to a manway nozzle for easy and safe access into the tank. Like dip pipes, ladders should be properly designed to withstand dynamic fluid forces due to agitation, and the blade pass frequency that develops cyclical loads and can cause fatigue failure.

4.4.9 Steady Bearings

Tall stirred tanks such as oxidation reactors and fermenters use long shafts supported by the top bearing(s). Because of intensive mixing, these long shafts experience vigorous lateral hydrodynamic forces, which can lead to periodic transverse movement of the overhung shaft, and consequently, damage to the bearings and mechanical seals. An option in addressing this issue uses a steady bearing at the tank bottom (see Figure 4.16), where the shaft bottom resides inside a bushing or bearing to prevent lateral shaft movement or vibration. A steady bearing provides additional support to an overhung (cantilever) shaft and increases its critical frequency. They can be particularly helpful for slender shafts whose diameter d_s is significantly smaller than its length l_s (i.e. $d_s \ll l_s$). Final steady bearing design depends on the process application. For certain sticky media, or media with a high solids content, steady bearings may need to be avoided completely.

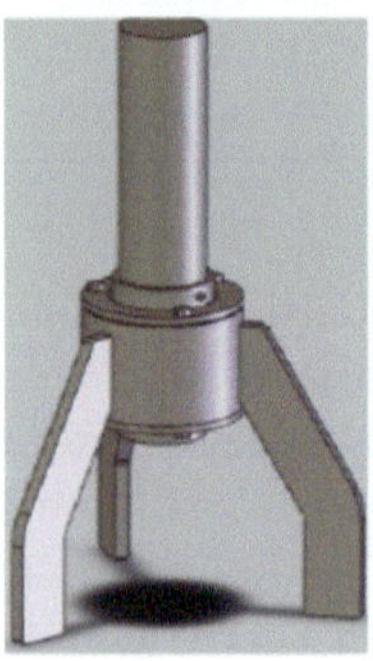

FIGURE 4.16 Image of a steady bearing.[23]

4.4.10 Limit Rings

In the case of slender shafts, limit rings are used to restrict lateral movement (wobbling/deflection) as the impellers experience unsteady fluid forces. Typical limit rings are represented in Figure 4.17 and may or may not include a bushing. Specifically, limit rings are often used in lieu of a steady bearing in tall biotech and slurry reactors.

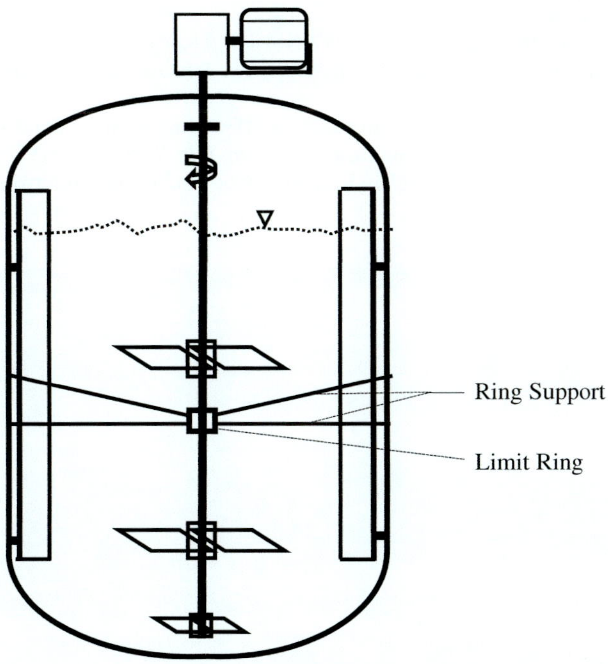

FIGURE 4.17 Image of a limit ring to limit a shaft's lateral support.

REFERENCES

1. ASME Boiler and Pressure Vessel Code (BPVC). www.asme.org/codes-standards/bpvc-standards/bpvc-2023
2. Section VIII, *Div. 1 (Rules for Construction of Pressure Vessels)*, National Board Inspection Code. www.nationalboard.org/SiteDocuments/NBIC/08errata.pdf
3. *What Is a Tank Head & How to Select the Correct One*, T Bailey, Inc., January 30, 2023. http://blog.tbailey.com/tank-heads
4. De Dietrich, *Glass Lined Vessel Brochure*, A Solution Ahead, 2004. www.ddpsinc.com/literature
5. Pfaudler, *RA, RT and RS Series Reactor Data Sheets*, 2012. www.gmmpfaudler.com/file/
6. www.spxflow.com/en/lightnin/pc-impellers
7. www.chemineer.com/products/chemineer.html#impellers
8. www.ekato.com/en-us/products/agitator-components/impellers/
9. www.dedietrich.com/en/solutions-and-products/reaction/mixing/glaslockr-system
10. www.pfaudler.com/en/products/mixing-components#lightbox
11. Kar, K. K., Cope, R. F., Sandor, S., Pennington, A., *Tickler for Slurry Reactors and Tanks*, US Pat. 6,955,461 B2, 2005.
12. Cope, R. F., Kar, K. K., KT3: A Novel Tickler for Solid Removal from Slurry Vessels, *Ind. Eng. Chem. Res.*, 48, pp. 4990–4997, 2009.

13. Kar, K., Piras, L., Monaghedu, M., Bai, H., *A Novel Slinger for Efficient Reflux Condensate Distribution in a Boiling Reactor, Presentation at North American Mixing Forum (NAMF) Mixing XXII*, NAMF, Victoria, British Columbia, Canada, June 2010.
14. Kar, K. K., Somasi, M., Cope, R. F., Adrian, D., Pressler, J., *Mechanical Gas Pocket Preventer/Breaker*, US Pat. 8,864,880, 2014.
15. https://en.wikipedia.org/wiki/Continuous_stirred-tank_reactor
16. Oldshue, J. Y., *Fluid Mixing Technology, Chemical Engineering*, McGraw-Hill Pub. Co., New York, 1983.
17. www.ddpsinc.com/de-dietrich-expertise/mixing/baffles#beavertail-baffles-1285249897488
18. www.pfaudler.com/en/products/mixing-components
19. https://cr4.globalspec.com/thread/18720/Vortex-Breaker
20. http://marineman.ir/wp-content/uploads/2015/03/ANSI-HI_9.8.pdf
21. Kar, K., Piras, L., *Impeller Draft Tube Agitation System for Gas-Liquid Mixing in a Stirred Tank Reactor*, WO/2001/041919, 2001.
22. Lueske, J., Kar, K., Piras, L., Pressler, J., Power Draw and Gas-Liquid Mass Transfer Characteristics of a Stirred-Tank Reactor with Draft Tube Configuration, *Chem. Eng. Technol.*, 38(11), pp. 1993–2001, 2015.
23. https://proquipinc.com/when-to-use-a-steady-bearing-in-your-industrial-mixer-shaft-design/

5 Hydrodynamics

5.1 INTRODUCTION

Many desirable mixing characteristics are observable in the flow patterns produced within a process fluid. Fluid distribution, a foremost mixing function, transports fluid elements to all regions of the vessel by bulk circulation or macroscale mixing behavior. These bulk macro-mixing features dissipate the imparted energy through collapsing macro-eddies that form smaller meso-eddies and then even smaller micro-eddies, a process known as dispersion. Dispersion, which is strongly dependent on the specific energy (ε. imparted into the fluid, is seen in the onset and continuation of turbulence. It leads to micro-mixing and micro-eddies whose length scale can approach that of λ_k, the Kolmogorov length scale of turbulence.

$$\lambda_k = \left(\nu^3 / \varepsilon\right)^{1/4} \tag{5.1}$$

In order to maximize mixing effectiveness, micro-mixing on a length scale even smaller than the Kolmogorov scale is essential. This is achieved, however, only by the molecular diffusion process known as reactive mixing. Since diffusion is typically among the slowest mixing sub-processes, it naturally limits the effectiveness and the time scale of turbulent mixing.

5.2 MIXING REGIMES

The impeller Reynolds number, *Re*, sheds light on whether a flow is viscous force dominated i.e., laminar, or inertial force dominated, i.e., turbulent. The following conditions specify the nature of flow within a stirred vessel, as depicted in Figure 5.1.

Laminar flow regime, $Re \leq 10$
Fully turbulent flow regime, $Re \geq 10^4$
Transition flow regime, $10 < Re < 10^4$

A log-log plot of the Power number, N_P, vs. Reynolds number, *Re*, curve (Figure 5.1a) shows three distinct regimes that depend on the dominant flow phenomena in the tank. The laminar flow regime is represented by a linear portion ($N_P = K_L / Re$ with negative slope) up to an impeller Reynolds number of ~10, followed by a transitional flow regime showing non-linear behavior at Reynolds numbers between 10 and ~10^4, and subsequently, by a turbulent flow regime in which the curve flattens to a fairly straight line ($N_P = K_T$) at Reynolds numbers beyond ~10^4. Values for constants characterizing laminar and fully turbulent flows, K_L and K_T, respectively, depend on the agitation system geometry.

Figure 5.1b shows circulation or flow number, N_{Fl}, and mixing time number, τ_m, as a function of Reynolds number, *Re*, in the laminar, transitional, and turbulent regimes.[1] Published K_L values for typical hydrofoils and standard pitched-blade turbines (PBT) are 27.4 and 43.2, respectively[2] and are independent of ($^D/_T$).

In the case of laminar flow, inertial forces quickly subside under the action of higher viscosity. The rotating impeller creates a large shear stress [$\tau = \mu\ (^{du}/_{dy})$] that continuously stretches and deforms fluid elements it is circulating within the working media. These fluid elements are continually thinning and elongating when passing through the high-shear regions until dimensionally too small to be further influenced by the velocity gradient ($^{du}/_{dy}$). The initial macro-mixing eddy

DOI: 10.1201/9781003606390-5

length scale, λ_l, is on the order of the impeller diameter, D. However, it gradually decreases to the Kolmogoroff micro-mixing length scale, λ_k, as flow energy, ε, dissipates within the vessel, i.e., $\lambda_l = \vartheta(D) \Rightarrow \lambda_k = \left(\nu^3 / \varepsilon\right)^{1/4}$. This is depicted schematically in Figure 5.2. At this scale, a slow interstitial molecular diffusion process occurs across disintegrating lamellar elements.

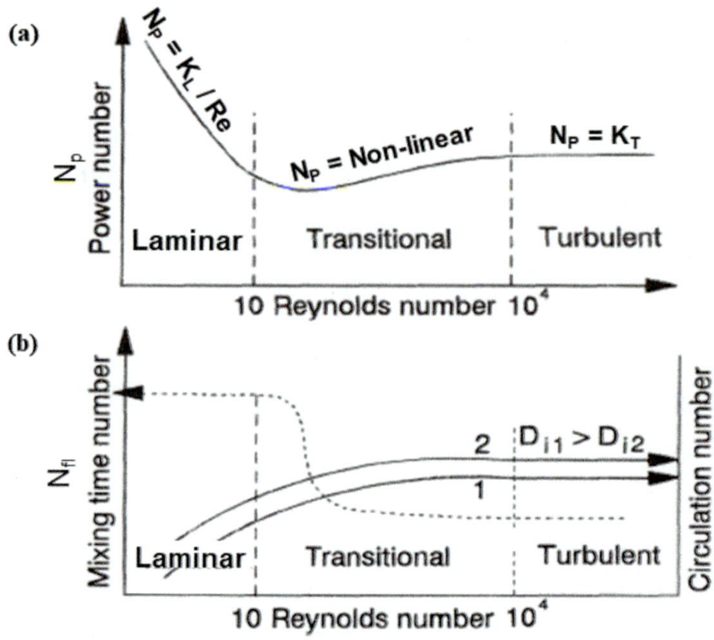

FIGURE 5.1 Log-log plots: (a) power number vs. Reynolds number for Newtonian fluid in standard baffled tanks, and (b) mixing time number and circulation number vs. Reynolds number.[1]

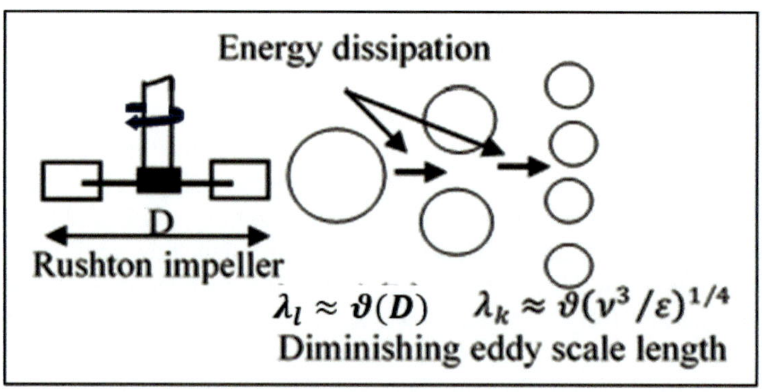

FIGURE 5.2 Laminar dissipation of freely produced vortices.[3]

For large Reynolds number (turbulent) flows, the smaller eddies are not influenced by impeller type. They are independent of macro-scale flow patterns and isotropic in nature. The larger turbulent eddies, on the other hand, depend on impeller type and the most energetic and anisotropic eddies. These larger eddies facilitate the blending process via convective mass transfer in a process tank.

Typical applications include both large and small eddies whose unique natures dictate that they be treated independently in computational fluid dynamics modeling, with larger eddies being modeled directly and smaller ones being solved using appropriate numerical methods. For all other practical purposes, however, the local turbulence intensity, ε_t, is determined based on the average specific power, ε. As expected, naturally, the maximum turbulence intensity ($\phi > 1.0$) is experienced near the impeller while the minimum intensity ($\phi < 1.0$) occurs farthest from the impeller and near the vessel wall. Figure 5.3a–b shows iso-energy contours inside stirred tanks with either a Rushton turbine or propeller.[3]

$$\varepsilon_t = \phi \cdot \varepsilon \tag{5.2}$$

where, ϕ represents the iso-energy contour parameter for an impeller.

For processes involving formation of fragile crystals during agitation, the Kolmogoroff length scale should be larger than the crystal size, thus the particles are less susceptible to breakage by micro-scale eddies. Since the kinematic viscosity ν of the working media remains constant, specific energy, ε, dictates the Kolmogoroff length scale, λ_k.

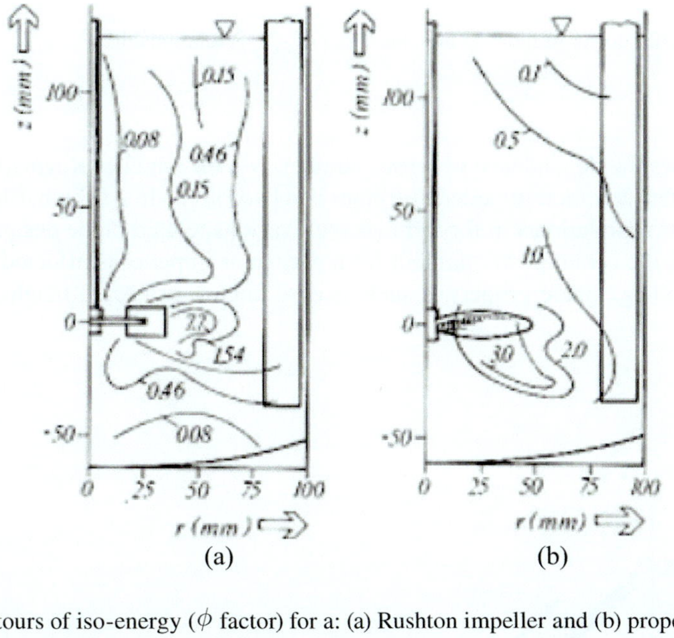

FIGURE 5.3 Contours of iso-energy (ϕ factor) for a: (a) Rushton impeller and (b) propeller.[3]

5.3 FLOW AND POWER CHARACTERISTICS

As described in Chapter 2, the non-dimensional power number, N_p, is the ratio of hydraulic resistance force to inertial force experienced by an impeller. Fluid resistance is primarily attributed to form drag and is a consequence of blade geometry, approach velocity and viscosity of the working media. Similarly, the dimensionless flow number, N_{Fl}, addresses the discharge flow from the impeller with respect to its diameter and velocity. The power, P, consumed by an impeller is related to its volumetric circulation rate, $\dot{Q}_f$, (pumping capacity) and the induced velocity head, ΔH, by

$$P = \dot{Q}_f \, \rho g \, \Delta H \tag{5.3}$$

Hence, power number, N_p, and flow number, N_{Fl} (N_Q), are related. Figure 5.4 shows this relationship for multiple axial impellers noted in the adjacent list.[4]

$$N_{Fl} = 0.77 N_P^{1/3} \tag{5.4}$$

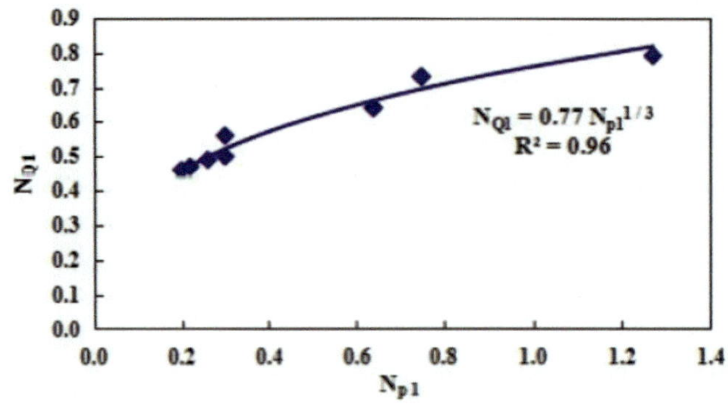

Type	N_{pl}	N_{Ql}
P-4	1.27	0.79
Lightnin A310	0.30	0.56
HE-3 D/T = 0.5	0.20	0.46
HE-3 D/T = 0.4	0.22	0.47
HE-3 D/T = 0.3	0.26	0.49
HE-3 D/T = 0.2	0.30	0.50
Lightnin A320	0.64	0.64
Lightnin A340	0.64	0.64
Lightnin A315	0.75	0.73
Lightnin A345	0.75	0.73

FIGURE 5.4 Impeller power number to flow number (N_p-N_{Fl}) relationship.[4]

Figure 5.5 shows the dependence of power number, N_p, on impeller Reynolds number, *Re*, and flow regime for a few common impellers (turbines and hydrofoils) in a fully baffled tank. Figure 5.6 presents the N_p vs. *Re* relationship for turbines with various related blade designs. Although these curves are typical, the actual power number for a particular impeller is affected by the fine details of its construction (e.g., blade properties such as cord angles, projected height, edge and surface finish, etc.).[5]

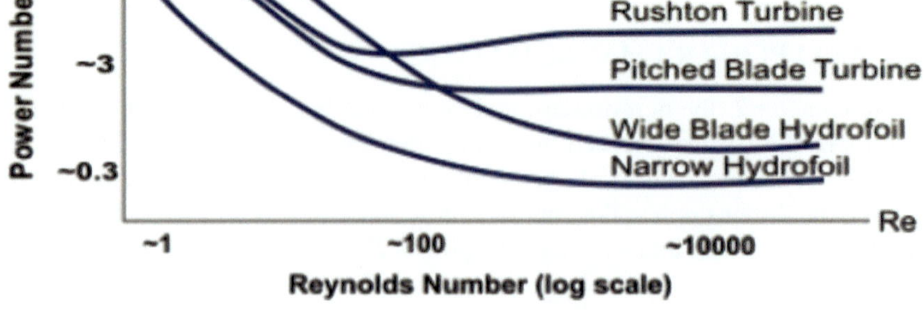

FIGURE 5.5 Power number impeller Reynolds number relationships for Rushtons, pitched blades and hydrofoils.[5]

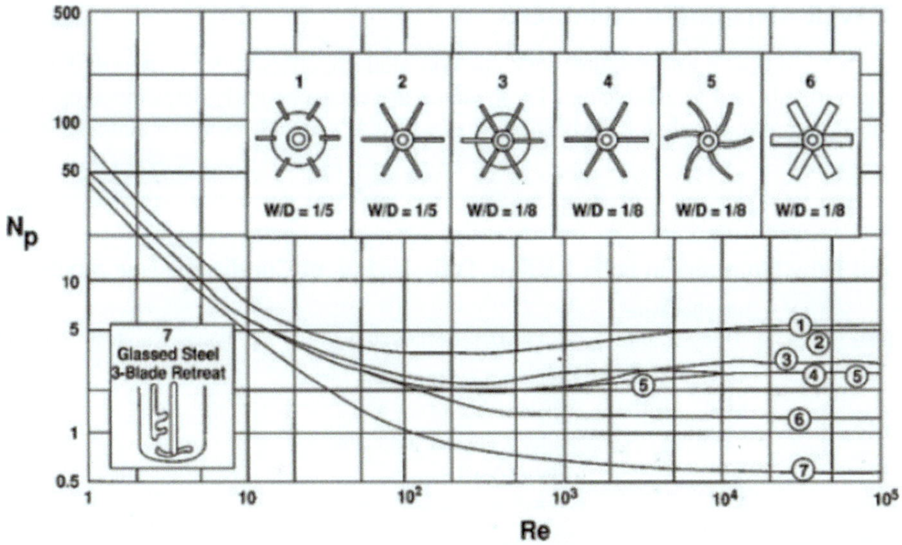

FIGURE 5.6 N_p-Re correlations for various impellers in standard baffled tanks with Newtonian fluids.[6]

As an illustration, impeller flow number to Reynolds number $\left(N_{fl} - Re\right)$ relationships for a pitched-blade turbine (fully turbulent flow Np = 1.27) in laminar, transition and fully turbulent conditions are represented in Figure 5.7. Flow numbers are constant in the laminar and fully turbulent regimes, but their values vary with D/T ratio. In the laminar zone, published flow numbers for a narrow-blade, high-efficiency hydrofoil and a standard PBT are 0.214 and 0.295, respectively.[2]

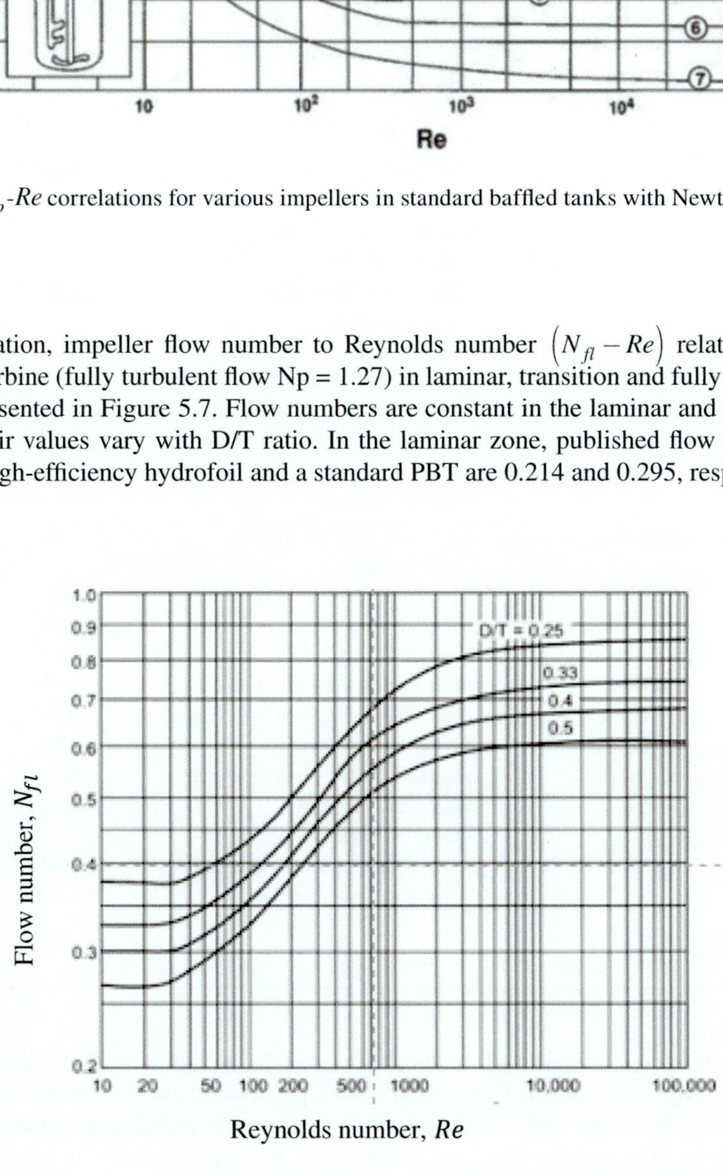

FIGURE 5.7 Flow number to Reynolds number relationships for PBT impellers with varying D/T ratio (0.25–0.5) in a baffled tank containing Newtonian fluids.[7]

5.4 POWER AND FLOW NUMBERS

Power number and flow number values for some common impellers under fully turbulent conditions are presented in Table 5.1 and Table 5.2. These values are experimentally determined using appropriate D/T ratios in multiple scaled, fully baffled tanks with Z/T and C/T fixed at 1 and 0.33, respectively.

TABLE 5.1
Power and Flow Numbers in Fully Baffled, Fully Turbulent Conditions (see mixit-io SS5.1)

Manufacturer	Impeller Type/ Designation	Description/Flow	Power Number	Flow Number
SPX Flow—Lightnin	A200	PBT (4 blades/45°)[4]—Mixed	1.27	0.79
	A100	Propeller (1.5 pitch)—Axial	0.87	0.77
	A310	Airfoil (32° chord angle) low shear– Axial	0.30	0.56
	A315	Hydrofoil (4 blades)[4]—Axial	0.75	0.73
	A320	Hydrofoil (3 blades)—Axial	0.65	0.68
	R200	Flat-blade Turbine (4 blades)—Radial	3.40	0.64
	R100	Rushton Turbine (6 flat blades)—Radial	5.75	0.72
	R130	Smith Turbine (6 concave blades)—Radial	3.20	0.80
NOV—Chemineer	P4	PBT (4 blades/45°)[4,10,11]—Mixed	1.30	0.79
	HE3	High efficiency low shear[11]—Axial	0.28	0.48
	Maxflo-W	Hydrofoil (4 blades)[11]—Axial	1.0*	0.8*
	D6	Rushton Turbine[4,10]	5.2	0.72
	BT6	Baaker Turbine[4,12](6 parabolic blades)—Radial	2.30	0.55
	CD6	Smith Turbine[4,12] (6 concave blades)—Radial	3.20	0.80
Ekato	Single Intermig	Clockwise rotation—Axial	0.45	0.26
	Dual Intermig	Clockwise rotation—Axial	0.66	0.19
	Viscoprop	25 deg—Mixed	0.26*	0.60*
Mixtech	HA720	High efficiency low shear—Axial	0.37	0.70
	HA 800	High efficiency low shear—Axial	0.30	0.60
	HA845	PBT (4 blades/45°)	1.60	0.85
PMSL	PH-A1	High-efficiency hydrofoil—Axial	0.55	0.72
Fluid Fusion	Cowles	High-shear sawtooth blade[10]—Radial	0.45	0.26
	RCI	Retreat curve impeller—Radial (low shear, 4 blades)[10]	2.52	0.74

*Approximate values

Practicing engineers are advised to contact the agitator manufacturers to obtain the N_p vs. Re curves of their proprietary or special impellers. Multiple review articles and classic books[8,9] address Power number to Reynolds number correlations of common radial and axial impellers used in industrial mixing. When using these data, however, care must be taken in identifying applicable geometrical ratios. Experience has shown that fibrous materials often hang up on impeller blades and significantly increase the effective impeller power number. While mixing under fully turbulent

TABLE 5.2
Power Numbers and Flow Numbers of a Few Common Glass-Lined Impellers (w/standard geometry and one flat/beavertail baffle) Under Fully Turbulent Flow Conditions

Manufacturer	Impeller Type/ Designation	Description/Flow	Power Number	Flow Number
Pfaudler	Cryo-Lock® RCI	Retreat curve impeller (3 blades)—Radial	0.40	0.23
	Cryo-Lock PBT	Pitch-blade turbine (4–45° blades)—Mixed	1.00	0.65
	Cryo-Lock FBT	Flat-blade turbine (4 blades)—Radial	2.70	0.65
	Cryo-Lock TBF	Turbofoil (4 blades)—Axial	0.35	0.48
DeDietrich	Glaslock®-45°	PBT (4 blades)—Mixed	0.65	0.34
	Glaslock-90°	FBT (4 blades)—Radial	1.22	0.39
	Glaslock-60°	Hydrofoil (3 blades)—Axial	0.83	0.46
	Glaslock-30°	Hydrofoil (3 blades)—Axial	0.35	0.20

conditions in a pilot reactor, the effective power number of a high-efficiency hydrofoil impeller was once doubled as cellulosic fibers got hung up on the blades.

In the case of an unbaffled tank, even under a fully turbulent flow condition, the power number of the mixing impeller is a function of the Reynolds number and continues to decrease, as shown in Figure 5.8.

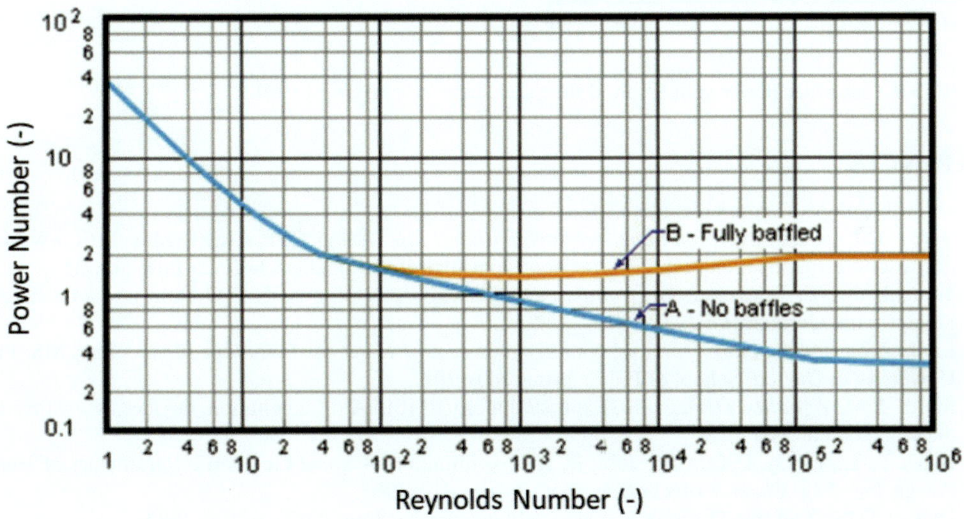

FIGURE 5.8 Power number vs. Reynolds number curve for fully baffled and unbaffled tanks.[13]

5.5 IMPELLER CIRCULATION VS. SHEAR

The mixing energy imparted into a working media is expended in two distinct forms: fluid circulation (flow) and shear. Both flow characteristics play critical roles in the mixing process. Their relative requirements vary significantly in different applications, however. For example, controlling crystal morphology in a crystallizer is promoted by faster circulation as opposed to higher shear, which often reduces crystal size. The operation of high viscosity resin mixers typically prefers high circulation over high shear to avoid thermal degradation caused by local viscous heating $\left(\dot{q}_v \ \alpha \ \dot{\gamma}^2\right)$. On the other hand, a viscous Bingham plastic slurry may need high shear rate for its flowability and/or solid dissolution in the stirred vessel.

Figure 5.9 is a qualitative graphical representation of the power split between flow and shear for different impellers.[14] Descriptions of basic impeller types are presented in a table by Oldshue (1983).[15]

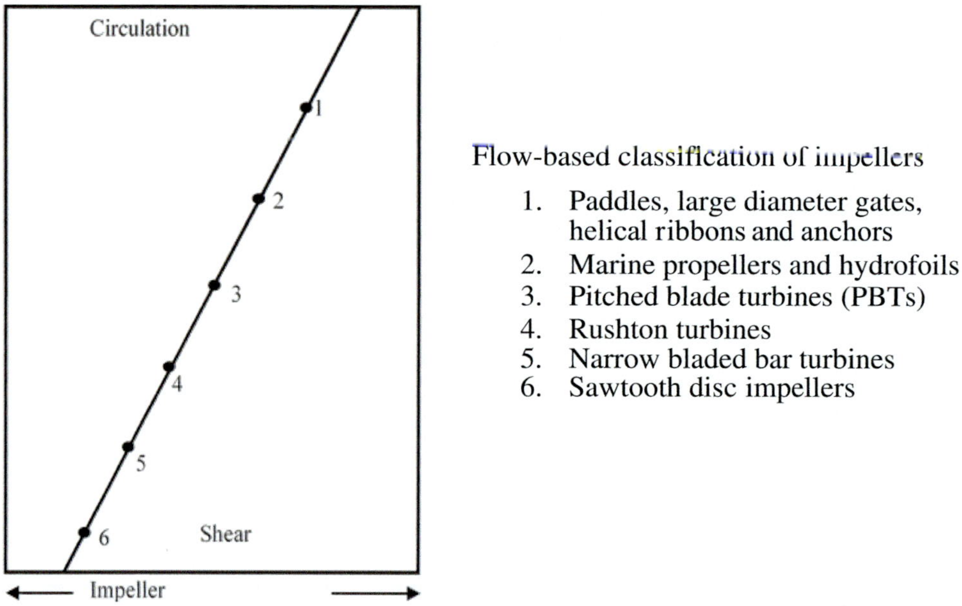

FIGURE 5.9 Impeller power split for fluid flow and shear in a mixing vessel.[12]

REFERENCES

1. www.thermopedia.com/cn/content/958/
2. Benz, G., *Consider Hydrofoil Impellers for Laminar-Flow Mixing*, CEP, AIChE, May 2018. www.aiche.org/resources/publications/cep/2018/may/consider-hydrofoil-impellers-laminar-flow-mixing
3. Todtenhaupt, E. K., Zeiler, G., *Handbook of Mixing Technology*, EKATO Ruhr-und Mischtechnik, Schopfheim, Germany GmbH, 1991.
4. Li, J., *Effect of Impeller Type and Number and Liquid Level on Turbulent Blend Time*, MS Thesis, University of Dayton, School of Engineering, May 2017.
5. Smith, J. M., *Agitation Devices*, 2011. https://doi.org/10.1615/AtoZ.a.agitation_devices, www.thermopedia.com/content/549/
6. Bates, R. L., Fondy, P. L., Corpstein, R. R., Examination of Some Geometric Parameters of Impeller Power, *Ind. Eng. Chem. Process Des. Dev.*, 2(4), p. 310, 1963.
7. Dickey, D. S., Tackling Difficult Mixing Problems, *Chem. Eng.*, 12(7), p. 102, 1984.
8. Uhl, V. W., Gray, J. B., *Mixing Theory and Practice*, Academic Press, 1966.
9. Nagata, S., *Mixing: Principles and Application*, John Wiley & Sons, Inc., New York, 1975.
10. www.fusionfluid.com/FusionFluidEquipmentLLC/html/mixing_impellers.html
11. Reeder, M., Fasano, J., *Achieving Mixing Efficiency with Intermediate Power Number Impellers, Paper no. 166c*, Presented at the Annual AIChE Meeting, Dallas, TX, November 1–5, 1999.
12. https://cercell.com/support/bactovessel-details/turbine-power/
13. www.bakker.org/dartmouth06/engs199/Lect1-Blend.pdf
14. McDonough, R. J., *Mixing for the Process Industries*, Van Nostrand, Reinhold, New York, 1992.
15. Oldshue, J. Y., *Fluid Mixing Technology*, pp. 58–64, McGraw-Hill Publications Co., New York, 1983.

6 Blending of Miscible Liquids

6.1 INTRODUCTION

Blending operations in mechanically agitated vessels seek to homogenize miscible components of dissimilar viscosity and/or density and/or concentration. The agitation system mixes by generating inter-reactor flows with intensities that can range from mild (i.e., laminar) to severe (i.e., fully turbulent). If the blending encompasses reaction, fluid motion in the reactor may help dissipate the reaction heat or transport it to and from heat exchange surfaces that may help promote and control the reaction. Typical blending applications are found in countless formulations of agricultural, food, household, healthcare and other products. Admittedly, blending is the simplest of all mixing processes employed in the chemical and allied industries. It can utilize both open impellers and proximity impellers, provided none of the system viscosities exceed the 50Pa-s range. Helical impellers and planetary agitation systems are ideally suited for extremely high viscosity media in the range of 50Pa-s to 1000Pa-s. Typically, pastes and dough-like media, such as caulks and sealants with a Bingham-pseudoplastic consistency of >0.5kPa-s, are blended using kneaders and extruders. While the blending of high viscosity working media (≥20Pa-s) is discussed in Chapter 14, Table 6.1 briefly describes appropriate agitation systems for a few typical blending applications.

TABLE 6.1
Common Impellers and Applications

Agitation System	Ideal Viscosity Range (Pa-s)	Examples of Applications
Axial flow open impellers:		Formulations:
(a) Hydrofoils	< 20	(a) Home, healthcare and personal care products
A315		(b) Agricultural products such as fertilizers,
A320		insecticides and pesticides
MaxFlo		(c) Paints
Viscoprop		
(b) Pitch-blade turbines	< 20	Homogeneous initiators and
A-200		catalyst with monomers
P-4		Food and pharma products
Glass-lined equivalents		
Anchors	20–50	Resins and gels
Helical ribbons	50–1,000	Sealants
Kneaders and extruders	>500	Caulks, pastes and doughs

6.2 FLUID MOTION AND BULK FLUID VELOCITY

Process requirements dictate the needed degree of agitation for blending and motion. In the early 1970s, Chemineer, Inc. introduced the well-known "Chem-Scale" to identify the degree of needed agitation[1] based on the viscosity and specific gravity ratios of the components subjected to the blending operation. The required dynamic response, i.e., the bulk fluid velocity, V_b, between 6–60ft/

DOI: 10.1201/9781003606390-6

min sets the agitation scale from 1–10, as shown in Table 6.2. The characteristic bulk fluid velocity in the tank is calculated as follows:

Effective pumping capacity of the impeller,

$$\dot{Q} = N_{Fl} N D^3 \tag{6.1}$$

where, N_{Fl} = impeller flow number
N = impeller rotational speed
D = impeller diameter.

Cross-sectional area of the tank,

$$A_{cs} = \frac{\pi}{4} T^2 \tag{6.2}$$

where, T = tank diameter.

bulk fluid velocity,

$$V_b = \dot{Q} / A_{cs} \tag{6.3}$$

TABLE 6.2
Chem-Scale for Degree of Agitation[1]

Scale of Agitation	Bulk Fluid Velocity, ft/min	Description
		Agitation levels 1 and 2 are characteristics of applications
1	6	requiring minimum fluid velocities to achieve the process result.
2	12	Agitators capable of level 2 will:
		• Blend miscible fluids to uniformity if specific-gravity differences are less than 0.1.
		• Blend miscible fluids to uniformity if the viscosity of the most viscous fluid is less than 100 times that of the other(s).
		• Establish complete fluid batch control.
		• Produce a flat, but moving, fluid-batch surface.
		Agitation levels 3 to 6 are characteristic of fluid velocities in most
3	18	chemical process industries agitated batches.
4	24	Agitators capable of level 6 will:
5	30	• Blend miscible fluid to uniformity if specific-gravity
6	36	differences are less than 0.6.
		• Blend miscible fluids to uniformity if the viscosity of the most viscous material is less than 10000 times that of the other(s).
		• Suspend trace solids (<2%) with settling rates of 2 to 4 ft/min.
		• Produce surface rippling at lower viscosities.
7	42	Agitation levels 7 to 10 are characteristic of applications
8	48	requiring high fluid velocity for the process result, such as in
9	54	critical reactors.
10	60	Agitators capable of level 10 will:
		• Blend miscible fluids to uniformity if specific-gravity difference is less than 1.0.
		• Blend miscible fluids to uniformity if the viscosity of the most viscous fluid is less than 100,000 times that of the other(s).
		• Suspend trace solids (<2%) with settling rates of 4 to 6 ft/min.
		• Provide surging surfaces at low viscosities.

6.3 SPECIFIC POWER

Shaft power, i.e., the energy imparted or delivered to the fluid by the agitator to accomplish blending of the constituents, is the most important mixing parameter. Specific power, ε, defined as power per unit mass of the working media, is often considered to scale up from laboratory or pilot scales to the production scale. It varies from low power 0.2W/kg (~1hp/1000gal for sp. gr. ≈ 1) for blending of low viscosity media to high power level, i.e., as much as 4W/kg (~20hp/1000gal), for blending extremely high viscosity substances such as pastes and doughs.

Specific power, ε, is calculated based on the shaft power, P, and the working media volume as follows:

$$P = \rho N^3 D^5 N_p \tag{6.4}$$

where, N_p = impeller power number.

$$\varepsilon = P/(\rho.V) = \frac{N^3 D^5 N_p}{V} \tag{6.5}$$

where, ρ and V are respectively the density and volume of the working media.

6.4 FLUID FLOW PATTERN

The fluid flow patterns generated by various impellers in the stirred tank are well-known in the literature. Figure 6.1 shows typical flow patterns of an axial, mixed flow, and radial flow turbines in a vessel equipped with four standard flat baffles. The top-to-bottom turnover is the most desired conditions for ideal blending operations. Often multiple impellers are used to avoid slow flow or dead regions in tall tanks. Swirling flow induced by radial impellers in unbaffled tanks must be avoided.

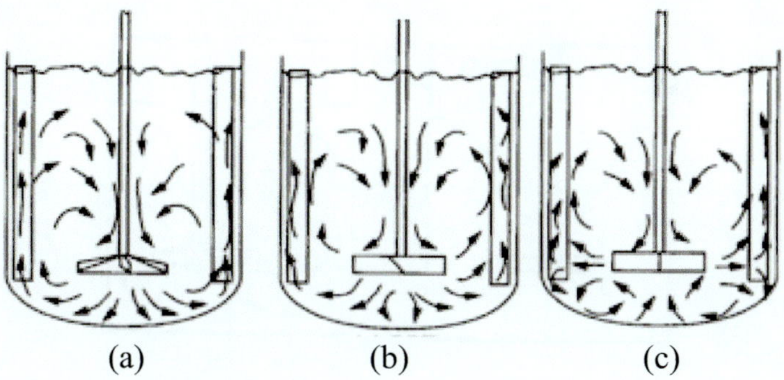

FIGURE 6.1A–C Schematic showing fluid flow patterns[2] generated by the (a) axial impeller (Hydrofoil), (b) mixed impeller (PBT), and (c) radial impeller (FBT).

Figure 6.2 shows positioning of the multiple impellers inside the tank. The off-bottom clearance, C, of the bottom impeller and the spacing, S, of the impellers with respect to one another, and the submergence, Z_s, of the upper impeller are critical parameters that must be taken into consideration for the design of the agitation system to ensure an ideal synergistic flow pattern in the tank.

It is hard to produce the desired axial flow pattern in an unbaffled tank using a centrally mounted agitation system like that shown in Figure 6.1. Furthermore, depending on the product of the Froude number and Reynolds number, i.e., $Fr \times Re$, the swirling flow may cause a surface vortex as depicted in Figure 6.3. Although a minor vortex may be desirable for addition and draw-down of a secondary

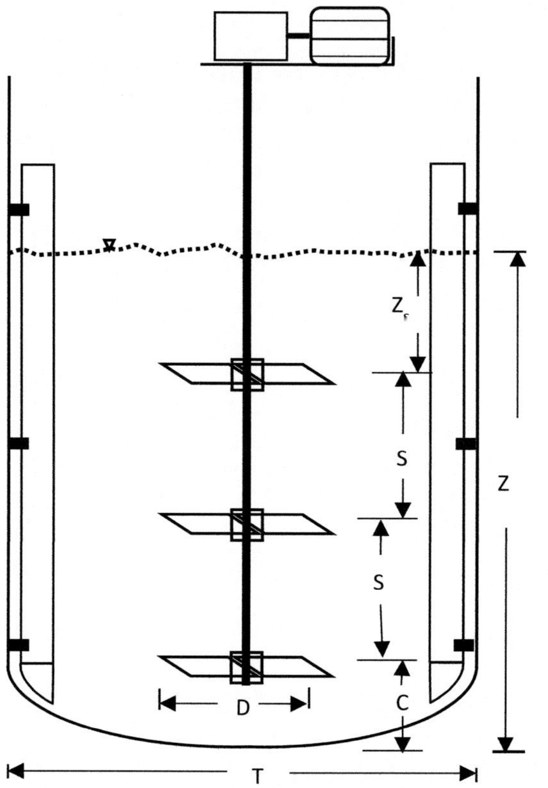

FIGURE 6.2 Schematic showing multiple impeller positions inside a stirred tank.

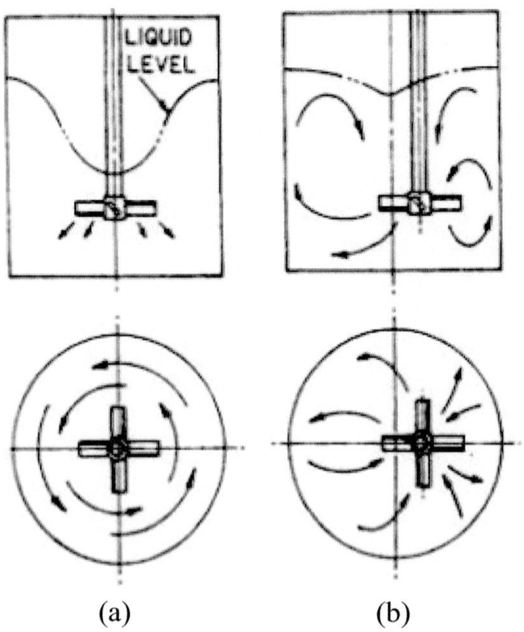

FIGURE 6.3 Agitator flow patterns: (a) axial or radial impellers without baffles produce a vortex and (b) an off-center location of the shaft reduces the formation of a vortex.[3]

component, a strong vortex is not acceptable for a couple of major reasons. It creates insignificant top-down circulation for an ideal blending operation, and since the vortex is often hydrodynamically unsteady, is detrimental to the mechanical sturdiness of the agitation system. Specifically, if the vortex touches or intersects the impeller blades, it significantly diminishes the longevity of the agitation system. Furthermore, it also creates uncertainty in the value of the impeller power number.

While blending can be accomplished with any of multiple unbaffled tank configurations, the three most common are depicted schematically in Figure 6.4. These are (a) vertical offset mounted, (b) angled offset mounted and (c) horizontal side mounted agitation configurations.

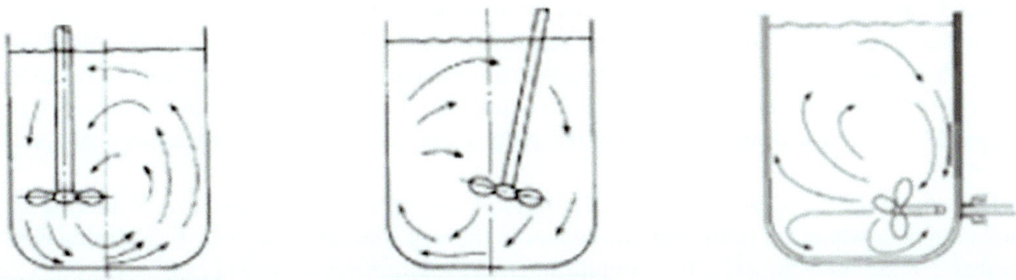

(a) Vertical offset mounted (b) Angled offset mounted (c) Horizontal side mounted

FIGURE 6.4A–C Agitation configurations in unbaffled tanks to prevent swirl flow pattern.[4]

6.5 MIXING TIME

Batch mixing time, θ, is the time difference between the instant a new component is added to the existing one and the time when the vessel contents have reached a specific degree of uniformity. Since the approach to 100% homogeneity is an exponential asymptotic function of time, mixing practitioners commonly report a lower uniformity level such as 95% (i.e., θ_{95}). The manner in which a representative concentration profile θ_{95} approaches a steady-state value is depicted in Figure 6.5.

If a higher order of uniformity is essential, i.e., mixing at a molecular level (micro-mixing) for reaction, θ_{99} is determined from the expression $\theta_{99} \approx 1.6\theta_{95}$. Since, as mentioned earlier, inhomogeneities generally decline exponentially, blend time, θ_x, to achieve the required degree of consistency, x, is expressed as[5]

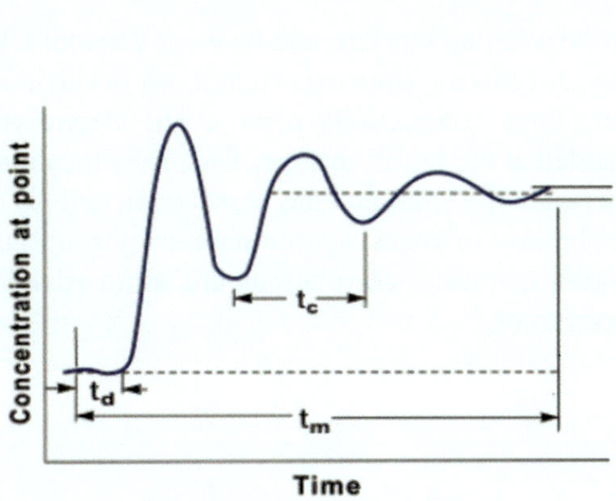

FIGURE 6.5 An idealized mixing time experiment.[6]

$$\theta_x = \theta_{95} \frac{\ln(1-x)}{\ln(0.05)} \tag{6.6}$$

Since mixing time is often considered the most critical parameter in assessing mixer performance, experiments are often carried out in geometrically scaled-down mixers using tracers (e.g., pH carriers, dyes and salt solutions) in real and simulated fluids. If the moment of tracer addition is taken as zero in Figure 6.5, t_d is dead time before that addition is first detected, t_c is a characteristic circulation time, and t_m is the 95% mixing time, defined by the last measurement whose non-uniformity was outside of the depicted 5% band of the total concentration change. Mixing time, t_m, or θ_{95} for an impeller in a fully baffled stirred tank containing a working media height that is equal to the tank diameter is sometimes used to quantify a mixer's blending efficiency. Numerous articles in the mixing literature address blend time correlations. An extensive review of the related literature prior to 1990 is presented in references 7 and 8. Authors typically followed FMP Consortium's recommended proprietary correlations[9] for their R&D and scale-up activities.

Table 6.3 shows a few well-established empirical correlations in literature for calculating θ_{95} for a single impeller where the liquid height is equal to the tank diameter, i.e., $Z = T$.[7,9] Mixing Reynolds number, Re_V, is characteristic of mixing throughout the entire vessel. It uses tank diameter, T, as the characteristic length and is defined as

$$\mathrm{Re}_V = \frac{T^2 \rho}{\mu \theta_{95}} \tag{6.7}$$

where, μ is the dynamic viscosity of the working media.

The dissertation[8] refers to the above mixing Reynolds number as the inverse of Fourier number, F_O, i.e., $Re_V = \frac{1}{F_O}$.

Blend time experimental studies with Newtonian and non-Newtonian (pseudoplastic) fluids were conducted in two transparent Perspex tanks (T = 0.305m and 0.609m) and in a single steel tank (T = 1.830m) with four viewing windows. All tanks had ASME dish bottoms, and each was equipped with four flat baffles. Baffle width and its clearance from the wall were T/12 and T/60. The three common types of impellers that were used were pitch-blade turbine (PBT; D = T/3, T/2), flat-blade turbine (FBT; D = T/2) and hydrofoils (Lightnin A310 and Chemineer HE3 impellers; D = T/2)). The value of impeller off-bottom clearances, C, was fixed at T/3.

The Newtonian and non-Newtonian working media were Versicol E11 (Acrylics) and Nitrosol™ 250MR (HEC), respectively. All mixing time experiments were carried out using an electrical conductivity technique utilizing three conductivity probes. The electrolyte tracer (nitric acid diluted with working media) was added at the liquid surface. The probe locations were (a) below the impeller blade tip, (b) halfway between the shaft and the vessel wall, and (c) behind a baffle.[8] Thus, these probes sampled fluids in the highest to lowest agitation intensity zones. Data processing and analysis details leading to the reported empirical correlations are summarized in Table 6.3 and presented comprehensively in the dissertation.[8]

TABLE 6.3
Mixing Time Correlations Based on Flow Regime ($Z = T$)[8–9]

Mixing Regime	Mixing Reynolds Number, Re_V	95% Blend Time, θ_{95} [sec]	Fluid Rheology Conditions	RSD*	Remarks
Turbulent	$\gtrsim 10^3$	$\frac{5.20\,(D/T)^{-2}}{N_p^{1/3}\,N}$	Newtonian	±10%	1. N_P obtained from the impeller Reynolds number, $Re = \frac{\rho N D^2}{\mu}$
		$\frac{5.47\,(D/T)^{-2}}{N_p^{1/3}\,N}$	Newtonian and non-Newtonian (pseudoplastic)	±13.6%	2. $\mu = \mu_W$ for non-Newtonian fluids** $\mu_W = k(\dot{\gamma}_W)^{n-1}$
Transitional and Laminar	$\lesssim 10^3$	$\frac{(183)^2\,\mu\,(D/T)^{-2}}{N_p^{\frac{2}{3}}\,N\,Re}$	Newtonian	±17.4%	3. Iteration needed due to dependency of Re_v on θ_{95} (see eq. 6.7)
		$\frac{(193)^2\,\mu\,(D/T)^{-2}}{N_p^{\frac{2}{3}}\,N\,Re}$	Newtonian and non-Newtonian (pseudoplastic)	±32.5%	

*RSD = Relative standard deviation of the proportionality constant; **viscosity of the non-Newtonian (pseudoplastic) fluid near the wall (see reference 8)

In the fully turbulent flow regime, $Re \gtrsim 10{,}000$, while reasonably assuming that mixing time is 4 times the circulation time ($t_m = 4\,t_c$), the following simple "rule-of-thumb" is commonly used to calculate θ_{95}.

$$\theta_{95} = \frac{4\,V}{\dot{Q}} \tag{6.8}$$

where, V is the working media volume and $\dot{Q}$ is the impeller pumping rate.

Mersmann[10] (1975) analyzed experimental results of multiple authors and proposed a simple correlation (see equation 6.9) that addresses mixing performance of many single stage mixers in baffled vessels having the liquid height equal to tank diameter ratio, i.e., $Z/T = 1$. Like the correlations in Table 6.3, in the turbulent flow regime, dimensionless mixing time, $N\theta$, depends on impeller power number, N_p, and the impeller diameter to tank diameter ratio, D/T.

$$N\theta_{95} = 6.7\left(\frac{D}{T}\right)^{-5/3} N_p^{-1/3} \tag{6.9}$$

Later experimental research by Ekato[11] (2000) suggested that equation 6.9 is appropriate for impellers with large diameter ratios $(D/T > 0.5)$ and proposed the following equation that accurately predicts non-dimensional blend times of axial pumping impellers with D/T ratios of 0.1–0.5 with an accuracy of ± 10%.[5]

$$N\theta_{95} = 5.5\left(\frac{D}{T}\right)^{-5/3} N_p^{-1/3} \tag{6.10}$$

A more recent thesis[12] by Strand (2017) suggests a different correlation coefficient for 95% blend time in mechanically agitated vessels with liquid height, Z, approximately the same as tank diameter, T, and flow in the fully turbulent regime.

$$\theta_{95} = \frac{7.5\,(D/T)^{-2}}{N_p^{\,1/3} N} \tag{6.11}$$

Strand conducted experimental investigations using SPX Flow-Lightnin's high-efficiency hydrofoils (A-310), standard PBTs (A-200), and six-bladed Rushton turbines (R-100) in 0.8m and 1.2m diameter Plexiglas® tanks with ASME-Dish bottoms. D/T and C/T ratios were varied from 0.2 to 0.5. Saturated sodium chloride solution fed at the liquid surface was used as the tracer media, and multiple conductivity probes were used to monitor conductivity levels as a function of time. The thesis attributes deviation in the turbulent blend time correlation coefficient to the combination of two factors, i.e., location of the farthest conductivity detection probes from the impeller (the "last-to-blend") and the definition of 95% homogeneity. Reference 8 placed the farthest probe at a distance of $T/3$ below the liquid surface, as opposed to Reference 12, where the probe was positioned behind the baffle at the liquid surface. The 95% homogeneity of the working media in Reference 9 is defined as the point at which the root mean square conductivity fluctuation was 5% of the total conductivity change. However, in the theses, 95% homogeneity was simply defined as the point at which 5% of the magnitude of the total conductivity change was remaining in targeted last-to-blend locations.[12] As expected, the off-bottom clearance ratio, C/T, of the impeller is one of the critical parameters that would affect θ_{95}. Its influence is presented in equation 6.12 as a correction factor[12] to equation 6.11.

$$\theta_{95} = \frac{7.5\,(D/T)^{-2}}{N_p^{\,1/3} N}\left[\left(\frac{C}{T} - \frac{1}{2}\right)^2 + 0.96\right] \tag{6.12}$$

The above mixing time correlations apply to systems where two or more components of approximately equal viscosity already present in the tank are subjected to agitation. However, the blend time for the con-added systems is much faster when the additives are added near the impeller region, wherein mixing is considered to be perfect. A typical addition location would be near the suction zone of an axial impeller or discharge region of a radial impeller, as shown in Figure 6.6. Ideally, the con-addition rate of the secondary fluid should be equal to or less than the impeller circulation (pumping) rate. This is especially important when the fluids have significantly different density and

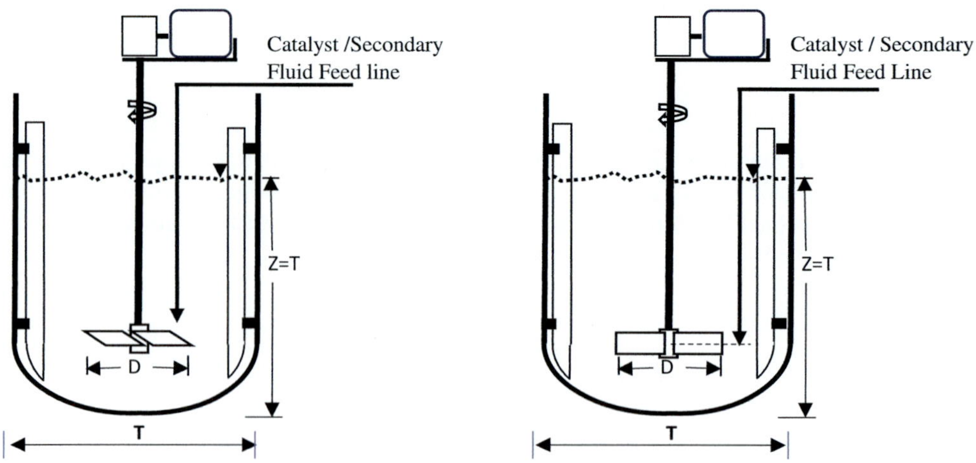

FIGURE 6.6 Ideal locations for a secondary feed line with an (a) axial and (b) radial impeller.

viscosity. The pumping rate of the impeller is calculated from the flow number, N_{Fl}, using the following expression. Note that similar to the power number, flow number is a function of the impeller Reynolds number, as shown in Figure 6.7a–b.

$$\dot{Q} = N_{Fl} N D^3 \tag{6.13}$$

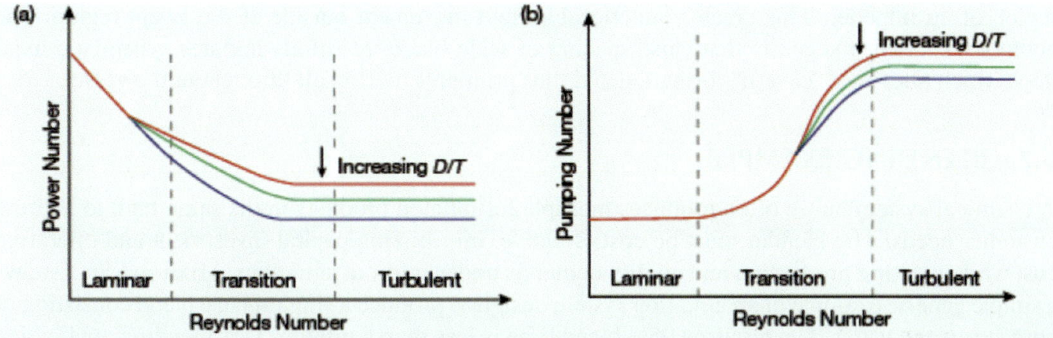

FIGURE 6.7 (a) Generic power number curve showing constant power number in fully turbulent flow, and (b) generic flow (pumping) number curves showing constant flow number in the laminar and turbulent flow regimes, but independent of D/T in the laminar region.[13]

6.6 DESIGN GUIDELINES

For blending applications, during scale-up the geometric similarity of the tank internals are maintained and the specific power remains constant. Table 6.4 shows typical sizing rules for the tank internals with respect to the tank diameter, liquid height, and working media viscosity.

TABLE 6.4
Rules of Thumb for Open Impeller Sizing and Locations

Number of impellers	n = 1 (Z/T $\lesssim$ 0.9); n = 2 (0.9 $\lesssim$ Z/T $\lesssim$ 1.7); n = 3 (1.7 $\lesssim$ Z/T $\lesssim$ 2.5)
Impeller dia/Tank dia	D/T = 0.30–0.56 (with open-impellers)
Off-bottom clearance	C = T/4 (2:1 SEL); C = T/6 (Dish bottom); C ≈ T/3 (Flat bottom and Z/T≈1)
Impeller spacing	S = 0.66 D–1.2 D [high (5–50Pa-s) low viscosity (≤ 5Pa-s)]
Liquid submergence	Z_s = 0.66–1.2 D (high low viscosity)
Baffle width	T/10–T/12 (high low viscosity)
Baffle clearance	T/48–T/64 (high low viscosity)
Number of baffles	n_b = 3–4 (Note: laminar flow regime i.e., Re $\lesssim$ 10, n_b = 0)

In the absence of a variable frequency drive (VFD), standard speeds recommended by the American Gear Manufacturers Association (AGMA) are practiced. The AGMA speeds are 25, 30, 37, 45, 56, 68, 84, 100, 125, 155, 190, and 230 RPM. Typical gear teeth machining allowance is ±1–3 rpm, depending on the speed. Occasionally, dual reduced gear systems are used to accommodate blending operations where flow regime is significantly altered due to changing working media viscosity, i.e., laminar to turbulent flow or vice versa.

Apart from the viscosity, physical properties and working media rheology may dictate operational and design constraints on the agitation systems. If possible, higher density medium must be con-added to the lower density medium to minimize stratification and improve mixing time. In the case of low surface tension fluids that are susceptible to foam, the agitation system should be designed to deliver minimum required specific power and minimum draw-down of headspace gas. Polyols, due to their low surface tension and hygroscopic nature, tend to foam. Thus, they should be blanketed with an inert gas and must experience mild agitation in storage tanks. A highly viscous but pseudo-plastic working media, such as Methocel™ (HPMC) solution, may form a cavern in the high-shear region of an impeller. This creates unnoticeable fluid movement outside of the swept region. The authors' experience suggests that close spacing of wide blade hydrofoils initiates synergistic axial flow, which alleviates "cavern" formation and thus promotes mixing all throughout the vessel.

6.7 BLENDING EXAMPLE

A chemical system house often produces multiple formulated products in the same tank to address customer needs. The blender must be cost effective, minimizing capital investment and operating cost while meeting production and quality demands under sold-out situations. Mixit-io SS6.1 shows a simple generic example of an agitation system design to produce a single-phase blend consisting of corn syrup and water. The required 95% blend time is less than a minute. This blending application spreadsheet can be used by practicing engineers to address blending of multiple components. Users are advised to read the comments in the spreadsheet cells for clarification on the calculations and on available design options that could be considered. Figure 6.8 represents a typical mechanically agitated tank with semi-elliptical 2:1 bottom head for simple blending applications. See mixit-io SS6.1 for details. Non-scaled dimensions are in inches.

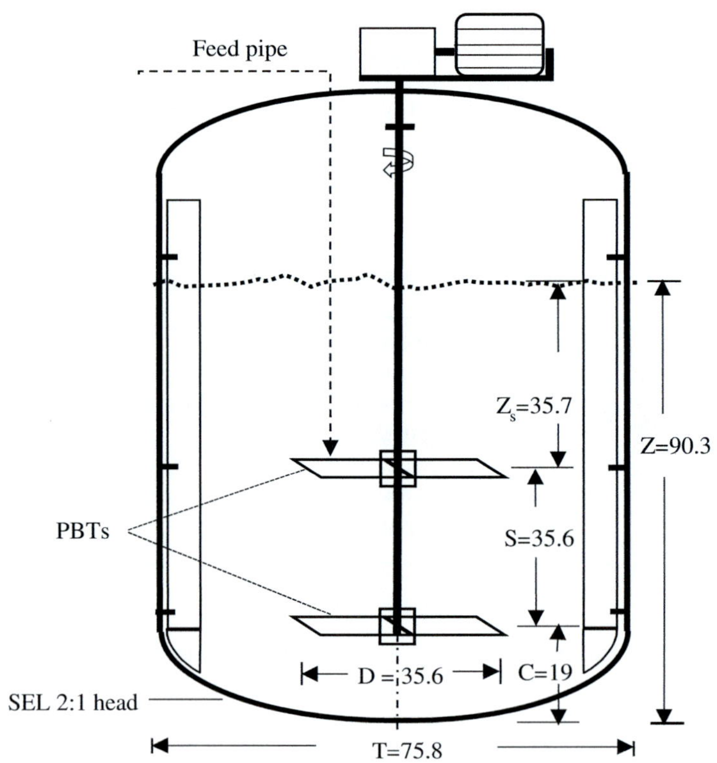

FIGURE 6.8 Configuration for ca. 2000gal vessel, V1. [non-scaled dimensions are in inches]

6.8 CATALYST ADDITION

Homogeneous or heterogeneous catalysts in the form of slurry are added to reactants in a reactor. For reactive components that are likely to undergo fast reaction, uniform catalyst dispersion within the working media must be rapid. The location of catalyst addition is critical to assure desired product yield and selectivity and to minimize byproduct formation. This is often an issue in reactors where a small amount (e.g., ppm) of catalyst "fairy dust" must be dispersed within a large vessel. The catalyst feed-pipe outlet should be located on the suction side of the axial impeller. Preferably, the distance from shaft centerline to discharge location should be about 0.8 times the impeller radius. A reasonable vertical distance between the top of the blade and the catalyst outlet should prevent the dip pipe from hindering the rotating blades. Placing the catalyst pipe near the suction side of the impeller serves multiple purposes, including (a) faster dispersion of the catalyst, (b) minimized back flow into the feedline, and (c) avoidance of dip-tube fouling after the catalyst stream is injected into the working media. Catalyst dip-pipe diameters are relatively small and often susceptible to plugging with product, especially if solids are present. Due to these issues, catalyst is most often added on the liquid surface. Thus, the ideal addition location should be within the down-pumping region induced by the impeller. Figure 6.9 shows a typical catalyst addition location, which must also avoid the immediate downstream side of the baffles.

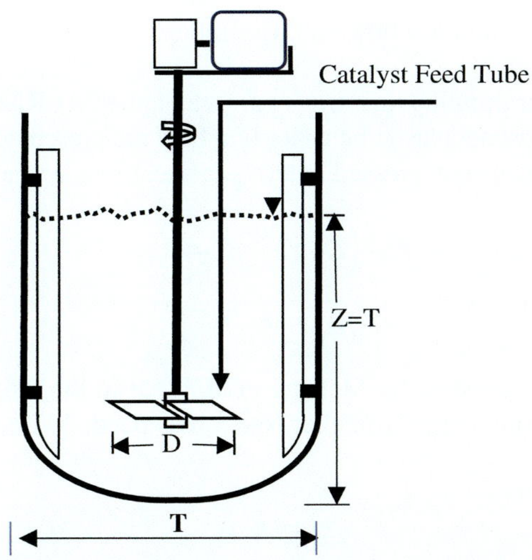

FIGURE 6.9 Catalyst addition location with respect to the impeller position.

6.9 MIXING OF STRATIFIED FLUID

At steady state, miscible working components are susceptible to stratification (density variations in the vertical direction) due to their density difference. The Richardson number, R_i, is a non-dimensional number expressing the ratio of a buoyancy term to the flow shear term.

$$R_i = \frac{Buoyancy}{Shear\,energy} = \frac{g\,\partial\rho / \partial z}{\rho\left(\partial u / \partial z\right)^2} \tag{6.14}$$

If the Richardson number is less than unity, buoyancy is unimportant in the flow. If it is much greater than unity, buoyancy is dominant because there is insufficient kinetic energy to homogenize the fluids. Thus, in order to disrupt stratification and achieve uniform density of stagnated fluid in a tank, agitation must be carried out to impart sufficient shear energy to overcome buoyancy effects. Axial or mixed flow impeller(s) must deliver enough specific power (energy), ε, to create bulk axial circulation of the working media.

A non-steady process to mix component A into an abundance of much-denser component B ideally locates an axial impeller barely below the component B surface and creates a minor vortex while continuously adding a low flow of component A. The addition rate should be less than the impeller pumping rate. As the working media level increases, agitation speed should increase to maintain a small vortex to draw down component A. Based on Z/T and the working media viscosity, design rules-of-thumb in Table 6.4 should be applied to determine how many impellers are required and where they should be placed on the shaft.

6.10 MIXING OF NON-NEWTONIAN LIQUID

In chemical and allied industries, liquids with non-Newtonian rheology are not uncommon. These are typically high-molecular-weight polymers (Polyols, CMC, Carbopol®, etc.), viscosifiers such as Kelzan® for agricultural formulations, and Xanthan gum for food and pharma products that exhibit high viscosity pseudoplastic behavior. The rheological relationship between shear stress, τ, and shear rate, $\dot{\gamma}$, is

$$\tau = K.\dot{\gamma}^n \tag{6.15}$$

where, K = consistency index and n = flow behavior index.

For pseudoplastic or shear thinning fluids, $n < 1.0$ (see Figure 1.1). Rheology instruments and measurement techniques of the pseudoplastic behavior of a fluid are presented in Chapter 1. The apparent viscosity, μ_a, of such fluids can be expressed in terms of shear rate, as shown in equation 6.17.

$$\tau = \left(K.\dot{\gamma}^{n-1}\right)\dot{\gamma} = \mu_a \cdot \dot{\gamma} \tag{6.16}$$

$$\mu_a = K.\dot{\gamma}^{n-1} \tag{6.17}$$

It is experimentally well established by Metzner et al.[14] that in the laminar regime ($Re \lesssim 10$), the average shear rate, $\dot{\gamma}$, is a function of impeller rotational speed, N. Local shear-rates can be much higher or lower.

$$\dot{\gamma} = k_s N \tag{6.18}$$

The shear rate constant, k_s, depends on the open-impeller geometry. However, in the case of proximity-impellers such as anchors and helical ribbons, it's also strongly influenced by the wall clearance ratios,[15,16] B_c / D or B_c / T. Table 6.5 presents measured k_s values for a few impellers. These values can be reasonably used to determine the impeller power for the design of agitation systems.

The Metzner-Otto Reynolds number for the impeller is derived as follows:

$$Re = \frac{\rho N D^2}{\mu_a} = \frac{\rho N^{2-n} D^2}{K\left(k_s\right)^{n-1}} \tag{6.19}$$

The power number equation for non-Newtonian fluids was expressed in Chapter 5, Figure 5.1a–b.

$$N_p = K_L / R_e \tag{6.20}$$

TABLE 6.5
Measured Average Shear Rate Constants, k_s, for the Open and Proximity Impellers[14–18]

Impellers	k_s	Conditions	References
A315	13		Anne-Archard et al.[17]
Propeller (w/3-blade)	10		Metzner[14]
Disc Turbine (w/6-blade)	11–13		Metzner[14]
Paddles	13		Metzner[14]
Ribbon	$34-114(B_C/D)$	$0.026<(B_C/D)<0.164$	
	30	$D/T=0.95, B_C/D=1$	Nagata[18]
Anchor	$33-172(B_C/T)$	$0.02<(B_C/T)<0.13$	
	25	$D/T=0.95$	Nagata[18]

$$N_p = K_L \left[\frac{\rho N^{2-n} D^2}{K (k_s)^{n-1}} \right]^{-1} \tag{6.21}$$

Shaft power, P, and specific power, ε, are determined using equations 6.4 and 6.5, respectively. These are essential quantities to appropriately scale-up and design of agitation systems for non-Newtonian fluid blending. Later, Hirata et al.[19] reported that in the laminar regime for highly shear thinning fluids, i.e., $n = 0.1 - 0.4$, the average shear rate constant, k_s, is also influenced by the fluid rheology. On the basis of this analysis, variation of k_s must be considered, and an improved understanding of hydrodynamics, particularly distributions in the sheared/unsheared region, would greatly enhance predictive abilities of system designers. Manufacturers' knowledge of experimentally determined (by LDA) and/or computationally determined (by CFD) average- and maximum shear-rate correlations (e.g., $\dot{\gamma}_{ave} = k_s N$ or $\dot{\gamma}_{max} = k_s N{\cdot}D$) for mixing impellers should be better leveraged.

In the case of fully turbulent flow ($Re \gtrsim 10^4$), inertia (turbulence) dominates over fluid viscosity. Thus, changing rheology due to turbulent shear plays an insignificant role, and varying shear zones near the impeller are nonexistent. For all practical purposes, power number is independent of Reynolds number, i.e., remains constant for a given type of impeller and geometry, as presented in Chapter 5, Figure 5.1a–b. The K_T values for a few standard impellers are given in Table 5.1.

$$N_p = K_T \tag{6.22}$$

In general, while designing agitation systems, practicing engineers must pay attention to the rheological behavior of the fluids at zero shear rate, particularly to the fluids exhibiting yield stress behavior, τ_y, prior to flow. Power imparted to the fluid must overcome the yield behavior. In other words, the impeller drive system (motor, gear box, bearings and shaft) should be designed to accommodate the required initial (starting) torque and the power. One must not rely on the apparent viscosity only. Installation of a variable frequency drive is suggested to compromise the required batch process variations. Pseudoplastic fluids with a low flow behavior index, n, at high shear rate are prone to create "caverns" in the impeller zone. Since fluids outside of the cavern remain almost unmixed, by all accounts, such "caverns" should be avoided. Caution is needed in the system design to control the shear rate by altering the agitation speed and/or the impeller geometry. In production plants, the problems are simply circumvented by installing a variable frequency drive and/or closely spaced high-efficiency hydrofoils or double helical ribbons.

6.11 CONTINUOUS STIRRED TANK REACTORS (CSTRS)

For completeness, process differences between a batch reactor and a continuous reactor are now briefly addressed. The latter is a reactor into which components continuously enter and intermediate or final product continuously exits, as shown in Figure 6.10. It is a steady-state operation with $\dot{Q}_{in} = \dot{Q}_{out}$. The mean residence time of the media in the vessel, i.e., $V / \dot{Q}$, must be at least 10 times more than the blend time, θ_b, of the media in the vessel.[20] Component inlet(s) and product outlet must be appropriately positioned relative to the impeller. Working media must be unable to bypass the impeller swept volume before exiting; dead media space in the vessel must be avoided. CSTR operations may include a cascade of multiple vessels as shown in Figure 6.11. Such batteries of serial stirred tanks blend multiple additives, including reactants, homogeneous/heterogeneous catalysts, and initiators that react in sequential stages to produce the desired conversion, prevent byproducts, and control thermal energy so as to avoid runaway reactions and a loss of control. Chapter 18 further discusses CSTR mixing time, θ_{95}, and residence time distributions, *RTDs*.

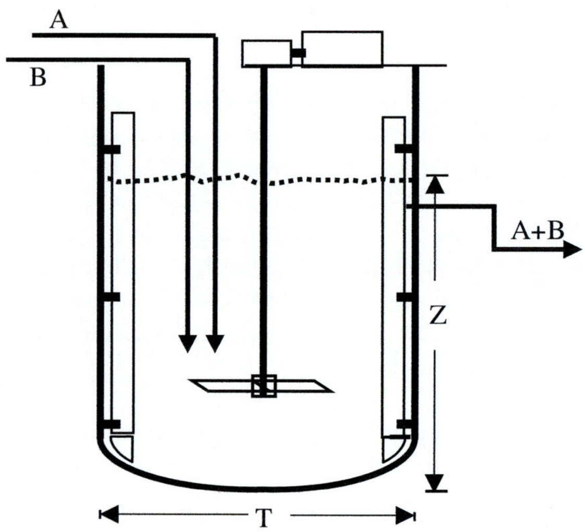

FIGURE 6.10 A continuous stirred tank reactor (CSTR) with Inlet A, Inlet B, and Outlet (A+B).

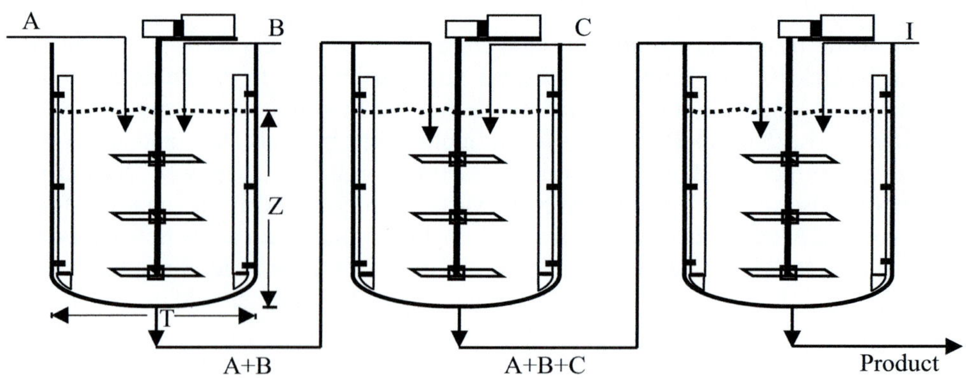

FIGURE 6.11 Three CSTRs with cascaded feeds (A = Reactant, B = Reactant, C = Catalyst, I = Initiator).

6.12 SCALE-UP AND PILOT PROCESS

This section addresses scale-up rules for reactive and non-reactive mixing processes, an essential step in reactor design. Commonly, the mixing of chemical monomers and catalyst is followed by reaction in the presence of initiators and/or appropriate thermal control. Laboratory chemists commonly conduct experiments in 20–1000mL round-bottom flasks with magnetic or overhead stirrers. Outcomes of using such small volumes are conversions and product yields that are mostly reaction controlled, since it is generally relatively easy to deliver sufficient specific mixing power, ε. Because shaft (hydraulic) power into the working media is low and of similar magnitude to the associated losses, it is difficult to accurately assess specific power needs from shaft torque or motor power measurements. This complicates the task of the R&D engineer who must scale up this process to the manufacturing plant while also ensuring that it is not mixing controlled. Caution must be taken, however, for the atypical case where excessive mixing and shear may be counterproductive and yield byproducts or hinder crystal growth.

Prior to the manufacturing scale, pilot operations are normally carried out to understand how fluid mechanics, heat transfer, and mass transfer impact yield. This is accomplished using smaller scale, i.e., typically 20gal to 500gal vessels made out of compatible materials. Although pilot scale should be related to the desired production scale, the size and shape of pilot reactors is often dictated by the available vessels in the existing pilot facility. Pilot processing plays a major role in manufacturing small volumes of new technology-based products, while helping to mitigate project risks and design unknowns until unexpected safety consequences are better understood. Pilot-scale testing generally provides the following essentials:

1. Assessment of product yield and quality
2. In-house product testing
3. Product samples for external evaluations by potential customers and regulatory authorities
4. Determination of scale-up design criteria for manufacturing process technology
5. Training of personnel for full-scale plant operations

Although pilot processing is desirable, it often becomes cost prohibitive. A current trend in the chemical and allied industries is avoidance of pilot processing to save money and/or time before market introduction.

Mostly, for the simple non-reactive blending operations of miscible fluids, the ideal scale-up rules from the laboratory scale to the production scale are: (a) use similar impeller(s), (b) maintain tank and agitation system geometric similarities such as $\frac{Z}{T}, \frac{D}{T}, \frac{S}{D}, \frac{C}{T}, \frac{B_w}{T}, \frac{B_c}{T}$, and (c) provide equal power per unit mass, ε. A comparable order of component addition and processing must be followed to overcome any significant differences in densities and viscosities. Using appropriate equations based on the fluid rheology, practicing engineers should determine the required agitation speed from the shaft power, P, i.e., $(\varepsilon.m)$, and confirm that the desired flow regime is maintained by comparing associated impeller Reynolds numbers, Re's. For design purposes, the required motor power, P_m, is ~20–25% more than the shaft power to account for friction losses in the drive system. If no VFD is present, the nearest but next higher AGMA speed must be used.

When scaling up a reactive system, however, simple power-per-mass scale-up rules may not be appropriate. Maximum allowable shear rate, $\dot{\gamma}_{max}$, which is proportional to the tip speed, πND, may be the dominating parameter for scale-up in crystallizers, fermenters, and reactors, producing polymers with a long chain and/or ultra-high-molecular-weight (UHMW, $M_n > 10^6$Da). Typical reaction product examples are the step-growth syntheses of traditional fiber precursors, ultra-high-molecular-weight polyethylene (UHMWPE) from its monomer ethylene, organometallic conjugate polymers for electronic applications, and carboxymethyl cellulose (CMC) for heavy metal remediation. Because these products are susceptible to mechanical shear degradation, their operations are often conducted in the laminar regime, and scale-up criteria include the maximum sustainable

TABLE 6.6
Mixing Scale-Up Parameters and Their Influence on Other Parameters

Scale-Up Quantity Volume, *V* [gal]	**50**	**5000**	**5000**	**5000**	**5000**	**5000**
Scale-up parameters		Specific power, $\varepsilon \alpha N^3D^2$	Maximum shear rate, $\dot{\gamma}_{max} \alpha ND$	Tank volume/ Circulation rate, $N_t \alpha N^{-1}$	Reynolds number, *Re*	Froude number, *Fr*
Scale factor, $\chi = V_{Pr} / V_{pi}$	1	100	100	100	100	100
Tank diameter, $T \alpha V^{1/3}$	1	4.64	4.64	4.64	4.64	4.64
D/T	0.4	0.4	0.4	0.4	0.4	0.4
D	0.4	1.86	1.86	1.86	1.86	1.86
Specific power, $\varepsilon \alpha N^3D^2$	**1**	**1**	1.23	1.51	2.28	0.81
Max. shear rate, $\dot{\gamma}_{max}$ $\dot{\gamma}_{max} \alpha ND$	**1**	0.54	**1**	1.86	1.86	0.54
Volume/Circulation time, $N_t \alpha N^{-1}$	**1**	3.46	1.86	**1**	3.46	1.86
Reynolds number, *Re* $Re \alpha ND^2$	**1**	0.08	0.54	3.46	**1**	0.16
Froude number, *Fr* $Fr \alpha N^2D$	**1**	1.35	1.36	1.36	2.53	**1**

shear rate observed in the lab or pilot scale. Because scale-up of such processes increases impeller diameter, processing speed must decrease. Consequently, production scale values of average shear stress and power per mass may be less than desired.

Table 6.6 summarizes the penalties associated with scaling based on one or other of the following parameters: specific power, maximum shear rate, or circulation rate (α mixing time). Also, for special applications, non-dimensional parameters such as Reynolds number and Froude number are used to control the flow regime and vortex formation. The following scale-up analysis assumes geometric similarities, $D/T = 0.4$, $Z/T = 1$, and fully turbulent flow, i.e., $Re \gtrsim 10^4$.

To clarify the numbers in Table 6.6, their derivations are presented based on the associated formulae and proportionality relationships.

1. Scale factor, $\chi = V_{Pr} / V_{pi}$, where V_{Pr} and V_{pi} are the working media volumes of the production and the pilot tanks, respectively. In this example, $\chi = \frac{5000}{50} = 100$.
2. Working media volume, $V = \frac{\pi}{4} T^2 Z = \frac{\pi}{4} T^3 \ \alpha \ T^3 \Rightarrow T \alpha V^{\frac{1}{3}}$, for $Z = T$.
   T is the tank diameter and Z is the working media height.
3. D/T ratios, $\kappa = D/T$ (constant for geometric similarity; in this example, $\kappa = 0.4$)
   It means similarity parameter, $\xi = \frac{D_{Pr} / T_{Pr}}{D_{pp} / T_{pp}} = 1$, where, D_{Pr} and D_{pp} are the impeller diameters; T_{Pr} and T_{pp} are the tank diameters of the production and the pilot plants, respectively.
4. Power, $P = N_P \rho N^3 D^5 \ \alpha \ N^3 D^5$, where N_P is the power number. It is a constant for an impeller in the fully turbulent flow regime. N is the rotational speed of the impeller.

5. Specific power, $\varepsilon = \frac{P}{m} = \frac{N_P \rho N^3 D^5}{\rho\left(\frac{\pi}{4}T^3\right)} = \frac{N_P \ N^3 D^5}{\left(\frac{\pi}{4}T^3\right)} = \frac{N_P \left(D/T\right)^3 N^3 D^2}{\frac{\pi}{4}} \alpha N^3 D^2$, for constant D/T.
6. Assuming the maximum shear rate, $\dot{\gamma}_{max}$, to occur in the swept zone. Also, assuming that it is proportional to the impeller tip velocity, v_t, $\dot{\gamma}_{max} \, \alpha \, v_t \; \alpha \; N \cdot D$.
7. Fluid circulation rate, $\dot{Q} = N_{Fl} ND^3 \; \alpha \; ND^3$. In the fully turbulent regime, for a specific impeller geometry, the flow number, N_{Fl}, is constant.

   Media volume in the tank, $V \alpha \; T^3 \Rightarrow V \alpha \; D^3$ as D/T is constant.

   The number of turnovers, N_t, defined as the media volume, V, to the circulation rate, $\dot{Q}$, is represented by, $N_t = \frac{V}{\dot{Q}} = \frac{D^3}{N_{Fl} ND^3} \alpha N^{-1}$.

   Thus, for fully turbulent flow, turnover ratio is proportional to the mixing time, θ.
8. Reynolds number, $Re = \rho ND^2 \; \alpha \, ND^2$
9. Froude number, $Fr = \frac{N^2 D}{g} \alpha N^2 D$

These analyses show the relative impeller diameter, D, and its speed, N, to be independent variables. Considering geometric similarity, the former is determined from the pilot and production tank capacities. Then the relative agitation speed, N, is calculated for each scale-up parameter. This is used to compute the rest of the parameters, as shown in Table 6.6.

As an example, the consequences of holding constant specific power as the scale-up criteria while also maintaining geometric similarity (i.e., $\kappa = D/T = 0.4$) is outlined briefly. The micro-mixing length scale, λ_k, remains constant, but maximum shear rate, $\dot{\gamma}_{max}$, and mixing time, θ, increase by 23% and 51%, respectively. Reynolds number increases by 128%, whereas the Froude number decreases by 20%. As a result, the implications of these non-dimensional parameters are higher turbulent intensity but a smaller vortex. If the maximum shear rate is held constant during scale-up, specific power would be half the pilot-scale value, but the mixing time and the Reynolds number will be doubled. One of the worst scenarios is when the mixing time is to be matched. As shown in Table 6.6, specific power and Reynolds number approximately triple while the shear rate doubles. In most instances with geometric similarity, the authors used specific power as the scale-up criteria, except in cases where shear degradation is intolerable, where maximum shear rate matching is considered. Unfortunately, in the scale-up process all parameters can't stay constant. In certain instances, therefore, compromise on the scale-up parameters is desired to optimize the power level for mixing. Shear and batch time address only the most important process needs, such as the shear rate in a crystallizer or a filamentous fermenter.

REFERENCES

1. Hicks, R. W., Morton, J. R., Fenic, J. G., *How to Design Agitators for Desired Process Response*, Chemical Engineering, p. 22–30, April 26, 1976.
2. http://users.rowan.edu/~hesketh/0906-309/Lectures/Mixing%20Notes%20rev2.pdf
3. Couper, J. R., Penney, W. R., Fair, J. R., Walas, S. M., *Chemical Process Equipment: Selection and Design*, Second ed., pp. 287–304, Butterworth-Heinemann, UK, 1995. www.slideshare.net/knowledge1995/agitaion-and-mixing
4. https://chemical.eng.usm.my/notes/AZMIER/notes/EKC212/EKC212_CH9_Solution%20WIP.pdf.
5. Himmelsbach, W., Houlton, D., Keller, W., Lovallo, M., *Mixing Systems: Design and Scale-Up*, pp. 46–53. Chemical Engineering, April 2006.
6. Smith, J. M., *Agitation Devices*. www.thermopedia.com/content/549/
7. Grenville, R. K., *Research and Industrial Review: Guidelines for Blending of Miscible Fluids*, FMP 051, October 1991

8. Grenville, R. K., Blending of Viscous Newtonian and Pseudo-Plastic Fluids, Doctor of Philosophy Cranfield Institute of Technology, 1992.
9. Grenville, R. K., Nienow, A. W., Blending of Miscible Liquids. In Chapter 9, Paul. E. D., Atiemo-Obeng, V. A., Cresta, S. M., Ed., *Handbook of Industrial Mixing*, Wiley, 2004.
10. Mersmann, A., Einenkel, W-D., Käppel, M., Auslegung und Maßstabsvergrößerung Von Rührapparaten, *Chem. Ing. Tech.*, 23, pp. 953–956, December 1975.
11. Todtenhaupt, E. K., Zeiler, G., *Handbook of Mixing Technology*, EKATO Ruhr-und Mischtechnik, Schopfheim, Germany GmbH, 1991.
12. Strand, A., Investigation of Blend Time for Turbulent Newtonian Fluids in Stirred Tanks, *Rochester Institute of Technology RIT Scholar Works Theses*, 2017.
13. Benz, G., *Consider Hydrofoil Impellers for Laminar-Flow mixing*, 2018. www.aiche.org/resources/publications/cep/2018/may/consider-hydrofoil-impellers-laminar-flow-mixing
14. Metzner, A. B., Otto, R. E., Agitation of Non-Newtonian Fluids, *AIChE J.*, 3(1), pp. 3–10, 1957.
15. Harnby, N., Edwards, M. F., Nienow, A. W., *Mixing in the Process Industries*, Second ed., pp. 143–144, Butterworths-Heinemann, Oxford, UK, 1992.
16. Oldshue, J. Y., *Fluid Mixing Technology*, p. 300, McGraw-Hill Publications Co., New York, 1983.
17. Anne-Archard, D., Marouche, M., Boisson, H. C., *Hydrodynamics and Metzner–Otto Correlation in Stirred Vessels for Yield Stress Fluids*, 2006. https://core.ac.uk/download/pdf/12044419.pdf
18. Nagata, S., *Mixing Principles and applications,* Kodansha Ltd., Tokyo, Wiley, New York, 1975.
19. Hirata, Y., Aoshima, Y., Formation and Growth of Cavern in Yield Stress Fluids Agitated Under Baffled and Non-Baffled Conditions, *Chem. Eng. Res. Des.*, 74(4), pp. 438–444, 1996.
20. Roussinova, V., Kresta, S. M., Comparison of Continuous Blend Time and Residence Time Distribution Models for a Stirred Tank, *Ind. Eng. Chem. Res.*, 47(10), p. 3532, 2008.

7 Solid–Liquid Mixing/ Dispersions

7.1 INTRODUCTION

A working media that contains suspended solids in a continuous liquid are commonly referred to as a slurry. Such systems are encountered in many chemical and allied operations. Typical non-reactive processes include simple gravity separations, strippings, and dissolution operations as well as such complex reactive operations as nucleation followed by precipitation. In many instances, the reaction products are solids that need to be uniformly suspended in the liquid phase. Typical process examples are suspension polymerization, solid catalyzed reaction, crystallization, and precipitation. Well-known processes for producing such basic chemicals as latex, ion-exchange resins, crude terephthalic acid (precursor to PTA/PET), high-density polyethylene (PE w/slurry process), and PVC/PVDC are solids suspensions within a solvent media. This chapter disseminates knowledge on the design of solid–liquid mixing and dispersion systems. The process objective must be understood to design a cost-effective agitation system.

7.2 DISPERSION OF SOLIDS

Acceptable levels of particle suspension and fragmentation must be achieved before the severity of an implemented agitation system can be deemed acceptable. Slurry dispersions are classified as presented previously in Table 1.1 (see Chapter 1). Commonly defined levels of particle suspension are depicted in Figure 7.1, as (a) partial or incomplete suspension, (b) off-bottom suspension, and (c) complete suspension with uniform distribution.

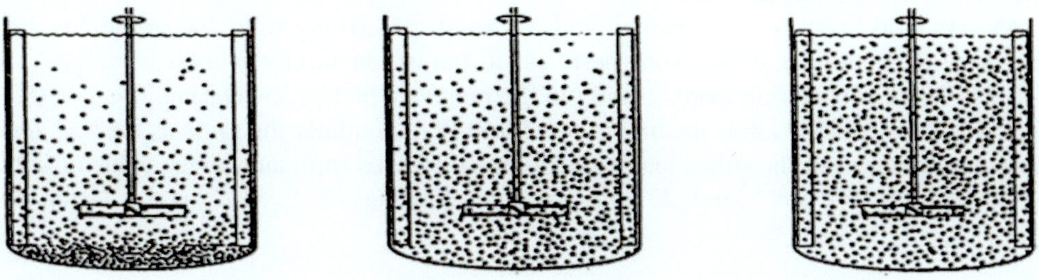

(a) partial or incomplete suspension distribution (b) off-bottom suspension, non-uniform distribution (c) complete suspension, uniform

FIGURE 7.1 Different levels of solids suspension and distribution in a stirred vessel.[1]

Zwietering (1958) proposed the idea of a "just suspension" speed, N_{js}, correlation that identifies the required agitation speed needed to, as a rule-of-thumb, keep all particles in motion and off of the vessel bottom for a minimum of 1–2 seconds.[2] Since then, multiple researchers[3–8] have critically analyzed and modified his correlation to overcome its drawbacks and prediction deficiencies. Zwietering's solid suspension experiments were carried out with sand (d_p = 125–850μm) and NaCl (d_p = 125–350μm) particles at 0.5–20 wt.% in relatively low viscosity (≤ 9cP) liquids (water, acetone, CCl_4, K_2CO_3 solution and oil), using fully baffled glass and Perspex tanks (T = 0.15–0.6m).

DOI: 10.1201/9781003606390-7

Most of the experiments used flat-bottom tanks. Multiple impellers, such as paddles, disk turbines, vaned disk, and propellers with D/T ratios ranging from 0.22–0.6, were used for solid suspension at impeller Reynolds numbers, $R_e > 3000$. A correlation constant, S, was determined experimentally for impellers based on their geometry, D/T, and off-bottom clearance, C/T.[2]

The Zwietering correlation, in equation 7.1, is an empirical correlation from a Buckingham-Pi type analysis that lacks underlying physics, and thus, appropriate rationale. It is widely used, however, to determine an approximate value of N_{js}.

$$N_{js} = S\nu^{0.1}\left[\frac{g_c}{\rho_l}(\rho_s - \rho_l)\right]^{0.45} \chi^{0.13} d_p^{\ 0.2} D^{-0.85} \tag{7.1}$$

where, N_{js} = just suspension speed [rps]
S = Zwietering constant (function of impeller type, C/T, D/T etc.) [—]
ν = kinematic viscosity [m^2-s^{-1}]
g_c = acceleration due to gravity [9.81m-s^{-2}]
ρ_l = density of liquid [kg-m^{-3}]
ρ_s = density of solid [kg-m^{-3}]
χ = mass ratio of solid to liquid × 100 [%]
d_p = particle diameter [m]
D = impeller diameter [m]
T = tank diameter [m]

"Just suspension speed" is considered the minimum speed required to expose the entire outer surface of all particles to the surrounding liquid media, for heat and mass transfer, and/or to initiate chemical reaction. Uniform suspension of solids, as shown in Figure 7.1c, is considered the ideal condition for catalyst suspension and uniform reaction. Typical values of the Zwietering constant, S, for axial and mixed flow impellers are noted[9] in Table 7.1. The table includes two different S values for each impeller at two different combinations of off-bottom clearance to tank diameter (C/T) and impeller diameter to tank diameter (D/T). A detailed description of the dependency of S on (C/T) and (D/T) is found in reference 9.

Meyers et al.[5] (2013) experimented with six solids at varying solid concentrations, up to 67%, and drew the following conclusions pertinent to the exponent of the solid loading, χ, in equation 7.1. When considering loadings only in the range studied by Zwietering, $0 < \chi < 18$, the averaged exponent over all solids loading is equal to 0.12, essentially the same as the 0.13 reported by Zwietering. However, the solids loading exponent increases from an average value of 0.097 at low solids loadings $(0 < \chi < 5)$ to 0.22 at intermediate loadings $(5 < \chi < 25)$ and 0.34 for the highest loadings $(25 < \chi < 67)$.[5]

TABLE 7.1
S Values for a Few Common Impellers[9] (flat-bottom vessels; $Z = T$)

Impeller Type	C/T	D/T	S
Standard PBT	1/4	1/3	6.2
	1/4	1/2	5.8
SPX Flow—A310	1/4	1/3	7.9
	1/4	1/2	6.0
NOV—HE3	1/4	1/3	7.2
	1/4	1/2	6.2
Ekato—Intermig (2 offset blades)	1/6–2/3	7/10	7.4

One of the latest N_{js} well-accepted correlations (see equation 7.2) is that proposed by Grenville et al.[7] (2015) for standard PBTs and hydrofoils, based on their dependency on respective power numbers. Unlike that of Zwietering, this correlation includes impeller off-bottom clearance with respect to its diameter, C / D, as an implicit independent term. The correlation is valid up to a slurry solids concentration of 20vol%. The experimental program of Grenville et al. is briefly described with Figure 7.2.

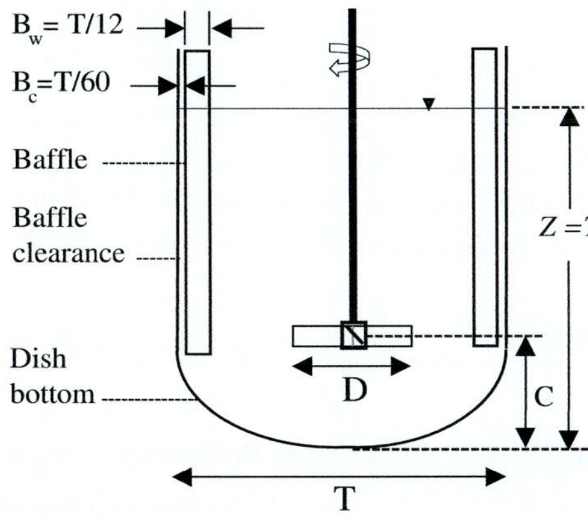

FIGURE 7.2 Schematic of experimental vessel and agitation system.[7]

[Dished bottom vessels of T = 0.30, 0.61, 1.00; pitch blade and hydrofoil impellers of $D = T/2, T/3$ and $C(\text{off - bottom clearance}) = T/3, T/4, T/6$]

Sand and urea formaldehyde are used as solid particles in dilute CMC solutions. The range of physical properties and flow Reynolds numbers are listed as follows:[7]

- d_p = 100, 180, 325, 360 and 605μm
- ρ_s = 1500, 2650kg-m^{-3}
- ρ_l = 2650kg-m^{-3}
- μ = 0.52 – 2000cP
- ϕ_v = 0.04 – 20.22%
- $Re = 3.9 \times 10^2 – 6.5 \times 10^5$

The Grenville, Mack and Brown (GMB) correlation is represented in equation 7.2:

$$N_{js} = z\left[\frac{g_c}{\rho_l}(\rho_s - \rho_l)\right]^{0.50} \phi_v^{\,0.154} d_p^{\,0.167} \left(\frac{C}{D}\right)^{0.10} N_p^{\,-1/3} D^{-2/3} \tag{7.2}$$

where, z = GMB correlation constant with values of:
(a) 1.528 for PBTs, with relative standard deviation of ±11%
(b) 1.213 for hydrofoils, with relative standard deviation of ±14%
N_p = impeller power number
ϕ_v = volume fraction (volume of particles/volume of slurry) × 100 [%]

Grenville et al. determined correlation constant, z, and exponents of ϕ_v, and ($^C/_D$) by regressing experimental data, while all other parameter exponents were adopted from the Davies' (1986) N_{js} model.[8]

Just suspension speed, N_{js}, calculations using the correlations of Zwietering (equation 7.1) and GMB (equation 7.2) are now illustrated while considering a working media (slurry) containing 6.0vol.% synthesized aluminum-silicate ($3Al_2O_3{\cdot}2SiO_2$) particles in an aqueous phase. For this illustrative example, physical properties of aluminum-silicate particles (with ~8wt.% of left-over ammonia content) and water are presented in Table 7.2.

TABLE 7.2
Physical Properties of Working Media (Slurry)

Components	Particle Size, $d_p = d_{90}$ [m]	Densities, ρ_s and ρ_l [kg-m^{-3}]	Viscosity [Pa-s or N-m^{-2}-s]	Kinematic Viscosity [m^2-s^{-1}]	$\frac{(\rho_s - \rho_l)}{\rho_l}$
$3Al_2O_3 \cdot 2SiO_2$	40×10^{-6}	2500	-	-	1.5
Water	-	1000	10^{-3}	10^{-6}	

In this example, particles are mesoporous (porosity ≈ 65%) with some nanopores. Water migration into the particles is neglected, and particle shape is assumed to be spherical. Total slurry volume in a mixing tank with 2:1 semi-elliptical base is 500gal (1.892m^3). Tank diameter, T, and liquid height from the tank bottom, Z, are 54in and 54.9in, respectively ($Z / T \approx 1$). Table 7.3 lists other mixer parameters.

TABLE 7.3
Vessel and Agitation System Design Parameters Pertinent to N_{js} Calculation

Parameters (see Figure 7.2)	Nomenclatures/Types		Units: m (inch)
Tank diameter	T		1.372m (54")
Type of impeller	Standard 4–45deg. PBT		
Imp dia./Tank dia. ratio*	D / T		0.46
Impeller diameter	D		0.631m (24.85")
Baffle width/Tank dia.*	B_W / T		1/12
Baffle width	B_W		0.114m (4.49")
Baffle clearance/Tank dia.*	B_C / T		1/48
Baffle clearance	B_C		0.029 (1.14")
Bottom head*	2:1 Semi-elliptical		
Bottom head depth/Tank dia.	h / T		0.25
Bottom head	h		0.343 (13.5")
Imp. blade width/Imp. dia.	b_W / D		0.2
Impeller blade width	b_W		0.126 (4.96)
Impeller off-bottom clearance* (imp. CL on tangent line	C		0.343 (13.5)
Imp. off-bottom clearance/Tank dia.	C / T		4
Imp. off-bottom clearance/Imp. dia.	C / D		0.544
Volume fraction, [%]	ϕ_v		6
Mass ratio**, [%] = $\{(0.06 \times \rho_s) / [(1-0.06) \times \rho_l]\} \times 100$	χ		16
Zwietering constant, S (from Table 7.1 for D / T and C / T)		5.8	
GMB correlation constant[5] for standard PBT, z		1.528	
Power number standard PBT[5], N_p		1.3	

Note: *Typical agitation system design criteria for solid/liquid suspensions. **Calculated values.

From values in Tables 7.2 and 7.3, just suspension speeds, in rpm, are calculated as follows:

(a) Zwietering correlation[2] (equation 7.3):

$$N_{js} = \left\{ S\, \nu^{0.1} \left[\frac{g_c}{\rho_l} (\rho_s - \rho_l) \right]^{0.45} \chi^{0.13}\, d_p^{\ 0.2}\, D^{-0.85} \right\} x60 \tag{7.3}$$

$$= \left[5.8\left(10^{-6}\right)^{0.1}\left(9.81\times1.5\right)^{0.45}\left(16\right)^{0.13}\left(40\times10^{-6}\right)^{0.2}\left(0.631\right)^{-0.85}\right]\times60 = 82rpm$$

(b) GMB correlation[7] (equation 7.4):

$$N_{js} = \left\{z\left[\frac{g_c}{\rho_l}\left(\rho_s-\rho_l\right)\right]^{0.50}\phi_v^{\ 0.154}\,d_p^{\ 0.167}\left(\frac{C}{D}\right)^{0.10}N_p^{\ -1/3}\,D^{-2/3}\right\}\times60 \tag{7.4}$$

$$= \left[1.528\left(9.81\times1.5\right)^{0.50}\left(6\right)^{0.154}\left(40\times10^{-6}\right)^{0.167}\left(0.544\right)^{0.10}\left(1.3\right)^{-1/3}\left(0.631\right)^{-2/3}\right]\times60$$

$$= 100rpm$$

The shape of the vessel bottom significantly influences the off-bottom suspension of solids. When axial flow impellers operate in semi-elliptical and dish bottomed tanks, rather than a flat-bottomed tank, S values are lower by 15–20%. A profiled up-flow pattern easily lifts up the solids, and unlike in the flat-bottom tank, leaves no deposits in the bottom periphery or corners. In designing vessels for solid–liquid mixing, practicing engineers should consider using dish and semi-elliptical bottomed tanks, while avoiding flat bottoms. Hemispherical bottom tanks should not be used. Large slurry tanks commonly use conical bottoms with a ~150-degree cone angle, as shown in Figure 7.3. Since lifting the particles into the cylindrical zone is challenging, using larger angled (with respect to tangent line) conical bottoms is not advisable.

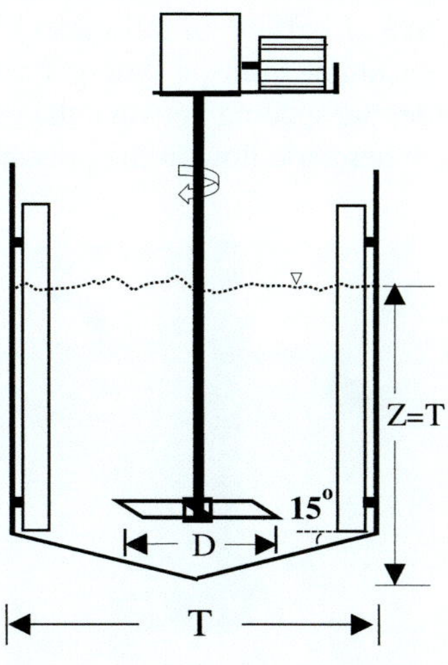

FIGURE 7.3 A typical 15-degree angled (w.r.t lower tangent line) conical bottom slurry tank.

If the tank has a bottom nozzle, its depth should be as shallow as possible, or more preferred, should be below a flush bottom shutoff valve, often referred to as a tank valve, angle bottom valve, bottom drain valve, flush-mount valve,[10] etc. As shown in Figure 7.4, this avoids dead space for particle settling and likely nozzle plugging before slurry drainage through the bottom nozzle. Several manufacturers offer flush bottom valves for reactors, transfer vessels, autoclaves, and storage tanks. The material of construction (MOC) of the valve typically matches that of the vessel, which must be chemically compatible with the working media.

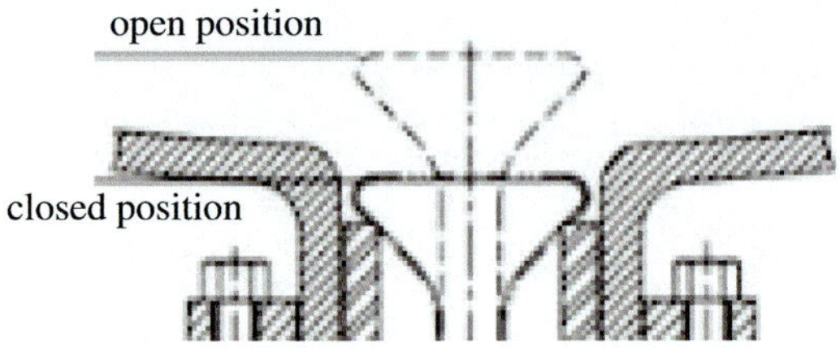

FIGURE 7.4 Cross-sectional view of a tank flush bottom valve.[10]

For example, the exposed flushing portion in the bottom of a glass-lined vessel is typically glass-lined. Pfaudler and De Dietrich are well-known leaders in the manufacture of glass-lined flush-mount valves. Fetterolf[11] and SchuF[12] are worldwide specialists in the design and manufacture of high-quality and complex industrial flush bottom valves for chemical process industries. Simple bottom nozzles with a relatively short depth-to-diameter ratio operate okay with light, non-sticky solids, and are less expensive alternatives to flush-mount valves.

7.3 AGITATION SYSTEM

As pointed out earlier, the working media flow pattern plays a significant role in lifting up the solid particles from the tank bottom. A synergistic downward flow in the central zone of the tank followed by an upward flow outside of the impeller zone provides the ideal circulation pattern for solid suspension. For axial impellers, the diameter ratio D/T must not be greater than 0.7, preferably 0.5. However, for mixed flow impellers such as the commonly used standard PBTs, the ratio must be <0.5 to avoid upward flow choking (throttling) outside of the impeller zone, preferably ≤0.46. The Ekato Intermig blades ($D/T \geq 0.7$) shown in Figure 7.5 are designed to provide a self-created forced circulation pattern irrespective of baffles. In tall tanks ($Z/T > 1.2$), where multiple impellers must be used for uniform suspension of solids, a synergistic circulatory flow pattern is desired. Hence, to be on the conservative side, the spacing between the impellers outlined in Chapter 6 may be reduced by up to 20% to obtain synergistic flow in the presence of solids.

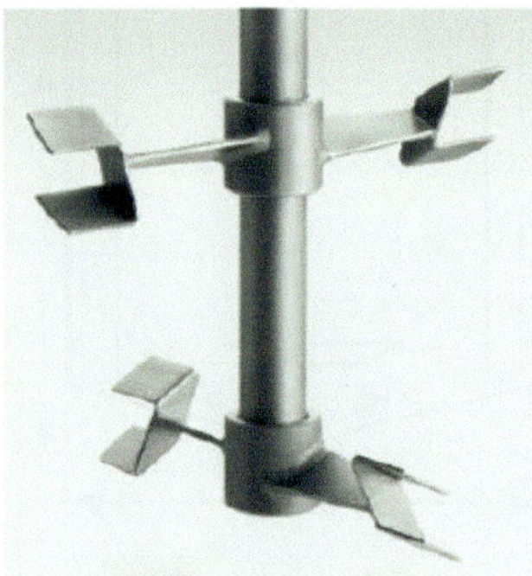

FIGURE 7.5 Ekato Intermig blades (two stages and two flow-directing elements).[13]

Special attention must be given to the design of agitation systems and baffles for crystallizers. Typically, oblong and needle-like crystal particles are susceptible to breakage due to mechanical shear experienced at the sharp edges of the blades, and hydraulic turbulent shear in the impeller swept region, including vortices trailing the impeller blades, or adjacent to the baffles. Particles may be subjected to attrition due to interaction with the baffles and tank wall. On the other hand, since in most cases complete dispersion is ideal for crystal precipitation, growth, and suspension, turbulent flow in the crystallizer is desired. Hence, collaboration between practicing engineers and crystallization researchers must optimize the agitation system to address these contradicting issues. Low shear impellers are preferred for such an application.

In days past, narrow-blade hydrofoil impellers such as the A310 (see Figure 7.6.a) and HE-3 were, like the Intermig, considered lower-shear impellers than traditional pitched-blade turbines. More recent results from Grenville et al.,[14] however, show that lower shear rates in primary flow immediately downstream from narrow-blade hydrofoils is accompanied by higher shear rates in vortices trailing each impeller blade. If shear in the trailing vortex is too high, a different impeller type should likely be used. Thus, testing should assess whether narrow hydrofoil impellers are suitable for a given crystallization.

A typical range of crystallizer power input would be 0.1–0.4 W/kg, depending on the working media viscosity. If breakage and attrition of crystals are not of primary concern, but control of particle size is, then agitation systems must create fully turbulent flow, i.e., $Re \geq 10^4$, to uniformly suspend particles, as shown in Figure 7.1.c. Wide-blade hydrofoils such as the A320 (see Figure 7.6.b), A315, Maxflo, and Viscoprop are ideal impellers. If the working media-height-to-tank-diameter ratio (Z / T) is at least 1.2, a multi-tier agitation system with at least two impellers and an impeller-spacing-to-tank-diameter ratio (S / T) not exceeding 0.8 is preferred. Note that impeller spacing is inversely proportional to the media viscosity. The middle of the bottom main impeller should ideally be at vessel tangent line. Since mixed flow, pitch-blade turbines (see Figure 7.6.c) are generic and widely available in plants, they are often used in slurry tanks, despite producing mixed axial and radial flow.

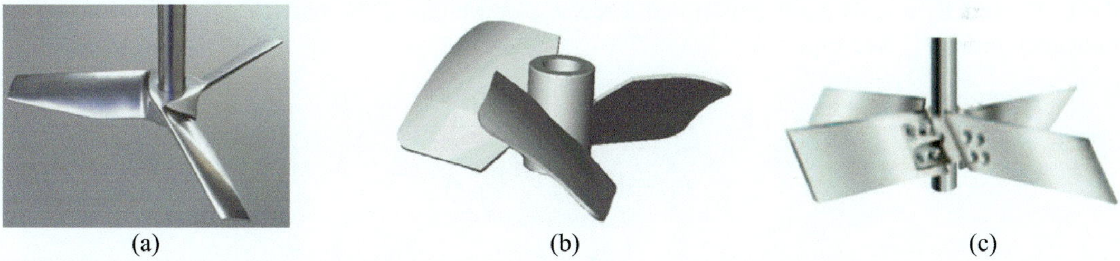

FIGURE 7.6 Photographs of SPX Flow-Lightnin Impellers: (a) narrow-blade hydrofoil, A310; (b) wide blade hydrofoil, A320; and (c) pitch-blade turbine, A200.[15]

Off-bottom clearance (C / T) of the impeller influences the fluid flow pattern, and hence, plays an important role in suspending solids. Even axial flow impellers create a radial flow pattern and pile solids in the dead zone under the hub, as seen in Figure 7.7, if the impeller is too low (e.g., way below the tangent line).

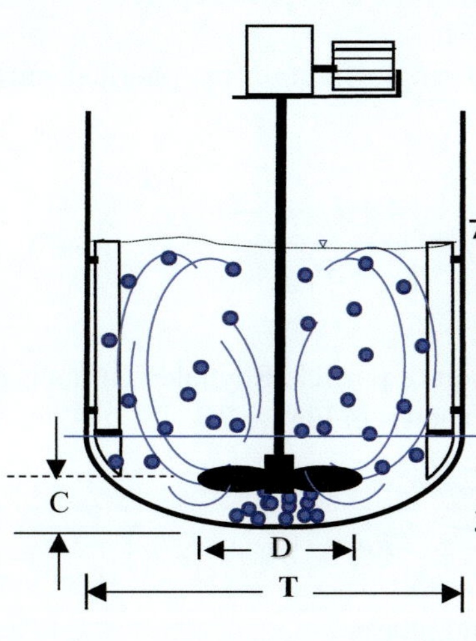

FIGURE 7.7 Schematic showing solids deposition under an impeller hub with low bottom clearance.

The authors' experience shows the Zwietering correlation to overpredict the required just suspension speed (N_{js}) in the dish-bottom tanks by 15–20%. Typically, uniform suspension of reasonably sized (10^2 – 10^3μm) particles of modest skeletal density (1.0–2.0 g/cc) requires an agitation speed of 1.25–1.5 × N_{js}.

In fully turbulent flow, power is directly proportional to the third power of agitation speed. Therefore, doubling the agitation speed increases the specific power, ε, by eight times. No matter how severe the agitation is, the top-most liquid layer will generally contain fewer particles than the rest of the tank, as depicted in Figure 7.1(c). The thickness of this layer, ideally less than 2–5 cm (1–2 in.), depends on the particle settling velocity, impeller locations, and agitation intensity. Solid particles are intermittently suspended in this layer, while below it, what may be a fluctuating combination of non-uniformly and uniformly suspended particles produce what is commonly identified as the particle cloud height. Researchers[16] at BHRG-FMP have developed correlations to predict cloud heights in stirred tanks. Cloud height is a function of the solid concentration, ϕ_v, impeller diameter to tank diameter ratio, i.e., D/T, and the impeller off-bottom clearance to the tank diameter ratio, i.e., C/T.[17]

7.4 PARTICLE SETTLING VELOCITY

Terminal velocity is the steady-state settling velocity of a single particle falling through quiescent liquid due to normal gravitational effects. Steady-state net force on the particle is zero since drag force, F_D, exactly equals net gravitational force, F_G, or weight of the particle, F_W, minus opposing buoyancy force, F_B. See Figure 7.8.

$$F_D = F_G = F_W - F_B \tag{7.5}$$

$$F_D = 1/2 \rho_l v_t^2 C_D A \tag{7.6}$$

$$F_G = (\rho_s - \rho_l) g V = (\rho_s - \rho_l) g \left(\frac{\pi}{6} d_p^3 \right) \tag{7.7}$$

where, ρ_s and ρ_l are densities of the solid and liquid [kg/m^3]
g acceleration due to gravity [m/s^2]
V volume of particle [m^3]
d_p is the diameter of the spherical particle [m]
A is cross-sectional area of the particle [m^2]
C_D is the drag coefficient for the flow past a sphere

Thus, settling velocity, v_t, of a single freefalling spherical particle in the quiescent liquid is expressed as

$$v_t = \left[\frac{4 g d_p (\rho_s - \rho_l)}{3 C_d \rho_l} \right]^{0.5} \quad [\text{m-s}^{-1}] \tag{7.8}$$

The drag coefficient, C_d is a function of particle Reynolds number, Re_p, particle shape and orientation, as presented in Table 7.4 and shown in Figure 7.9.

By definition, Re_p is

$$Re_p = \left(\rho_l \; v_t \; d_p / \mu \right) \tag{7.9}$$

where, μ is the viscosity of liquid phase.

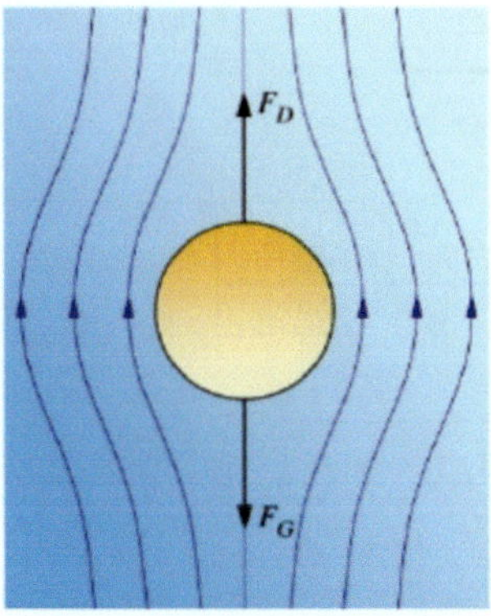

FIGURE 7.8 Forces acting on a spherical particle under steady-state settling condition.[18]

TABLE 7.4
Equations Representing C_d for Spherical Particles as a Function of Re_p[20]

Range of Re_p	C_d	Settling Principles
$Re_p < 2$	$24 / Re_p$	Stokes' eqn. (laminar)
$2 < Re_p < 500$	$18.5 / Re_p^{0.6}$	Allen's eqn. (intermediate)
$500 < Re_p < 2x10^5$	0.44	Newton's eqn. (turbulent)

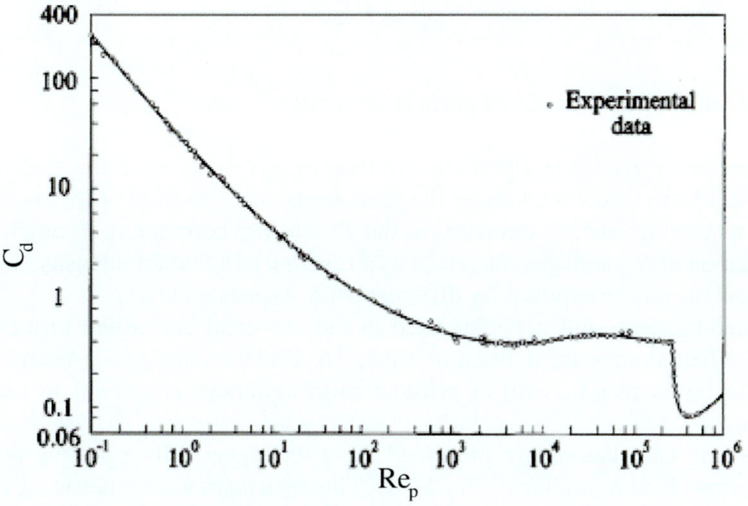

FIGURE 7.9 Flow past a sphere: experimental curve of drag coefficient vs. particle Reynolds number.[19]

Typically, particles with settling velocity, v_t, below 5mm/s are considered "easy to suspend," while those exceeding 30mm/s are considered "difficult to suspend."

In the case of Newtonian flow, using the C_d vs. Re_p relationships for spherical particles in Table 7.4, terminal velocities are determined from equation 7.8.

$$v_t = \frac{g d_p^{\,2} (\rho_s - \rho_l)}{18\mu} \qquad Re_p < 2 \qquad (7.10)$$

$$v_t = \left[0.072 \frac{g d_p^{\,1.6} (\rho_s - \rho_l)}{\rho_l^{\,0.4} \mu^{0.6}} \right]^{0.72} \qquad 2 < Re_p < 500 \qquad (7.11)$$

$$v_t = \sqrt{3.03 \frac{g d_p (\rho_s - \rho_l)}{\rho_l}} \qquad 500 < Re_p < 2x10^5 \qquad (7.12)$$

7.5 PARTICLE MORPHOLOGY

Actual solid particles are typically not spheres with a fixed diameter as commonly assumed. Hence, practicing engineers must consider particle morphology and determine an appropriate shape factor.[16] Considering dynamic response to the particle settling in the working media, an equivalent particle diameter, d_t, based on terminal velocity, v_t, may be used in the N_{js} calculation. The dynamic diameter is defined as the diameter of a sphere that has the same density and the same free-falling velocity in any given fluid as the particle in question. It can be determined from particle settling experiments ($Re_p \leq 2$) under simulated conditions. The following implicit equation for determining d_t is derived from equations 7.10 substituting 7.9 for Re_p.

$$d_t = \frac{Re_t \, \mu}{\rho \, v_t} \qquad (7.13)$$

$$Re_t = \frac{\rho \, v_t \, d_t}{\mu}$$

where Reynold number Re_t is based on particle terminal velocity.

Furthermore, solids may have a particle size distribution. If so, from the particle size distribution (psd) curve, ideally one would target the particle size d_{90} (90vol.% of the solids are below the diameter) for specifying particle diameter in the Zwietering correlation. Example mixit-io SS7.1 presents calculation of N_{js} and specific power as a function of agitation intensity. Table 7.5 presents approximate specific power required for different solid suspension levels.

As mentioned earlier, solid particles used in the chemical and allied processes are mostly non-spherical. Their shapes are defined in Table 7.6. Understanding long-term settling of these particles in the liquid media, with or without mild agitation, is critical to their storage and transportation.

Settling velocity calculations are presented for a few spherically isotropic bodies in Stoke's creeping (very slow) flow conditions,[22] $Re_p \leq 0.05$. These particles are isometric, i.e., all edges are of equal length, and the last three in Table 7.7 are regular polyhedral, with all faces equal. The sphericity, ψ, for a non-spherical particle is defined as the ratio of the area of a sphere, A_s, of the same volume as the particle, divided by the area of the particle itself, A_{ns}.

$$\psi = A_s \, / \, A_{ns} \qquad (7.14)$$

TABLE 7.5
Agitation Speed for Different Levels of Solid Suspension

Levels of Solid Suspension	State of Solids in the Liquid	Process Applications	Mix Speed, N [1/sec]	Required Power ε [W/kg]
Partial (incomplete) suspension	Some solids remain on vessel bottom	Dissolve highly soluble solids	$N \le N_{js}$	0.1–0.2
Off-bottom suspension	All solids lifted up from the vessel bottom	Minimum condition to dissolve and mix, uncontrolled reaction	$N \approx N_{js}$	0.2–0.4
Uniform (complete) suspension	Solids off-bottom and uniformly dispersed	Crystallize and mixing controlled reaction in slurry	$N \ge N_{js}$	0.4–1.2

TABLE 7.6
Definition of Particle Shape[21]

Shape Name	Shape Definition
Acicular	needle-shaped
Angular	sharp-edged or having roughly polyhedral shape
Crystalline	freely developed in a fluid medium of geometric shape
Dendritic	having a branched crystalline shape
Fibrous	regularly or irregularly thread-like
Flaky	plate-like
Granular	having approximately an equidimensional irregular shape
Irregular	lacking any symmetry
Modular	having rounded, irregular shape
Spherical	global shape

Table 7.7 shows sphericity of various particle shapes. Terminal velocity, v_t, is presented in equation 7.15.

$$v_t = \frac{K g d_p^{\,2} \left(\rho_s - \rho_l\right)}{18 \mu} \tag{7.15}$$

TABLE 7.7
Sphericity of Particle Shapes[22]

Particle Shape	Sphericity, ψ
Sphere	1
Cube-octahedron	0.906
Octahedron	0.846
Cube	0.806
Tetrahedron	0.670

$$K = 0.843 log_{10} \frac{\psi}{0.065} \tag{7.16}$$

Mixit-io SS7.2 shows how to calculate the sphericity of short rod-shaped particles with two different aspect ratios. The crystal morphology in a process is often platelets or needles. Such orthotropic objects may be considered as oblate and prolate spheroids, respectively. Where fluid motion is parallel to the axis of symmetry, we may define the length to diameter ratio as $\varphi = a / c$. The variables a and c are the length of the particle and that of its semi-axis, respectively. Table 7.8 presents values of "equivalent radius of a sphere," R, for discoid and needlelike objects based on their φ values.[23]

TABLE 7.8
Equivalent Radius of Needlelike Particles[23]

Needlelike Objects (prolate spheroids), $\varphi > 1$	
Length/Diameter: a / c	**Equivalent Radius/Actual Radius: R / c**
1.0	1.0
2.0	1.204
5.0	1.785
10	2.647
20	4.172
50	8.117
100	13.895
Discoid Objects (oblate spheroids), $\varphi > 1$	
Length/Diameter: a / c	**Equivalent Radius/Actual Radius: R / c**
10^{-6}–10^{-3}	0.8488
10^{-2}	0.8489
10^{-1}	0.8525
$2x10^{-1}$	0.8615
$4x10^{-1}$	0.8888
$8x10^{-1}$	0.9606

7.6 PARTICLE SAUTER DIAMETER

The volume equivalent sphere diameter, d_v, of a given particle is equal to the diameter of a sphere with the same volume as that particle, V_p, i.e.,

$$d_v = \sqrt[3]{\left(\frac{6}{\pi} V_p\right)} \tag{7.17}$$

The surface equivalent sphere diameter, d_s, of a given particle is equal to the diameter of a sphere with the same surface as that particle, S_p, i.e.,

$$d_s = \sqrt{\left(\frac{1}{\pi} S_p\right)} \tag{7.18}$$

The volume-surface diameter of a particle is known as the Sauter diameter, d_{32}, and is widely used in chemical engineering. It is defined as the ratio of the cube of volume equivalent diameter to the square of the surface equivalent, i.e.,

$$d_{32} = d_v^{\,3} / d_s^{\,2} \tag{7.19}$$

$$d_{32} = 6V_p / S_p \tag{7.20}$$

Thus, d_{32} is inversely proportional to the surface density (surface area per unit volume), S_v, or specific surface area (surface area of the particle per unit mass), S_m, i.e., S_v / ρ_p.

$$S_v = \frac{K_{sv}}{d_{32}} \tag{7.21}$$

The constant of proportionality, K_{sv}, is called a shape factor. Values of K_{sv} for spheres and platonic solids are listed in Table 7.9. In chemical engineering, particle surface density, S_v, is often used as a key transfer-surface parameter in dissolution, diffusion, and reaction equations. This is further discussed in Chapter 8.

TABLE 7.9
Shape Factors K_{sv} for Spheres and the Platonic Solids (regular polyhedral)[24]

Shape	K_{sv}
Sphere	6
Tetrahedron	8.94
Octahedron	8.06
Cube	7.44
Dodecahedron (pentagonal faces)	6.59
Icosahedron (triangular faces)	6.39

7.7 HINDERED SETTLING

A great many chemical applications encounter concentrated suspension/slurry systems, where settling velocity is a function of particle concentration and morphology. Surface interactions between a particle and its neighbors and its container wall significantly alter settling behavior and normally reduce the free-falling velocity. This phenomenon is known as hindered settling, a significantly different situation from individual settling spheres in dilute slurry systems, as modeled by equation 7.8. To properly model hindered settling, practicing engineers should consider published semi-empirical sedimentation equations for polydisperse systems derived from laboratory simulated experiments.

Particle-particle interaction due to surface energy (adhesion) and mechanical entanglement could lead to faster-settling agglomerated particles, as typical in polymerization reactors with semi-solid sticky particles. This behavior is well understood from experiments in small-scale STRs that maintain similar flow behavior (laminar or turbulent) to that expected in the production vessel.

7.8 SOLIDS DRAW-DOWN AND DISPERSION

The bulk density and wettability of solid particles, relative to the liquid phase, determine their settling or floating behaviors. As an example, powdered solids such as the wheat flours fed to fermenter microorganisms tend to float before being gradually drawn down into the liquid media.

Initially, lumps of the flour are uniformly distributed in the fermenter and then gradually broken up into individual particles that remain suspended. Thus, in most instances, sparingly wetted solids are sucked into the liquid as agglomerates. These are subjected to gradual de-agglomeration while passing through local high-shear zones near the blade tips and then additional turbulent shear along the flow circulation path.

Dry solids must be safely added to a vessel of agitated liquid to avoid static charges and associated explosion and fire hazards. Although the preferred addition method uses a hopper and metered chute, reality is that solids are often poured uncontrollably into a tank top nozzle, directly from the sack. Ideally, the solid addition rate must be less than the impeller pumping rate ($\dot{Q}_s < \dot{Q}_l = N_q N D^3$), so the solids are drawn down into the liquid as soon as they reach the liquid surface. Most instances use at least two axial pumping impellers to create a synergistic circulation pattern. The upper impeller assists in the draw-down process by creating a downward suction flow, while the lower impeller helps disperse the solids into the bulk liquid media. If the solid draw-down rate must be increased to minimize the process batch time, it is accomplished by creating a minor surface vortex. One of the techniques for vortex creation is to reduce the baffle height as shown in Figure 7.10.

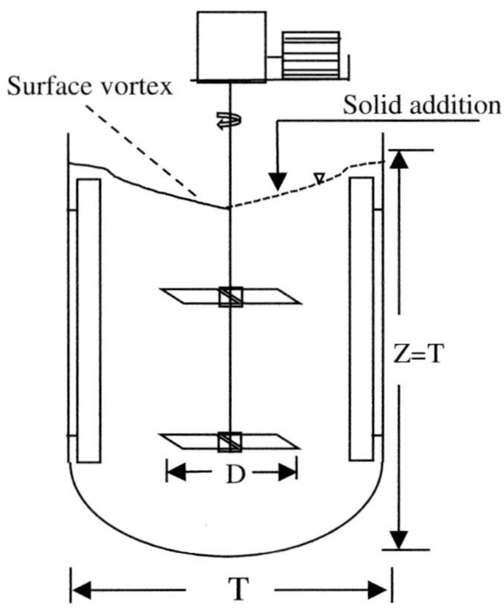

FIGURE 7.10 Schematic showing surface vortex formation with a reduced baffle height for solids draw-down.

In the case of non-wetting solids with respect to the solvent media, chemically non-participating wetting agents, also called as surfactants, are introduced to reduce the surface tension and weaken solid–liquid cohesiveness relative to solid-solid adhesion. Thus, solid particle boundaries adjacent to the liquid surface are more easily wetted, so that particle pull-down into the liquid requires less effort. By definition, the solid–liquid contact angle is the angle created right at the interface between a liquid droplet and the solid surface. A smaller contact angle is indicative of reduced surface tension and superior wettability, as illustrated in Figure 7.11. If the contact angle is greater than 90 degrees, the solid is considered non-wettable. As mentioned previously, adding surfactant or dispersant to the liquid or the solid surfaces is vital to the process. The commercial instruments and measurement techniques of continuous-phase surface tension and solid–liquid contact angle are briefly mentioned in Chapter 1. In order to facilitate breakage of the solid agglomerates and flocculants, a dispersant may be used. The wetting agents are often designed to both de-agglomerate and disperse. Note that the surface tension of the continuous phase must not be significantly lowered in

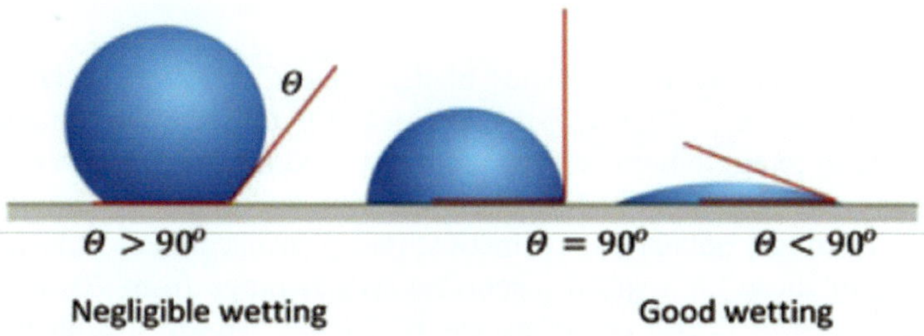

FIGURE 7.11 A schematic showing the contact angle of the liquid drops on a solid surface.[25]

order to avoid foaming of the working media. Wetting agent concentration must be optimized for the given application, to avoid adverse effects on the product performance.

In solid–liquid mixing batch processes, attention to the component addition sequence is needed to prevent solids clumping in the tank bottom and plugging in the overhead feed chute. In one instance, when fine solids were added to a tank containing methanol, the chute frequently plugged due to methanol vapor saturating and reacting with solids in the chute. To circumvent the issue, although atypical, solids were added to the tank first and slowly agitated while liquid was added through a dip pipe. Thus, use of a complicated compartmental chute and automated plunger design were avoided.

7.9 PARTICLE BREAKAGE

In solid–liquid contacting processes such as crystallization and precipitation, the design target is often avoiding particle breakage by creating a synergistic circulation flow with minimal turbulence. The objective of this controlled turbulence is to keep microscale eddies (i.e., sized near the Kolmogorov length scale, λ_k) larger than the desired particle size. Particle size, d_p, is set by knowingly maintaining the relationship between kinematic viscosity, ν, and specific mixing energy, ε_{loc}, as shown in equation 7.22.

$$\lambda_k = \left(\nu^3 / \varepsilon_{loc} \right)^{1/4} \tag{7.22}$$

On the other hand, dissolution can be expedited and mass transfer enhanced by increasing both fluid circulation and turbulence to expose the solids to greater nascent liquid surface area. Although the authors recommend open impellers with a high-efficiency profiled blade (e.g., A310, HE3 and Intermig), for crystallizers, they prefer higher power number axial-flow impellers (e.g., pitch-blade turbines and hydrofoils) when dissolving and reacting particles in the continuous phase.

7.10 BAFFLES AND SURFACE FINISH

Careful consideration of baffle shape and baffle clearance from the tank wall can help minimize breakage and/or attrition of fragile crystals. In lieu of flat baffles, the tank may be configured with round and profiled baffles. Attrition of expensive heterogeneous catalysts such as palladium and platinum are avoided with appropriate surface finish of the process vessel walls. One may consider vessel walls that are glass-lined carbon steel or dairy finished (surface roughness average, Ra range 0.5–0.8 μm) stainless steel. Such vessels are widely used in the pharmaceutical and food processing industries where sanitation is critical.

7.11 DISPERSERS

High-shear radial dispersers such as Cowles blades[26] and Conn & Co's ITT (Intensive Type with Teeth) blades,[27] shown in Figure 7.12, are ideal for uniform suspension of non-fragile solid particles that tend to agglomerate. They are often used in catalyst slurry tanks for uniform feeding of slurry into the reactors. Other deagglomeration and dispersion applications are practiced in the paint, coatings, adhesives, cosmetics, and pharmaceutical industries. The desired tip speeds (πDN) of these high-speed dispersers (HSD) range from 20–30 m/s (4000–6000 fpm), while specific power draws are in excess of 2W/kg. Motor specification as "rules-of-thumb" are ~1hp/10gal. The optimum design D/T and C/T ratios are approximately 1/3 and 1/2, respectively, and full holding capacity of the vessel is typically 1.3 times the batch size. The normal operating viscosity range is 1,000–25,000 cP and up to 50,000 cP max.[28] The ITT impeller design with multiple louvers on top of the disc surface (see Figure 7.12.b) provides minor axial flow in addition to radial flow. In the fully turbulent flow regime, power numbers of HSDs (Morehouse, Ross, SPX Flow-Lightnin sawtooth design) and Conn & Co. ITT blades are reported to be between 0.3 to 0.45.

(a) (b)

FIGURE 7.12A–B High-shear dispersers: (a) Morehouse Cowles blade[26] ($N_p = 0.45$ and $N_{Fl} = 0.26$, in a fully baffled tank) and (b) Conn & Co. ITT dispersion blade[27] ($N_p = 0.45$).

7.12 TICKLER IMPELLERS

For solid suspension in the bottom head of the vessel, a smaller impeller ($D/T = 0.15 - 0.2$) known as a tickler or kicker is often used (see Figure 7.13). The purpose of the tickler is twofold: (a) assist the main impeller in suspending heavy solids at the bottom, and (b) keep the particles suspended during the draining process after the slurry level recedes below the main impeller. While a tickler is commonly a down- or up-pumping standard PBT, its close proximity to the tank bottom nozzle ($C/T \leq 0.10$) creates radial flow as the slurry level drops below the influence of the main impeller. A typical radial flow pattern throws the solid particles towards the wall and may present an adverse pressure gradient on the draining operation, depending on the tickler's rotational speed and direction of pumping. Consequently, the draining process typically slows down and solids begin to settle in the tank bottom. Encountering such scenarios in slurry processes motivated the authors to innovate in creating a new tickler impeller design,[28] known as the KT-3, to minimize solids loss and enhance the draining process. After the slurry level recedes below the main impeller, the KT-3 tickler blades generate a downward swirling flow pattern that directs the slurry towards the bottom discharge nozzle. Slurry inertia maintains this action temporarily, even after the slurry level is below the tickler. The turbulent ($Re \geq 10{,}000$) power number, N_P, of the KT-3, available for purchase from SPX Flow, is 0.6. For further insight into the KT-3's performance and design details, please see references 29–30.

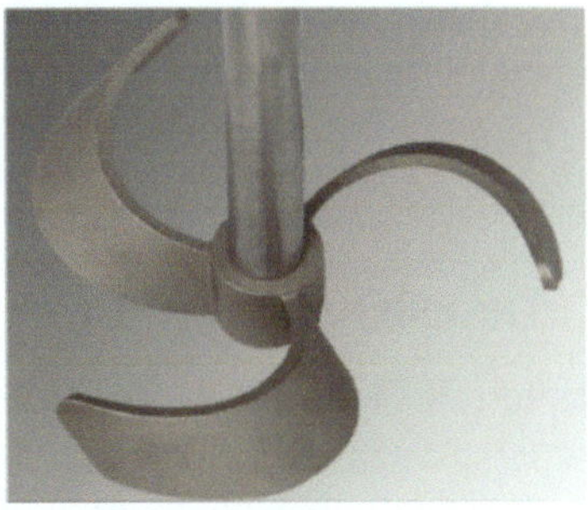

FIGURE 7.13 CAD drawing of a SPX Flow KT-3 tickler impeller.[30]

7.13 SLINGERS

Boiling reactors are often used to remove excessive heat from a working media that is experiencing a highly exothermic reaction. For example, CTA slurry reactors contain 30–40vol.% fine solids (100–300 μm) in a sacrificial solvent of acetic acid. Energy extracted by the acetic acid vapor is removed in overhead condensers, and then the condensate is re-introduced into the tank. Unfortunately, the high face velocity of the acid vapor entrains a small fraction of the fine CTA solids into the condenser, where they cause plugging, reactor shutdown, and tremendous loss of plant productivity. To mitigate this plugging, a torrent of acetic acid droplets is dispersed throughout the reactor vapor space to knock down entrained solids before they reach the condenser inlet. A preferred means of creating these acid droplets is a slinger impeller mounted on the rotating agitation shaft, below the height of the condenser. Typical slingers use 4–8 radially straight vanes that are spaced uniformly atop a flat disk. Directing the stream of returning acetic acid condensate onto this rotating slinger (via a dip pipe) theoretically disperses it across the reactor as liquid droplets that are shed from the edges of the rotating vanes. Unfortunately, droplet sheets from straight vane slingers covered less than the full 360-degree reactor cross section, and thus, failed to consistently mitigate condenser plugging. Consequently, the authors designed and implemented the novel curved vane KS-8 slinger,[31] shown in Figure 7.14. Its rotating curved blades retain liquid longer than needed for a liquid sheet (umbrella), spanning the full 360-degree reactor cross section. Thus, it more completely disengages CTA particles entrained in the acetic acid vapors.

FIGURE 7.14 Photograph of the KS-8 Slinger.[31]

7.14 EXAMPLE

A 2m ASME dish-bottom slurry tank contains 30wt.% of rod-shaped calcite crystals in a non-polar solvent (hexane) at room temperature and atmospheric pressure. Working media height to the tank diameter is approximately 1.0. The particle diameter and length are approximately 100 and

500 µm, respectively. Skeleton or true density of the particle is 2.71gm/cc. The tank is equipped with a standard A310 high-efficiency impeller located on the tank tangent line. Determine the required Zwietering just suspension speed, N_{js}, a preferred complete dispersion speed, N_c, and required specific power, ε. See mixit-io SS7.1.

7.15 NON-STR S-L MIXING/DISPERSION DEVICES

In a few special instances, such as in railcars, large wastewater treatment tanks, and remediation ponds, implementation of jet mixer(s) are a cost-effective approach to suspending minor levels of light solids. In the case of continuous operations with addition of fine light solids (powder suspensions), sometimes high-shear rotor-stators such as the Silverson—Flashmix® powder/liquid mixers (see Figure 7.15) are used to suck in and mix the solids with a continuous liquid stream. Later jet mixing and high-shear mixing topics are covered in Chapters 11 and 13, respectively.

FIGURE 7.15 Silverson Flashmix® powder-liquid mixer.[32]

Special multi-shaft and/or planetary agitation systems such as those shown in Figure 7.16.a and Figure 7.16.b uniformly disperse fine solids into polyurethanes, polyesters, and epoxies for preparation of adhesives, sealants, and putties, etc. These are highly solid loaded, extremely high viscosity liquids, often in the range of millions centi-Poise. An example would be carbon black in polyurethane for the manufacture of automotive windshield sealant. This topic is discussed later in the high viscosity mixing chapter.

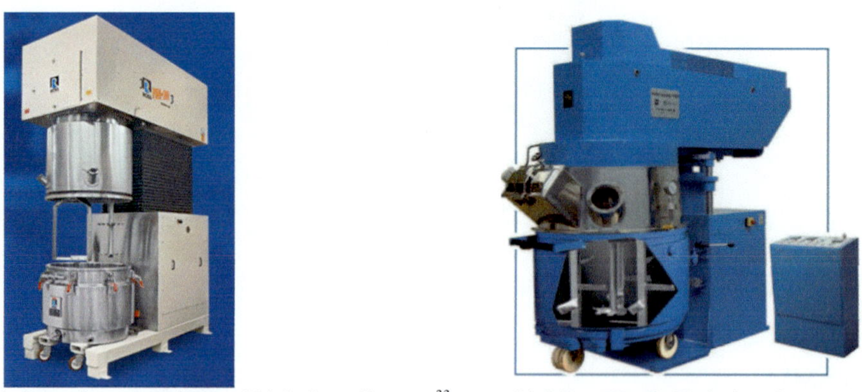

(a) Ross planetary mixer and high shear disperser[33]

(b) Oliver+Batlle Hydrobat planetary mixer[34]

FIGURE 7.16 High viscosity mixing equipment.

7.16 SOLID SUSPENSION IN TALL TANKS

It is normal practice to use standard solid–liquid mixing tanks with a liquid height-to-tank diameter aspect ratio of ≤ 1.2. For specific applications needing a larger batch size, a higher overall heat transfer rate due to an increased surface area-to-volume, or a smaller reactor footprint, however, "tall" tanks (H/T ≥ 2.0) with multitier axial flow impellers are often used. The authors encountered a slurry process that demanded laboratory investigation of solids suspension uniformity in a "tall" tank, prior to a process scale-up.[35] Experiments were carried out with ~300μm Saran® particles (density = 1.4gm/cc) suspended in water by a 0.44m ID Plexiglas® tank with H/T ≥ 2.0, as depicted in Figure 7.17. Observations concluded that an agitation system of purely axial flow wide-blade hydrofoils (A320s) consumed 25% less specific power than the mixed flow PBTs while also producing a similarly uniform distribution of solids within the tall tank. Significantly, higher energy or agitation speeds improved the solid distribution insignificantly, partly due to choking of the upward return flow near the vessel wall, which was more pronounced in case of PBTs. The results suggested that once enough specific power had been supplied to produce the uniform particle distribution, a lower sustained power level could maintain that distribution due to hindered particle settling.

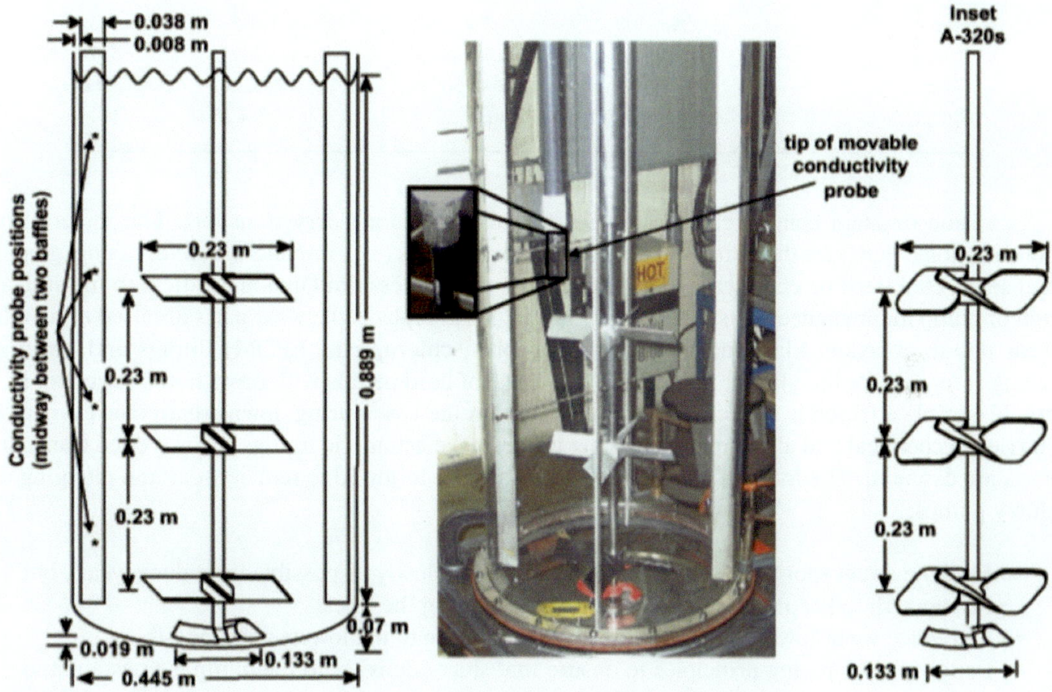

FIGURE 7.17 Experimental set-up with multi-tier impellers in a tall tank (H/T = 2.0).[35]

7.17 SOLID–LIQUID MIXING/DISPERSION PROCESSES

The objectives of typical solid–liquid mixing operations can be categorized as:

1. simple suspension and distribution of solids as slurries
2. dissolution of solids
3. reaction of solids with the liquid media

7.17.1 Slurries

As mentioned earlier, uniform suspension and distribution of solids in a stirred tank is essential to obtain unvarying downstream transport into a storage tank or a reactor. An example would be to uniformly feed a polyol reactor with a heterogeneous catalyst for controlled reaction. In many instances, transporting varying concentrations of solids into pipe lines from the mixing tank causes line plugging and unwanted shutdown, with lost productivity in the downstream operations. A steady flow regime in which turbulent eddy drag forces are sufficient to suspend and distribute the particle mass is desirable. In Table 7.10, the authors outline rules-of-thumb for maintaining sufficient velocity to uniformly suspend slurry particles in a pipeline while avoiding solids settling and plugging. The minimum desired slurry velocity in the pipe primarily depends on the type of solid and its size (Mesh number).[36]

TABLE 7.10
Downstream Piping—Slurry Velocity to Avoid Solids Settling and Plugging[36]

Type of Solids	Size of Solids, d_p		Minimum Flow Velocity	
	Mesh No.	Microns	m/s	ft/s
Fine	Over 200	<74	1–1.5	3–5
Sand	200–20	74–841	1.5–2	5–7
Coarse	20–4	842–4760	2–3.25	7–11
Sludge	Semi-solid (mushy) slurry		3.25–4.25	11–14

Abrasion-resistant centrifugal pumps are typically used for slurry transport. The material of construction (MOC) for the pump's major wetted components (e.g., impellers, volute, casing, glands and seals, etc.) must be chemically compatible with the process fluid and are critical to minimization of pump maintenance costs. Sacrificial materials and replaceable wear parts are used to handle wear due to abrasion. Elastomer (e.g., natural rubber, chloroprene, EPDM) linings and ceramic facings are used for the glands. Static parts are made of hard metals with or without the rubber liners. If particle attrition is unacceptable, which is often the case during downstream conveyance of slurries in chemical and allied manufacturing processes, elastomeric linings of the wetted transport parts are essential. The following basic guidelines may be followed to reduce wear and plugging of slurry pumps:[37]

- Find the sweet spot that allows the pump to run as slowly as possible (to reduce wear), but fast enough to keep solids from settling and clogging the lines.
- To reduce wear, lower the pump's discharge pressure to the lowest point possible.
- Follow proper piping principles to ensure that slurry delivery to the pump is constant and uniform.

In the chemical and allied industries, transfer of slurry through shell-and-tube heat exchangers (S-T Hex) is quite common. When flowing a concentrated slurry through a horizontally oriented heat exchanger, caution must be especially taken to avoid solids deposition in either the inlet plenum and solids plugging in the tubes. The authors invented a tube-side, sequentially pulsable-flow, shell-and-tube heat exchanger[38] (KHX) to avoid such chronic issues in production plants. The KHX uses an impeller in the entrance head to keep particles suspended in the plenum and to create a pulsatile flow in the heat transfer tubes as the blades pass by their apertures. Pulsating flow disturbs the velocity and thermal boundary layers, and thus, mitigates particle adherence to the wall while enhancing heat transfer.

7.17.2 Solid Dissolution

Simple non-reactive solid dissolution, known as solvation, is a solid–liquid mass transfer dominated process wherein solute molecules diffuse into the surrounding liquid, known as solvent. Exposure of the particle surface to the unsaturated liquid media invigorates the process of dissolution and helps to create a homogeneous media. Hence, turbulent flow within the tank is essential to create uniform solid suspension and to maximize exposure of interfacial surfaces. It must also maintain transience of the solid–liquid boundary layer and enhance the rate of solute (i.e., solid) dissolution (mass transfer) until the solvent's saturation limit is reached. At the solubility limit (i.e., solvent saturation), the competing processes of solvation and precipitation are in equilibrium, and the solvent is incapable of dissolving more solute. For example, the solubility limits of table salt and sugar in room temperature water are 35 and 100 grams/100mL, respectively.

For readily dissolving solids, integral mass transfer coefficient, $\bar{k}$, is experimentally measured at the laboratory scale. It is used for production reactor scale-up and maintains specific power and geometric similarities to predict expected dissolution time.[39] The intrinsic mass transfer coefficient for readily dissolving solids is calculated as follows:

$$\bar{k} = 3\rho_s / \theta A_p \left(C^* - C_\infty\right) \tag{7.23}$$

where ρ_s = density of solid [kg-m^{-3}]
θ = dissolution time from initial particle diameter, d_p to 0 [sec]
A_p = specific surface area of the initial particle [m^{-1}]
C^* = species concentration at the solid–liquid interface [kg-m^{-3}]
C_∞ = species concentration in the bulk liquid phase [kg-m^{-3}]

Based on Kolmogorov's transport theory, the bulk diffusion mass transfer coefficient, k, is determined from the following general form of mass transfer equation:

$$Sh = A + B\,Re_p^{\ n}\,Sc^m \tag{7.24}$$

where, $Sh = k d_p / \vartheta_L$ and $Sc = \upsilon / \vartheta_L$ are respectively, the Sherwood and Schmidt numbers
ϑ_L = diffusion coefficient [m^2-s^{-1}]
υ = kinematic viscosity [m^2-s^{-1}]

The modified Froessling equation[40] for stirred tank is represented by

$$Sh = 2 + 0.72\,Re_p^{\ 0.5}\,Sc^{0.33} \tag{7.25}$$

where particle Reynolds number, $Re_p = \left(\epsilon_T\, d_p^{\ 4} / \upsilon^3\right)^{1/3}$

For very small particles, Sh $\rightarrow$ 2, dissolution time is independent of agitation conditions, and the mass transfer coefficient, k, is simply given by

$$k = 2\vartheta_L / d_p \tag{7.26}$$

7.17.3 Solid–Liquid Reaction

Reaction may occur on the solid surface—typical of a heterogeneous catalytic reaction, or in the bulk liquid phase—associated with molecular transport leading to the solvation (dissolution) process. As mentioned earlier, solids suspension preferably should be uniform to promote reaction and to avoid the mixing-controlled rate limit. However, in a few cases, solvation and reaction rates must be optimized to circumvent exothermic reactions, leading to the formation of excessive heat and/or undesired byproducts. Typically, reaction on the solid surface (e.g., during crystallization, ion exchange, heterogeneous catalytic reaction, etc.) is not significantly affected by agitation conditions. Figure 7.18 shows a catalyst particle in a well-mixed reactor. Concentrations of reacting species are C_∞ in the bulk fluid and C_i at the solid–liquid interface.

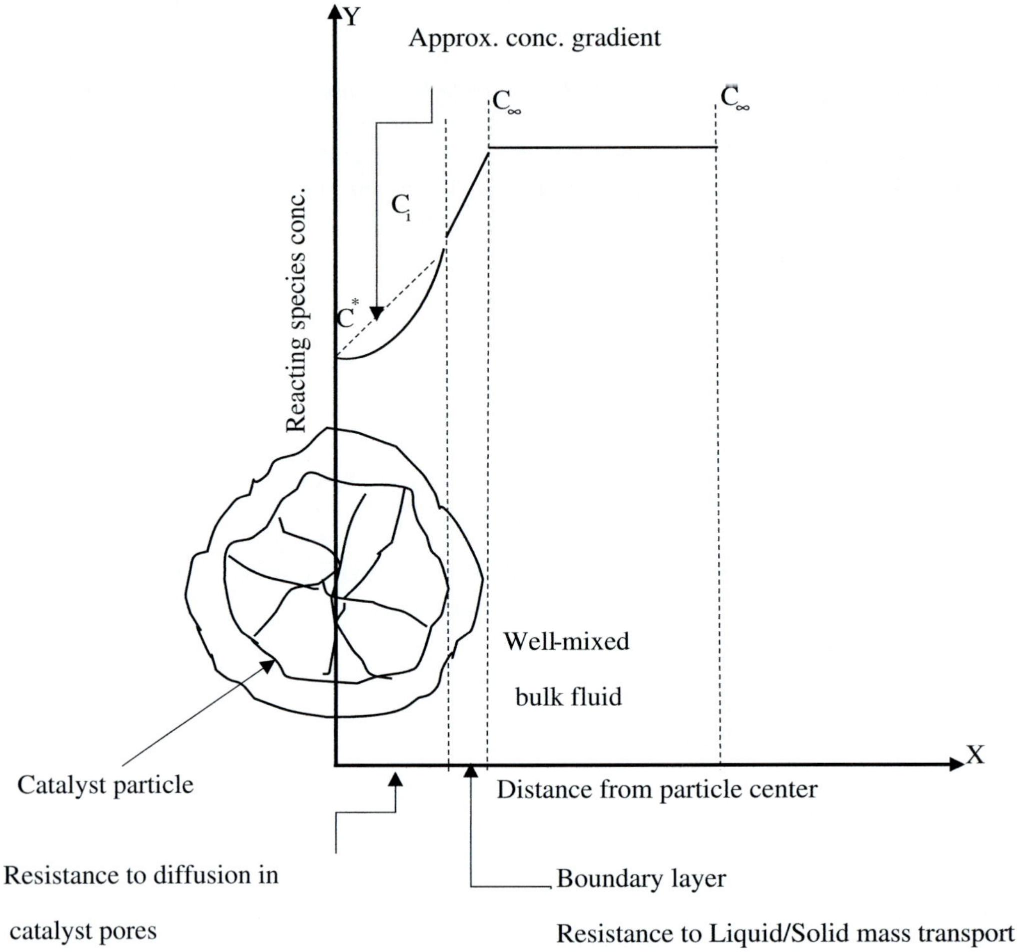

FIGURE 7.18 Idealized mass transfer and reaction two-step model for a catalyst within an STR.

Reactions occurring entirely in the bulk liquid phase are characterized by the mass transfer coefficient, k [m²/s], which is a strong function of agitation system hydrodynamics. During physical dissolution of the solid, only bulk diffusion is important so the reaction rate, R_G, equals the rate of dissolution, R_D, as represented in equation 7.27. A is the particle surface area exposed to the liquid phase.

$$R_D = R_G = k\,A\left(C_\infty - C_i\right) \tag{7.27}$$

However, surface reaction rate, R_G, includes in-pore diffusion and is characterized by the mass transfer coefficient, k_r. C* is the species solubility/saturation limit, and n' is the order of reaction.

$$R_G = k_r\, A\left(C_i - C^*\right)^{n'} \tag{7.28}$$

Based on these rate constants, the solid–liquid reactive mixing processes are identified as

(a) reaction limited for $k_r \ll k$
(b) mass transfer limited for $k_r \gg k$
(c) balanced if $k_r \approx k$

Although k may be approximately determined using appropriate correlations, the reaction rate constant, k_r, and the order of reaction, n', are measured in a stirred laboratory vessel under a no-mixing-limitation condition.

REFERENCES

1. Solid-liquid mixing in agitated vessels (Just Suspended Speed). www.cheresources.com/invision/blog/4/entry-511-solid-liquid-mixing-in-agitated-vessels-just-suspended-speed/
2. Zwietering, TH. N., Suspending of solid particles in liquid by agitators, *Che. Eng. Sci.*, 8, pp. 244–253, 1958.
3. Chapman, C. M., Nienow, A. W., Cooke, M., Middleton, J. C., Particle-gas-liquid mixing in stirred vessels. I: Particle-liquid mixing, *Chem. Eng. Res. Des.*, 61, pp. 71–83, 1983.
4. Bujalski, W., Konno, M., Nienow, A. W., In *Proceedings of the 6th European Conference on Mixing*, pp. 389–398. BHRA, Cranfield, 1988.
5. Myers, K. J., Janz, E., Fasano, J. B., Effect of solids loading on agitator just-suspension speed, *Can. J. Chem. Eng.*, 91(9), pp. 1508–1512, September 2013.
6. Ayranci, I., Kresta, S. M., Critical analysis of Zwietering correlation for solid suspension in stirred tanks., *Chem. Eng. Res. Des.*, 92, p. 413, 2014.
7. Grenville, R. K., Mak, A. T. C., Brown, D. A. R., Suspension of solid particles in vessels agitated by axial flow impellers, *Chem. Eng. Res. Des.*, 100, pp. 282–291, 2015.
8. Davies, J. T., Particle suspension and mass transfer rates in agitated vessels, *Chem. Eng. Proc.* 20, pp. 175–181, 1986.
9. Harnby, N., Edwards, M. F., Nienow, A. W., *Mixing in the Process Industries*, 2nd Edition, p. 374. Butterworth Heinemann Ltd., 1992.
10. www.piping-engineering.com/flush-bottom-valve-special-purpose-valve.html
11. www.fetterolfvalves.com/pdfs/Valve%20News%20Tank%20Drain%20Installation%20Guide.pdf
12. www.schuf.com/pdf/schuf-pharmaceutical-and-fine-chemical-industries-2015-rev.1
13. www.ekato.com/en-us/products/agitator-components/impellers/intermig/
14. Grenville, R. K., Giacomelli, J. G., Padron, G., Brown, D. A. R., *Impeller Performance in Stirred Tanks*, p. 46–55. Chemical Engineering, August, 2017.
15. www.spxflow.com/en/lightnin/pc-impellers
16. Eagles, W. P., *Solid-Liquid Mixing Design Guide*, FMP Report 053 v3, January 2006.
17. Hicks, M. T., Meyers, K. J., Bakker, A., Cloud height in solids suspension agitation, *Chem. Eng. Comm.*, 160(1), p. 137, 1997.
18. Dey, S., Ali, S. Z., Padhi, E., Terminal fall velocity: The legacy of Stokes from the perspective of fluvial hydraulics, *Proc. R. Soc. A*, 475, p. 20190277, 2019.
19. Schlichting, H., Gersten, K., *Boundary-Layer Theory*, 9th Edition, Springer-Verlag Berlin, 1979.
20. Allen, T., *Particle Size Measurements*, Chapman & Hall, 1990.
21. Fan, L., Zhu, C., *Principles of Gas-Solid Flow*, Cambridge University Press, 1998.
22. Pettyjohn, E. A., Christiansen, E. B., Effect of particle shape on free-settling of isometric particles, *Chem. Eng. Prog.*, 44(2), p. 157, 1948.
23. Hipple, J., Brenner, H., *Low Reynolds Number Hydrodynamics*, Martinus Nijhoff Publishers, Dordrecht, 1983.
24. Pabst, W., Gregorová, E., Characterization of particles and particle systems, *The Institute of Chemical Technology (ICT)*, Prague, 2007.
25. https://fmps.fbk.eu/contact-angle-platform
26. https://fusionfluid.com/products/mixing-impellers/cowles-disperser-multi-blade-stx
27. www.connblade.com/cb/blades.asp

28. Mixing technology insight #95, Charles Ross & Son Company. www.mixers.com/insights/mti_95.pdf
29. Cope, R. F., Kar, K. K., KT-3: A novel tickler for solids removal from slurry vessels, *Ind. Eng. Chem. Res.*, 48(10), pp. 4990–4997, 2009.
30. www.spxflow.com/lightnin/products/kt-3-low-level-mixing-impeller/
31. Kar, K., Piras, L., Monaghedu, M., Bai, H., A novel slinger for efficient reflux condensate distribution in a boiling reactor, *Presentation at North American Mixing Forum (NAMF) XXII*, Victoria, British Columbia, Canada, June, 2010.
32. www.findtheneedle.co.uk/companies/silverson-machines-ltd/products/flashmix-powder-liquid-mixers.
33. www.planetarymixers.com/
34. https://oliverbatlle.com/en/prod/mixing-equipment-planetary-hidrobat/
35. Cope, R. F., Somasi, M., Kar, K. K., Solid distribution in tall tanks, *Can. J. Chem. Eng.*, 89(5), pp. 1104–1111, August 2011.
36. www.engineeringtoolbox.com/slurry-transport-velocity-d_236.html
37. Simpson, J., A beginner's guide to pumping slurry, *Crane Engineering*, January 2015. https://blog.craneengineering.net/a-beginners-guide-to-pumping-slurry
38. Kar, K. K., Cope, R. F., Yuan, Q., Somasi, M., Morgan, B. M., *Tube-Side Sequentially Pulsable-Flow Shell-and-Tube Heat Exchanger Apparatus, System, And Method*, US Pat. 9068782, Dow Global Technologies LLC., June 2015.
39. Nienow, A. W., Miles, D., The effect of impeller/tank, configurations on fluid-particle mass transfer, *Chem. Eng. J.*, 15, pp. 13–24, 1978.
40. Rowe, P. N., Claxton, K. T., Lewis, J. B., Heat and mass transfer from a single sphere in an extensive flowing fluid, *Trans. Inst. Chem. Eng.*,43, p. T14, 1965.

8 Immiscible Liquid–Liquid Mixing/Dispersion

8.1 INTRODUCTION

Mixing immiscible liquid components under a reasonable shear strain produces dispersions like that shown in Figure 8.1. Dispersion processes may involve non-reactive or reactive systems. Formulations and extractions/stripping are non-reactive processes. Home-care and health-care products, as well as pesticide preparations (micro-emulsions), are simple formulations. Complex formulations include biphasic solvent systems that use a suitable solvent to extract a soluble fraction from within an insoluble matrix. Reactive systems include suspension and emulsion polymerizations, while well-known products such as polyvinylchloride (PVC) and ion exchange resins (e.g., Dowex®) are created using suspension polymerization processes. Common acrylic and styrene-butadiene latexes used in paints and paper coatings, respectively, are produced using emulsion polymerization processes. Typically, these polymerization processes generate insoluble particles ranging from colloidal to micron-sized in their respective continuous phases. Figure 8.2–Figure 8.4 show examples of common products obtained from some of the non-reactive and reactive immiscible liquid systems described earlier.

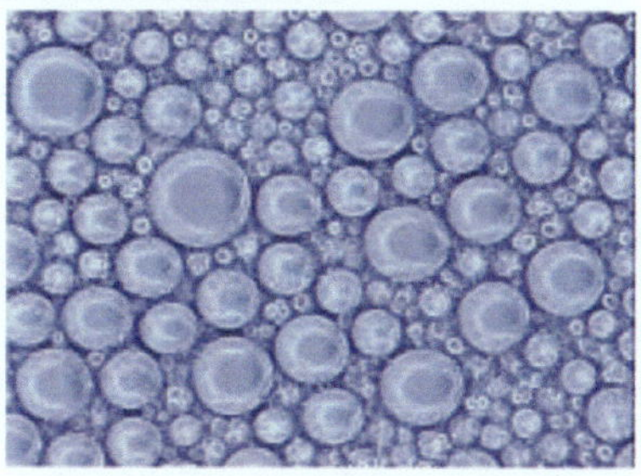

FIGURE 8.1 Photograph showing dispersion of a liquid in a continuous media.

FIGURE 8.2 Photograph of a common household non-reactive micro-emulsion product (Pine-Sol®).

DOI: 10.1201/9781003606390-8

FIGURE 8.3 Schematic of suspension polymerization process and photograph of a product (PVDC–Saran®).

FIGURE 8.4 Schematic and photograph of latex emulsion polymerization process[1] and product (paint).

8.2 EMULSION PROCESS

Generally, emulsions are thermodynamically unstable systems of at least two immiscible phases, one dispersed as globules in the other, and stabilized by a third substance called an emulsifying agent. The droplets are called the dispersed (internal) phase, and the liquid in which the droplets are dispersed is the continuous (external) phase. Microemulsion, an isotropic and thermodynamically stable class of emulsion, contains oil whose droplets are <0.1μm. Since the size of the droplets is less than 25% of the wavelength of visible light (~0.4μm), microemulsions are transparent, i.e., appear to be clear (see Figure 8.2).

Figure 8.5 shows examples of a couple of commonly used macro- and micro-emulsion products, which are respectively kinetically and thermodynamically stable systems.

FIGURE 8.5 Photographs of macro-emulsion (salad dressing) and micro-emulsion (face cream) products.

8.3 SURFACTANT CHEMISTRY

Acrylic is produced in a polymerization process that micro-emulsifies acrylic acid monomer droplets (oil), and surfactants, in a continuous water phase. Certain water-soluble polymers such as polyvinyl alcohols (PVA) and hydroxyethyl celluloses (HECs) are used to stabilize emulsions. Surfactant chemistry that dictates the functioning of surfactants and emulsifiers is a subject that is extensively studied and applied by chemical formulators. Without trying to fully describe this subject, a few meager statements are offered.

Figure 8.6 is schematic diagram of a micelle of oil in aqueous suspension, such as might exist in an emulsion of oil in water. In this example, the oil-soluble tails of the surfactant molecules project into the oil, while the water-soluble ends remain in contact with the water phase to create a stable oil droplet.

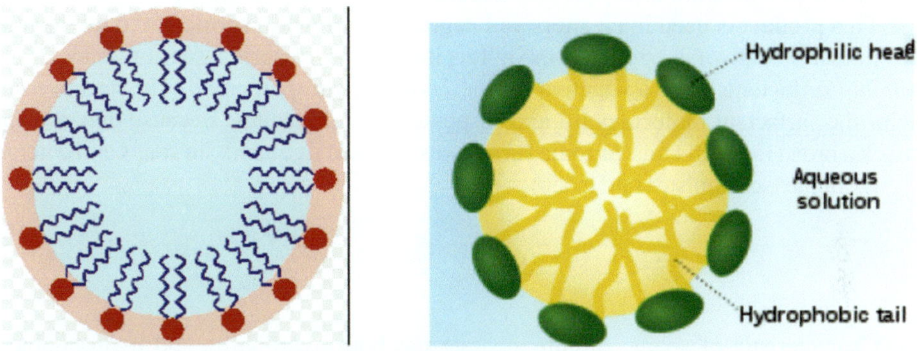

FIGURE 8.6 Schematic representations of droplet formation.

A vast number of commercial surfactants are manufactured by specialty chemical companies. These are tailored for use in such markets as health, home, food, pharmaceutical, and industrial products. Surfactants are typically classified based on the nature of their heads, i.e., hydrophobic or hydrophilic, as schematically shown in Figure 8.7. The hydrophilic-lipophilic balance (HLB) is an important parameter in choosing the surfactant for each specific application.

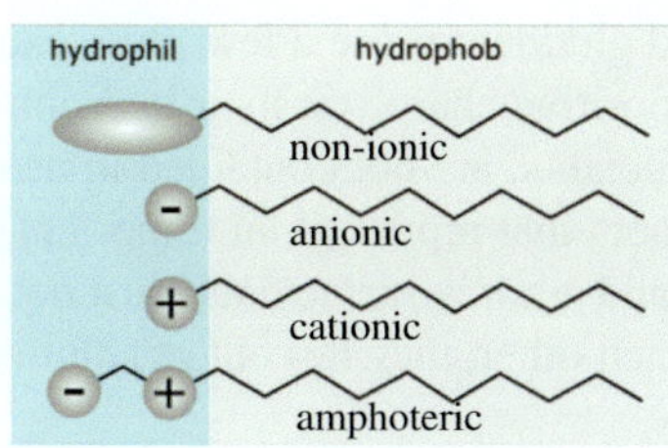

FIGURE 8.7 Classification of surfactants based on their head charges.

A few commonly used surfactants are represented as follows with their respective molecular structures.

(a) *Nonionic* surfactants are neutral, having no charge on their hydrophilic end. Nonionic surfactants are very good at emulsifying oils in the aqueous phase.
   (i) Polyvinyl alcohol (PVA—non-ionic)

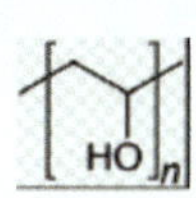

Polyvinyl alcohol is most commonly used to stabilize emulsions caused by the formation of relatively small-sized particles with uniform size distribution. It is an excellent nanoparticle stabilizer, and is widely used to stabilize agricultural microemulsions enroute to granular products.

(ii) Tergitol® nonylphenol ethoxylate (non-ionic)

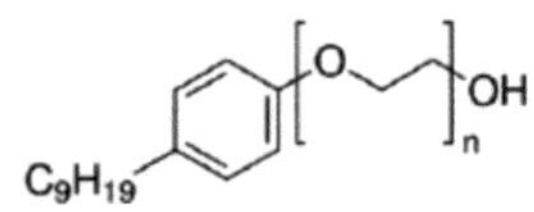

Tergitol® NP-10 is a nonylphenol ethoxylate-based surfactant manufactured by Dow Inc. It is non-ionic and provides excellent detergency, rinse-ability, and low odor. This product is used in cleaners and degreasers, prewash spotters and metalworking fluids.[2] For diverse applications of Tergitol® NP-4-15 series, see reference 2.

(b) *Anionic* surfactants have a negative charge on their hydrophilic end. The negative charge helps the surfactant molecules lift and suspend soils in micelles. Because they are able to attack a broad range of soils, anionic surfactants are used frequently in soaps and detergents.
Example: sodium stearate (soap)

Characteristic of soaps, sodium stearate has both hydrophilic and hydrophobic portions, namely, a carboxylate and a long hydrocarbon chain, respectively. It is also used in the pharmaceutical industry as a surfactant to solubilize hydrophobic compounds in producing mouthwashes.

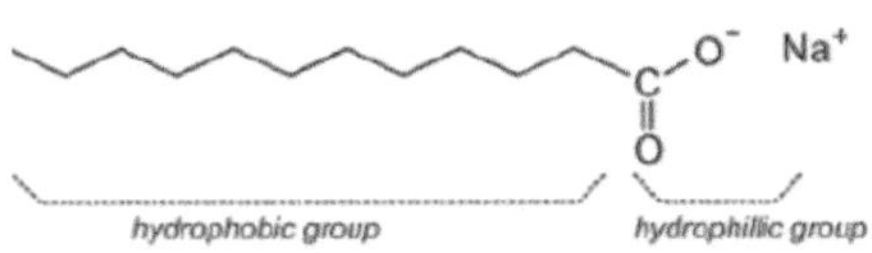

a soap

Sulfates, sulfonates, and gluconates are a few more examples of anionic surfactants.

(c) *Cationic* surfactants have a positive charge on their hydrophilic end. Alkyl ammonium chlorides, common cationic surfactants, are not contained in usual cleaning products but are the active agents in fabric softeners that represent an important group of detergent products. As might be expected, cationic and anionic surfactants must not be used in the same formulation because, after neutralizing each other, they fall out of solution and are no longer effective.

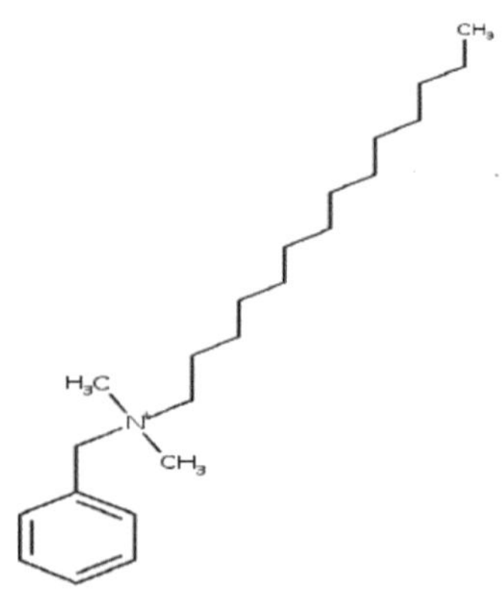

(d) *Amphoteric* surfactants simultaneously carry anionic and cationic hydrophilic groups with a structure that simultaneously contains hermaphroditic ions capable of forming a cation or anion depending on ambient conditions (e.g., pH). Because amphoteric surfactants become positively charged in acidic solutions, they behave like cationic surfactants. In alkaline solutions, they develop a negative charge, and behave like anionic surfactants. All polyethylene glycols (PEGs) are amphiphilic and soluble in water as well as in many organic solvents: the higher the molecular weight, the lower their solubility in water.

Polyethylene glycols

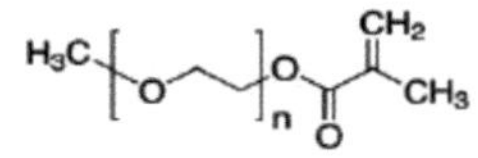

Amphoteric surfactants are often used in personal care products such as shampoos and cosmetics. Examples of some frequently used amphoteric surfactants are betaines and amino oxides.

Alkyl dimethyl butane

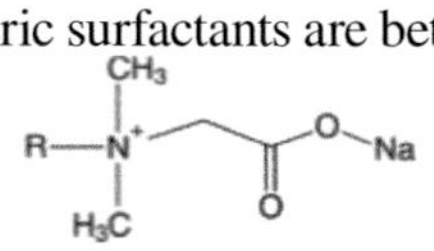

A few other common surface-active agents and their usages are listed as follows:

(a) Fatty acid sulfonates, e.g., sodium lauryl sulfate (SLS), are anionic surfactants used as an emulsifying cleaning agent in household cleaning products (laundry detergents, spray cleaners, and dishwasher detergents).
(b) Ethoxylated alcohol compounds, e.g., ethoxylated propylene glycol, are used as detergents, wetting agents, emulsifiers, degreasers, and emollients in many lines of commercially available products and industrial practices, including paints and coatings, agricultural, home and personal care products, oil and gas production, and industrial and institutional cleaning products.

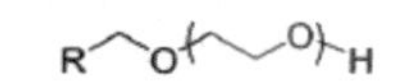

(c) Short-chain starches (polygluconates) are cationic surfactants. A positive charge makes them useful in anti-static products such as fabric softeners, while their antimicrobial properties prompt their use in disinfectants.
(d) Quaternary ammonium fatty acid adducts (ammonium quats)

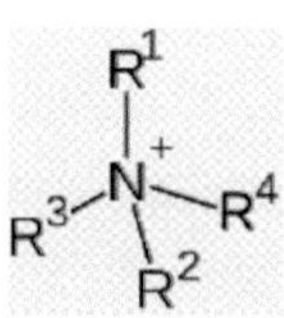

Quaternary ammonium salts are used as disinfectants, surfactants, fabric softeners, and as antistatic agents (e.g., in shampoos). In liquid fabric softeners, the chloride salts are often used.
(e) Lignin sulfonates (lignosulfonates) are water-soluble anionic polyelectrolyte polymers used to disperse pesticides, dyes, carbon black, and other insoluble solids and liquids into water. They are used in tanning leather. These are typically generated as byproducts from the production of wood pulp using sulfite pulping.

Benzyl sulfonic acid

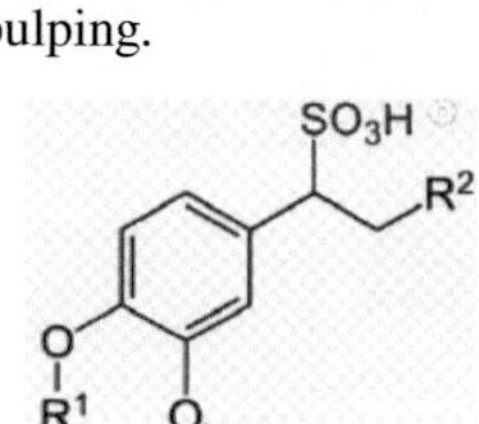

(f) Organo-silicone surfactants are amphiphilic, but the non-silicone-soluble group can be oil- or water-soluble. The silicone surfactant, if composed of silicone- and water-soluble groups, will lower the surface tension of water to around 20 dynes/cm. If the silicone surfactant has oil- and silicone-soluble groups, the surface tension of the oil will be lowered to 20 dynes/cm. Adding silicone surfactants to agricultural chemicals greatly reduces solution surface tension and enhances their wettability, spreadability, and adhesion on the plant or pest surfaces, thereby improving their efficacy.[3]

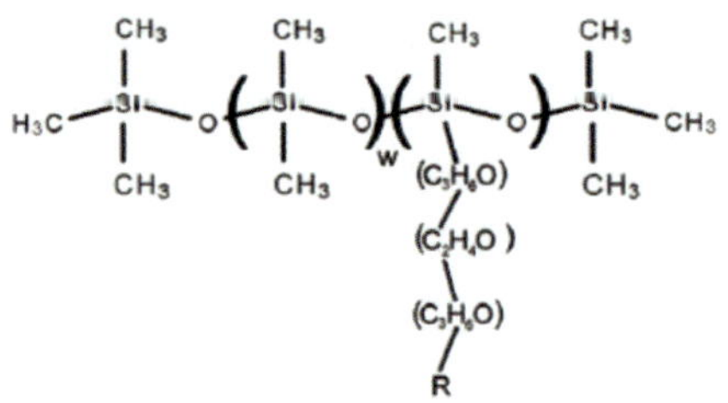

A general organo-silicone surfactant structure (R = acetyl, w = 0 with a variable polyethoxylated tail)[4]

8.3.1 Hydrophilic/Lipophilic Balance (HLB)

Ranging in arbitrary units of 1–20, the HLB of a nonionic surfactant can be calculated and used to determine the propensity of a compound to work effectively in a given solution of oil and water. Lower HLB values (<10) are commonly used for oil-rich solutions while surfactants with higher HLB values (>10) are typically most efficient in oil-in-water emulsions. For example, length of the polyoxyethylene component (i.e., the hydrophilic portion) of alcohol ethoxylate provides this class of compounds with a wide assortment of water solubilities and detergency properties. Increasing the amount of ethylene oxide on the ethoxylated alcohol typically increases its water solubility and also increases the hydrophilic/lipophilic balance (HLB) of the compound. Ethoxylated alcohol products can vary widely in their HLB values, offering numerous options for chemical formulators. Readers are advised to view the "Surfactants" publications of Oxiteno, a manufacturer of specialty chemicals and surfactants.[5]

8.3.2 Reactive Surfactants

A reactive surfactant is an amphiphilic molecule with an additional functionality that provides it with chemical reactivity. An important class of reactive surfactants are those that act in radical polymerizations as co-monomer (Surfmers), as initiator (Inisurfs), or as chain transfer agent (Transurfs).[6] In products represented schematically from the silicone industry (Figure 8.8), reactive surfactant produces silicone rubber from nano-scale silicone oil microemulsions in aqueous hydroxy-ethyl-methacrylate (HEMA) monomer, known as room temperature vulcanizing (RTV).

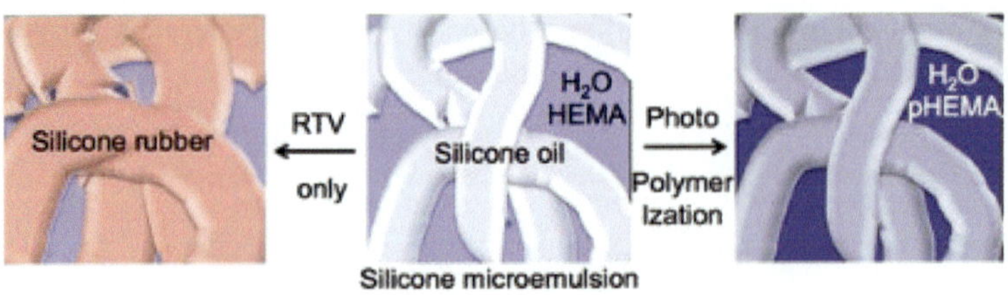

FIGURE 8.8 Schematic of silicone micro-emulsions with HEMA and morphological changes.[7]

8.4 LIQUID–LIQUID DISPERSION CLASSIFICATION

In Figure 8.9, liquid–liquid dispersions are classified based on droplet size of the dispersed phase. Emulsions are characterized as having drop sizes ≤ ~ 1μm, while macro-emulsions or suspensions have typical drop sizes greater than ~1 μm.

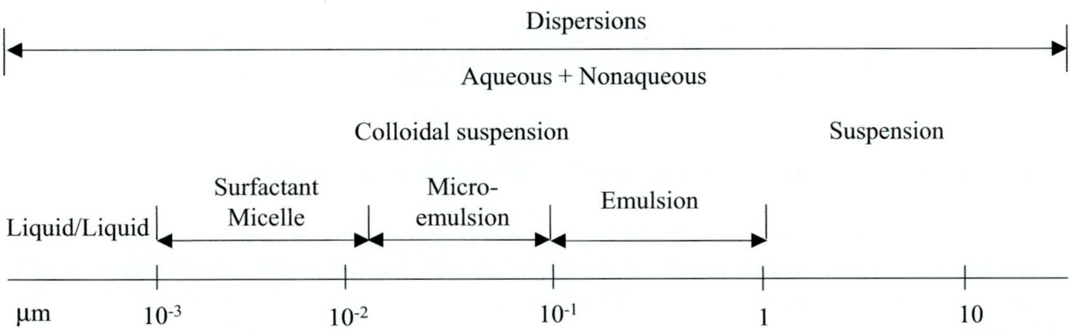

FIGURE 8.9 Dispersion classification criteria based on dispersed-phase drop size.[8]

Types of dispersions are categorized based on components in the dispersed phase and the continuous phase. If the non-aqueous phase is dispersed in an aqueous phase, it is generalized as an oil/water (o/w) emulsion, while its opposite is called a water/oil (w/o) emulsion. Consider three common dairy products:

(a) Milk is an oil-water emulsion where fat globules are dispersed in water. Proteins in the form of micelles are dispersed as colloids.
(b) Butter is a water-oil emulsion where water is emulsified in milk fat and then chilled to provide the required consistency.

The food and pharma industries typically prepare o/w and w/o emulsions using edible surfactants, high-intensity shear mixers, and stabilizers that prolong preservation before ultimate consumption. For example, mayonnaise is an oil-water emulsion that is stabilized by the lecithin in egg yolks.

Parameters influencing the desired droplet size and dispersion quality are identified as follows:

(a) Physical Parameters

• Component viscosities	μ_c and μ_d	[Pa-s]
• Working media viscosity	μ_m	[Pa-s]
• Solubility (mass fraction solute to solution)	χ	[-]
• Interfacial tension	σ_i	[N/m]

The solubility of a component within a given mixture, χ, is the relative proportion of that component that saturates the mixture at a specified temperature and pressure. The term “saturated” implies equilibrium, and no mass transfer of the component between phases.

(b) Process Parameters
- Phase volumetric fraction, φ
- Influence of additives
- Geometry (hydrodynamics) of the dispersing system
- Influence of superimposed chemical reaction

Dispersion physical properties are determined from the following formulae:

(a) Mixture density:

$$\rho_m = \rho_c - (\rho_c - \rho_d)\,\varphi_d \tag{8.1}$$

(b) Mixture viscosity:[9]

$$\mu_m = \frac{\mu_c}{(1-\varphi)}\left[1 + \frac{\varphi(1.5\mu_d)}{\mu_c + \mu_d}\right] \tag{8.2}$$

where, subscript c = continuous phase, and subscript d = dispersed phase.

The mixture viscosity, μ_m, of light emulsions containing a suspension of spherical liquid droplets is given by the formula in equation 8.3. This classic formula was derived by Taylor[10] and is used to obtain order of magnitude viscosities for emulsions containing less than 2–3 volume fraction (φ) of the dispersed phase.[11]

$$\mu_m = \mu_c\left[1 + 2.5\varphi\left(\frac{\mu_d + \frac{2}{5}\mu_c}{\mu_d + \mu_c}\right)\right] \tag{8.3}$$

8.5 DISPERSION MECHANISM

Figure 8.10 illustrates droplet stabilization mechanisms in the quiescent and colliding states. The latter is most common in mechanically agitated and hydrodynamically sheared environments.

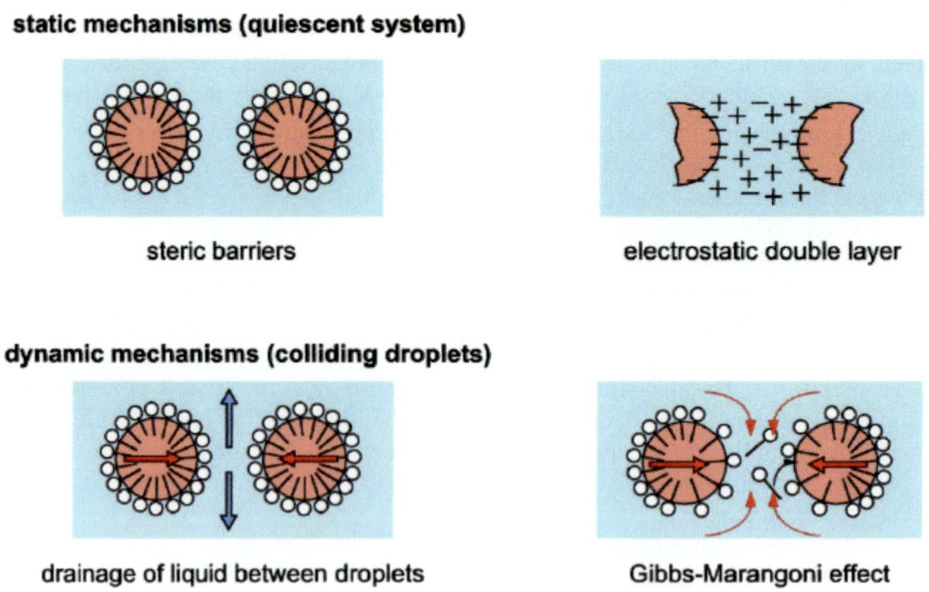

FIGURE 8.10 Schematic illustrations of droplet stabilization mechanisms.[12]

Successively subjecting a fluid element to environments of shear, followed by extension, are vital to dispersing stabilizable droplets of one immiscible phase within a second continuous phase. The deformation and breakup of droplets in steady flow[13] are represented in Figure 8.11.

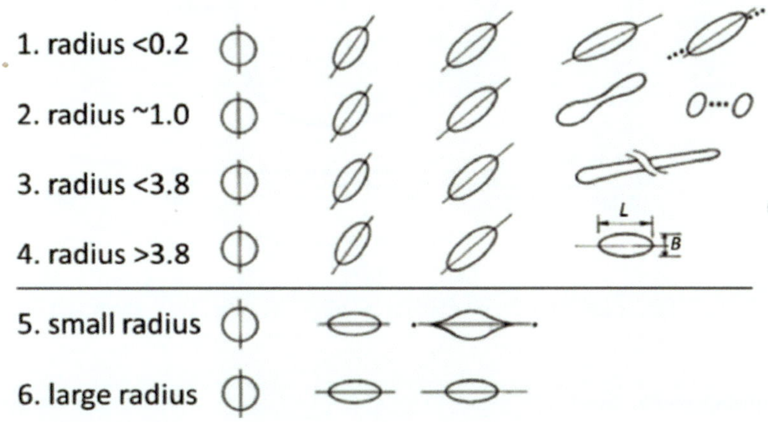

FIGURE 8.11 Possible drop deformation patterns and breakage (1–4 shear, 5–6 extension).[13]

Photographs in Figure 8.12a–b were captured by researchers at BHR Group to vividly show the effects of shear and elongation on dispersed-phase droplets.[14] This hydro-mechanical phenomenon is typically dictated by such non-dimensional parameters as shear flow Weber number, We, and dispersed phase-to-continuous phase viscosity ratio, *R*, as shown in Figure 8.13, where the separation between drop deformation and breakup zones are delineated with a dashed line for extensional irrotational shear flow and with a solid line for simple shear (Couette) flow.[15]

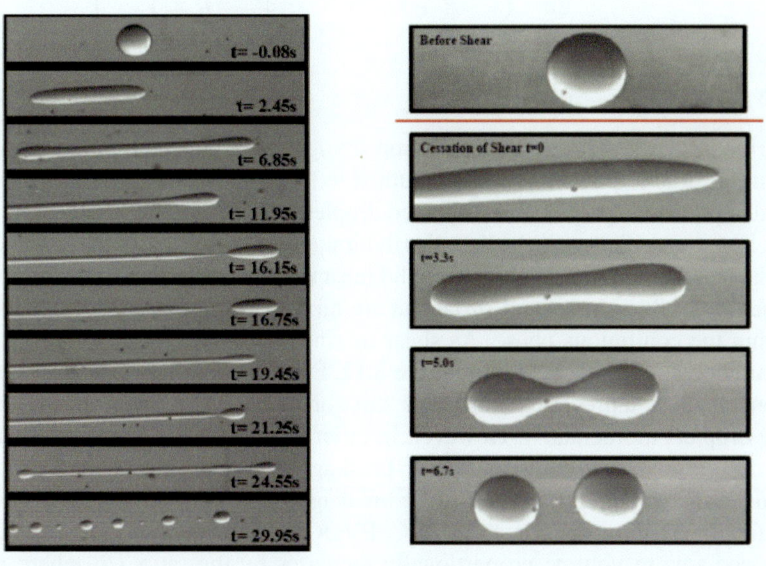

FIGURE 8.12A–B Photographs illustrating the influence of elongation and shear forces on drop formation.[14]

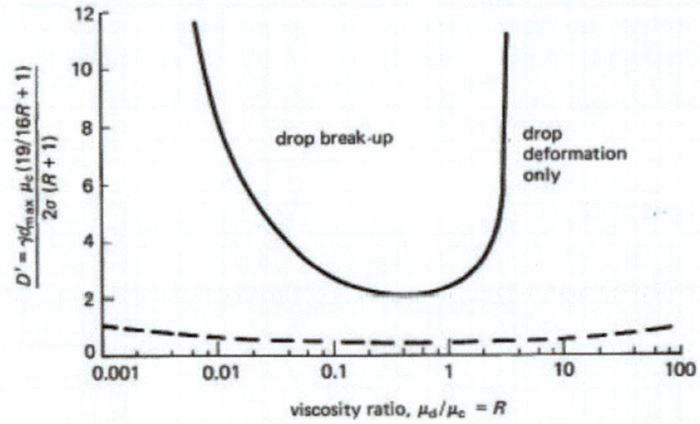

FIGURE 8.13 Critical shear rate, $\dot{\gamma}$ for drop break-up vs. viscosity ratio, R (—shear and - - - -extension).[13,15]

(a) Weber number (shear flow),

$$\mathrm{We} = \frac{\text{Deforming shear stress}}{\text{Restoring Laplace stress}} = \frac{\mu_c \gamma}{(4\sigma / d_d)} \quad \text{or} \quad \mathrm{We}^* = \frac{\mu_c \gamma d_d}{2\sigma}$$

where, We^* is based on the droplet radius, $d_d / 2$, as its characteristic length.

(b) Viscosity ratio, $R = \dfrac{\textit{Viscosity of dispersed (drop) phase}}{\textit{Viscosity of continuous phase}} = \dfrac{\mu_d}{\mu_c}$

(c) According to Taylor, the deformation D' of an ellipsoidal drop derived from the relationship between fluid shear and Laplace surface forces is presented as follows:[13,15]

$$D' = \frac{\dot{\gamma} d_d \mu_c}{2\sigma} \left(\frac{19/16R + 1}{R + 1} \right) = We \left(\frac{19/16R + 1}{R + 1} \right)$$

8.6 DROP SIZING

Here are described the two primary stages of drop sizing, which depend on mixing intensity as indicated by specific hydraulic power, which is assumed to be proportional to shear or strain rate. Since most industrial applications polymerize dispersed droplets to form particles, successful drop sizing is crucial to meeting a specified particle size distribution (psd). In this case, the volumetric shrinkage factor of the dispersed liquid to the polymeric solid must be considered during the drop sizing process.

Initially, the immiscible phases and surfactant are mixed to create coarse dispersions of the internal phase within the continuous phase. As shear mixing intensity increases, macro drops undergo deformation, as shown in Figure 8.11 and Figure 8.12. Extension stretches the drops until they break up due to turbulent shear, a process identified as "first-order breakage," with the surfactants typically stabilizing the droplets momentarily. However, in order to further size the droplets, intensive shear mixing for a desired period follows first-order breakage to create the required drop size distribution (dsd). This is recognized as "second-order breakage," a critical process step in manufacturing most industrial polymeric products such as PVC, PVDC, and ion exchange resins. As the drop size decreases, surface area to volume proportionally increases by the ratio *1/r*, where *r* is the droplet radius. Hence, the specific energy, ε_m, imparted to the working media must significantly increase, as illustrated in Figure 8.14. This is analogous to the comminution process in mining industries. Furthermore, in order to maintain the size of dispersed droplets, the required quantity of stabilizer/surfactant increases to cover the increased surface area and suppress coalescence. It is to be noted that while immiscible media experience second order breakage, the duration of agitation eventually has minor influence on drop size reduction,[13] as determined experimentally and presented in Figure 8.15.

Generally, end users specify the required particle size distribution (psd) to satisfy their applications or extruder processing conditions. Manufacturers of polymeric materials must tailor the mixing and reaction conditions to abide by the customer specifications. A classic procedure is to assume a normal (Gaussian) distribution of particles with the desired 10% and 90% tail ends, and to establish a correlation between the volume median drop size (d_{vm}), specific energy (ε_m), duration of sizing operation (t), and the surfactant/stabilizer concentration (c_s). Preferably the correlation is in appropriate non-dimensional forms. Experimental data are generated using a reasonably scaled-down version of the R&D reactor, normally 10gal or more. In order to attain the desired psd, the authors in many instances pursued this experimental correlation approach *a priori*, and successfully set the physical and chemical process parameters for the pilot and production reactors to achieve the desired psd.

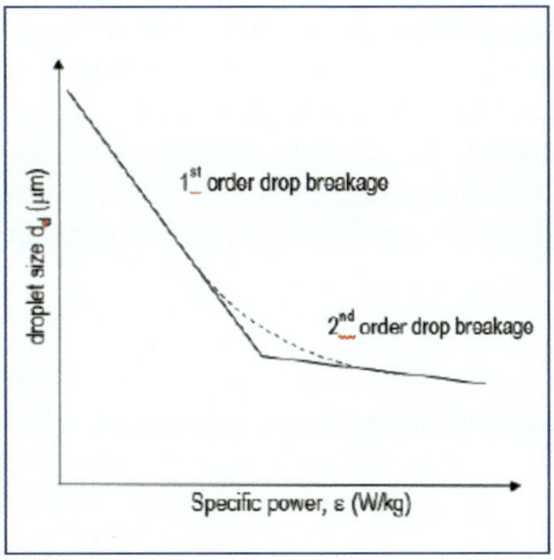

FIGURE 8.14 Illustration of first- and second-order drop breakage phenomena.

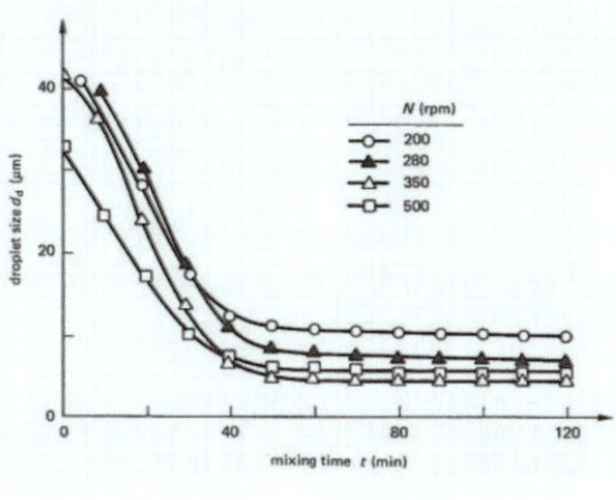

FIGURE 8.15 Experimental data showing drop size as a function of mixing time.[13]

8.7 MICRO-TURBULENCE AND DROPLET SIZE

A purely theoretical approach to predict drop size distribution as a function of the specific power is addressed based on micro-turbulence eddy scale and interfacial tension. As noted earlier, turbulent shear stress creates sufficient droplet deformation against the stabilizing effects of the interfacial area stress, so that extension and splitting of the droplet occurs. The ratio of the two stresses is expressed by the Weber number (inertia force/interfacial tension):

$$We = \frac{\rho N^2 D^3}{\sigma_i} \tag{8.4}$$

The size of micro-vortices is represented by

$$\lambda_k = \left(\nu^3 / \varepsilon_{loc} \right)^{1/4} \tag{8.5}$$

Typically, the drop size range is three to five times the Kolmogorov eddy length scale, λ_k. For example, if a typical dilute dispersed phase of oil-water media ($\nu \approx 1cSt.$) is subjected to a local specific power, ε_{loc}, of 1W/kg, the expected drop size would be 100–150μm. There are multiple established correlations between the Sauter mean diameter, d_{32}, and the Weber number, We. In the case of dilute liquid–liquid dispersions experiencing turbulent flow while viscous forces are insignificant, drop breakup is influenced by the turbulent energy dissipation, and drop sizes are restored by the interfacial tension. If the average diameter is larger than the Kolmogorov length scale, and d_{32} / d_{max} is constant, then average stable drop size can be related to specific energy,[16] as shown in equation 8.6.

$$d_{32} = C_o \cdot (\sigma_i / \rho_c)^{\frac{3}{5}} \varepsilon^{-2/5} \tag{8.6}$$

Substituting for $\varepsilon = P / m$, where P is the hydraulic power and m is the working media mass, yields

$$d_{32} = C_1 \cdot (\sigma_i / \rho_c)^{\frac{3}{5}} \left[\frac{N_P \, \rho \, N^3 D^5}{\rho (\pi D^3)} \right]^{-2/5}$$

$$d_{32} \propto (\sigma_i / \rho_c)^{\frac{3}{5}} \left[\frac{N^3 D^5}{D^3} \right]^{-2/5}$$

and the common non-dimensional representation of d_{32} with respect to the stirrer diameter, D, is presented in equation 8.7.

$$\frac{d_{32}}{D} = C \cdot We^{-0.6} \tag{8.7}$$

$$\text{or, } d_{32} \propto \varepsilon^{-0.4}$$

Thus, for a given product, the same droplet size distribution should result by scaling up the stirred tank based on the constant power per mass criteria. The constant, C, in equation 8.7 varies from 0.09 to 0.15 depending on impeller type and power number, N_P.[17]

In the case of a stirred tank, multiple physical parameters influence dispersion. Volume fraction, c_v, of the dispersed phase affects dispersing and coalescing behaviors of the media significantly. As expected, damping of turbulence occurs with increasing volume concentration, C_v [%]. Hence, the following correction to the Sauter mean diameter has been proposed:[18]

$$d_{32-cv} / d_{32} = 1 + C.C_v \text{ up to } C_v \le 10\% \text{ } (C \approx 0.07 \text{ for standard PBT}) \tag{8.8}$$

An increase in viscosity of the dispersed phase can result in considerably larger droplets. The following correlation addresses the viscosity effect:[18]

$$d_{32} / D \propto \left(Sc.N.D^2 / \mu_d\right)^{-0.75} \tag{8.9}$$

where the turbulent Schmidt number, $Sc = \dfrac{eddy\, viscosity}{eddy\, mass\, diffusivity}$.

Initiating coarse dispersion depends on such impeller characteristics as type, tip speed, and location. Macro-vortices deform the phase that is to be dispersed, drawing it into the continuous phase, perhaps in layer-by-layer fashion, until it has been totally dispersed. The minimum required tip speed for axial impellers to initiate coarse dispersion[18] is

$$v_{tip} = A(\rho_c / \rho_d)^{3/16} [g(\rho_c - \rho_d) / \rho_c]^{5/16} D^{1/8} \tag{8.10}$$

where, the constant A depends on the impeller type.

Although axial (hydrofoils) and mixed flow (PBT) impellers exhibit a low sensitivity to their location with respect to the interface, the preferred location for axial impellers in a down-pumping agitation system is $\sim 0.5D$ below the interface to avoid a significant instantaneous pool of the lower density phase in the higher density phase. The impeller submergence decreases in higher viscosity media. The goal is to easily draw the lighter floating phase down into the heavier phase to initiate coarse dispersion, while avoiding either undesirable foaming or phase inversion of the system.

8.8 MIXING DEVICES USED FOR L-L DISPERSION

Figure 8.16 graphically represents maximum stable drop size, d_{max-s}, as a function of local power per mass, ε_{loc}, for various mixing devices.[19] Note that $\varepsilon_{loc} \gg \varepsilon$ and represents the energy dissipation in the swept volume or impeller region.

Stirred tank reactors are common in most polymerization processes that normally target larger volume median particle size (d_{vm} = 10–10^{3}µm). Their agitation systems typically include a variable frequency drive (VFD) in order to impart different specific energy levels simply by altering the impeller speed. Stirred tank reactors are ideal for reaction-controlled processes. In several instances, drop sizing ($d_{vm} \le$ 10µm) is conducted using high-shear devices (see Chapter 13), but the working media is still subjected to an STR for the particle formation reaction to proceed. The agitation system must be designed to overcome the deficiency of particle settling, unless the desirable outcome is separation of solids from the continuous media. Hence, a variable-speed agitation system may be essential to keep the particles uniformly suspended.

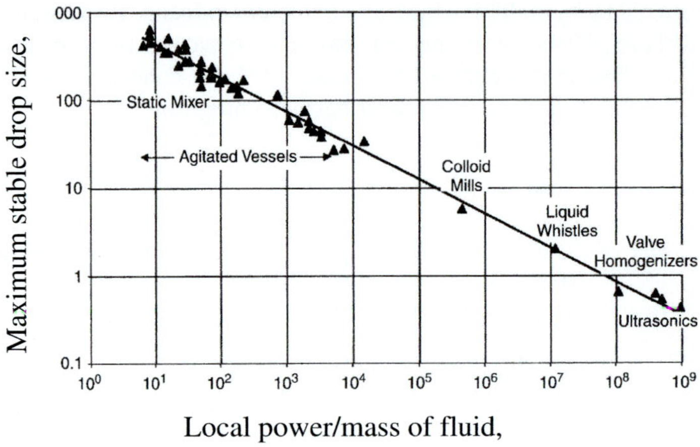

FIGURE 8.16 Graph of drop size as a function of local power per mass for various mixing devices.[19]

Table 8.1 presents a number of mixing devices used to create liquid–liquid dispersions with the desired drop size. Description of the non-stirred devices and their performance characteristics will be addressed later in separate chapters.

TABLE 8.1
Typical Mixing Devices Used for Drop Sizing

Drop Size	Comments	Equipment/Agents
<0.5μm	Stabilized by Brownian motion; non-settling dispersion	Emulsifiers, ultrasonic devices, rotor-stator mixers and high-pressure homogenizers; surface active agents usually required
0.5–3.0μm	Marginally stable; can cream and separate	Rotor-stator, impingement mixers, and static mixers
>3.0μm	Usually unstable; coalescence and phase separation common when agitation ceases	Static mixers, inline mixers, and stirred vessels

8.9 DROP DISPERSION

Skelland and Seksaria[20] (1978) pursued an extensive experimental program and established an empirical correlation to predict minimum impeller speeds, N_{cd}, for liquid–liquid dispersions in baffled vessels. The experimental program used four types of impellers (propeller, pitched-blade turbine, flat-blade turbine, and curved-blade turbine) in five equal-volume (vol. fraction, $\varphi = 0.5$) binary phase working media such as Dow Corning 200 Fluids (5, 10 and 15cSt.), benzaldehyde, and ethyl-acetate in water. All of the tests were performed using a 0.213m-diameter, flat-bottom glass vessel, as shown in Figure 8.17. Liquid height to tank diameter ratio, Z/T, was maintained at unity, except for two cases involving ½ and 2/3. Single impellers were located at three different off-bottom clearances, C_i of $Z/4$, $Z/2$, and $3Z/4$. Dual impellers were located only at $C_i = Z/4$ and $3Z/4$, i.e., midway between the denser and lighter immiscible fluids, shown as follows.

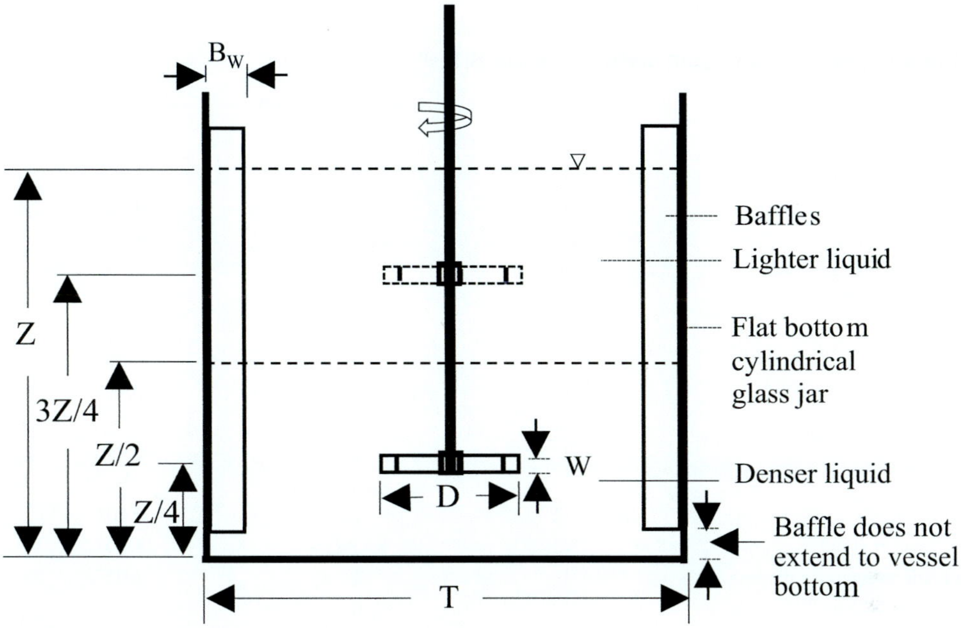

FIGURE 8.17 Schematic of experimental vessel.[20]

[Impeller diameters, D ≈ 0.1, 0.075, 0.06m, baffle width, Bw = 0.190m ≈ T/11]

Minimum impeller rotational speed for complete liquid–liquid dispersion in agitated vessels is a function of impeller type, dispersed-phase volume fraction (φ), impeller diameter to tank diameter ratio (D/T), and impeller off-bottom clearance to liquid height (C_i/Z). Physical properties of the dispersed and continuous phases, such as densities (ρ_d, ρ_c), viscosities (μ_d, μ_c) and interfacial tension, σ etc., are also important. Since the initial correlation proposed by Skelland and Seksaria,[20] several groups[21–23] have investigated minimum required dispersion speed, N_{cd}, for complete suspension of liquid droplets.

Based on the data from three different sources, Skelland and Ramsay[22] (1987) later developed the following empirical correlation to calculate minimum impeller speed, N_{cd}, needed for complete liquid–liquid dispersion, in fully baffled vessels.

$$N_{cd} = C\left(\frac{T}{D}\right)^{\alpha} \frac{g^{0.42}\Delta\rho^{0.42}\mu_m^{\ 0.08}\sigma^{0.04}\varphi^{0.05}}{D^{0.71}\rho_m^{\ 0.54}} \tag{8.11}$$

Values of the constant, C, and (T/D) exponent, α, for the various impellers based on their locations with respect to the liquid height, Z, and the liquid height-to-tank diameter ratio, Z/T, are presented in Table 8.2. Limitations to the current application of the above N_{cd} correlation include experimenting in a flat-bottom tank with no baffle clearance and using atypical impellers for liquid–liquid dispersion.

TABLE 8.2
Constants C and α Associated with Equation 8.11[20]

Impeller Type	Z/T	Impeller Location	C	α
Square pitch, down-thrusting propeller (three blades)	1	$Z/4$	4.38	0.67
	1	$3Z/4$	2.76	0.95
	1	$Z/2$	4.33	0.79
	1	$Z/4, 3Z/4$	1.46	1.33
Down-thrusting pitched-blade turbine (six blades)	1	$Z/4$	1.95	1.44
	1	$3Z/4$	1.96	1.17
	1	$Z/2$	0.84	1.97
	1	$Z/4, 3Z/4$	0.94	1.27
Flat-blade turbine (six blades)	1	$Z/4$	0.91	2.02
	1	$3Z/4$	*	*
	1	$Z/2$	0.95	1.38
	1	$Z/4, 3Z/4$	*	*
	1/2	$Z/2$	0.70	1.24
	3/2	$Z/2$	1.10	1.70
Curved-blade turbine (six blades)	1	$Z/4$	1.03	1.86
	1	$3Z/4$	*	*
	1	$Z/2$	1.34	1.20
	1	$Z/4, 3Z/4$	1.20	0.94
Disc turbine (six blades)	1	$Z/2$	0.53	1.70

Note: (a) * indicates insufficient data due to splashing; (b) average deviation of correlation = 12.7%, (c) $10^4 < \mathrm{Re} < 10^5$ (d) for scale-up, typical equal power per volume, i.e., ε_v, although incorrect, is on the safe side to use.[20]

Following is an illustration of the critical speed, N_{cd}, calculation using equation 8.11 for complete suspension of a dispersed phase in a continuum media. Consider the instance where styrene monomers are mechanically dispersed in water at 20°C prior to their suspension polymerization in the presence of monomer- soluble initiator at 90°C. The polymerization reactor (90" ID x 125" height) has 2:1 semi-elliptical heads, and contains 2000gal of working media. The volume fraction of styrene monomer is 50%.

TABLE 8.3
Physical Properties of Immiscible Components and Their Mixture at 20°C

Components and Mixture Parameters	Density, ρ [kg-m^{-3}]	Viscosity, $\propto$ [N-m^{-2}-s]
Styrene monomer (dispersed phase)	909	0.762×10^{-3}
Water (continuous phase)	998	1.0×10^{-3}
Interfacial tension (styrene/water), σ	35.5×10^{-3} N-m^{-1}	
Mixture density at $\varphi = 0.5$ (eqn. 8.1)	$\rho_m = \rho_c - (\rho_c - \rho_d)\,\varphi = 955$ kg-m^{-3}	
Mixture viscosity at $\varphi = 0.5$ (eqn. 8.2)	$\mu_m = \dfrac{\mu_c}{(1-\varphi)}\left[1 + \dfrac{\varphi(1.5\mu_d)}{\mu_c + \mu_d}\right] = 2.65 \times 10^{-3}$ N-m^{-2}-s	

TABLE 8.4
Polymerization Reactor and Agitation System Parameters

Parameters	Nomenclature	Values/Remarks
Reactor diameter	T	2.286m (90")
Working media height	Z	2.032m (80")
Media height/Reactor dia.	Z/T	0.89 < 0.9 ⇨ a single impeller is required
Impeller	6-bladed PBT	Atypical, but selected as info., available in ref. 20
Impeller diameter/Tank diameter	D/T	0.46 (mixed flow impeller)
Impeller diameter	D	1.052m (41.42")
Impeller location from tank bottom*	$C_i = T/4$	0.571m (20.5") i.e., at bottom tangent line
Correlation constant (Table 8.2), $C_i \approx Z/4$	C	1.95
Correlation exponent (Table 8.2), $C_i \approx Z/4$	α	1.44

Note: *assumption $C_i = T/4 \approx Z/4$ i.e., 0.571 vs. 0.508 with 12% difference

Substituting the predetermined parameters from Tables 8.3–8.4 in equation 8.11 yields the minimum required complete dispersion speed, N_{cd}, in rpm.

$$N_{cd} = C\left(\frac{T}{D}\right)^{\alpha} \frac{g^{0.42}\Delta\rho^{0.42}\mu_m^{\ 0.08}\sigma^{0.04}\varphi^{0.05}}{D^{0.71}\rho_m^{\ 0.54}} \times 60$$

$$= 1.95\left(\frac{2.29}{1.05}\right)^{1.44}\left[\frac{(9.81)^{0.42}(998-909)^{0.42}\left(2.65\times10^{-3}\right)^{0.08}\left(35.5\times10^{-3}\right)^{0.04}(0.5)^{0.05}}{(1.05)^{0.71}(955)^{0.54}}\right]\times 60.$$

$$= 77 rpm$$

8.10 DROP STABILIZATION

8.10.1 Destabilizing Phenomena of a Dispersion

Many processes subject the newly formed dispersion to light agitation or a static state to destabilize the droplets and develop the following desirable phenomena. Such stirred vessels are commonly known as mixer-settler units, typical of chemical manufacturing plants. Figure 8.18 illustrates different droplet destabilization conditions.

(a) *Crémage*: Droplets phase, if lower density, floats and accumulates near or above the surface of the continuous phase.
(b) Sedimentation: Heavier droplets settle under gravity and create a sedimentation layer
(c) Flocculation: Droplets are intentionally clustered by flocculants that migrate to the liquid surface. This process is typical of mining operations where the metal particles are attached to the clusters and subsequently recovered therefrom.
(d) Coalescence: Droplets are subjected to coalescence for faster separation and a cleaner distillation process.

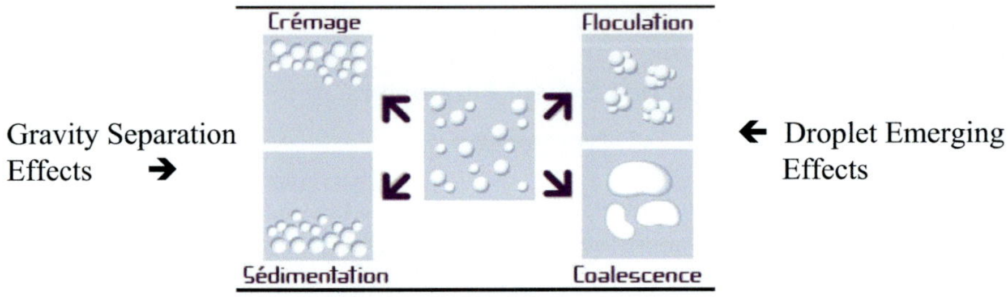

FIGURE 8.18 Schematic of drop destabilization effects.[24]

Decantation time in mixer-settler units is often critical to the liquid/liquid extraction processes. Shorter settling/separation times are of decisive advantage for high-throughput production rates and to minimize secondary reactions. Typically, the influence of polar complex agents, low interfacial tension, and high dispersion viscosity of the working media are responsible for coalescence-inhibited systems. Also, the local energy dissipation rate of the agitation system in the mixer-settler unit is a key parameter influencing the decantation rate as shown in Figure 8.19. Hence, understanding of local energy, $\varepsilon_{tip} / \varepsilon_{av}$, and tip velocity, v_{tip} ($\propto$ max shear rate), is critical to the quality of drop stabilization. According to the *Ekato Handbook of Mixing*,[18] flat-blade disk turbines lead to a considerably increased settling time compared to equally energy dissipating impellers.

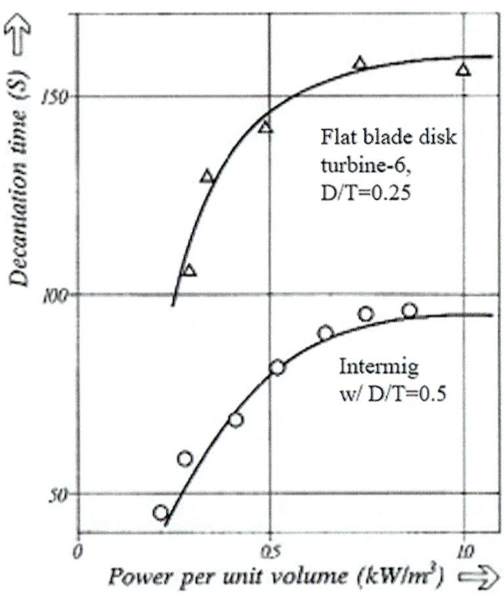

FIGURE 8.19 Graph showing decantation time in mixer-settler unit as a function of specific energy.[18]

8.10.2 Stabilization of Droplets

Unless the dispersion is a Brownian-motion-controlled micro-emulsion, the droplets are normally stabilized using suspending agents to prevent coalescence. As mentioned earlier, a critical amount of agent is required for surface coverage. Leng and Quarderer (1982) proposed several models to predict largest surviving drop size, d_{l-s}, based on their research with liquid–liquid dispersions and suspension polymerizations. In non-coalescing systems, d_{l-s} is a function of impeller geometry, speed, and such physical properties as the type of monomer and the stabilizer concentration.[25] A dispersion must be stable, i.e., resistant to property changes over time,[26] to be useful in most food, pharma, healthcare, and ag emulsion products.

The mixture viscosity of suspensions containing stabilized spherical droplets is a strong function of the drop size distribution (dsd) and dictates their packing behavior. In spite of the shear forces, dilute ($\phi \leq 0.05$) and low concentration ($\phi \leq 0.2$) dispersed phase droplets thrive under unrestricted mobility in the continuous phase. Although the mixture viscosity is influenced by droplet concentration, as depicted in equations 8.2 and 8.3, it remains Newtonian. At higher concentrations ($0.2 < \phi < 0.65$), drop mobility incrementally diminishes and mixture viscosity increases due to droplet collisions and surface interactions. Thus, mixture viscosity starts trending towards non-Newtonian behavior. The theoretical random packing density of mono spheres is limited to a volume fraction of ~64%, but their reported 3D packing density is around 74%. This is the transition domain where dispersion rheology significantly deviates from its non-Newtonian behavior, and its viscosity tends to increase exponentially, as shown in Figure 8.20.

As mentioned, while discussing phase inversion, dispersed-phase drops typically have a Gaussian or a log-normal distribution. Unlike solids, they are inherently deformable. Thus, the transition volume fraction, φ_{cr}, is not clearly evident. Past the transition, either the dispersion encounters phase inversion (with high shear rates and/or lack of stabilizer for surface coverage) or behaves as a solid. Figure 8.20 graphically illustrates the five typical stages of stabilized mono-spherical drops suspended in a continuous media, and their influence on mixture viscosity.

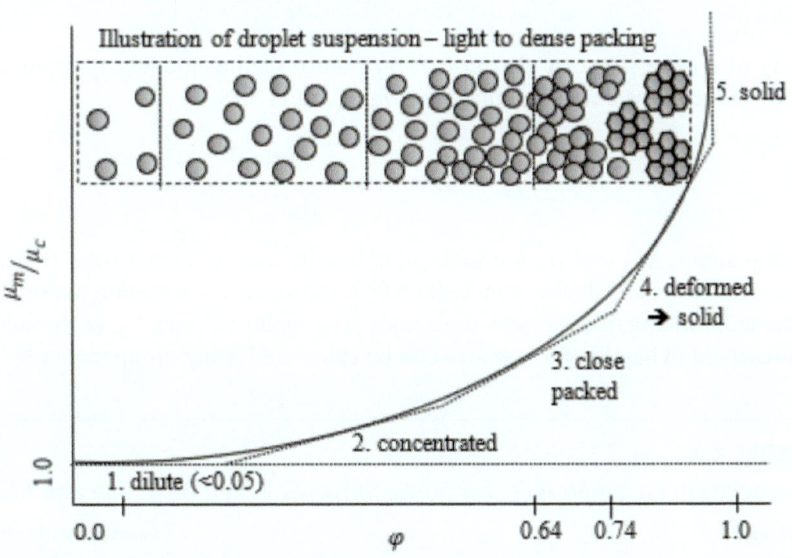

FIGURE 8.20 Graphical representation of the increasing mixture viscosity of a typical liquid–liquid dispersion, as a function of drop volume fraction, ϕ.

8.11 MASS TRANSFER PHENOMENA

8.11.1 Diffusivity

Mass transport affects reactor productivity (reaction yield, selectivity) and downstream product recovery operations. Solute mobility in the continuous phase is governed by diffusivity (diffusion coefficient, D), which is a function of viscosity, temperature, concentration of solute, and type of continuous phase.

Self-diffusivity, (D_{AA}), or diffusion within itself, is defined as diffusivity of component A within a continuous phase consisting of A. Mutual diffusivity, (D_{AB}), on the other hand, is the diffusivity of component A within a continuous phase B. Typically, D_{liquid} is 10^{-2}–$10^{-4} \times D_{gas}$. Table 8.5 presents typical diffusivities for different material phases.[20]

TABLE 8.5
Order of Magnitude of Self-Diffusivity, D_{AA}, of Different Phases[27,28]

Continuous Phase	Diffusivity, m²/s
gas (atmospheric pressure)	10^{-5}
liquid	10^{-9}
liquid held in a solid matrix	10^{-10}
polymers	10^{-12}
solid	10^{-14}

Diffusivity in solvents, polymer solutions, and gels are briefly outlined as follows.[27] The diffusivity of a small solute A (molar volume $V_A \leq 0.5$ m³/k-mol) diffusing in a dilute liquid B (Wilke-Chang[29]) is determined using equation 8.12.

$$D_{AB} = 1.173 \times 10^{-16} \frac{(\xi M_B)^{0.5} T}{\mu_B V_A^{0.6}} \tag{8.12}$$

The diffusivity of a large solute A ($V_A >> 0.5$ m³/k-mol), such as globular protein, diffusing in a dilute liquid B (Stokes-Einstein) is calculated using equation 8.13.

$$D_{AB} = 9.96 \times 10^{-16} \frac{T}{\mu_B V_A^{1/3}} \tag{8.13}$$

In these two equations, M_B and μ_B are molecular weight and viscosity of the solvent, while T is the absolute temperature, and ξ (listed in Table 8.6) is the solvent association parameter, which is a measure of the polar interaction between molecules. The molar volume, V_A, of the solute at its boiling point is presented in handbooks and also can be calculated using group theory.[30]

TABLE 8.6
Association Parameters, ξ, for Some Solvents Based on Wilke and Chang[29]

Solvent	Association Parameter, ξ
water	2.6*
methanol	1.9
ethanol	1.5
propanol	1.2
benzene	1.0
ether	1.0
heptane	1.0
Other unassociated solvents	1.0

*A value of 2.26 better described the data, as shown in later publications.

The dilute phase diffusivities of some common solutes, when paired with various solvents at different temperatures, are presented in Table 8.7.

Additionally, Table 8.8 lists the diffusivity of a few large bio-pharma molecules (globular proteins) in dilute aqueous solutions.

TABLE 8.7
Diffusion Coefficients of Various Solutes/Solvents in Dilute Phase[31]

Solute	Solvent	Temperature (°C)	D_{AB} (×10^9 m²/s)
acetic acid	water	25	1.26
acetic acid	benzene	25	2.09
acetone	water	25	1.28
benzoic acid	water	25	1.21
carbon dioxide	water	20	1.50
carbon dioxide	water	25	2.00
chloroform	benzene	20	2.11
chloroform	ethanol	20	1.23
ethanol	water	25	1.24
glucose	water	20	0.60
glycerol	water	20	0.83
methanol	water	15	1.26
nitrogen	water	20	1.64
oxygen	water	20	1.80
oxygen	water	25	2.41
phenol	water	20	0.84
phenol	ethanol	20	0.80
sodium chloride	water	20	1.35
sucrose	water	20	0.45
urea	water	25	1.38
water	ethanol	25	1.13

TABLE 8.8
Diffusion Coefficients of Large Bio-Molecules into Water in Dilute Phase[32]

Protein	Molecular Weight (kg/k-mol)	Temperature (°C)	D_{AB} (×10^{11} m²/s)
Bovine serum albumin	67,500	25	6.81
Urease	482,700	25	4.01
Human serum albumin	72,300	20	5.93
Human fibrinogen	339,700	20	1.98

8.11.2 Mass Transfer Coefficients

The process of mass transfer across an interface, or across a virtual surface in the bulk of a phase, is the result of a chemical potential driving force. This driving force is usually expressed in terms of concentrations of the species, or partial pressures in the case of gas phases.

Table 8.9 presents non-dimensional mass transfer coefficient correlations (Sherwood number, Sh) comprehended inside the mobile or oscillating drops in liquid–liquid dispersion. Note that the Reynolds number, Re, is based on drop diameter and relative velocity between phases. All other properties are for the drop phase unless otherwise stated. The density and viscosity in Re_C are for the continuous phase. Subscripts C and D refer to continuous and dispersed phases, respectively.[27]

Table 8.10 contains some continuous-phase mass transfer coefficient correlations for liquid–liquid dispersions.

TABLE 8.9
Mass Transfer Coefficients as a Function of Reynolds Number[27]

Correlation	Range
$Sh \approx 16.7$	$Re < 50$
$Sh = 0.32\,Re^{0.68}\left(\dfrac{\sigma^3 \rho_C^2}{g\mu_C^4 \Delta\rho}\right)^{0.10}\left(\dfrac{4D_D t_c}{d_D^2}\right)^{-0.14}$	$Re > 150 - 200$
$Sh = 0.32\,Re^{0.63}\,Sc^{0.50}\left(1+\dfrac{\mu_D}{\mu_C}\right)^{-0.5}$	$10 \le Re \le 1000$, spherical drops
$Sh = 7.5x10^{-5}\,Re_C^{2.0} Sc^{0.56}\left(1+\dfrac{\mu_D}{\mu_C}\right)^{-0.5}$	$100 \le Re_C \le 1500$, larger oblate drops

TABLE 8.10
Mass Transfer Coefficient Correlations for Stationary and Mobile Drops[27,31,32]

Small, noncirculating drops	$Sh = 2 + 0.79Re^{0.5}\,Sc^{0.33}$
	$Sh = 2 + 0.76Re^{0.5}Sc^{0.33}$
	$Sh = 0.562Re^{0.5}Sc^{0.33}$
Larger, mobile drops	$Sh = \dfrac{2}{\sqrt{\pi}}\left[1 - Re^{-0.5}\left(2.89 + 2.15\left(\mu_D / \mu_C\right)^{0.64}\right)\right]^{0.5}\left(Re.Sc\right)^{0.5}$
When $Re < 120, \left(\mu_D / \mu_C\right) \le 2$ and $0 \le \left(\rho_D / \rho_C\right) \le 4$	$Sh = -126 + 18Re^{0.5}\,Sc^{0.33}$
When $8 \le Re \le 800$	$Sh = -178 + 3.62Re^{0.5}Sc^{0.33}$

To fully conceptualize the process of liquid–liquid dispersion and the ultimate product goal, various rate coefficients must be understood, including the ones relating to homogeneous or heterogeneous reaction. Table 8.11 briefly describes the definitions of these rate coefficients.[33]

TABLE 8.11
Various Transfer Rates, Corresponding Equations, and Their Flow Dependency[33]

Effect	Basic Equation	Rate	Force	Coefficient
Mass transfer	$N_i = k\Delta c_i$	Flux per area relative to an interface	Difference of concentration	Mass transfer coefficient, $k[L/t]$ is a function of flow
Diffusion	$-j_i = D\Delta c_i$	Flux per area relative to the volume average velocity	Gradient of concentration	Diffusion coefficient, $D[L^2/t]$ is a physical property independent of flow
Dispersion	$-\overline{c_i'v_i'} = E\Delta\overline{c_i}$	Flux per area relative to the mass average velocity	Gradient of time average concentration	Dispersion coefficient $E[L^2/t]$ depends on flow
Homogeneous chemical reaction	$r_i = k_i c_i$	Rate per volume	Concentration	Rate constant $k[L/t]$is a physical property independent of flow
Heterogeneous chemical reaction	$r_i = k_i c_i$	Flux per interfacial area	Concentration	Rate constant $k[L/t]$ is a surface property often defined in terms of a bulk concentration

Generally, a rate process is described as:

$$\text{Flow or Rate} = \frac{\text{Driving Force}}{\text{Resistance}} = \frac{\text{Driving Force}}{1/\text{Conductance}}$$

Mass transfer uses the following representations for various parameters:

(a) mass flux = N_A / A [kg-moles of A per hr-m^2]
where, A is the interfacial area between two immiscible phases available for mass transfer
(b) concentration difference driving force = Δ_{c_A} [kg-moles of A / m^3]
(c) conductance or over all mass transfer coefficient = k [m-hr^{-1}.]

Overall, mass flux and mass flow are represented in equations 8.14–8.15.

$$\text{mass flux, } N_A / A = \frac{\Delta c_A}{1/k} \tag{8.14}$$

$$\text{mass flow, } N_A = k A \Delta c_A \tag{8.15}$$

High values of driving force (i.e., concentration gradient Δc_A) interfacial area, A, and overall mass transfer coefficient, k, are desirable to increase the mass flux, N_A.

8.12 LIQUID–LIQUID DISPERSION AND SPECIES TRANSPORT

This section outlines the mean and fluctuating velocity components associated with turbulence in stirred tanks, followed by transport mechanisms of the dispersing species. Generally, mean velocity and fluctuating velocities are derived from LDA/PIV data or discrete random tracking of statistically meaningful samples.[34]

Turbulent fluctuations in the flow field are represented by breaking the instantaneous fluid velocity, u_i, into mean, $\bar{u}_i$, and transient, u_i', components.

$u_i = \bar{u}_i + u_i'$, where, u_i' is derived from the local turbulence parameter and is defined as
$u_i' = \varsigma\sqrt{2k/3}$, with ς as a normally distributed random number,
k as turbulent kinetic energy, i.e., $k = {}^3\!/_2(\bar{u}_i I)^2$, and I as turbulent intensity.

Since models are often complex (containing multiple terms to account for the interphase mass transport mechanism) and numerical solution is time consuming, use of the correlations in Table 8.10 and Table 8.11 are preferred. The transfer of momentum by turbulent eddies is regularly modeled with an effective eddy viscosity similarly to the way momentum transfer is modeled with molecular diffusion (i.e., friction) with an effective molecular viscosity.

To facilitate the numerical analyses used in computational fluid dynamics (CFD), the concept of eddy viscosity $k-\varepsilon$ (turbulent energy and its dissipation rate) and liquid–liquid dispersion species transport mechanisms are described briefly. Details of the CFD models, and an example, are presented in Chapter 16.

The species transport is modeled as follows (while assuming constant density and incompressible flow):

$$\frac{\partial c}{\partial t} + \frac{\partial}{\partial x_i}(u_i c) = \frac{\partial}{\partial x_i}\left(\mathcal{D}\frac{\partial c}{\partial x_i}\right) + \mathcal{S} \tag{8.16}$$

where c = concentration of the chemical species, u_i = velocity, $\mathcal{D}$ = diffusion coefficient, and $\mathcal{S}$ = source term

Equation 8.16 is solved numerically using the discretized form in Figure 8.21. It utilizes neighboring cells (defined by compass coordinates) to calculate species transport and associated local concentrations.

The important terms in equation 8.16 are briefly mentioned as follows:

(a) Unsteady term: $\frac{\partial c}{\partial t}$ = change in local concentration as a function of time
(b) Convection term: $\frac{\partial}{\partial x_i}(u_i c)$ = convective transport of species due to fluid velocity
Flux from one cell to the next = normal velocity x concentration x area

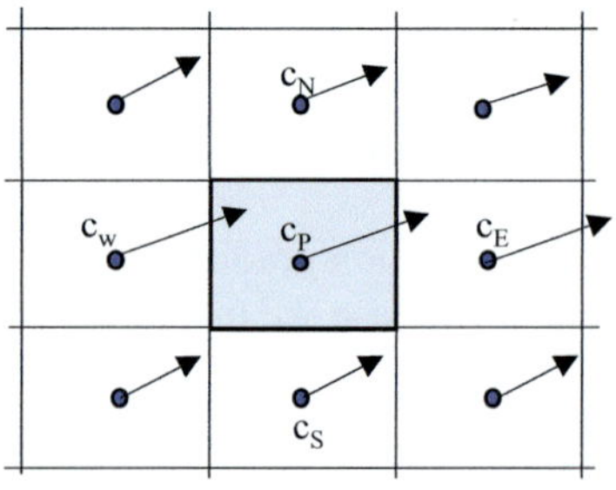

FIGURE 8.21 Schematic representation of the computational cell discretizing principle used in CFD.[34]

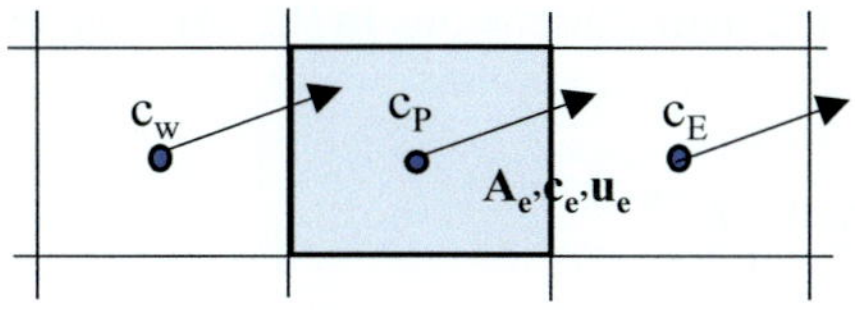

FIGURE 8.22 Schematic representation of the discretizing principle demonstrating convective flux.[34]

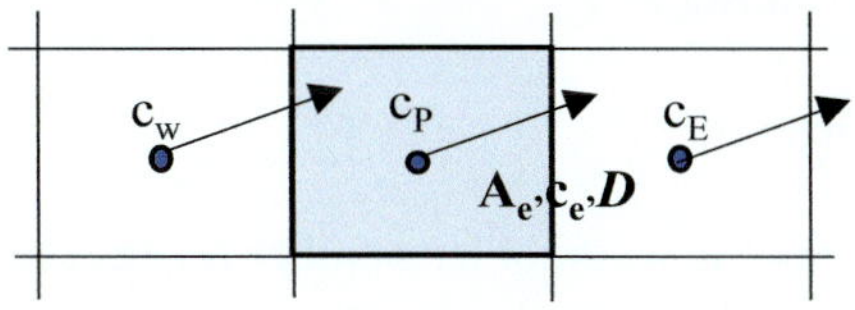

FIGURE 8.23 Schematic representation of the discretizing principle showing diffusion flux.[34]

(c) Diffusion term: $\frac{\partial}{\partial x_i}\left(\mathcal{D}\frac{\partial c}{\partial x_i}\right)$ = Diffusive transport resulting from a concentration gradient.

$$\mathcal{D}\, A_e \frac{\partial c}{\partial x}\Big|_e$$

Diffusion flux from one cell to the next = diffusion coeff. x conc. gradient x area, as seen below.

(d) Source term: $\mathcal{S}$ = creation or destruction of species due to reaction

Two types of diffusion mechanisms are encountered in liquid–liquid dispersion and must be considered as deemed appropriate. These are molecular, i.e., mass diffusion due to the concentration gradient and/or transport diffusion due to the mixing action of turbulent velocity fluctuations. Mass diffusion coefficients for molecular transport are approximated as

(a) the constant dilute approximation: same constant for all species
(b) the dilute approximation: different constant for each species
(c) the multi-component: a separate binary diffusion coefficient $\mathcal{D}_{ij}$ for each i, j species combination.

The value of the turbulent diffusion coefficient, $\mathcal{D}_t$, is calculated from the turbulent viscosity, μ_t.

$$\mathcal{D}_t = \frac{\mu_t}{\rho\, Sc_t} \tag{8.17}$$

where Sc_t = the turbulent Schmidt number, a model constant, that has one of the following values
~ 0.7 for an eddy viscosity model ($k-\varepsilon$ model),
~ 1.0 for Reynolds stress model (RSM).

Mechanically agitated vessels are often used for physical separation of the unreacted/reacted dispersed phase (usually products) from the continuous phase. In this case, liquid–liquid extraction process advantages are high separation efficiency, ability to handle liquid–liquid systems with high interfacial tension, and low susceptibility to flooding. In order to have optimum control over dispersion and coalescence, the vessel agitation system should have a variable speed motor. Table 8.12 identifies the conventional non-dimensional parameters widely used in mixing, and their applications to various processes, with emphasis on the importance of the Damköhler number to address reaction rate and species transport.[26]

The symbols in Table 8.12 and their dimensions in the MLT system are as follows:

D = diffusion coefficient, L^2/T
g = acceleration due to gravity, L/T^2
K = mass transfer coefficient, L/T
l = characteristic length, L
v_o = fluid velocity, L/T
α = thermal diffusivity, L^2/T
k = first-order reaction rate constant, T^{-1}
υ = kinematic viscosity, L^2/T
$\Delta\rho/\rho$ = fractional density change [-]

In its most commonly used form, the Damköhler number relates the reaction timescale to the convection time scale, i.e., volumetric flowrate through the reactor for continuous (plug flow or stirred tank) or semi-batch chemical processes. Damköhler number (D_a) is a useful ratio for determining whether convection rates or reaction rates are more 'important' for defining a steady-state chemical distribution over the length and time scales of interest. Hence the definition:

TABLE 8.12
Non-Dimensional Groups and Their Applications[33]

Non-Dimensional Group	Formula	Physical Meaning	Application
Sherwood number, *Sh*	$K.l/D$	$\frac{mass\ transfer\ velocity}{diffusion\ velocity}$	usual dependent variable
Stanton number, *St*	K/v_o	$\frac{mass\ transfer\ velocity}{flow\ velocity}$	occasional dependent variable
Schmidt number, *Sc*	υ/D	$\frac{diffusivity\ of\ momentum}{diffusivity\ of\ mass}$	correlations of gas or liquid data
Lewis number, *Le*	α/D	$\frac{diffusivity\ of\ energy}{diffusivity\ of\ mass}$	simultaneous heat and mass transfer
Prandtl number, *Pr*	υ/α	$\frac{diffusivity\ of\ momentum}{diffusivity\ of\ energy}$	heat transfer included here for completeness
Reynolds number, *Re*	$l.v_o/\upsilon$	$\frac{inertia\ force}{viscous\ force}$	forced convection
Grashof number, *Gr*	$l^3.g(\Delta\rho/\varrho)/\upsilon^2$	$\frac{buoyancy\ force}{viscous\ force}$	free convection
Peclet number, *Pe*	$l.v_o/D$	$\frac{flow\ velocity}{diffusion\ velocity}$	correlations of gas or liquid
Damköhler number, *Da*	$k.l/K$	$\frac{reaction\ velocity}{mass\ transfer\ velocity}$	correlations involving reaction and convection
Second Damköhler number, Da_{II}	$k.l^2/D$	$\frac{reaction\ velocity}{diffusion\ velocity}$	correlations involving reaction and interphase diffusion

$$D_a = \frac{reaction\ rate}{convective\ mass\ transport\ rate} = \frac{convection\ time}{reaction\ time}$$

In reacting systems that include interphase mass transport, the second Damköhler number (Da_{II}) is defined as the ratio of the chemical reaction rate to the mass transfer rate.

$$D_{aII} = \frac{\textit{reaction rate}}{\textit{diffusive mass transfer rate}} = \frac{\textit{diffusion time}}{\textit{reaction time}}$$

D_a *and* D_{aII}, in principle, are associated with the characteristic convection or diffusion and reaction times. If $D_a \gg 1$, the process is said to be diffusion limited, i.e., mass transport characteristics dominate and the reaction will be instantaneously in equilibrium. If $D_a \ll 1$, diffusion occurs much faster than the reaction, so diffusion reaches an 'equilibrium' state well before the reaction is at equilibrium.[33]

In general, D_a is also defined as the ratio of the characteristic fluidic and chemical time scales. Since the reaction timescale is determined by the reaction rate, the exact formula for the Damköhler number varies according to the rate law equation. For a general chemical reaction, A → B of *nth* order, the Damköhler number for a convective flow system is defined as:

$$D_a = kC_0^{\,n-1}\tau \tag{8.18}$$

where, k= kinetic reaction rate constant, c_0 = initial concentration, n = reaction order, and τ = mean residence time or space time

On the other hand, the second Damköhler number, D_{aII}, is defined as:

$$D_{aII} = \frac{kC_0^{\,n-1}}{k_g\, a} \tag{8.19}$$

where, k_g is the global mass transport coefficient, and a is the interfacial area.

The value of D_a provides a quick estimate of the degree of conversion that can be achieved. As a rule of thumb, when D_a is less than 0.1, a conversion of less than 10% is achieved, and when D_a is greater than 10, a conversion of more than 90% is expected. The limit $D_a \to \infty$ is called the Burke–Schumann limit.[35,36]

8.13 PHASE INVERSION

Phase inversion in the liquid–liquid dispersion processes refers to the transformation of a disperse phase to a continuous phase and vice versa. This phenomenon occurs due to change in the thermodynamic condition or imposition of ultra-high flow-induced deformation (extension) of the dispersed phase.

(a) State of interfacial thermodynamics: When significant reduction in drop size occurs due to excessive specific energy for a long duration, starvation of emulsifier (surfactant and stabilizer system) at the droplet interface alters its surface thermodynamics that lead to phase inversion. If phase inversion is undesired, the process recipe must have excess emulsifier concentration to accommodate potential change in the drop size. Simply, the thermal state of the working media and stripping of the solvent may also influence the interfacial state of thermodynamics to cause phase inversion. The phase inversion diagram of the recipe should be studied.

(b) Mixing system flow induction: When concentration of the dispersed phase exceeds the particle packing parameter ($c_v \gtrsim 0.6$), it is subjected to excessive flow induced stretching that leads to deformation of droplet surface curvature. Phase inversion may occur as illustrated in Figure 8.24.

The occurrence of phase inversion in a stirred tank is a nuisance, often considered a catastrophic process failure that must be avoided. However, there are a few processes, such as those synthesizing nanoparticles for drug delivery products, colloidal polymeric epoxide resins, and ABS latex coagulation, etc., where flow- and/or thermodynamically induced phase inversion are intentionally carried out. As mentioned earlier, location of the impeller near the initial o/w interface and the energy imparted to the working media should be considered to avoid phase inversion in a stirred tank. The latter may create an unwanted solid polymeric mass in the tank that will be a nightmare to dislodge. Such unforeseen incidents are, unfortunately, not uncommon. This topic is further discussed in Chapter 13, "Inline Dynamic Mixer."

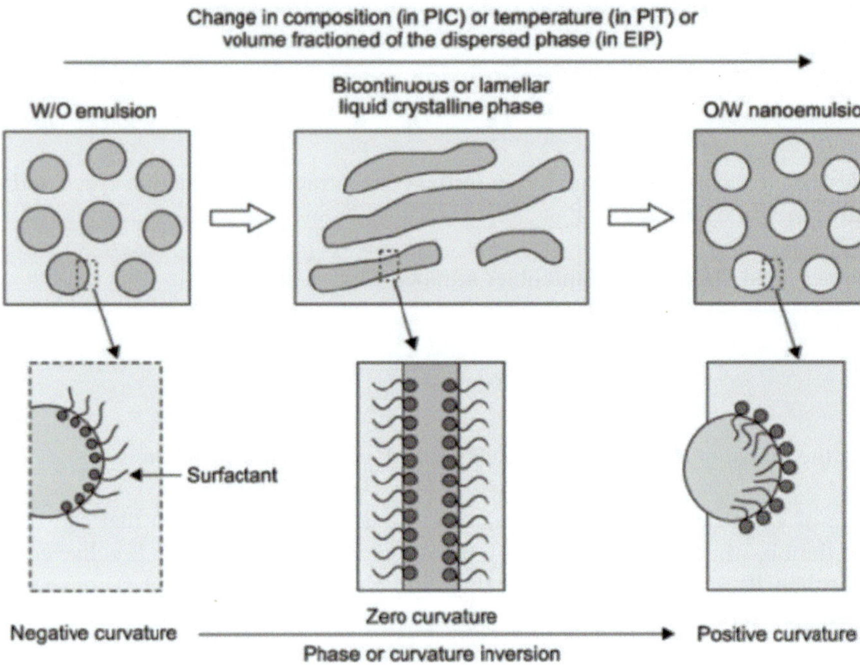

FIGURE 8.24 Schematics of phase inversion due to droplet deformation leading to curvature alteration.[37]

8.14 EXAMPLE: AGITATION SYSTEM FOR L-L DISPERSION

As an example, the design of a 1500gal. stirred tank and its agitation system are described for dispersion of an immiscible liquid (divinylbenzene organic monomer) in water. Prior to thermal polymerization with the initiator, the required organic phase drop size distribution (dsd) must have a volume median, d_{vm}, and Sauter mean drop size d_{32} of approximately 400µm and 240µm, respectively. Scale-up is based on a 100gal. pilot-scale reactor that used approximately 4hp/1000gal to produce the desired droplet size. Refer to MS Excel® spreadsheets SS8.1-2 for the reactor design based on geometric and dynamic similarities. Figure 8.25 shows a schematic of the reactor with its pertinent features; Table 8.13 presents relevant dimensions per mixit-io SS8.1. Mixit-io SS8.2 is used to determine the proportionality constant, C, from the pilot reactor data, applied to equation 8.7 for d_{32} calculation.

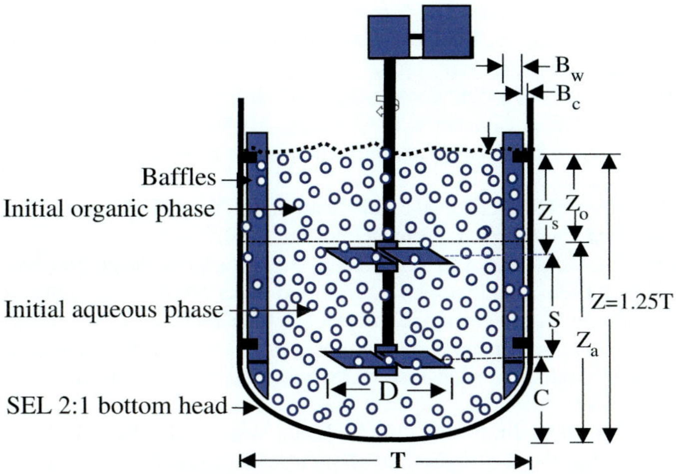

FIGURE 8.25 Pertinent features and design parameters for a liquid–liquid dispersion reactor.

TABLE 8.13
Important V-1 Reactor Design Results from mixit-io SS8.1

Important Characteristics	Symbol	Type/~Value
Tank bottom head	-	SEL 2:1
Type of baffles and number	-	Flat—4
Impeller(s)	-	Standard 45 deg. 4-blade PBT
Tank diameter	T	72.3″
Working media height	Z	90.4″
Working media height/tank diameter	Z/T	1.25
Impeller diameter	D	33.3″
Initial aqueous phase height	Z_a	623″
Initial organic phase height	Z_o	28.1″
Lower impeller of-bottom clearance	C	18.1″
Impeller spacing	S	33.3″
Upper impeller submergence	Z_s	39.0″
Baffle width	B_w	6.0″
Baffle clearance	Bc	1.51″

Note: To avoid possibility of phase inversion, Zs should be greater than Zo

REFERENCES

1. https://en.wikipedia.org/wiki/Emulsion_polymerization.
2. Dow surfactants alternatives to Alkyl Phenol Ethoxylate (APE, APEO), Dow Inc. Form No. 119-02307-0414MM. www.dowsurfactants.com
3. An overview of agricultural silicone surfactants. www.scribd.com/document/355715559/An-Overview-of-Agricultural-Silicone-Surfactant
4. A general organo-silicone surfactant structure. www.nature.com/articles/srep40499/figures/1
5. Surfactants, *Surfactants and Specialty Chemicals for North America*. www.oxiteno.us.com
6. Tauer, K., Reactive surfactants. In: Asua, J. M. (eds.), *Polymeric Dispersions: Principles and Applications*, p. 335, NATO ASI Series (Series E: Applied Sciences). Springer, Dordrecht, 1997.
7. Whinton, M., Hughes, T. C., Peng, S., Brook, M. A., Silicone microemulsion structures are maintained during polymerization with reactive surfactants, *Langmuir*, 34(14), pp. 4374–4381, March 2018.

8. Hao, T., *Electrorheological Fluids: The Non-aqueous Suspensions*, p. 2. Elsevier, 2005.
9. Vermeulen, T., Williams, G. M., Langlois, G. E., Interfacial area in liquid-liquid and gas-liquid agitation, *Chem. Eng. Prog*., 51(2), pp. 85F, 1955.
10. Taylor, G. I., The viscosity of a fluid containing small drops of another fluid, *Proc. R. Soc. Lond. Ser. A-Contain. Pap. Math. Phys. Character*, 138(834), pp. 41–48, 1932.
11. Govier, G. W., Aziz, K., *The Flow of Complex Mixtures in Pipes*, p. 98. Von Nostrand Reinhold, New York, 1982.
12. Schubert, H., Karbstein, H., Mechanical emulsification. In: Yano, T., Matsuno, R., Nakamura, K. (eds.), *Developments in Food Engineering*. Springer, Boston, 1994.
13. Harnby, N., Nienow, A. W., Edwards, M. F., *Mixing in the Process Industries*. Butterworth-Heinemann, 1992.
14. Padron, G. A., BHRG, Liquid-liquid dispersion in stirred tanks. Short course lecture. In: *Mixing XXIV, 24th Biennial North American Mixing Forum's Conference*. NAMF (AIChE), Lake George, NY, USA, June 2014.
15. Grace, H. P., Dispersion phenomena in high viscosity immiscible fluid systems and application of static mixers as dispersion devices in such systems, *Chem. Eng. Comm*., 14, pp. 225–277, Gordon Beach Science Publisher, 1982.
16. Leng, D.E., Calabrese, R.V., Immiscible liquid–liquid systems. In: Paul, E. L., Atiemo-Obeng, V. A., Kresta, S. M. (eds.), *Handbook of Industrial Mixing: Science and Practice*, pp. 639–753. John Wiley & Sons, Inc., Hoboken, NJ, USA, 2004.
17. Pacek, A.W., Chamsart, S., Nienow, A. W., Bakker, A., The influence of impeller type on mean drop size and drop size distribution in an agitated vessel. *Chem. Eng. Sci*., 54, pp. 4211–4222, 1999.
18. Todtenhaupt, E. K., Zeiler, G., *Handbook of Mixing Technology*. EKATO Ruhr-und Mischtechnik, Schopfheim, Germany GmbH, 1991.
19. Davies, J. T., A physical interpretation of drop sizes in homogenizers and agitated tanks, including the dispersion of viscous oils, *CES*, 42, pp. 1671–1676, 1987.
20. Skelland, A. H. P., Seksarla, R., Minimum impeller speeds for liquid-liquid dispersion in baffled vessels, *Ind. Eng. Chem. Proc. Des. Devel*, 17(1), pp. 56–61, 1978.
21. Baldi, G., Conti, R., Alaria, E., Complete suspension of particles in mechanically agitated vessels, *Chem. Eng. Sci*., 33(1), pp. 21–25, 1978.
22. Skelland, A. H. P., Ramsay, G. G., Minimum agitator speeds for complete liquid-liquid dispersion, *Ind. Eng. Chem. Res*., 26(1), p. 77, 1987.
23. Armenante, P. M., Huang, Y-T, Li, T., Determination of the minimum agitation speed to attain the just dispersion state in solid-liquid and liquid-liquid reactors provided with multiple impellers, *Chem. Eng. Sci*., 47(9–11), pp. 2865–2870, June 1992.
24. Dispersion stability. https://en.wikipedia.org/wiki/Dispersion_stability
25. Leng, D. E., Quarderer, G. J., Drop dispersion in suspension polymerization, *Chem. Eng. Commun*., 14, pp. 177–201, 1982.
26. McClements, D. J., *Food Emulsions: Principles, Practices and Techniques*, 3rd ed., Chapter 9, CRC Press, Taylor & Francis Group, London, UK.
27. Mass Transfer, Ed. Kirk-Othmer, *Kirk-Othmer Encyclopedia of Chemical Technology*, Vol. 15, John Wiley & Sons, Inc., 2005.
28. Bin, K. A., Mass transfer to the free interface in stirred vessels, *Chem. Eng. Comm*., 31, p. 155, 1984.
29. Wilke, C. R., Chang, P., Correlations of diffusion coefficients in dilute solutions, *AIChE J*., 1, p. 264, 1955.
30. Geankoplis, C. J., *Transport Processes and Unit Operations*, 3rd ed., Prentice Hall, Englewood Cliffs, NJ, 1993.
31. Pratt, H. R. C., In: Lo, T. C., Baird, M. H. I., Hanson, C. (eds.), *Handbook of Solvent Extraction*, pp. 91–123. John Wiley & Sons, Inc., New York, 1983.
32. Schugerl, K., Solvent extraction. In: *Biotechnology: Recovery of Primary and Secondary Metabolites*, Springer-Verlag, New York, 1994.
33. Cussler, E., Fundamentals of mass transfer. In: *Diffusion: Mass Transfer in Fluid Systems*, (Cambridge Series in Chemical Engineering, P. 237–273), Cambridge University Press, 2009.
34. Bakker, A., *Lecture on Modeling Reacting Flows in Stirred Tanks: Lecture II-Material/Species Transport*, 2006. https://www.bakker.org/Lectures-Reacting-Flows.pdf
35. Fogler, H. S., *Elements of Chemical Reaction Engineering*, 4th ed. Pearson Education International, Upper Saddle River, NJ, 2006.
36. https://en.wikipedia.org/wiki/Damk%C3%B6hler numbers
37. Kumar, M., Bishnoi, R. S., Shukla, A. K., Jain, C. P., Techniques for formulation of nano-emulsion drug delivery system: A review preventive nutrition and food science, *Prev. Nutr. Food Sci*., 24(3), pp. 225–234, September 2019.

9 Gas–Liquid Contactors

9.1 INTRODUCTION

Gas–liquid contacting operations, which transfer one or more components between a gas phase and a liquid phase, are important to numerous industrial chemical processes. Their significance is reflected in the abundance of different contactor designs and review articles.[1,2] The importance of these operations to the chemical industry is affirmed by their global prevalence and involvement in annually producing hundreds of millions of tons of basic chemicals. The various gas–liquid contactor designs attempt to optimize controlling parameters or such specific domains as the gas–liquid interface or continuous-phase residence time.

9.2 BACKGROUND

To illustrate the various gas–liquid contactor designs, consider worldwide phenol production, which totaled 11 million tonnes in 2020.[3] Bubble columns are the gas–liquid contactor typically used in a common phenol process that oxidizes cumene (at ~7bar and ~110°C) to produce intermediate cumene-hydroperoxide. The compressed air introduced into the base of the oxidizer provides oxygen for chemical reaction, as well as the agitation needed to keep the alkali and organic phases intimately mixed.

Another process example, that of terephthalic acid (TPA), the precursor to polyester, had a worldwide production of 79 million tonnes in 2020.[4] Stirred tank reactors (STRs) operating at 14 bar and 200°C are a preferred gas–liquid contactor in at least one continuous, multi-step TPA process. This particular liquid-phase air blown process oxidizes p-xylene to crude terephthalic acid (CTA) in the presence of acetic acid, a bromine promoter, and a cobalt catalyst. Heat from the highly exothermic oxidation is removed by condensing and refluxing solvent, p-xylene, and water vapor. Another gas–liquid contactor located downstream, a packed bed hydrogenation reactor, uses hydrogen to remove color impurities (4-CBA) from aqueous CTA. Hydrogen in the reactor headspace is absorbed into a thin film of liquid pooled atop the catalyst packing. As this liquid film moves down through the packing, the absorbed hydrogen reacts with the trace liquid impurities.

More fundamentally, gas–liquid systems promote reaction in the liquid phase by utilizing the following three possible modes of gas–liquid contact: 1) gas bubbling through a liquid (as in bubble columns and stirred tanks); 2) gas contacting a thin film of flowing liquid (as with packed beds or columns); and 3) liquid droplets dispersed in gas (as in spray columns and venturi scrubbers).

The best possible mode of gas–liquid contact for a given process depends upon a combination of effects, including hydrodynamics, mass transfer, and chemical kinetics. In treating this combination, a dimensionless parameter β has been defined as the ratio of "total volume of the liquid phase" to "volume of the liquid diffusion layer." Krishna and Sie[5] reported general values of β to be 10–40 for thin liquid films and liquid sprays, and 10^3–10^4 for gas bubbles within a continuous liquid. The relative rates of mass transfer and chemical reaction show whether high values or low values of β best utilize available reactor volume.

For example, when mass transfer limits the overall reaction rate, reaction occurs near the gas–liquid interface to make significant liquid volume unnecessary. To minimize the value of β, bulk liquid volume is minimized, while interfacial area is maximized. As intense mixing and turbulence throughout a large liquid volume would be of limited benefit, the gas–liquid contactor of choice is a packed bed or a spray column, as in the earlier TPA hydrogenation reactor example.

DOI: 10.1201/9781003606390-9

Conversely, if chemical reaction in the liquid phase limits the overall reaction rate, a large value of β suggests minimizing the gas–liquid interface and maximizing both liquid volume and turbulent mixing. The best contactor would disperse discrete bubbles in a continuous liquid phase. Both the bubble column mentioned in the phenol example and the stirred reactor in the CTA example utilize this gas–liquid contacting mode.

9.3 SCOPE

After introducing the theoretical basis of gas–liquid contacting, the remainder of this chapter focuses on gas–liquid contact through gas bubbles dispersed within a continuous liquid, i.e., the most common mode of gas–liquid contact within the chemical industry. Finally, the gas–liquid contactor design procedure is presented, followed by an example involving an industrial-scale STR. Further discussion of bubble columns, packed beds, thin films, and venturi scrubbers is found elsewhere.

9.4 GAS–LIQUID CONTACTING IN AN STR

STRs create the intimate gas–liquid contact needed for mass transfer in aerobic fermentation, hydrogenation, phosgenation, neutralization, chlorination, organic oxidation, and numerous other chemical processes. Extensive experimental and computational (e.g., computational fluid dynamics) effort has been expended to develop and apply gas–liquid contacting competencies to manufacturing practices. STR design significantly affects bubble dispersion, bubble size distribution, bubble surface transients, and consequently the coefficient of mass transfer. In gas dispersion applications, an ideal agitation system performs the following tasks:

- Increases the interfacial area for mass transfer by dispersing the gas as small bubbles.
- Decreases mass transfer resistance by providing shear and turbulence to thin films and constantly renews the liquid film around the bubbles.
- Distributes the gas bubbles throughout the liquid by creating synergistic axial flow.
- Achieves homogeneous temperature and concentrations by blending the liquid.

9.5 GAS–LIQUID CONTACTOR THEORY

Film theory says that the gas–liquid interface is separated from the bulk gas and liquid phases by a gas film on one side and a liquid film on the other. Although Figure 9.1 shows this specifically for bubble or droplet systems, it is easily extended to liquid film contactors as well. Resistance in both films decreases the concentration of species diffusing across the gas–liquid interface irrespective of whether that diffusion is from gas to liquid (as depicted in Figure 9.1) or from liquid to gas.

Because diffusion in the liquid phase is much slower than in the gas phase, the liquid film presents most of the resistance to mass transfer. Gas–liquid film theory defines the mass transfer rate (MTR) in terms of the liquid film, as follows:

$$MTR = k_L a\left(C^* - C_L\right)$$

$$= k_L a\left(C_g H_e - C_L\right) \qquad [\text{mol/m}^3\ \text{s}] \tag{9.1}$$

where k_L is the liquid film mass transfer coefficient [m/s], a is the interfacial area per unit volume [m^{-1}], C^* is the saturation concentration at the gas–liquid interface [mol/m^3], C_L is the bulk concentration in the liquid phase [mol/m^3], C_g is the transfer gas concentration in the bubble [mol/m^3], H_e is Henry's law constant [(= C^*RT/P_g), R is the universal gas constant (~8.314 J/mol K), T is the absolute temperature at the gas–liquid interface [K], and P_g is the transfer gas pressure in the bubble [Pa]].

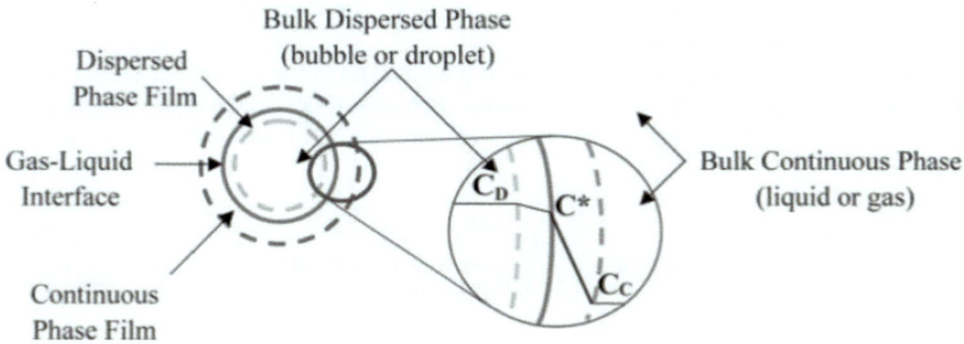

FIGURE 9.1 Gas-liquid film theory for diffusion from the dispersed phase to the continuous phase (with maximum diffusion resistance in the continuous-phase film).

Values of k_L typically average 0.02 cm/s ± 0.02 depending on diffusivity of the dissolved gas and relative velocities of the gas and liquid phases.[5] The value of "*a*" can vary by several orders of magnitude, especially in bubble systems, where it is inversely proportional to bubble size. Because of difficulties in measuring k_L and "*a*" individually, the lumped parameter or volumetric mass transfer coefficient k_La (i.e., k_L times "*a*") is typically used to determine MTR. The value of k_La can be raised by increasing superficial gas velocity and power input. A higher k_La can cut capital costs by producing a comparable MTR with a smaller vessel, a shorter batch process time, or a higher continuous process throughput.

9.6 EXPERIMENTAL DETERMINATION OF K_LA

Values of k_La can be evaluated for a diffusing gas (e.g., O_2) as a function of operating conditions by using an appropriate dissolved gas probe, and either a transient or a steady-state experimental measurement method. While these methods apply to all diffusing gas species, they will be confined to oxygen for the sake of this discussion. The transient method of determining k_La for oxygen at a given set of operating conditions begins by sparging an inert gas (e.g., N_2) into the liquid to purge it of dissolved oxygen (DO). After the DO probe reading indicates essentially zero concentration, the flowing N_2 is turned off and replaced with air. The increasing concentration of DO in the liquid phase "*C(t)*" is recorded as a function of time. As the gas dissolution process can be described by equation 9.2, the corresponding value of k_La is obtained by regressing the experimental data as suggested in equation 9.3:

$$\ln\left[\frac{C^* - C_L}{C^* - C(t)}\right] = k_L a.t \tag{9.2}$$

$$C(t) = C^* + \left(C_L - C^*\right)\exp(-k_L a.t) \quad [\text{mol/m}^3] \tag{9.3}$$

Although these equations include interfacial concentration of the DO (C^*), its value need not be known to determine k_La from the "*C(t)* vs. t" data regression shown in equation 9.3. Accurate determination of k_La requires a DO probe with adequate response time.

The steady-state k_La method uses chemical reaction in the liquid phase to generate oxygen at a "steady-state" rate. Oxygen is transferred from the bulk liquid into the gas bubbles, an exact reversal of the transfer direction employed in the transient k_La measurement method. For example, the NEL/Hickman[6] steady-state method for measuring oxygen mass transfer continuously feeds hydrogen peroxide solution (H_2O_2) and dry air into the liquid at known rates. Bovine catalase in the liquid decomposes H_2O_2 into H_2O and O_2.

The DO concentration in the liquid is monitored with a DO probe and recorded after it reaches a steady-state (i.e., constant) value. The oxygen generation rate (OGR) in the liquid equals the oxygen transfer rate (OTR), and is related to the hydrogen peroxide mass feed rate ($\dot{Q}_{H_2O_2}$), as shown in equation 9.4:

$$\text{OGR} = \left[\frac{\dot{Q}_{H_2O_2}}{2 \cdot MW_{H_2O_2}} \cdot \frac{1}{V_L}\right] = \text{OTR}$$

$$= k_L a\left(C_L - C^*\right) \qquad [\text{mol/m}^3\ \text{s}] \tag{9.4}$$

where $MW_{H_2O_2}$ is the H_2O_2 molecular weight (34g/mol), V_L is the volume of liquid phase (corrected for H_2O_2 addition) [m^3], and C^* is the average saturation concentration at the gas–liquid interface [mol/m^3].

The H_2O_2 feed rate is divided by 2 because of the stoichiometry of oxygen formation. As liquid temperature and salinity change during a measurement, oxygen concentrations in the liquid must be corrected accordingly. In addition, use of this method requires the following assumptions:

- Gas phase is ideal gas.
- Liquid phase is well mixed.
- Oxygen concentration at the gas–liquid interface is in equilibrium with that in the bulk gas.
- Oxygen partial pressure in air is fixed at 0.2095atm.

The presence of H_2O_2 and salt in the liquid phase generally elevates the associated k_La values above those obtained by the transient method.[7] A slight modification of this steady-state method allows the authors to use ferric chloride instead of bovine catalase in large-scale experiments.

9.7 MASS TRANSFER COEFFICIENT

The mass transfer coefficient, k_La, is correlated with power per mass, superficial gas velocity, and media viscosity as follows:

$$k_La = \alpha \cdot \varepsilon^{\beta} \cdot \left(V_{GS}\right)^{\gamma} \left(\eta_r\right)^{\kappa} \quad [\text{s}^{-1}] \tag{9.5}$$

$$k_La_{20} = k_La.\left(1.024\right)^{(20-t)} \quad [\text{s}^{-1}] \tag{9.6}$$

where ε is the specific energy (shaft power per unit mass) under gassed conditions [W/kg], V_{SG} is the superficial gas velocity (vol. gas flowrate [m^3/s]/tank cross-sectional area [m^2]) [m/s], η_r is the relative viscosity, i.e., viscosity of working media/viscosity of water, and k_La_{20} is the mass transfer coefficient at 20°C.[8]

Example values of the coefficient α and the exponents β, γ, κ are shown in Table 9.1[9] for an air–water system that was extended to higher viscosities as noted.

The value of α can change significantly for different working media. For example, $k_La_{salt\ water}$ = *1.5–2* × $k_La_{pure\ water}$. Furthermore, k_La is less dependent on specific power at higher levels of power input, such as $\varepsilon > 3$W/kg (see Table 9.2). In practice, appropriate values for each variable in the k_La correlation should be determined for specific process or simulation fluids of interest.

TABLE 9.1

Fitting Parameters for a Coalescing Gas-Liquid System *(ε < 3W/kg)***

Source	Working Media	Interfacial Characteristics	Viscosity Ratio	α	β	γ	κ
Van't Riet/ Dow	air-water	72 dynes/cm	1	0.412	0.4	0.5	−0.46

* Extended correlation using k_La data from non-coalescing 5cP glucose broth, 10-300cP penicillin cultures, and 2000cP polyglycol.

9.7.1 Effect of Surface-Active Agents

The presence of low-molecular-weight surfactants in the working media reduces interfacial tension and resulting bubble sizes. Consequently, k_La increases significantly. In one instance, the authors' research at Dow showed the presence of 1wt% formic acid (free surface ions) in an agricultural intermediate, to surprisingly yield a 34% increase in k_La, as seen in Table 9.2. High concentrations of high-molecular-weight surfactants, however, tend to decrease k_La by increasing the viscosity of liquid film surrounding the bubble.

TABLE 9.2

Effect of Surface-Active Agents (free surface ions) on k_La (ε >3W/kg, at 40°C)

Working Medium	α	β	γ
Dilute slurry + SARAN particles	0.945	0.05	0.47
Dilute slurry + SARAN particles + formic acid	1.265	0.05	0.47

9.7.2 Effect of Solids on the Mass Transfer Coefficient

In gas–liquid mixing applications, solids are often present as a catalyst or a product. Inherently having small solids present on the bubble surface can diminish the effective intrinsic bubble surface area and significantly decrease the mass transfer coefficient. Although others report just the opposite effect, an illustrative example from the authors' research on prediction of the oxygen transfer rate for the catalytic oxidation in a stirred batch reactor at Dow is shown in Table 9.3.

TABLE 9.3

Effect of Solid SARAN Particles on k_La (0.4 <ε < 3W/kg, at 22°C)

Working Medium	α	β	γ
Slurry	0.476	0.52	0.46
Dilute slurry + SARAN particles	0.378	0.53	0.44

9.8 GAS HOLDUP

Besides the mass transfer coefficient, k_La, other parameters help quantify gas–liquid contactor performance in bubble systems. One such parameter is gas holdup (ε_g), which is the ratio of gas volume within the media to the mixture volume, as shown in equation 9.7.

$$\varepsilon_g = \frac{V_g}{\left(V_g + V_L\right)} \tag{9.7}$$

where V_g is the gas-phase volume [m³] and V_L is the liquid phase volume [m³].

Sometimes gas holdup information is expressed as the fluid volume expansion factor, or $1/(1-\varepsilon_g)$. Gas holdup is correlated with power per mass (ε_g) and superficial gas velocity (V_{SG}) as follows

$$\varepsilon_g = \alpha_1 \cdot \varepsilon^{\beta_1} \cdot \left(V_{SG}\right)^{\gamma_1} \tag{9.8}$$

For the air–water (coalescing) system at specified conditions[8], $\alpha = 0.65$, $\beta = 0.20$, and $\gamma = 0.55$.

Gas holdup is an important parameter for fill volume calculations and reactor sizing. In the case of aqueous coalescing systems, gas holdup normally ranges between 10% and 20%. This is also typical of most aerobic bioreactors. In aqueous non-coalescing systems or viscous systems, gas holdup can be appreciably higher (e.g., 50%). For example, gas holdup in an ~12% acetic acid solution is in the range of 45–50% depending on power level and superficial gas velocity. Significantly, a typical coalescing medium such as water may often be converted into a non-coalescing medium with the addition of salt or other electrolytes.

Higher gas holdup generally produces significant increases in the volumetric mass transfer coefficient, k_La. An exception, however, is high viscosity media, where initially many tiny (< 2mm) non-coalescing bubbles and a few relatively large (50–300 mm) coalescing bubbles form. As viscous forces dominate over local turbulence (inertial forces), the tiny bubbles are trapped in the liquid phase and have little-to-no relative velocity with respect to the bulk liquid motion (i.e., the two-phase media behave as a single liquid phase). This phenomenon limits bubble surface transience and gas disengagement, and consequently, the volumetric MTR. Simultaneously, the large bubbles quickly disengage (burp) from the media with little contribution to the mass transfer.

9.9 AGITATION SYSTEM

A modern high-efficiency gas–liquid STR (minus an unseen gas sparge) is shown in Figure 9.2. Such systems commonly have a radial gas-dispersing impeller mounted below one or more axial-flow impellers.

Table 9.4 shows the recommended numbers of impellers to use as Z/T increases. Impeller spacing, S, tends toward the high end of the indicated ranges in low viscosity working media. For high viscosity media, impeller spacing normally must decrease to the lower end of each Z/T range.

For media of lower viscosities (<200cP), axial impellers are commonly standard 45° four-bladed pitch-blade turbines (PBTs), although high-solidity three-blade hydrofoils are also used in such systems. Examples of the latter, including the Lightnin A-320 and Chemineer Maxflo, are suggested axial impellers for high-viscosity working media. Possible radial impellers include the Rushton and concave-blade Smith turbines (see Figure 9.3), and more advanced parabolic-blade turbines like the Chemineer BT-6 and Lightnin R135 (not shown). Smith turbines are recommended over Rushtons for normal impeller-to-tank diameter ratios of 0.4–0.54. The BT-6 and R135 are recommended when the system and related financials are suitable.

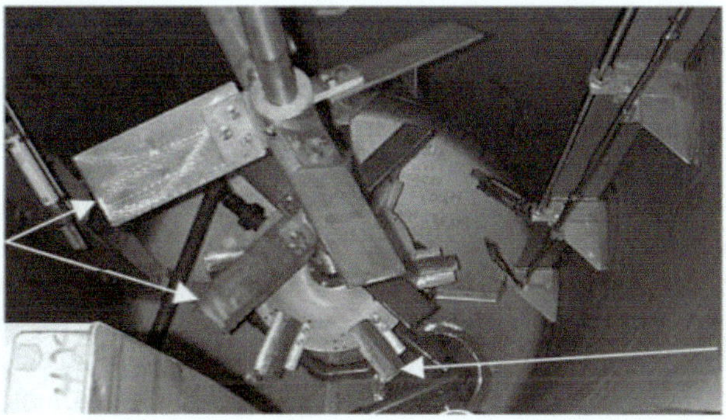

FIGURE 9.2 A typical high efficiency gas–-liquid mixing system.

TABLE 9.4
Multiple Impeller Arrangements as a Function of Liquid Height (Z/T ratio)

Working medium gassed height/tank diameter (Z/T):	<0.9	0.9–1.7	1.7–2.5	2.5–3.3
Numbers of impellers:	1	2	3	4

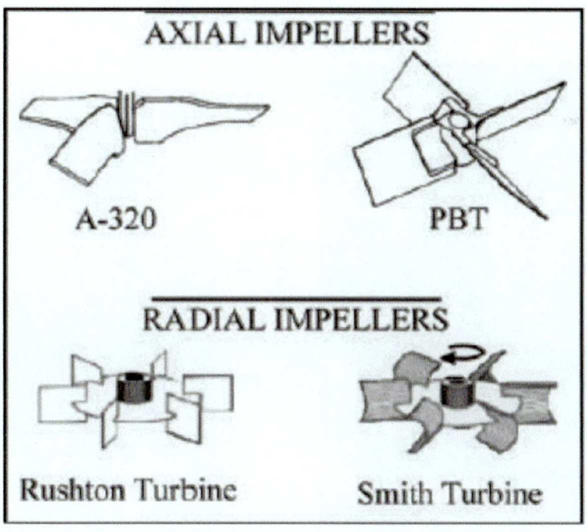

FIGURE 9.3 Examples of common axial-flow and radial-flow impellers.

9.9.1 Down-Pumping vs. Up-Pumping

Some recently installed gas–liquid agitation systems use up-pumping axial flow impellers, instead of down-pumping axials.[10] The literature contains findings from such systems, although they appear to be somewhat system dependent. Sardeing et al. found that up-pumping axials improved gas dispersion and k_La over values obtained in their down-pumping system because an improved bulk-flow pattern increased circulation and gas–liquid contact time.[11] Zhu et al. say that switching from down-pumping to up-pumping axial impellers reduced observed decreases in both impeller flow numbers and turbulent power numbers (thus, P_g/P_{ug}), as gas flow was increased.[12] Strand and Courtemanche found that up-pumping produced a 26% faster average blend time, but also a 14% lower k_La value.[13] Benz says, however, that fermenters with up-pumping axial flow impellers generally gives higher gas hold-up and shorter blend times. If the relative P_g/P_{ug} ratio is also increased, up-pumping could draw more power at high gas flows, and thereby, deliver higher k_La.[14] Even with these observations for upper up-pumping axial-flow impellers, at least one unnamed mixer manufacturer still recommends down-pumping axials unless other requirements such as high maximum gas flow and/or intolerance for headspace gas draw-down make up-pumping a more attractive alternative.

The authors conducted experiments in both low-viscosity (water) and high-viscosity (polyglycol) media using either the up-pumping configuration of Figure 9.4a or the corresponding down-pumping configuration. Results from these particular experiments (Figure 9.4b) showed the down-pumping k_La to be 1.2 times the corresponding up-pumping value.

The superior mass transfer performance of down-pumping vs. up-pumping in these experiments, at various locations in the tank, was presented at the NAMF Mixing XIII Conference.[15] Since the up-pumping axials performed below the down-pumping axials, which are considered most effective mixing technology (METs) for gas–liquid mixing, no further details are discussed from the cited articles[11–14] regarding systems in which up-pumping was apparently advantageous over down-pumping. The cited up-pumping references provide an introduction to the topic for readers interesting in seeking further insight into their operations.

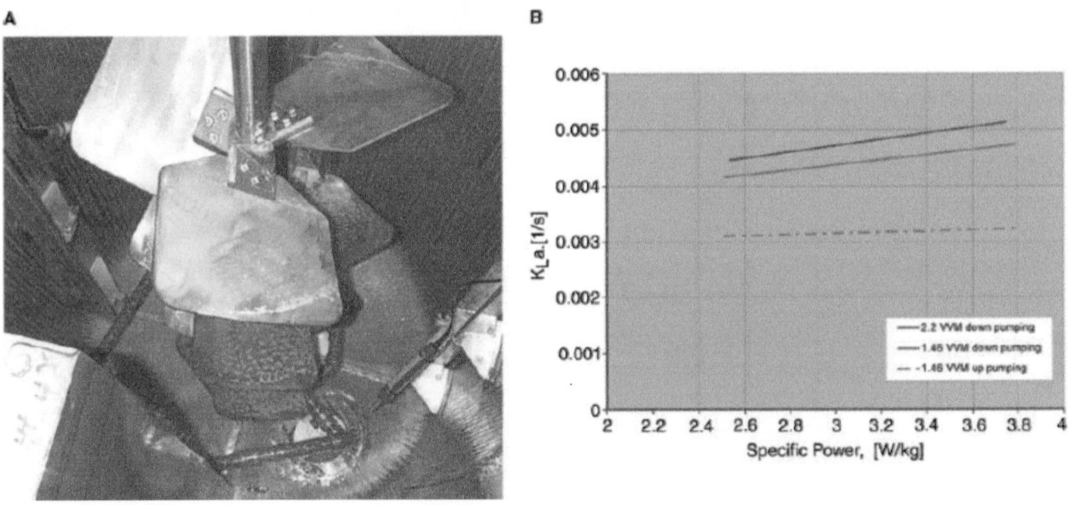

FIGURE 9.4 Up-pumping (A-320) agitation system (a) photograph (counterclockwise rotation), and (b) experimental k_La results in a polyglycol (2000cP) medium.

Conceptually, most gas–liquid contacting applications maximize gas-to-liquid mass transfer by retaining gas bubbles in the liquid for as long as possible before their release into the vessel headspace. Since up-pumping agitation necessarily transports gas bubbles upward at a faster rate, this seemingly opposes maximizing retention time and mass transfer of gas-phase species within the liquid. If, as reported by Strand and Courtemanche, up-pumping increases gas holdup while decreasing k_La, it is implied that the average diameter of resulting gas bubbles has undesirably increased. In addition, since up-pumping creates downward fluid thrust on the drive system bearing(s), conventional tapered roller bearings that sustain upward fluid thrust (created in downward pumping), must be altered to sustain a downward fluid thrust.

9.9.2 Power Draw

The power draw characteristics of open impellers are expressed in terms of power number (N_P) vs. Reynolds number (N_{Re}). By definition,

$$N_p = P/(\rho \cdot N^3 \cdot D^5) \tag{9.9}$$

$$N_{Re} = \rho \cdot N \cdot D^2/\mu \tag{9.10}$$

where P is the shaft (hydraulic) power [W], ρ the density of the working media [kg/m^3], and μ the viscosity of the working media [kg/m-s or Pa-s].

Impeller power number is fairly constant for the fully turbulent ($N_{Re} > 10^4$) flows encountered in most production-scale vessels working with a low-viscosity media. For various reasons including synergy between impellers, the N_P vs. N_{Re} relationship approaches an asymptote more slowly in multiple-impeller systems than in single-impeller systems. Figure 9.5 shows power number vs. Reynolds number characteristics for two different dual-impeller systems, i.e., Rushton turbine + PBT and concave disk turbine + PBT. Slopes of curves in Figure 9.5 flatten as $N_{Re} > 10^6$ (beyond the fully turbulent flow regime) and increase significantly where $N_{Re} < 75{,}000$ (the transition flow regime). The un-gassed fully turbulent flow power numbers for the standard Rushton, Smith, and PBTs are 4.6, 3.2, and 1.27, respectively.

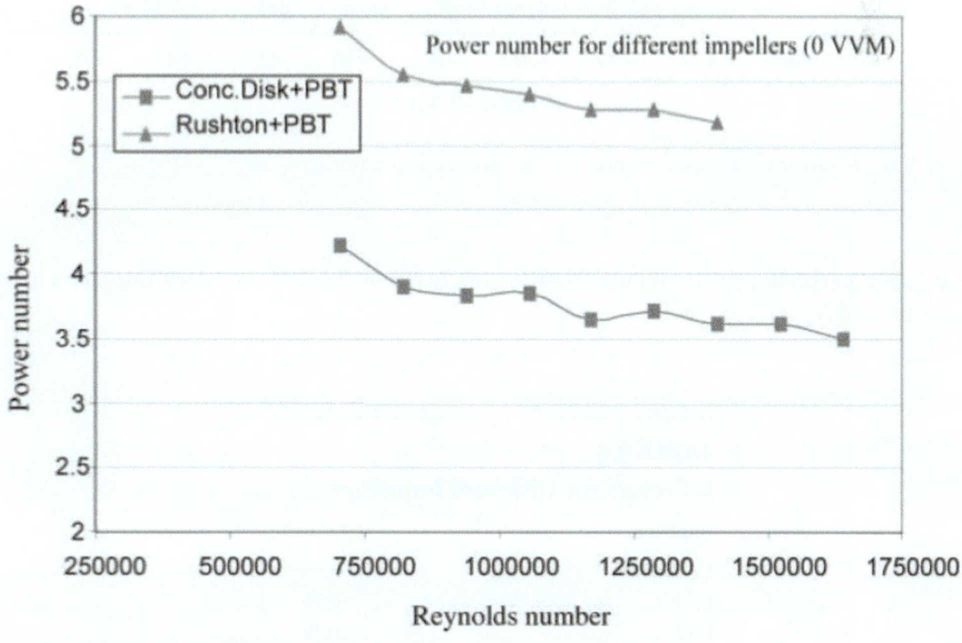

FIGURE 9.5 Power number vs. Reynolds number for two agitation systems.

9.9.3 The k-Factor

The *k*-factor is the ratio of the power drawn under gassed conditions to that under ungassed conditions, i.e., P_g/P_{ug}. Its value is generally less than unity and varies with changes in such operating conditions as gas flowrate, impeller type, and impeller speed. The k-factor decreases with increasing aeration number, N_a, as shown in Figure 9.6. Although lower power draw under gassed conditions is generally not considered a problem, it becomes one if the system motor power is sufficient under gassed conditions but insufficient during system startup under ungassed conditions. In the limit of fully turbulent gassed conditions, the power draw of all the gas-dispersing impellers approaches a similar value.

Figure 9.6 shows k-factor vs. aeration number curves for the two dual agitator combinations discussed previously. As the power draw (and the power number) is higher under ungassed conditions, agitators typically operate at lower speeds when ungassed. The change in power draw with increasing gas flow is managed by a two-speed motor or a variable frequency drive (VFD) in the drive system.

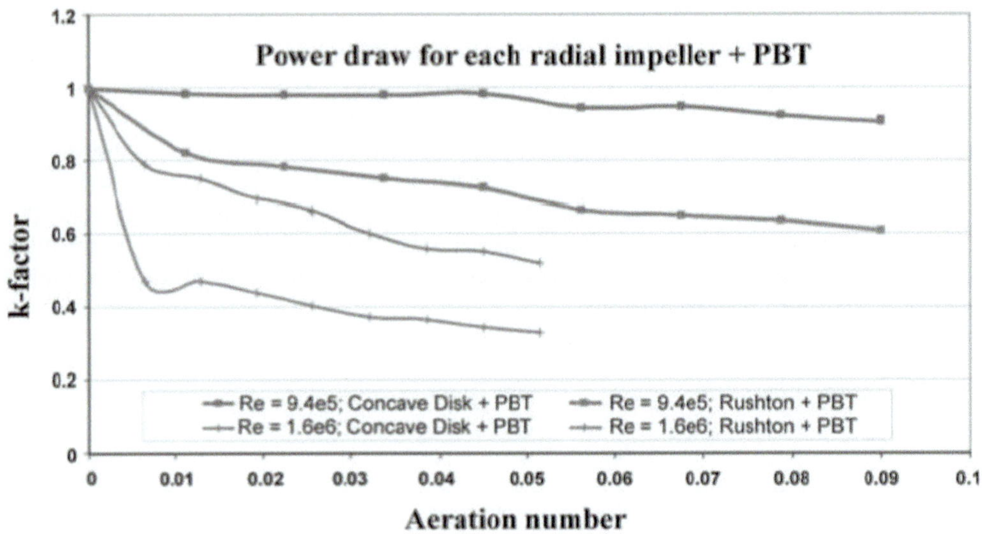

FIGURE 9.6 *k*-Factor vs. aeration number for the two dual agitation systems.

For design purposes, one could use the k-factors listed in Table 9.5 for open impellers operating at $N_{Re} > 10^5$ and $N_a > 0.03$.

TABLE 9.5
***k*-Factors for Different Impellers**

Impellers	*k*-Factors
Rushton	0.35
CD-6	0.70
PBT	0.90
A320	0.90

9.9.4 Criteria for Impeller Flooding

The Froude number (N_{Fr}) characterizes gas flow regimes using the following ratio of inertial forces to gravitational forces:

$$N_{Fr} = N^2 D / g \tag{9.11}$$

where g is the acceleration due to gravity (9.81 m/s^2).

The aeration number (N_a) is the ratio of gas flowrate to impeller discharge flow and is written as

$$N_a = \dot{Q}_g / ND^3 \tag{9.12}$$

where $\dot{Q}_g$ is the actual gas flowrate [m^3/s], N is the agitator speed [s^{-1}], and D is the gas-dispersing impeller diameter [m].

For a given gas flowrate, the dispersion pattern of gas bubbles in the working media depends on the interplay between the isothermal expansion energy of the gas and the hydraulic power of the impeller. One of three bubble dispersion scenarios is typically expected, i.e., flooding, dispersion in the upper region only, or complete dispersion. Figure 9.7[15] illustrates each of these flow regimes.

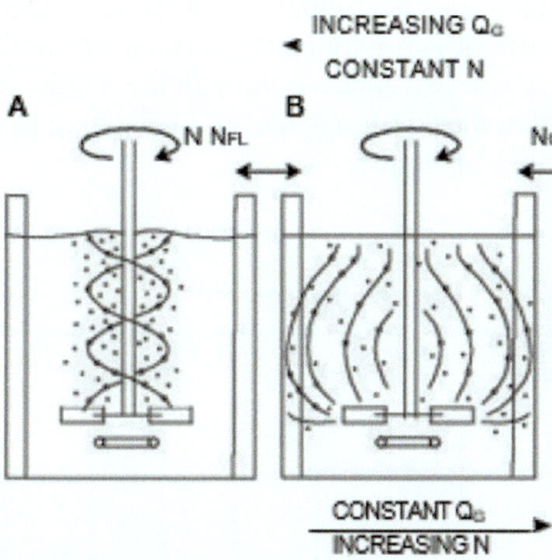

FIGURE 9.7 Gas bubble dispersion in an STR.[16]

In water, the maximum aeration number at which flooding occurs ($N_{a\,FL}$) with one of several different impellers is correlated[16] as:

$$N_{a\,FL} = C_{FL}\, N_{Fr}\, (D / T)^{3:5} \tag{9.13}$$

where for Rushton $C_{FL} = 30$ and for Smith turbine $C_{FL} = 70$.

In the same water system, gas is completely dispersed at transition aeration numbers ($N_{a\,CD}$) given by the following correlation:

$$N_{a\,CD} = C_{CD}\, N_{FR}^{0.5}\, (D / T)^{0.5} \tag{9.14}$$

where for Rushton $C_{CD} = 0.2$ and for Smith turbine $C_{CD} = 0.4$.

A rule of thumb is that one can expect complete dispersion when the desirable gassed shaft power is greater than three times (minimum two times) the isothermal gas expansion power (defined in equation 9.15). A design check for adequate shaft power is absolutely necessary to avoid an undesirable flooding situation:

$$P_{gas} = p_1 \dot{Q}_1 \ln(p_1 / p_2) \tag{9.15}$$

where P_{gas} is the power delivered by the isothermally expanding gas [W], P_1 is the gas pressure (hydrostatic + back pressure) at the sparger [Pa], p_2 is the back pressure [Pa], and $\dot{Q}_1$ is the actual volumetric gas flowrate at the sparger [m^3/s].

9.9.5 Gas Sparger

The gas sparger is a critical internal component of STRs used in gas–liquid applications. The type of sparger used depends on the ratio of volumetric gas flowrate to liquid volume (VVM) and the particle size.

Figure 9.8 shows the different types of spargers used in STRs. For systems containing large (>50 microns) settling particles, tangential sparger nozzles are more appropriate than common ring spargers. Open pipe spargers provide adequate gas delivery under low (<0.2 min^{-1}) VVM conditions. For high gas flowrates (high VVM, i.e., >2.0 min^{-1}), plate spargers are typical, although an alternative approach uses a draft-tube impeller agitation system (see Chapter 10). Use of Ring spargers are quite common. Their basic design parameters with an illustration are described in the "Fermenter" chapter. Sometimes, gas is self-induced through the rotation of hollow shafts and blades.[17] Disadvantages of self-induction include susceptibility to plugging in sticky polymers and variation of gas flow with hydrostatic head pressure.

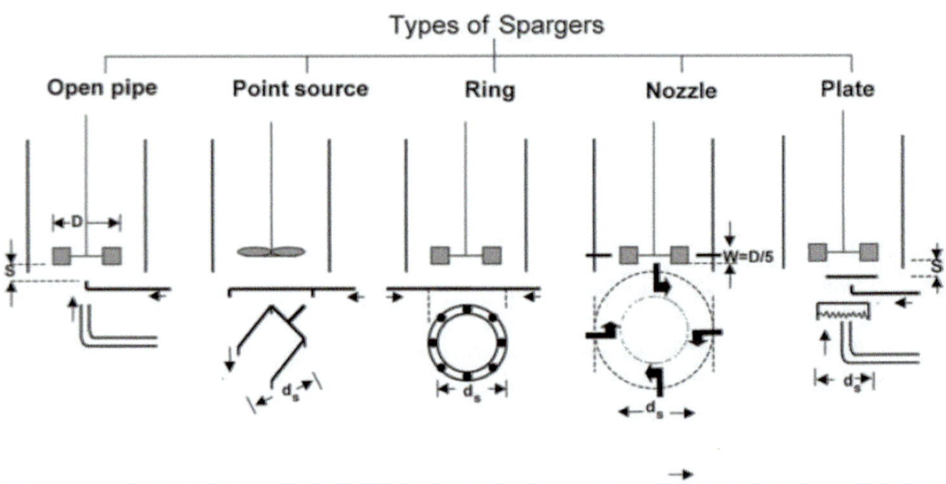

S = 0.1 D	0.1 D	0.2 - 0.3 D	- D/20	0.1 D
d_s = -	1.3 - 1.4 D	0.7 - 0.8 D	1.1 D	0.7 D
V_p = 11 - 23 (~18)	11 - 23	11 - 23	11 - 23	11 - 23
EX: Low VVM	Axial Impellers + Solids	+ Fine light solids	+Solids	High VVM

V_p = Gas velocity in the pipe, m/s

FIGURE 9.8 Sparger types and design rules.

9.9.6 Tickler (Kicker)

A tickler is a relatively smaller impeller placed at close proximity to the vessel bottom. In gas–liquid mixing systems that contain some solids, an axial (up- or down-pumping) or a radial tickler is often used to overcome the solid suspension deficiency of radial gas-dispersing impellers. Pitched-blade or flat-blade turbines are normally used as ticklers. In the case of gas-liquid-solid systems where the slurry needs to be drained through a central nozzle, the authors recommend the Dow-designed KT-3 tickler mentioned in Chapter 7. Use of this tickler prevents pump starvation and facilitates solid removal.

9.9.7 Steady Bearing

Gas–liquid mixing systems typically operate at high power levels, i.e., 1–4 W/kg, and often involve tall vessels (H/T > 2) with long mixing shafts. Most tall vessels incorporate a steady bearing or a limit ring to prevent the shaft deflection caused by severe transient loading of the agitation system. Unfortunately, steady bearings wear out (requiring regular maintenance), are susceptible to fouling in sticky polymers, and are often recognized as a contamination source in bioreactors. Thus, caution must be exercised and vendors should be consulted during the consideration and mechanical design of steady bearings.

9.9.8 Baffles

In stirred tanks for gas–liquid mixing, baffles are essential to create axial flow and maintain overall mixture uniformity. Normally, four simple flat baffles are recommended. The baffle width and baffle clearance from the wall are usually in the range of 1/12–1/8 and 1/64–1/48 times the tank diameter, respectively. Profiled baffles are preferred for sticky high viscosity or shear-sensitive materials (see Chapter 4).

9.10 STIRRED TANKS DESIGN METHODOLOGY

As discussed earlier, stirred tanks are desirable when the overall reaction rate is limited by liquid-phase reaction. Bubbling gas through well-mixed liquid in a stirred tank should minimize the thickness of the gas–liquid interface while maximizing the availability of the liquid volume ($\beta = 10^3$–10^4). Consequently, process conditions that achieve this objective are the first consideration in designing stirred tank gas–liquid contactors.

The optimal configuration depends on numerous factors, including the required gas transfer and mixing rates, and an acceptable range of shear rates. Some factors may be mutually contradictory, as in a crystallizer where the required gas–liquid transfer rate is best achieved by increasing turbulence and shear; yet the crystals are shear sensitive. In these cases, successful design carefully balances the contradictory factors.

This section describes the design of a stirred tank contactor for a given range of operating variables and includes the selection of geometric parameters. Important correlations and formulae are presented and followed by an example design case. The design process actually requires multiple iterations to simultaneously satisfy both geometric and mass transfer requirements, but such iteration has become almost trivial when using tools such as spreadsheet-based "goal-seek" computations.

Process conditions that must be considered early in the design process include:

- Desired batch size (required tank dimensions) based on required yield
- Physical and thermal properties of the working media
- Volumetric gas demand based on reaction kinetics

Gas demand influences mass flow meter selection, as indicated in the following calculations:

- Actual cubic meters per minute (ACMM) of the reacting gas
- ACMM of (reacting + carrier) gas based on percentage of reacting gas consumed
- VVM (gas volumetric flowrate per minute/volume of liquid)
- Standard cubic meters per minute (SCMM) in selecting a mass flow meter

Design of the gas sparger and agitation system can proceed at this point.

1. Sparger:
   a. Choose the sparger type.
   b. Determine superficial gas velocity from the required ACMM at the sparger.
2. Agitation system (radial + axial impellers):
   a. Calculate C* and required bulk mean gas concentration in the liquid phase.
   b. Determine k_La from the gas transfer rate.
   c. Calculate the overall transfer rate OTR = k_La (C^*—C_L) or k_La (ΔC_{lm}).
   d. Predict gassed power per mass using k_La correlation (α, β, γ, and κ).
   e. Determine gassed power, P_g.
   f. Determine ungassed power per mass using k-factor.
   g. Calculate gas isothermal expansion energy.
   h. Check the P_g/P_{gas} ratio (>2 as a minimum; >3 to ensure full dispersion).
   i. Check the flooding conditions for the gas dispersing impeller.
   j. Check for gas holdup and recheck tank dimensions.
   k. Specify agitation system:
      i. Motor power, P_m
         Motor and gear box design normally assume 80% drive system efficiency ($\eta_m = 0.8$)

$$P_m = \frac{1}{\eta_m} P_{ug}$$

      ii. Gearbox
      iii. VFD or two-speed motor
      iv. Operating conditions

The optimal stirred tank design is identified through iterating until both geometric and mass transfer requirements are simultaneously satisfied.

9.10.1 Design Example

The design of an STR for a fast-reacting oxidation process is illustrated in the following section. Operational parameters include a process oxygen demand of 250mmol l^{-1} hr^{-1} and a required batch size of 45m.[3] Reactor absolute pressure and temperature are 1.32bar and 30°C, respectively. Minimum required oxygen concentration in the liquid is 1.0mol m^{-3}. Physical properties of this Newtonian working media are: viscosity of 1×10^{-3}Pa-s; density of 1000kg m^{-3}; and interfacial tension (gas–liquid) of 7.1×10^{-2}N m^{-1}.

Design Worksheet

Based on product batch requirements, preliminary vessel dimensions are listed in Table 9.6. Gas flowrate (ACMM) is determined to obtain the mass transfer required by reaction kinetics. A ring sparger is selected as discussed in conjunction with Figure 9.8. Table 9.7 summarizes the calculation of the required specific power and gas holdup based on mass transfer and the desired oxygen concentration in the liquid. It must be confirmed that vessel dimensions accommodate the aerated liquid height. An appropriate agitation drive system is recommended in Table 9.8. Figure 9.9 is a schematic of the reactor in this design example.

TABLE 9.6
Specification of Reactor Dimensions, Sparger Type, and Mass Flow Meter

Tank diameter [m]	3.80
Tank height: tangent-tangent [m]	4.66
Nominal tank volume [m^3]	60
Working media volume [m^3]	45
Unaerated liquid height [m]	4.3
Liquid height/tank diameter	1.13
Volumetric gas flowrate [m^3s^{-1}]	0.21
Midpoint pressure [Pa]	1.57×10^5
Pressure at the sparger [Pa]	1.78×10^5
Required ACMM of oxygen [m^3min^{-1}]	3.0
Required ACMM of air [m^3min^{-1}]	14.4
VVM [min^{-1}]	0.37
Required SCMM [m^3min^{-1}]	23.6, specify mass flow meter
Superficial gas velocity at the sparger [m^3s^{-1}]	0.031
Type of sparger	Ring

TABLE 9.7
Design of Agitation System

Property	Value	Note
C^* [mol m^{-3}]	1.72	
C_L [mol m^{-3}]	1.0	
Henry's law constant [mol m^{-3} Pa^{-1}]	1.26×10^{-5}	
Volumetric mass transfer coefficient [s^{-1}]	0.12	eqn. 9.4
Power per mass [W kg^{-1}]	1.82	eqn. 9.5 and 9.6
Gassed power [kW]	82	eqn. 9.9
k-Factor [—]	0.7	Table 9.5
Ungassed power [kW]	117	Motor power ≈ 150kW
Gas isothermal expansion power [kW]	14.4	eqn. 9.15
Gassed agitation power/gas expansion power [—]	5.7	>3, Check
Aeration number for flooding conditions [—]	2.10	eqn. 9.11–9.14, Check
Gas holdup [—]	0.11	eqn. 9.8
Aerated liquid height [m]	5.0	Recheck tank dimensions

TABLE 9.8
Specifications of Agitation System

Number of impellers	2	
Radial impeller: Smith turbine	1	Np = 3.2
Axial impeller(s): four-blade-45° PBT	1	Np = 1.27
Impeller diameter [m]	1.7	D/T = 0.50
Off-bottom clearance [m]	0.85	C/T = 0.25
Impeller spacing [m]	1.7	S/T = 1.0
Shaft speed under gassed condition [rpm]	74 (AGMA = 84)	eqn. 9.9
Shaft speed under ungassed condition [rpm]	66 (AGMA = 56)	eqn. 9.9

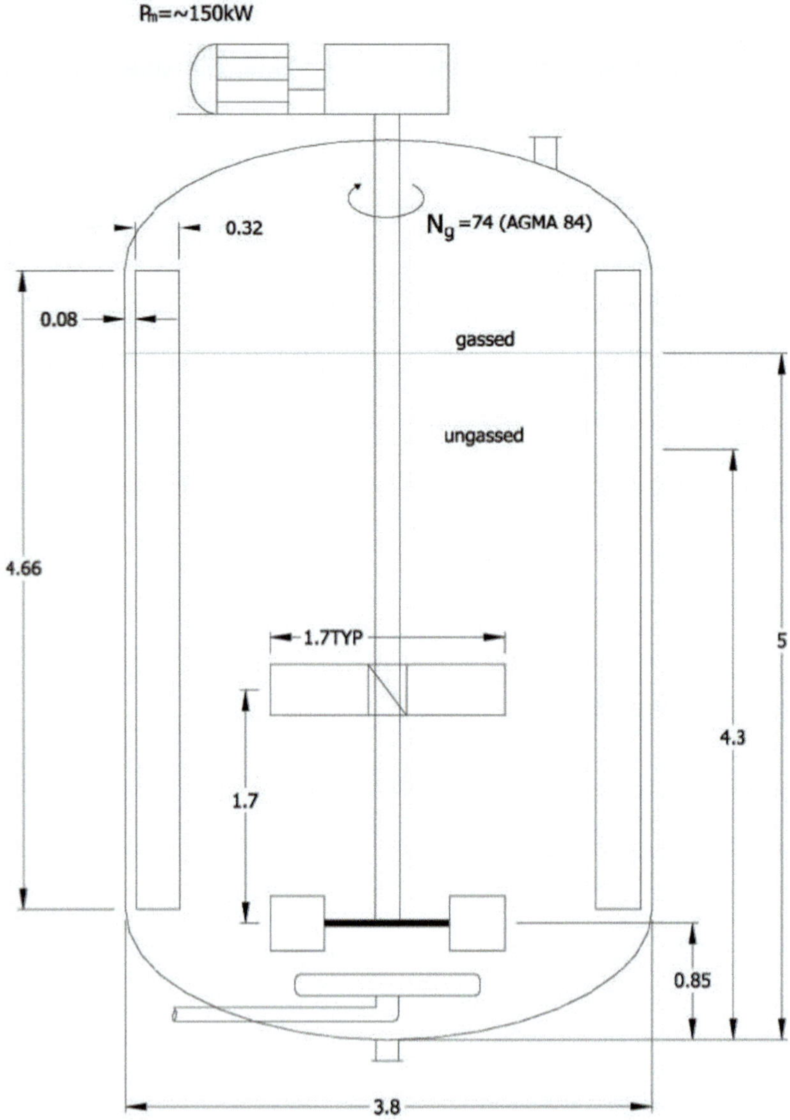

FIGURE 9.9 Schematic of the production stirred tank reactor (dimensions in meters).

9.11 CONCLUSIONS

This chapter briefly presents the mass transfer fundamentals of gas–liquid contacting followed by a series of experimental correlations for the design of STRs. Discussions are intended to familiarize practicing engineers with how to design, optimize, and troubleshoot stirred tank gas–liquid contactors. To illustrate the design methodology, the design of a typical industrial-scale stirred tank oxidation reactor is presented in a worksheet. Practicing engineers are strongly advised to collaborate with the equipment manufacturers for detailed mechanical design aspects of the agitation system components, and their safety such as shaft, gearbox, critical speed etc. Gas–liquid contactors such as bubble columns and pneumatically or jet agitated airlift reactors are discussed in Chapter 10.

9.12 NOMENCLATURE

a	Gas–liquid interfacial area per unit volume [m^{-1}]
c	Impeller off-bottom clearance [m]
C	Concentration of diffusing gas [mol m^{-3}]
C^*	Saturation concentration at the gas–liquid interface [mol m^{-3}]
C_L	Bulk concentration in the liquid phase [mol m^{-3}]
C_g	Transfer gas concentration in the bubble [mol m^{-3}]
k_L	Liquid film mass transfer coefficient [m/s]
D	Impeller diameter [m]
g	Acceleration due to gravity [m s^{-2}]
H	Tank height [m]
H_e	Henry's law constant ($H_e = C^*RT/p_g$)
MW	Molecular weight [g mol^{-1}]
N	Impeller speed [rpm]
p	Pressure [Pa]
P	Power [W]
$\dot{Q}$	Volumetric gas flowrate [m^3 s^{-1}]
R	Universal gas constant (= 8.314) [m^3 Pa mol^{-1} K^{-1}]
S	Impeller spacing [m]
t	Time [s]
T	Temperature at the gas–liquid interface [K]
T	Tank diameter [m]
V	Volume [m^3]
V_{sg}	Superficial gas velocity [m s^{-1}]

Greek Letters

ε	Power per mass [W kg^{-1}]
ε_g	Gas holdup
ρ	Density [kg m^{-3}]
μ	Dynamic viscosity [Pa s]
v	Kinematic viscosity [m^2 s^{-1}]
ζ_r	Viscosity ratio of the liquid phase to that of water

Subscripts

g	Gassed or gas phase
gas	Gas expansion
ug	Ungassed
L	Liquid phase

9.13 ACKNOWLEDGMENT

The authors would like to thank the publisher, Taylor & Francis, New York, for their permission to reproduce the entry "Gas-Liquid Contactors" by Kar, K. K., Cope, R., Lueske, J., and Ed. S. Lee, In *Encyclopedia of Chemical Processing*, Vol. 1., 2005.

REFERENCES

1. Middleton, J., Smith, J., Gas–liquid mixing in turbulent systems. In *Handbook of Industrial Mixing: Science and Practice*, Paul, E. L., Atiemo-Obeng, V. A., Kresta, S. M., (Eds.). John Wiley and Sons, New York, pp. 585–638, 2003.
2. Lee, S., Tsui, Y., Succeed at gas/liquid contacting, *Chem. Eng. Prog.*, pp. 23–49, 1999.

3. www.statista.com/statistics/979265/global-phenol-market-volume/, Global phenol market volume 2015–2026, Published by Lucía Fernández, October 19, 2021.
4. www.statista.com/statistics/1245249/purified-terephthalic-acid-market-volume-worldwide/Global purified terephthalic acid market volume 2015–2026, Published by Lucía Fernández, October 19, 2021.
5. Krishna, R., Sie, S. T., Strategies for multiphase reactor selection, *Chem. Eng. Sci.*, 49, pp. 4029–4065, 1994.
6. Cooke, M., Dawson, M. K., Nienow, A. W., Moody, G. W., Whitton, M. J., Mass transfer in aerated agitated vessels: Assessment of the NEL/Hickman steady state method. In *Proceedings of Seventh European Mixing Conference*, Bruxelmane, M., Froment, G., (Eds.). KVIV: Belgium, pp. 409–418, 1991.
7. Coyne, J., Kaufman, P., Post, T. A., Key parameters for consideration in up-pumping technology in fermentation. In *Proceedings INTERPHEX East*. ISPE, Philadelphia, PA, March 19, 1998.
8. Stenstrom, M. K., Gilbert, R. G., Effects of alpha, beta, theta factors upon the design, specification and operation of aeration systems, *Water Res.*, 15, pp. 643–654, 1981.
9. Van't Reit, K., *Turbine Agitator Hydrodynamics and Dispersion Performance*, PhD thesis, Delft University of Technology. The Netherlands, 1975.
10. Kaufman, P., Post, T. A., Preston, M., Up-pumping mixing technology: Breaking the mold of traditional systems, *Chem. Process*, 61(10), pp. 63–66, 1998.
11. Sardeing, R., Aubin, J., Xuereb, C., Gas-liquid mass transfer: A comparison of down- and up-pumping axial flow impellers with radial impellers, *Chem. Eng. Res. Des.*, 8(12), pp. 1589–1596, 2004.
12. Zhu, H., Nienow, A., Bujalski, W., Simmons, M., Mixing studies in a model aerated bioreactor equipped with an up- or a down-pumping 'Elephant Ear' agitator: Power, hold-up and aerated flow field measurements, *Chem. Eng. Res. Des.*, 87(3), pp. 307–317, March, 2009.
13. Strand, A., Courtemanche, N., *Impact of Agitator Design on Bioreactor Performance*. Pharmaceutical Technology: Biologics and Sterile Drug Manufacturing, pp. 20–24, 2018.
14. Benz, G.T., *Agitator Design for Gas-Liquid Fermenters and Bioreactors*. Wiley, pp. C89–C92, 2021.
15. Gigas, B., Kar, K., Lueske, J., Piras, L., Geiger, T., *Mass Transfer Capability Study of Multiple Up-pumping High-efficiency Hydrofoil Impellers*, Paper 12.7 presented at NAMF Mixing XVIII, Conference, June 24–29, 2001.
16. Bakker, A., Smith, J. M., Myers, K. J., How to disperse gases in liquids, *Chem. Eng.*, 101, pp. 98–104, 1994.
17. Saravanan, K., Mundale, V. D., Joshi, J. B., Gas inducing type of agitated contactors, *Ind. Eng. Chem. Res.*, 33(9), pp. 2226–2241, 1994.

10 Fermenters

10.1 INTRODUCTION

Fermentation and separation processes are vital unit operations in the production of biological products using microbial and mammalian cells in the pharmaceutical, food, and chemical industries. Although primary metabolites leading to such industrial chemicals as alcohols and acids often reach yields close to their theoretical maximum, the fermentation of secondary metabolites such as antibiotics and enzymes yields significantly lower values. In recent years, while microbiologists have focused their research on new strain developments, biochemical engineers have devoted their research to the advancement of fermentation and separation technologies to address productivity and environmental or regulatory issues.

10.2 BACKGROUND

Industrial fermentation processes are either aerobic, where aeration is a must for the growth and production of the microorganisms, or anaerobic, where aeration is irrelevant to the process. For example, in the presence of the Acetobacter species, acetic acid is produced by the oxidation of ethanol at a concentration of up to 15% (wt./vol) [g/100 ml]. Full aeration is required; interruption of the air supply for only 30 seconds can kill acetic-acid-producing bacteria. On the other hand, lactic acid is the end product of dextrose glycolysis using species of the genus Lactobacillus, especially *L. delbrueckii*. The specific rate of lactic acid synthesis is strongly dependent on the media pH but independent of aeration. In this case, adjustment of media normality is essential. Localized low-pH zones seriously hinder the bacterial growth and consequently the yield. Fast dispersion and dissolution of the pH adjusters [alkali—$Ca(OH)_2$] in the glucose media are critical factors for successful production of lactic acid. Because the design of lactic acid (anaerobic) fermenters is akin to that of typical large-scale stirred tank reactors, this chapter will be limited to the design of aerobic fermenters only.

Biochemical engineers are often challenged when scaling up fermentations demonstrated by microbiologists on a laboratory scale. Those skilled in the art have realized the difficulty and ambiguity involved in scaling up multiphase mixing systems from a biochemist's 1-liter flask to a cubic-meter scale. Hence, tests in an intermediate pilot-scale process are usually completed to understand process feasibility and to identify parameters that control yield. Pilot samples of the therapeutics are submitted for pharmacological and toxicological studies during preclinical and clinical trials. Business strategies such as speed to market, early cash flow, and capturing value by extending sales under patent protection often necessitate engineering-only authorization or toll manufacturing while the clinical studies by the regulatory authorities are under way. Irrespective of the approach, biochemical engineers play a crucial role in the design of pilot plants, production fermenters, and a web of associated peripherals.

Multipurpose batch fermenters are typical installations in the toll manufacturing and pharmaceutical industries. The batch fermentation process provides economic and operational advantages because a single fermenter can produce multiple products. Although most fermentation uses fed-batch processes, biological wastewater treatment commonly utilizes large-scale continuous processing to treat a vast amount of water. In this case, a series of configurations of continuous stirred tank fermenters possesses advantages such as sequential treatment of multiple substrates, reasonable residence times for high conversion efficiency under flow-through conditions, and improved volumetric reaction rates due to high substrate concentration.

DOI: 10.1201/9781003606390-10

10.3 TYPES OF FERMENTERS

Mechanically and pneumatically agitated fermenters are commonly used in the bioprocessing industries. Figures 10.1–10.5 show the schematics of these bioreactors, while Table 10.1 shows their specific usage based upon the fermentation criteria that is outlined by Storhas.[1]

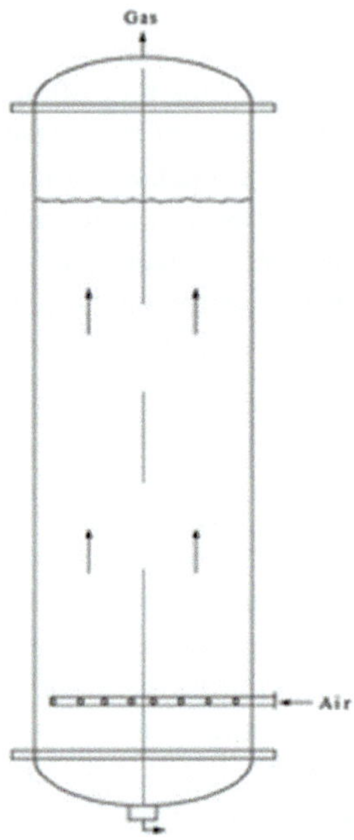

FIGURE 10.1 Bubble column fermenter.

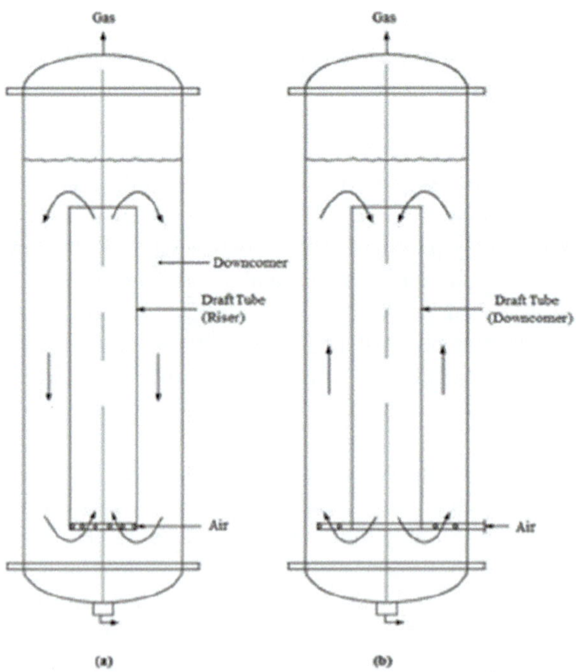

FIGURE 10.2 Concentric tube airlift fermenters: (a) draft-tube sparged and (b) annulus sparged.

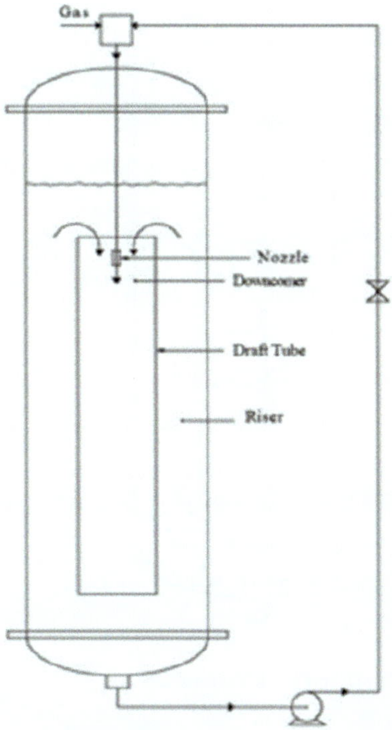

FIGURE 10.3 External loop, jet nozzle airlift fermenter.

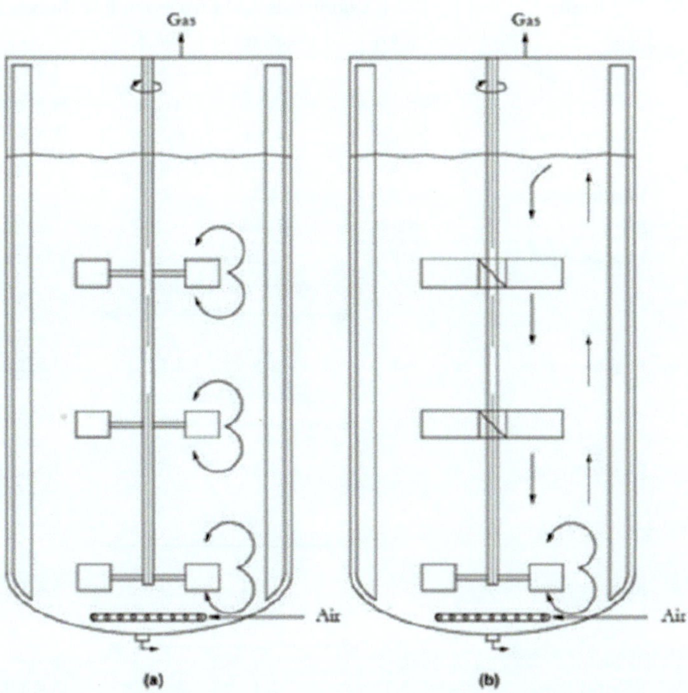

FIGURE 10.4 Stirred tank fermenters: (a) with multiple radial impellers and (b) with axial and radial impellers.

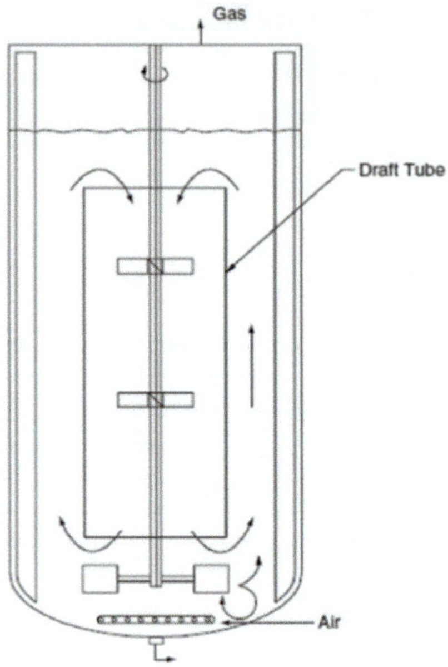

FIGURE 10.5 Draft-tube–impeller stirred tank fermenter.

TABLE 10.1
Critical Matrix for Fermenter Selection[1]

Selection Criteria	Units	Conditions and Corresponding Bioreactor Types					
Media viscosity	Pa-s	> 2.0	<2.0	<0.4	<0.1		
	Reactor types*	4,5	4,5,3	4,5,3,2	5,4,3,2,1		
Ease of pumping		Very poor	Poor	Good	Very good		
	Reactor types	4,5	4,5,3	4,5,3,2	5,4,3,2,1		
Susceptibility to shear		No	Little	Yes			
	Reactor types	4,5,3	4,2	1,2			
Max. reactor size m^3	m^3	>500	<400	<300	<100	<10	<5
	Reactor types	1	4,1	5,4,2	5,4,3,2	5,4,3,2	4,3,2
Media solid content			Very high ⟷	Very low			
	Reactor types	4	4,5	4,5,2,3	4,5,2,3,1		
Tendency to foam			Very strong ⟷	Very weak			
	Reactor types	4	4,5	4,5,2,1	4,5,2,1,3		
Difficulty in achieving media homogeneity			High ⟷	Low			
	Reactor types	5,4	5,4	5,4,3,2	5,4,3,2,1		
Mass and heat transfer			High ⟷	Low			
	Reactor types	5,4	5,4	5,4,3,2	5,4,3,2,1		
Sterilization required (Cleaning-in-process)			High ⟷	Low			

TABLE 10.1 ***(Continued)***
Critical Matrix for Fermenter Selection

	Reactor types	1,2	1,2,3	1,2,3,4	1,2,3,4,5	
Biological safety			High Low ◄────►			
	Reactor types	1,2	1,2,3	1,2,3,4	1,2,3,4,5	
Example case						
Criteria	0.4–2.0 Pa-s	<300m^3	30% solids		High mass transfer	High sterility
Bioreactor types	4,5,3	5,4,2	4,5		5,4	2,1
Preferred bioreactor	4 or 5					

* Refer to the bioreactor types shown in Figures 10.1–10.5.

10.3.1 Mechanically Agitated Fermenters

A stirred tank fermenter consists of a centrally mounted agitation system inside a cylindrical vessel. Typically, the agitation system is composed of either multiple radial flow impellers (see Figure 10.4a) or a combination of radial and axial flow impellers as shown in Figure 10.4b). A gas sparger is located below the radial gas-dispersing impeller. The role of the bottom radial impeller is to break the sparged gas stream into smaller bubbles. Axial, down-pumping impellers establish a synergistic flow pattern with the radial impeller that includes upward flow near the wall and downward flow in the core region of the tank. Multiple radial flow impellers create undesirable compartmentalized flow regions in the tank.

Stirred tanks are commonly used as the most efficient and reliable gas–liquid contactors for mass-transfer-limited aerobic processes in the biochemical industries. More than 90% of today's bioproducts are produced in stirred tank fermenters. Typical applications range from the large-scale production of such industrial enzymes as amylases and lipases, antibiotics or amino acids to the smaller-scale production of value-added specialty recombinant DNA therapeutic proteins and vaccines. The key advantages of stirred tank fermenters are enhanced mass transfer due to intimate gas–liquid turbulent mixing, improved heat removal through external heat transfer jackets and internal heat exchangers, and shear-induced thinning of the pseudoplastic viscous media. Stirred tank fermenters are easily scalable from the pilot to the large industrial scale. However, stirred tanks are not ideal for shear-sensitive microorganisms and cannot be aerated at high rates because of impeller flooding. Poorly designed, tall, stirred tank fermenters may create regions of slow flow or stagnation, which lead to zones of poor oxygen transfer.

For low-viscosity media, the draft-tube–impeller combination in a stirred tank achieves better mass transfer performance than a conventional stirred tank fermenter. A co-current flow of gas and liquid is induced down through the draft tube and exits at its bottom. This flow pattern provides intimate gas–liquid contact and a high gas–liquid mass transfer rate. Consequently, bioreactors can consist of a tank, a draft tube, and an agitation system with axial and radial flow impellers as shown in Figure 10.5. Dual axial flow impellers are located inside the draft tube and pump the fluid down toward the radial flow impeller. The radial gas-dispersing impeller is located above the gas sparger ring. This configuration confines aeration to the annular zone and establishes a highly directional flow pattern in the tank. An optimized design draft-tube–impeller agitation system allows higher aeration rates before the onset of flooding.

10.3.2 Pneumatically Agitated Fermenters

The airlift fermenter relies on air as the principal means of achieving various transport requisites. It is composed of a vertical cylindrical vessel and a concentric draft tube into which air or other suitable oxidation gases are injected (see Figure 10.2). Reduced bulk density causes the contents of the central riser to move upward. This action displaces the contents of the surrounding downcomer and moves them downward to complete the fermenter circulation cycle.

Airlift bioreactors have the advantages of generating liquid mixing and gas transfer without the use of a mechanical agitator. The special characteristics of airlift fermenters make them most suitable for some processes, especially those cultivating shear-sensitive cultures such as mycelial fungi, filamentous bacteria, and certain mammalian cell lines. The system fluid dynamics ensure satisfactory mass and heat transfer together with an even distribution of shear stress. This offers advantages for both high-volume, low-cost processes and low-volume, high-value-added processes like those producing animal and cell cultures. Airlift fermenters are basically simple, low-capital-investment units.

There have been many simple modifications to air-lift bioreactors for specific applications. For example, a novel airlift loop fermenter (schematic unavailable in the literature) utilizes a side arm. The external loop in this integrated system overcomes the problem of ethanol inhibition by continuously stripping ethanol from the fermentation broth and recovering it by condensation. This is suitable for the simultaneous production and recovery of ethanol.[2]

Mechanically stirred hybrid airlift reactors (see Figure 10.6) are well suited for use with shear-sensitive fermentations that require better oxygen transfer and mixing than is provided by a conventional airlift reactor. Use of a low-power axial flow impeller in the downcomer of an airlift bioreactor can substantially enhance liquid circulation rates, mixing, and gas–liquid mass transfer relative to operation without the agitator. This enhancement increases power consumption disproportionately and also adds other disadvantages of a mechanical agitation system.

Figure 10.3 shows a modified airlift fermenter consisting of an external loop where gas is injected into the recycled broth through a gas-dispersing nozzle. In the working media inside the downcomer,

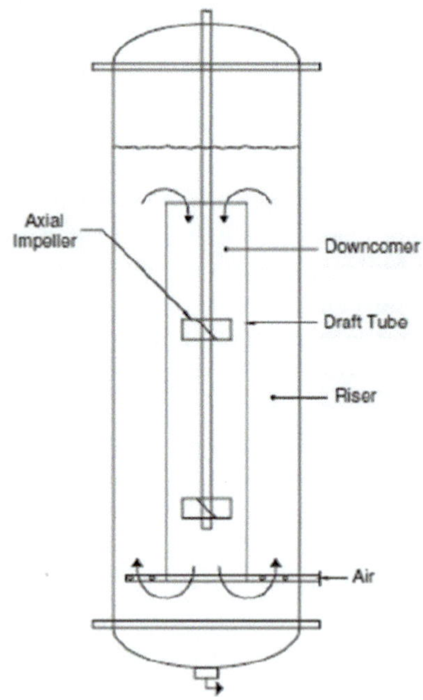

FIGURE 10.6 Mechanically agitated hybrid airlift fermenter.

two-phase discharge from the nozzle creates a jet of microbubbles. Jet-nozzle airlift fermenters provide better mass transfer, but the accompanying high-shear bubble breakup is detrimental to the growth of most microorganisms.

The bubble column in Figure 10.1 is the simplest pneumatically agitated bioreactor. Typical installations provide low-cost oxygen transfer in low-viscosity media where mass transfer is not limiting. Bubble columns are ideal for extremely shear-sensitive microorganisms because their simplicity makes sterilization and maintenance of an aseptic environment simple. Because most bioprocessing applications are known to be mass transfer limited, bubble columns are rarely used for fermentation despite their obvious advantages.

Tubular fluidized and fixed bed fermenters are deviations from the simple bubble column fermenter. Often utilized in producing beer and ciders, these fermenters contain immobilized microorganisms or microbial films on support surfaces. Microbes lost with the product are continuously replenished by adding fresh microorganisms into the packed bed fermenters. In the fixed bed case, slow downward flow of the medium significantly reduces the shear removal (mobilization) of the microbes from the support materials and increases the residence time in the packed column. This is a typical characteristic of the trickle bed fermenter for continuous operation. Readers are referred to the packed bed reactor entry elsewhere for a more comprehensive understanding of packed bed fermenter operations, as they are not discussed here.

10.4 SCOPE

The focus of this entry is limited to the design of aerobic fermenters, i.e., stirred tank and concentric tube airlift fermenters, which are commonly utilized in the bioprocessing industries. Design principles and the basic calculations are described with a couple of industrial examples.

10.5 DESIGN PRINCIPLES

10.5.1 OXYGEN TRANSFER RATE

Gas–liquid mass transfer is commonly modeled in terms of a gas film (between the bulk gas and interface) and a liquid film (between the interface and bulk liquid). Hindrance to mass transfer causes soluble gas (e.g., O_2) concentrations to decrease across these films. The highest mass transfer resistance usually exists in the liquid film; therefore, it controls the overall oxygen transfer rate (OTR). In aerobic fermentation, an effective fermenter design achieves an efficient OTR through intimate gas–liquid contact. OTR is described in terms of oxygen concentration and characteristics of the gas–liquid interface, as follows:

$$OTR = k_L\mathrm{a}\ \Delta c = k_L\mathrm{a}\ (c^* - c_L) \qquad (10.1)$$

In this equation, k_L is the mass transfer coefficient in the liquid film, a is the gas–liquid interfacial area per unit liquid volume, c^* is the oxygen concentration near the gas–liquid interface, and c_L is the oxygen concentration in the bulk liquid.

Henry's law (i.e., $c^* = H\,p_G$) relates the equilibrium oxygen solubility in the liquid (c^*) to oxygen partial pressure in the gas (p_G) by the corresponding Henry's law constant (H). Because the liquid film mass transfer coefficient, k_L, is difficult to measure independently, k_La (i.e., k_L times a) is used. It is a lumped parameter known as the volumetric mass transfer coefficient to characterize the overall mass transfer rate.

Because c^* is a function of hydrostatic head, a log-mean concentration difference, Δc_{lm}, generally replaces the Δc term used in the earlier equation:

$$OTR = k_L\mathrm{a}\ \Delta c_{lm} \qquad (10.2)$$

$$\Delta c_{lm} = \frac{\left(c_1^* - c_L\right) - \left(c_2^* - c_L\right)}{ln\left[\left(c_1^* - c_L\right) / \left(c_2^* - c_L\right)\right]} \tag{10.3}$$

The average logarithmic concentration difference is calculated with the oxygen gas concentrations in the bulk liquid phase and saturation conditions at the inlet (1) and outlet (2). In this application, the value of c_L is commonly assumed to be 0 ($c_L << c^*$).

Several simple theories of interaction at the gas–liquid interface are pertinent to the design of pneumatically and mechanically agitated fermenters, and are now discussed.[3]

The first is the penetration theory of Higbie (1935). If the liquid immediately adjacent to a rising bubble is assumed to rise with the bubble, i.e., the relative velocity between the bubble and the liquid is 0, then the mass transfer conditions are those of non-steady-state molecular diffusion. The mathematical solution of this problem leads to

$$k_L = 2\sqrt{D_L / \pi\theta} \tag{10.4}$$

where k_L is the liquid film mass transfer coefficient over the bubble lifetime (θ), and D_L is the coefficient of molecular diffusion.

The second theory is that of surface renewal by Danckwerts (1951). If turbulent mixing occurs in liquid near a bubble, then it is likely that unsteady molecular diffusion will take place in individual eddies. In these conditions, mathematical analysis leads to

$$k_L \propto \sqrt{D_L s} \tag{10.5}$$

where S is the rate of surface renewal, i.e., the rate at which eddies sweep the interfacial boundary layer. Because the surface renewal rate incorporates the eddy lifetime (θ), it increases with increased turbulence intensity, just as θ does in Higbie's penetration theory.

Because of the absence of mechanical agitators, airlift fermenters are generally unable to produce the turbulence necessary for high mass transfer rates. Superficial gas velocity and liquid circulation rate in the risers determine the relative velocities of rising bubbles and adjacent liquid and, in turn, the volumetric mass transfer coefficient, k_La. Increasing the gassing rate increases the liquid circulation rate indirectly by increasing hydrostatic pressure on top of the downcomer. Decreasing downcomer pressure drop by increasing the ratio of downcomer to riser cross-sectional area also increases the circulation rate. Unfortunately, other challenges arise if higher liquid circulation rates shift liquid conditions from bubbly flow to churn flow. Although this transition contributes to turbulence and bubble surface transience, bubble surface renewal also increases coalescence and bubble streaming along the wall, thus reducing specific interfacial area, a, and k_La. Particularly in concentric draft-tube airlift fermenters, designers must maintain bubbly flow in the riser by optimizing the gassing rate and liquid circulation rate. Typically, industrial fermenters have an a_r/a_d ratio of 0.8–1.2 (preferably $d_t/d_c = 0.7$) and a superficial gas velocity, u_g, of 0.02–0.10 m/s.

The gas–liquid interfacial area per unit volume, a, increases as bubble size decreases. Hence, the design and location of gas spargers (aerators) are critical to the success of aerobic fermenters, particularly those with pneumatic agitation. Ring spargers, rake spargers, and porous plates are used as aerators in airlift fermenters. In the case of stirred tank fermenters, energy dissipation from the gas-dispersing radial impellers normally controls bubble size. Ring spargers are typical aerators in mechanically agitated fermenters. Irrespective of the sparger holes, bubble size distribution depends on the media rheology, interfacial tension, and turbulent intensity. Increased broth viscosity reduces diffusivity, increases bubble size (i.e., under turbulent conditions Sauter mean bubble diameter, d_{b32}, is proportional to $\mu_L^{0.1}$), and decreases bubble-specific surface area, a.[4] Therefore, it significantly decreases k_La. Generally, k_La decreases with decreases in pressure and temperature and with increases in gas–liquid interfacial tension and liquid viscosity.

Designing fermenters that produce high yields with such viscous media as polysaccharides is a challenge. In batch fermenters, for example, agitation initially disperses the sparged gas as tiny, well-distributed bubbles. Viscous media stabilize these bubbles and extend their lifetime beyond that required for oxygen mass transfer. The extended lifetime of these bubbles hinders subsequent oxygen transfer, and thus fermenter productivity, by dampening agitation and its ability to continue dispersing fresh sparged gas. The resulting mass transfer coefficient falls off dramatically with continued batch time. A much higher energy input is needed to generate fine bubbles from the sparged gas and prevent them from coalescing.

10.5.2 Microbial Metabolic Heat

In some cases, heat added to the initial broth promotes the desired microbial activity. During metabolism, however, thermodynamics of the overall microbial activity generates heat in such quantities that, if not removed, it would raise broth temperature beyond the system optimum. Elevated temperatures affect the fermenter by producing everything from a slight productivity decrease to widespread microbe death.

Metabolic heat evolution is determined from small-scale laboratory experiments and used for the appropriate design of heat transfer elements. When data are not available, heat evolved during bacterial fermentation can be estimated using the following correlation:[5]

$$\dot{Q}_f \approx 0.12 \times OUR \tag{10.6}$$

where $\dot{Q}_f$ is heat evolution [kcal/L/hr] and OUR is the oxygen uptake rate [mmol O_2/L/hr].

Addition and removal of heat is achieved by such means as a fermenter jacket, internal heat transfer coils (horizontal or vertical), and/or an external heat exchanger. A fermenter jacket provides insufficient heat removal by itself, especially in large microbial fermenters. Internal cooling parts may be particularly necessary in slender tanks despite their impedance to cleaning, sterility, and intermixing. Appropriately designed vertical heat transfer coils often serve as baffles in mechanically agitated fermenters, thus minimizing extraneous structures inside the vessel. Sterilization difficulties and asepsis also make external heat exchangers with multiphase pumping a last resort for heat removal. When an external heat exchanger is implemented, its design and materials of construction must allow it to withstand thermal stress cycling and chemical cleaning.

In the case of stirred tank fermenters, heat that must be dissipated includes not only that generated by microbial metabolic activity but also that evolved from agitation power (i.e., 2500 Btu/shp-hr), expansion of sparged gas, and excess heat in the liquid or gaseous feeds. This can lead to scale-up problems because vessel volume is proportional to vessel diameter cubed, while heat transfer area is proportional only to vessel diameter squared.

10.5.3 Sterilization and Asepsis

For most aerobic fermentation processes, maintenance of absolute sterility is critical to cell growth and biomass production. Culture growth rate determines susceptibility to culture contamination. Mammalian cells divide in a day, whereas microbials divide in an hour or less. The difference in growth rate makes slower-growing mammalian cell cultures more susceptible to contamination than faster-growing bacteria or yeast cultures, which can out-compete contaminating microorganisms.

Substrates that include such heat-labile nutrients as vitamins or enzymes are often added to the media as sterilizing agents and growth enhancers.[6] Prior to fermenter operation, standard operating practice usually includes a caustic wash and/or superheated steam cleaning of the vessel and gas sparger.

Typical sources of contamination include the steady bearing, mechanical seal, sparger, and fasteners. Because bushing interfaces and the shaft wear-sleeve of a steady bearing are prone to biomass deposition, their design is given special attention. Use of steam-purged or dry seals minimizes biomass contamination. Type 316L stainless steel is the most common construction material for bioreactor applications. Because micro-crevices are known to harbor bacteria and contaminants, the surface of the tank walls and internals are electro-polished to a minimum surface finish of 0.6μm (average roughness). To prevent the influx of air, which can compromise aseptic conditions, fermenters normally operate with a positive internal pressure. A high level of positive pressure also has the advantage of increased oxygen partial pressure and, consequently, mass transfer rate.

10.5.4 Turbulence and Shear Tolerance

Prior knowledge of cell sensitivity to shear and turbulence is critical in designing mechanically agitated fermenters. Morphology dictates the tolerance of microorganisms to turbulence intensity and shear stress, although most microorganisms are too small (typical sizes: bacteria 1–5 μm) to be destroyed by turbulent microscale eddies or local shear forces. Secondary metabolites with spherical, ellipsoidal, and rod-like structures are not susceptible to mechanical degradation, whereas filamentous microbes such as *Spinosad mycelia* and fungi are degraded by shear. Mammalian cells, because of their lack of cell wall and their larger size (> 5 vs. 1 μm), are more shear sensitive than microbial cells. Hydrodynamic shear stresses or shear rates also affect the production of extracellular microbial polysaccharides. During pilot-scale fermentation, potential cell damage due to high specific energy must be considered. Intense shear fields can disrupt even the most robust microorganisms, but the likelihood of cell damage can be safely ignored if the Kolmogoroff length of a micro vortex is greater than the effective size of the microbe in question.

According to Kolmogoroff, the size of the micro-vortex, λ', can be determined as follows:

$$\lambda' = \left(\frac{\nu_L^3}{\varepsilon_{local}} \right)^{\frac{1}{4}} \quad (10.7)$$

where ν_L is liquid kinematic viscosity [m^2/s], and ε_{local} is local power per mass [W/kg].

10.5.5 Media Rheology

Flow properties of biofluids and slurries fall into two categories:

1. Newtonian fluids such as water, and most bacterial and yeast fermentation broths
2. Non-Newtonian media such as polysaccharide fermentations and broths of *Streptomyces*, *Aspergilli*, and *Penicillia*

The viscosity of Newtonian suspensions of yeast and bacteria growing as individual cells in a water-like medium can be estimated from the following empirical equation proposed by Thomas:[7]

$$\mu_{susp} = \mu_L \left(1 + 2.5\, c_s + 10.05\, c_s^2 + 0.00273 e^{16.6 c_s}\right) \quad (10.8)$$

where, c_s is the volume fraction of solids.

While working media can be Newtonian or non-Newtonian, most high-viscosity non-glucose-based broths containing microorganisms demonstrate pseudoplastic or shear thinning behavior. For fungal suspensions (2–10 kg/m^3 dry biomass) such as *Aspergillus niger*, in the power law model, $\tau = k\,\lambda^n$, the flow behavior index n varies from about 0.6 to 0.16, while the consistency index, k, increases from 0.17 Pa-s$^{0.6}$ to 9.1 Pa-s$^{0.16}$.[8] While *Acetobacter* fermentation is highly aerobic, the fermentation broth itself is viscous, even viscoelastic in certain instances. Although the fluid's turbulence does not seem to significantly affect the biomass or its productivity, the quality of the polymer produced is distinctly shear stress dependent. To ensure sufficiently long polymer microfibrils, the shear must be kept low.[9]

For high-viscosity applications, lack of turbulence adjacent to the bubble, and the general difficulty in pumping around the draft tube (riser), prevent the use of simple pneumatically agitated airlift fermenters. Stirred tank fermenters are typically used for media with a viscosity exceeding 100cP. As noted, an increase in viscosity results in a substantial decrease in the mass transfer coefficient, all other parameters being equal.

Figure 10.7, a graph of k_La vs. apparent viscosity for a penicillin slurry, illustrates this trend.[10] To illustrate further, achieving a reasonable mass transfer coefficient (>0.1 s^{-1}) typically requires two to three times more specific power in a high-viscosity broth (>1000cP) than in low-viscosity (<100cP) media. In viscous broth fermentation, yield is often limited by oxygen starvation.

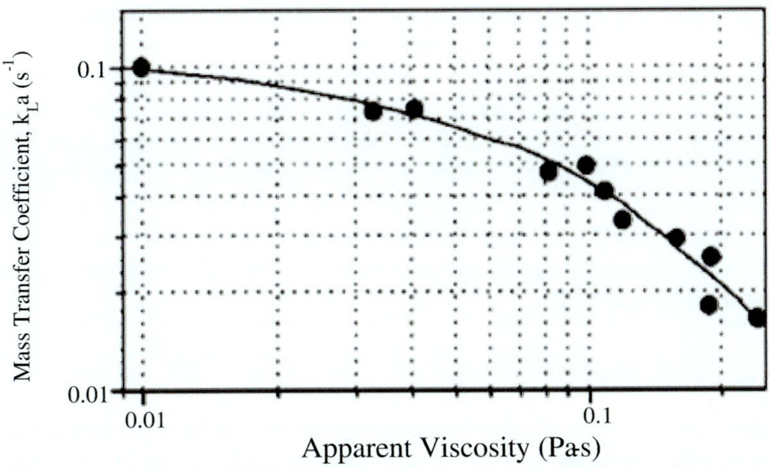

FIGURE 10.7 Mass transfer coefficient (k_La) as a function of apparent viscosity of penicillin broth at a specific energy of 3 W/kg.[10]

10.5.6 Blending and Homogeneity

Uniformly distributing air bubbles, nutrients, and microbes within a broth are of paramount importance for the successful cultivation of microorganisms. Airlift fermenters are ideal for achieving homogeneity in broths with a low solids content, while mechanically agitated fermenters are suitable for high solids (slurry) broths. Despite the possibility of localized non-homogeneity due to slow flow regions in stirred tanks, mechanically agitated fermenters are preferred over pneumatically agitated ones. An appropriately designed draft-tube–impeller agitation system (shown in Figure 10.5) overcomes the uniform blending deficiency of the standard stirred tank fermenter. Configuration and location of tank internals such as agitators, baffles, feed pipes, and heat transfer coils/plates are critical for blending and efficient heat transfer. Fermentation yield depends strongly on the homogeneity of pH and temperature in the working media.

Fed batch fermentation is widely practiced in the industry. Substrates including nutrients such as glucose are continuously fed through dip pipes as the fermentation proceeds. For instantaneous or fast dispersion, these ingredients must be added in high-turbulence regions in the vessel. It must be noted that a local surplus of glucose can cause excessive biomass or by-products to be produced, which can negatively impact the yields of the desired products, e.g., antibiotics. In this regard, selection of the broth, its feed point, and the feeding rate strategy are important.

10.5.7 Gas Holdup and Foam Formation

In vessel design work, size and geometry must account for the fact that aerobic fermenters experience significant gas holdup owing to finely dispersed bubbles in the broth. Gas holdup can be in the range of 10–30% depending on the media viscosity and the presence of surface-active substrates. Because of much higher turbulence intensity, mechanically agitated fermenters have a much higher gas holdup than do pneumatically agitated fermenters. As with the mass transfer coefficient, gas holdup in a given media depends on the specific power input, ε and the superficial gas velocity, u_g.

From the mass transfer point of view, airlift fermenters should be designed and operated in such a way that carryover of air from the riser into the downcomer is kept as low as possible. Gas in the downcomer liquid contributes little to oxygen transfer. It reduces the effective density difference between the contents of the riser and downcomer, however, which reduces the liquid circulation rate and also impairs the mixing performance of the fermenter. In fermentations where even a momentary lack of oxygen can very seriously affect productivity, some air is essential to maintain aerobic conditions and sustain fermentation in the downcomer.

Foaming is a major concern in aerobic cell cultures because excessive foaming significantly reduces broth volume and impedes bioreactor productivity. Normally, silicon-based or oil-based antifoam agents (e.g., corn oil, silicone antifoam emulsions such as Momentive SAG™ 471, Dow Inc. XIAMETER™ AFE-1510 etc.) are added to suppress foaming. In choosing a foam inhibitor, one must be aware of possible adverse impact on the cells, products, or downstream processes and on pertinent regulatory (Food and Drug Administration) and/or kosher certification requirements. High-molecular-weight antifoam agents, when used in the low-viscosity broth, typically reduce the gas–liquid mass transfer coefficient.

10.5.8 Biological Safety

In the fermentation plant, the vent gas is commonly controlled by a backpressure regulator that maintains positive pressure in the bioreactor. The vent is usually piped to a scrubber, which scavenges microbes from the fermenter off-gas. In addition to the scrubber, fermenters are equipped with an overhead safety relief device for potential loss of containment. With the advent of genetic engineering and the targeting of proteins from mammalian cells, biological safety standards have been extended to overcome unforeseen health issues. In this case, all U.S. large-scale bioreactors must comply with the Good Large-Scale Practice (GLSP) guidelines set forth by the U.S. National Institutes of Health (NIH). When using an organism covered by Biological Safety Level (BSL) classifications, BSL-1 or BSL-2, it is necessary to ensure that the cells are contained within the bioreactor's hygienic envelope. BSL-1 containment commonly requires inactivation of the organism by means of heat or chemicals (pH kill) in a separate tank.[5,6]

10.6 DESIGN GUIDELINES

The design basis is the most important consideration when determining the size of a fermenter. Factors to consider are the final product, the growth rate and oxygen requirement of the microorganism, the product expression concentration or titer, and the nature of the product expression.

Choosing an optimal bioreactor configuration for a given fermentation depends on a number of factors including oxygen transfer and mixing requirements, and the acceptable range of shear rates. Some of these factors are mutually contradictory. For example, while a given fermentation may require a very high gas–liquid OTR, which is generally increased by increasing turbulence and shear, its tolerance to such shear may be low. In such cases, successful bioreactor design obviously involves a careful balance of these contradictory factors.

The two sections that follow describe the designs of a stirred tank fermenter and a concentric tube airlift fermenter. Each design is for a given range of operating variables and includes the selection of geometric parameters. Important correlations and formulae are presented and followed by an example design case. The design process actually requires multiple iterations to simultaneously satisfy both geometric and mass transfer requirements. Such iteration has become almost trivial, however, with the advent of such tools as spreadsheet-based "goal seek" computations.

10.6.1 Stirred Tank Fermenter

This, the first of two fermenter design examples, deals with a stirred tank fermenter. Non-dimensional parameters that are applied in this example are defined as follows.

Non-dimensional parameters

N_P	power number, $P/(\rho N^3 D^5)$	Φ	gas holdup, $V_g/(V_g + V_L)$
N_a	Aeration number, V_g/ND^3	μ_{sr}	suspension viscosity ratio, μ_{susp}/μ_w
Fr	Froude number, N^2D/g	Nu	Nusselt number, $\alpha_w T/\lambda$
K	k-factor, P_g/P_{ug}	Re	Reynolds number, $N_g D^2/v_L$
R	diameter ratio, D/T	Re_o	orifice Reynolds number, $u_o d_o/v_g$
S	impeller spacing ratio, s/D	Pr	Prandtl number, $\mu c_p/\lambda_w$
C	impeller bottom clearance ratio, c_{btm}/D		

Mass transfer: The relationship between the volumetric mass transfer coefficient, specific aerated shaft power, superficial gas velocity, and suspension viscosity ratio is given as follows:

$$k_L a_{20} = \alpha \varepsilon_g^{\beta} u_g^{\gamma} \left(\frac{\mu_{susp}}{\mu_w} \right)^{\kappa} \tag{10.9}$$

The following model parameters are specific to proprietary glucose-based fermentation media that were correlated with the above $k_L a$ model for $1.5 < \varepsilon_g < 3.5$ W/kg:

$$\alpha = 0.258,\ \beta = 0.562,\ \gamma = 0.232 \text{ and } \kappa = -0.333$$

Using equation 10.8, the viscosity of the suspension is determined from the broth viscosity and volumetric concentration of solids.

The $k_L a$ value determined with the above model is valid at a constant temperature of 20°C. The mass transfer coefficient at actual operating temperatures is given by

$$k_L a = k_L a_{20} \left(\frac{1}{1.024^{(20-t)}} \right) \tag{10.10}$$

Stirrer power under gassed conditions is as follows

$$P_g = \varepsilon_g m_{broth} = \varepsilon_g \rho_{broth} V_{broth} \tag{10.11}$$

Gas holdup: Similar to the mass transfer model, gas holdup may also be expressed as a function of the specific power, ε_g, the superficial gas velocity, u_g, and the viscosity ratio. Model parameters (α, β, γ, and κ) obtained from the gas holdup expression are neither related to nor interchangeable with those obtained in the mass transfer model:

$$\Phi_g = \alpha\, \varepsilon_g^{\beta}\, \bar{u}_g^{\gamma} \left(\frac{\mu_s}{\mu_w} \right)^{\kappa} \tag{10.12}$$

The following parameters are specific to the broth used in fitting the above gas holdup model. In this example, these parameters have the following values:

$$\alpha = 0.500,\ \beta = 0.143,\ \gamma = 0.197 \text{ and } \kappa = -0.333$$

The k-factor: A common way to determine the agitator power under gassed conditions is to use the k-factor, sometimes called the relative power demand, defined as the ratio of gassed to ungassed shaft power.

$$\text{k-factor: } K \equiv \frac{P_g}{P_{ug}} \tag{10.13}$$

For a stirred system with multiple impellers of different types, the overall k-factor can be experimentally determined as in Figure 10.8 (k-factor vs. aeration number at different Reynolds numbers), or it can be approximated by adding weighted k-factors for the individual impellers, as in equation 10.14.

$$K = \frac{N_{P_R}}{N_P} K_R + \frac{N_{P_A}}{N_P} K_A \tag{10.14}$$

The first term on the right represents the radial, bottom impeller and the second term on the right applies to axial, down-flow impeller(s). The following relationships are used in the design:[11]

$$K_R = 1 - (b' - a'\mu_{b'}) Fr_R^{d'} \tanh(c' Na_R) \tag{10.15}$$

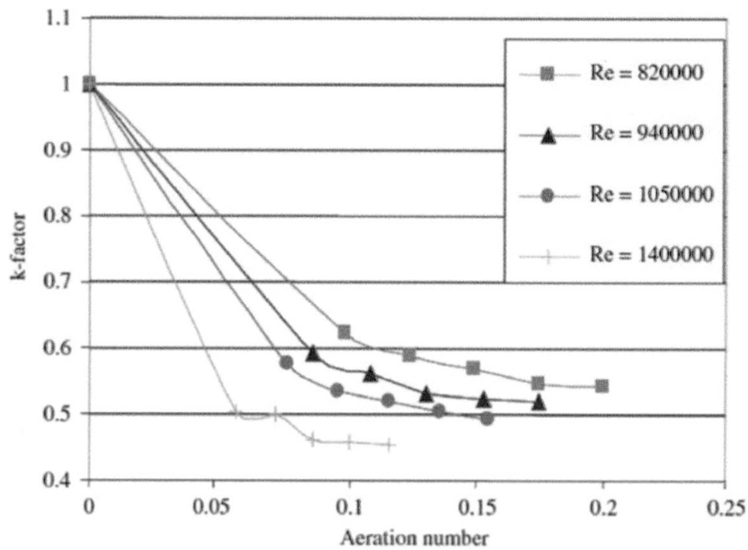

FIGURE 10.8 k-Factor of a multiple impeller system (Rushton + PBT) as a function of aeration number.

Values of the constant parameters a', b', c', and d', listed in Table 10.2, depend on the radial impeller type.

TABLE 10.2
Constants for Equation 10.15

Blade Type	a′	b′	c′	d″
Flat	0.72	0.72	24	0.25
Concave	0.12	0.44	12	0.37

$$K_A = 1 - \left(a'' + b'' Fr_A\right) Na^{c'' + 0.4 Fr_A} \tag{10.16}$$

The parameters a'', b'', and c'' in the K_A relationship are not constant, but depend on the impeller to tank diameter ratio D/T as follows:

$$a'' = 5.3 e^{-5.4(D_A/T)},\ b'' = 0.47\left(D_A/T\right)^{1.3},\ c'' = 0.64 - 1.1\left(D_A/T\right)$$

Equations 10.15 and 10.16 are valid for the following ranges of D/T ratio, Froude number, and Aeration number:

$$0.40 < D/T < 0.65,\ 0.50 < Fr < 2.0,\ 0.05 < Na < 0.35$$

$$P_{ug} = \frac{1}{K} P_g \tag{10.17}$$

Motor and gear box design normally assumes an 80% drive system efficiency ($\eta_m = 0.8$).

$$P_m = \frac{1}{\eta_m} P_{ug} \tag{10.18}$$

At a fixed agitation rate, the gassed power, P_g, is generally less than the ungassed power, P_{ug}. Consequently, the ungassed agitation rate is typically lower than the gassed agitation rate, i.e., $N_{ug} < N_g$. If gas feed suddenly stopped while operation continued at the gassed agitation rate, then the mixer drive system would likely be overloaded and damaged.

Gas flooding: A stirred tank fermenter has power introduced through both the agitation system and the expanding gas. With the assumption of fast temperature equalization, an isothermal expansion of the gas (air) takes place. The gas power is:

$$P_g = \dot{m}_{gas} \frac{RT}{M_{gas}} ln\left(\frac{p_1}{p_2}\right) \tag{10.19}$$

where R is the gas constant (8.314 kJ/kmol-K), T is the absolute temperature at operating conditions, and M_{gas} is the molecular weight of gas. The pressure ratio is given at inlet (1) and outlet (2) tank locations. For high aeration values, gas expansion power, P_{gas}, can reach the magnitude of the aerated agitator power, P_g. If gas expansion power is greater than or equal to the stirrer power, there is the likelihood of impeller flooding, which limits gas dispersion to the central region of the tank.

To avoid potential flooding, it is recommended that

$$\left(P_g / P_{gas}\right)_{design} \geq 3 \tag{10.20}$$

Design example: The design of a stirred tank fermenter for the production of an industrial enzyme at an annual rate of 800 ± 5% mt/yr. is illustrated as follows. Product recovery efficiency is 80%, and the expected yield (product concentration) is 75kg/m^3. Maximum oxygen uptake demand is 185 mmol O_2/L/hr. Operational parameters and media physical properties are listed as follows:

Batch length – 3 days
Fermenter turnaround rate = 2 days
Number of operating days per year = 320
Annual fermenter output = required production/recovery efficiency = 800 mt/yr./0.8 = 103 mt/yr.
Required batch = (annual output / titer) × (total batch duration / year) = (106/75) (5/320) = 208 m^3
Number of bioreactors = 2
The required batch size per fermenter = 208/2 = 104 m^3
Assuming a tank volume utilization of 80% the size of the bioreactor = 130 m^3
The nominal tank size is 140,000 L

Broth properties of Newtonian fluids:
Effective glucose broth (liquid) viscosity = 5×10^{-3} Pa-s
Effective media (suspension) viscosity = 15.3×10^{-3} Pa-s
Density = 993 kg/m^3
Interfacial tension (gas–liquid) = 7.1×10^{-2} N/m
Thermal conductivity = 0.617 W/m-K
Specific heat capacity = 4.118 kJ/kg-K
O_2 broth diffusion coefficient = 4.3×10^{-9}m^2/s

Additional conditions for fermentation and heat removal, as determined at pilot scale, are presented in Table 10.3. State-of-the-art aerobic fermenter agitation system design recommendations are shown in Table 10.4.

TABLE 10.3
Production Fermenter Design Parameter

Batch operation parameters	
Temperature [°C]	32
Head space pressure [bar (gage)]	0.689
VVM i.e., v_g/v_L (standard conditions) [1/min]	1.3
Solid concentration (vol%)	30
Respiration ratio (mol CO_2/mol O_2)	0.8
Heat transfer conditions	
Reaction heat liberation [kJ/mol O_2]	500
Chilled water supply temperature [°C]	12
Max. chilled water flow [m^3/h]	650
Thermal conductivity [W/m-K]	16
Fouling coefficient on product side [m^2-K/W]	0

TABLE 10.4
Agitator Design Recommendations

Impeller type	Radial	Axial
Description	Gas-dispersing disk turbine	Pitched-blade turbine
Number of impellers	1	2 (0–3)
Impeller diameter, *D/T*	0.4–0.6	0.4–0.6
Impeller spacing, *s/D*	—	1.0–1.5
Impeller bottom clearance, c_B/T	0.25	—
Material of construction	SS-316	

Based on required fermentation batch size and agitator design recommendations in Table 10.4, fermenter dimensions and internals are summarized in Table 10.5 and Table 10.6, respectively. Drive system specifications for satisfactory OTR are calculated using pertinent design equations and correlations (see Table 10.7).

TABLE 10.5
Fermentation Tank Data

Material of construction	SS-316
Outside diameter [m]	4.30
Height (cylindrical section) [m]	8.25
Wall thickness [m]	0.025
Bottom head shape	2:1 semi-elliptical
Total fill volume [m^3]	130
Nominal tank volume [L]	140,000
Number of baffles (flat)	4
Baffle width [m]	0.43
Baffle clearance to tank wall [m]	0.09
Baffle height [m]	7.425

TABLE 10.6
Agitation System Design Data

Number of radial flow impellers (bottom)	1
Impeller type	Smith turbine
Diameter [m]	2.15
D/T ratio [-]	0.50
Number of axial flow impellers second to fourth	2.00
Impeller type	Hydrofoil A-320
Diameter [m]	2.15
D/T ratio [-]	0.50
Distance between bottom and middle impellers [m]	2.58
Distance between middle and top impellers [m]	2.58
Distance between top impeller and ungassed liquid [m]	1.30
Distance between top impeller and gassed liquid [m]	2.45
Critical speed (Ncr > 1.25Ng)	Contact manufacturer
Shaft diameter	Contact manufacturer

TABLE 10.7
Results of Stirred Tank Fermenter Design Calculations

Variable	Equation	Value
Required k_La [1/s]	10.2	0.12
VVM [1/min]		1.30
Standard volume flow [m^3/hr.]		8124
Superficial gas velocity [m/s]		0.072
Required aerated power per mass [W/kg]	10.9	2.3
Aerated power [kW]	10.11	238
k-Factor (P_g/P_{ug})	10.14–16	0.75
Ungassed power [kW]	10.17	320
Shaft power (aerated power) [kW]	10.11	238
Gassed speed (from ungassed power) [rpm]		69/AGMA-68
Ungassed speed (from shaft power) [rpm]		63/AGMA-56
Motor power [kW]	10.18	400
Gas expansion power [kW]	10.19	80
Shaft power/gas power (check)	10.20	>3.0
Aeration number		0.091
Gas holdup [%]	10.12	13.80
Ungassed height [m]		7.55
Gassed height [m]		8.70

Figure 10.9 shows a schematic of the designed production fermenter. Details of the mechanical design are found in the publication[12] by Charles and Wilson.

Gas sparger design: Some basic guidelines for sparger design are presented in the "Gas–Liquid Contactor" chapter. The recommended sparge ring to impeller diameter ratio is 0.65–0.8, while the ring sparger orifice diameter is in the range of 2–10 mm, and the gas velocity through the orifice is between 15 and 30 m/s.[13]

Pressure drop across a perforated disperser or orifice is estimated by the conventional orifice equation:

$$\Delta P_o = \rho(v_o/c_v)^2/2 \tag{10.21}$$

Equation 10.21 neglects velocity of approach to the orifice. Typically, the coefficient of velocity, c_v, is assumed to be 0.8.

To ensure uniform flow through all of the orifices or holes, rules of thumb for the gas sparger design call for satisfaction of the following criteria.[14] First, the pressure drop ratio between the total pressure loss along the pipe, ΔP_p, and the pressure loss across each hole, ΔP_o, should be:

$$\Delta P_p / \Delta P_o < 0.1 \tag{10.22}$$

$$\Delta P_p = [(4 f L_p / 3 d_p) - 1] (\rho u^2_{in} / 2) \tag{10.23}$$

with f being the Fanning friction factor.

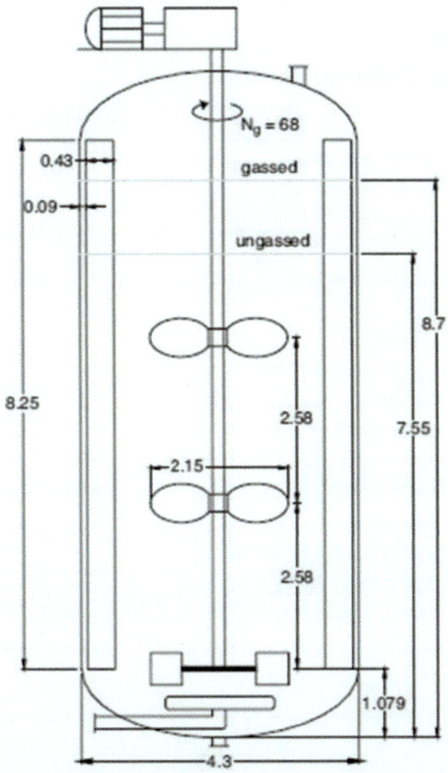

FIGURE 10.9 Schematic of the production fermenter (dimensions in meters).

Second, the ratio of gas kinetic energy at the pipe inlet, and the pressure drop across each hole should be

$$KE_{in} / \Delta P_o < 0.1 \tag{10.24}$$

$$KE_{in} = 1.05(\rho u^2_{in} / 2) \tag{10.25}$$

The ring sparger design is summarized in Table 10.8. The sparger schematic is shown in Figure 10.10.

Heat transfer calculations: The first step in the design calculations is to determine the amount of heat that has to be removed to maintain isothermal conditions in the tank. As mentioned earlier, a gas–liquid stirred tank bioreactor has three sources for heat generation: (1) the gas expansion power, (2) the agitation power under gassed conditions, and (3) the heat liberated by the bioreaction (assuming that gas entering the system is already at system temperature).

The amount of heat to be removed can be expressed as

$$\dot{Q} = P_{gas} + P_g + \dot{Q}_f \tag{10.26}$$

TABLE 10.8
Gas Sparger Design Results

Variable	Equation	Value
Diameter of sparge ring [m]		1.7
Sparge pipe inside diameter [m]		0.20
Sparger-to-bottom impeller diameter ratio		0.79
Sparger bottom clearance [m]		0.55
Number of holes		570
Hole diameter [mm]		8.00
Velocity through each orifice [m/s]		36.55
Number of drain holes		16
Size of drain holes [mm]		16
Pressure drop per orifice [Pa]	10.21	4500
Pressure drop in the pipe ($f = 0.005$) [Pa]	10.23	127
Sparge pipe velocity [m/s]		16.03
Sparge feed pipe inside diameter [m]		0.33
Sparge feed pipe velocity [m/s]		11.86
Pressure drop ratio, pipe to hole	10.22 (check)	0.03

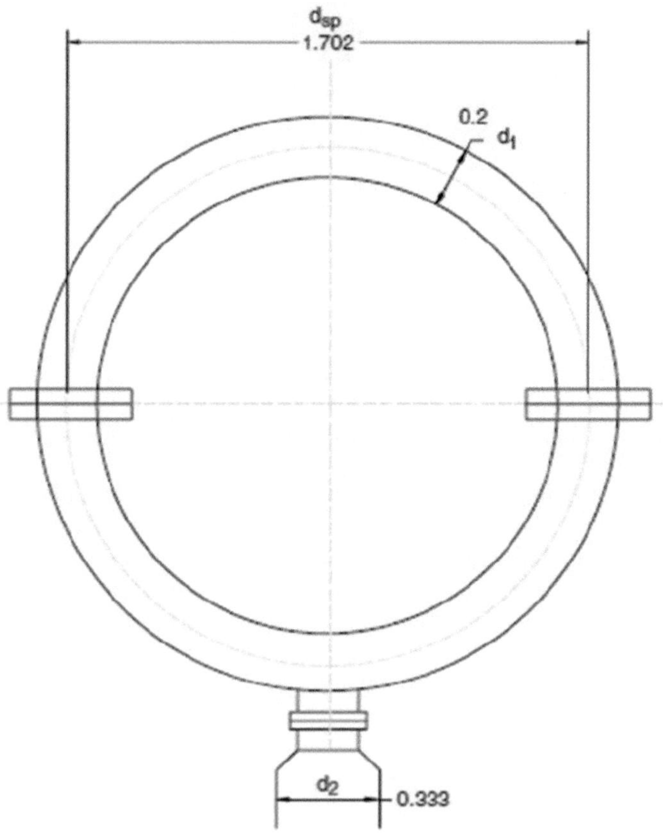

FIGURE 10.10 Schematic of sparger ring (dimensions in meters).

The last term, $\dot{Q}_f$, is the heat of formation in the fermenter. It is directly proportional to the OUR and normally much higher than the gas expansion and agitation power. For enzyme production, the relationship between heat of formation and oxygen demand, $\dot{n}_{O_2}$, is:

$$\dot{Q}_f[kW] \approx 500\,\Delta\dot{n}_{o_2}\ [\text{kmol/s}] \tag{10.27}$$

Whether the heat duty can be removed depends on the heat resistances discussed in the following section. The heat duty is described as

$$\dot{Q} = U\,A\,\Delta T_{lm} \tag{10.28}$$

U is the overall heat transfer coefficient, A is the heat transfer area, and ΔT_{lm} is the logarithmic mean temperature difference defined as

$$\Delta T_{lm} = \frac{(\Delta T_1 - \Delta T_2)}{ln(\Delta T_1 / \Delta T_2)} \tag{10.29}$$

ΔT_1 is inlet and ΔT_2 is outlet temperature differences, respectively, of broth and utility fluid.

The overall heat transfer coefficient, U, is determined as follows:

$$\frac{1}{U} = \frac{1}{\alpha_p} + f_p + \frac{\delta_w}{\lambda_w}\frac{A}{A_w} + f_j\frac{A}{A_j} + \frac{1}{\alpha_j}\frac{A}{A_j} \tag{10.30}$$

where, α_w is the wall thickness, α_p and A_p are the heat transfer coefficient and area of the process side, and α_j and A_j are the heat transfer coefficient and area of the jacket side, respectively. Heat transfer calculations for large-scale fermenters having multiple impellers must combine separate correlations for each impeller, with appropriate weighting, since experimental correlation developers likely used Newtonian liquids and a single, standard impeller. For example, the following ungassed liquid phase Nusselt number correlation for heat transfer at the tank wall, in different vessel geometries with four standard baffles and a Rushton turbine, was developed by Strek.[15]

$$Nu_{pw-ug} = \left(\frac{1}{T}\right)^{0.13}\left(\frac{c_{bot}}{T}\right)^{0.12} Re^{2/3}Pr^{1/3}\left(\frac{\mu_L}{\mu_w}\right)^{0.14} \tag{10.31}$$

where μ_w represents process-side locations near the wall (i.e., heat transfer surface). Under gassed conditions with a ring sparger in a similar system, Kurpiers and Steiff developed the following correlation:[16]

$$Nu_{pw-g} = 0.1\left\{\xi\left(\frac{T}{D}\right)\left(\frac{T^3}{V_{broth}}\right)N_{P,g}Re^3\left[1-2\left(\frac{D}{T}\right)\right.\right.$$
$$\left.\left.\left(\frac{D}{H_{ug}-H_{sp}}\right)N_{P,g}^{0.5}Fr\right]\right\}^{0.25} Pr^{0.4}\left(\frac{\mu_L}{\mu_w}\right)^{0.23} \tag{10.32}$$

where the constant, ξ, changes linearly from 0.2 (T/D = 3) to 0.37 (T/D = 1.9). Equation 10.32 is used to determine the internal heat transfer coefficient (α_{pw-g}) under gassed conditions in a jacketed stirred tank. After appropriate corrections, it can also be used to approximate the process-side heat transfer coefficient (α_{p-g}) in the presence of an internal heat exchanger. A multiplying correction factor, the ratio of measured heat transfer coefficients under ungassed conditions ($\alpha_{p-ug} / \alpha_{pw-ug}$), is used as indicated in equation 10.33. In this case, $\alpha_{p-ug} = \alpha_{pw-ug}$.

$$\left(\alpha_{p\text{-}g}\right) = \left(\alpha_{pw\text{-}g}\right)\frac{\left(\alpha_{p\text{-}ug}\right)}{\left(\alpha_{pw\text{-}ug}\right)} \tag{10.33}$$

The biggest challenge in using equation 10.28 is the lack of reliable values of the overall heat transfer coefficient, U. It depends on a number of factors, including rheology of the broth, thermal properties of the broth and utility fluid, aeration and agitation levels, and the nature of the heat transfer surfaces (e.g., jacket, dimpled jacket, coils, or tube bundles). Because so few applicable correlations exist for aerated fermenters, experience and a conservative approach are essential for their design.

Design guidelines for a typical half-pipe jacket are discussed here. Based on the fermentation heat duty (see Table 10.9), half-pipe coil design results are summarized in Table 10.10.

Heat transfer in a coiled half-pipe jacket: According to Stein and Schmidt, a few adaptations make it possible to use correlations for horizontal tube coils to estimate heat transfer in a vessel with a coiled half-pipe jacket.[17] Necessary adaptations include using the thermic diameter $d_{th} = (\pi / 2)\, d_i$ instead of the tube inner diameter d_i to calculate the Reynolds and Nusselt numbers, and replacing the bending ratio (d_{be} / D_b) with the ratio of $d_i / 2(T + 2\delta_w)$.[1,2]

The geometry with the main dimensions of the coil is shown in Figure 10.11. Here, *d* is the inside diameter of the tube coil and *h* is the pitch of the coil. D_w is the average diameter of the coil. Another characteristic parameter used to describe the coil geometry is the bending ratio d_{hc} / D_b, with D_b as the average bending diameter of the coil. The calculation of D_b and that of the heat transfer coefficient are described next. The average turning diameter D_t of a coil can be determined for a tube with length *l* and *n* turns of slope *h*:

$$D_t = l / n\pi \tag{10.34}$$

With a right triangle configuration of the coil, the average diameter of the coil (shown in Table 10.10) is:

$$D_w = \sqrt{D_t^2 - \left(\frac{h}{\pi}\right)^2} \tag{10.35}$$

TABLE 10.9
Fermentation Heat Duty

Variable	Equation	Value
Agitation power [kW]	Table 10.7	238
Gas expansion power [kW]	Table 10.7	80
Reaction heat liberation [kJ/mol. O_2]		500
Reaction heat liberation [kW)]	10.27	2676
Total heat to be removed [MW]	10.26	3

TABLE 10.10
Results of Half-Pipe External Heat Transfer Jacked Design

Variable	Equation	Value
Half-pipe jacket design		
Vessel wall thickness [m]		0.032
Thermal conductivity tank wall [W/m-K]		16
Number of jacket coils [-]		3
Nominal tube size [m, in.]		0.1, 4
Tube outside diameter [m]		0.114
Schedule		Sch. 10
Tube thickness [m]		0.003
Tube inside diameter [m]		0.108
Number of coil-turns [-]		16
Coil tube pitch [m]		0.149
Height of coil section [m]		2.377
Distance between coil section [m]		0.114
Total surface area of half pipes [m^2]		71.20
Velocity [m/s]		2.0
Total chilled water (CW) flow [m^3/hr]	10.46	100.0
Reynolds number [-]		2.94 x 10^5
Zeta [-]	10.41	0.0183
Nusselt number [-]	10.40	1,922
Process-side heat transfer coefficient [W/m^2-K]	10.31–33	1,764
Utility side heat transfer coefficient [W/m^2-K]		10,608
Fouling process side [m^{2}K/W]		0.00000
Fouling utility side [m^{2}K/W]		0.00017
Overall heat transfer value [W/m^2-K]	10.30	355
Process-side temperature [°C]	10.32	
Log mean temperature difference [K]	10.28–29	18
Effective heat transfer area [m^2]	10.42–43	71.2
Actual heat duty [MW]	10.26	0.455
CW supply temperature [°C]		12.0
CW return temperature [°C]	10.43–44, 10.46–47	15.9

The average bending diameter, D_b, of the coil that is used in calculating the bending ratio, d_{hc}/D_b, is:

$$D_b = D_w \left[1 + \left(\frac{h}{\pi D_w} \right)^2 \right] \tag{10.36}$$

In contrast to a straight pipe, the critical Reynolds number for the transition from laminar to turbulent flow in a coiled pipe increases with increasing bending ratios.

For the critical Reynolds number, Schmidt found

$$Re_{crit} = 2300 \left[1 + 8.6 \left(\frac{d_{hc}}{D_b} \right)^{0.45} \right] \tag{10.37}$$

Based on the value of the critical Reynolds number or flow regime, different heat transfer correlations are applied. For laminar flow with $Re < Re_{crit}$, the Nusselt number is correlated to

$$Nu_{lam} = \left\{3.66 + 0.08\left[1 + 0.8\left(\frac{d_{hc}}{D_b}\right)^{0.9}\right]\right\} Re^m Pr^{1/3} \left(\frac{Pr}{Pr_w}\right)^{0.14} \tag{10.38}$$

$$m = 0.5 + 0.2903\left(\frac{d_{hc}}{D_b}\right)^{0.194} \tag{10.39}$$

Previous definitions have been slightly altered, so: $Nu = (\alpha\, d_{hc} / \lambda_{cool})$, $Re = w\, d_{hc}\, \rho_{cool} / \mu_{cool}$, and $Pr = \mu_{cool} \cdot Cp_{cool} / \lambda_{cool}$.

For heat transfer in the turbulent flow regime, i.e., Re > 2.2 × 10[4], the Nusselt number is determined from the following equation:

$$Nu_{turb} = \frac{\xi / 8 \cdot Re \cdot Pr}{1 + 12.7\sqrt{\xi / 8} \cdot \left(Pr^{2/3} - 1\right)} \left(\frac{Pr}{Pr_w}\right)^{0.14} \tag{10.40}$$

$$\xi = \frac{0.3164}{Re^{0.25}} + 0.03\left(\frac{d_{hc}}{D_b}\right)^{0.5} \tag{10.41}$$

The installed area of half-pipes can be calculated as:

$$A_{inst} = n_{coil}\, d_o\, L_{coil} = n_{coil}\, d_o\, n_{tc}\, \pi\,(T + 2\delta_w) \tag{10.42}$$

where, d_o is the outside diameter of the half-pipe.

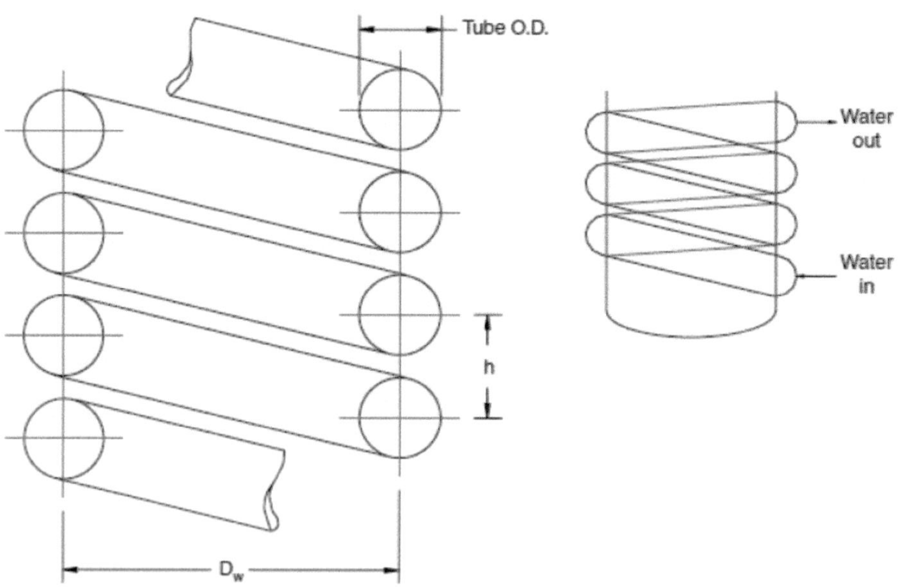

FIGURE 10.11 Schematic of a coiled half-pipe jacket.

Effective heat transfer area is linearly proportional to gassed height in the cylindrical portion of the tank:

$$A_{eff} = A_{inst}\ [(H_g—H_{btm}) / n_{coil} \cdot h_{coil}] \qquad (10.43)$$

The heat removal duty from the tank is:

$$\dot{Q} = U\,A_{eff}\ \Delta t_{lm} = U\,A_{eff}\left[\frac{(t-t_1)-(t-t_2)}{ln\left[(t-t_1)/(t-t_2)\right]}\right] \qquad (10.44)$$

$$\dot{Q} = m_{cool}\ c_{p,cool}\ (t_2—t_1) \qquad (10.45)$$

The cooling medium volumetric flowrate is:

$$\dot{V}_{cool} = w\left(\frac{\pi}{4}d_i^2\ n_{coil}\right) \qquad (10.46)$$

where, w (assumed to be 2 m/s) is the utility fluid velocity in the half-pipe.

The cooling medium return temperature t_2 is:

$$t_2 = \left(\frac{U\,A_{eff}\ \Delta t_{lm}}{\dot{m}_{cool}\ c_{p,cool}}\right) + t_1 \qquad (10.47)$$

10.6.2 Concentric Tube Airlift Fermenter

This, the second of two fermenter design examples, deals with a concentric tube airlift fermenter. Non-dimensional parameters applicable in this example are defined as follows.

Non-dimensional parameters are:

R	diameter ratio, d_c / d_t
S	slenderness ratio, l_t / d_c
M	bubble separation group, $d_s / 4d_c$
X_t	draft-tube bottom clearance ratio, $c_{t\text{-}b} / d_t$
Y_t	draft-tube top clearance ratio, $c_{t\text{-}t} / d_t$
F_g	gas holdup, $V_g / (V_g + V_l)$
DR	disengagement ratio, $(d_c / d_s)\ f(R_2–1) / R_g$
Eu'	modified Euler number, $\Delta_P / \rho u_g^2$
Fr	Froude number, $u_g / (gd_c)^{0.5}$ [also defined as $u_g^2 / (gd_c)$]
Ga	Galileo number, $gr^2\ Ld_c^3 / m^2$
Gr	Grashof number, $g\ \rho_L\ (\rho_L—\rho_g)\ d_B^3 / \mu_L^2$
Sc	Schmidt number, $\mu_L / (\rho_L D_L)$
Sh	Sherwood number, $k_L a d_c^2 / D_L$
We_o	Weber number for the orifice, $\upsilon_o^2\ d_o\ \rho_o / \sigma$

Mass transfer correlations: The following non-dimensional mass transfer correlation, represented by the Sherwood number, is valid for the draft-tube sparged concentric tube airlift fermenter.[18]

$$Sh = 17.3 \times 10^3\ Fr^{0.9}\ M^{-3.4}\ Ga^{0.13}\ X_t^{-0.07}\ Y_t^{-0.18} \qquad (10.48)$$

Apparent viscosity, represented by Ga, reflects the expected decrease in mass transfer rate with increased viscosity. Both draft-tube bottom and draft-tube top clearances, i.e., X_t and Y_t, affect mass transfer rate only modestly, whereas bubble separator-to-column diameter ratio, M, obviously has the greatest influence.

Gas holdup correlations: These correlations are used to calculate gas holdup in the riser, downcomer, and separator, respectively.

$$\Phi_r = 1.5\ Fr^{0.87}\ M^{-0.4}\ X_t^{-0:19}\ Y_t^{-0:2} \quad (10.49)$$

$$\Phi_d = 4.76\ Fr^{1.3}\ Ga^{-0.09} M^{-3.8} X_t^{\ 0.65} \quad (10.50)$$

$$\Phi_{to} = \Phi_s = 0.29\ Fr^{1.05}\ M^{-2.5} X_t^{0.1}\ Y_t^{-0.07} \quad (10.51)$$

The main variable is Fr, which represents the influence of gas input rate (and also the energy input rate). Its effect on holdup is much stronger in the downcomer than in the riser, and the balance between the two gives an exponent of almost 1 (1.05) for the total holdup. Interestingly, the behavior of holdup in the gas separator is very close to that of total holdup. As draft-tube bottom clearance $c_{t\text{-}b}$ increases, Φ_r decreases.[18]

Dynamic pressure drop correlations: According to Merchuk et al., the following correlation satisfactorily predicts pressure drop in concentric tube airlift fermenters over a wide range of physicochemical, geometrical, and operating variables.[18]

$$Eu' = 0.858\ Fr^{-1.51}\ Ga^{0.06}\ X_t^{-1.1}\ M^{\ 4.2} \quad (10.52)$$

Substituting the definitions of Eu' and Fr into this expression shows pressure drop to be a function of $u_g^{0.5}$. The bottom clearance ratio, X_t, and the bubble separation group, M, have a strong influence on the pressure drop because of their effect on liquid velocity. Note that all of these correlations (mass transfer, gas holdup, and dynamic pressure drop) are valid only within the following operating ranges:

$$6 \times 10^{-4} < Fr < 3.5 \times 10^{-2}$$

$$3 \times 10^{7} < Ga < 6 \times 10^{11}$$

$$0.04 < X_t < 0.4$$

$$0.22 < M < 0.34$$

Shear rate correlations: The rate of shear in the airlift bioreactor remains the subject of speculation. The disparity in "average shear rate" $\dot{\gamma}$, as calculated with several available correlations, is shown by Chisti.[8] Several investigators (Nakanoh and Oshida) have accepted the following conservative expression for use in airlift bioreactor:[19]

$$\dot{\gamma} = 5000\ u_{gr} \text{ for } 0.04 \le u_{gr} \le 0.1 \text{ m/s} \quad (10.53)$$

Schumpe and Deckwer, and Moo-Young et al. propose the following for determining average shear rate:[20,21]

$$\dot{\gamma} = 2800 u_{gr} \text{ for } u_{gr} \le 0.08 \text{ m/s} \quad (10.54)$$

$$\dot{\gamma} = 1000 u_{gr} \text{ for } u_{gr} \le 0.04 \text{ m/s} \quad (10.55)$$

These expressions show shear rate dependence on superficial gas velocity, but not as a function of the specific energy that is responsible for turbulence. In an airlift bioreactor, pneumatic power input is[22]

$$P_g/V_L = \rho_L\, g\, u_{gr}[a_r / (a_r + a_d)] \tag{10.56}$$

Incorporating this definition of pneumatic power into equations 10.53 and 10.55 produces the following modified and corrected shear rate correlations:[23]

$$\dot{\gamma} = [5000\, P_g / (V_L\, \rho_L\, g)] / [(a_r + a_d) / a_r] \text{ for } u_{gr} \geq 0.04 \text{ m/s} \tag{10.57}$$

$$\dot{\gamma} = [1000\, P_g / (V_L\, \rho_L\, g)] / [(a_r + a_d) / a_r] \text{ for } u_{gr} < 0.04 \text{ m/s} \tag{10.58}$$

Sparger design equations: For satisfactory gas distribution from a ring sparger,[24] Weber number for the orifice should be

$$We_o = (\upsilon_o^2\, d_o\, \rho_o / \sigma) > 2 \tag{10.59}$$

In the case of gas emerging from a simple submerged orifice at very low flowrates, bubbles periodically form at the orifice, grow to a certain size, and break away. For media with a reasonably low viscosity, bubble diameter is a function of the interfacial tension and density as indicated in equation 10.60:

$$d_b = 1.82\, [(d_o\, \sigma g_c) / (\rho_L - \rho_g) g]^{1/3} \tag{10.60}$$

This relationship holds for "slow" bubbling in the "constant volume" region, although the exponent may be nearer 1/4 when static surface tension data are used. Bubble size is independent of flowrate, whereas bubble frequency is proportional to it. At higher gas flowrates, the frequency levels off and bubble volume increases. Finally, a point is reached where distinct bubble formation is replaced with a chain bubbling process. This undesirable phenomenon is avoided by keeping overall gas rates below those shown in Table 10.11.[24]

The turbulent churn flow regime typically sets in as superficial gas velocity, u_g, exceeds 0.08 m/s. As cited by Bhavaraju, Russel, and Blanchard, Davidson obtained the following correlation for bubble diameter using orifices ranging in diameter from 0.001 to 0.01 m:[25]

$$d_b = 0.19\, d_o^{0.48}\, Re_o^{0.32} \tag{10.61}$$

TABLE 10.11
Maximum Gas Flow Rates for Separate Bubbling

Orifice Diameter, $d_o \times 10^3$ [m]	Orifice Area, $a_o \times 10^6$ [m²]	Gas Rate/Orifice, $q_o \times 10^6$ [m³/s]
0.25	0.05	0.275
0.51	0.20	0.471
1.02	0.82	0.864
2.54	5.06	1.807
5.08	20.26	3.300
10.16	81.03	6.679

Design example: The design of a concentric tube internal loop airlift fermenter for the production of a biocatalysis enzyme at an annual rate of 8000 ± 5% kg/yr. is illustrated as follows. Product recovery efficiency is 80% and the titer (product concentration) is 20 g/L. Maximum oxygen uptake demand is 10 mmol O_2/L/hr. Operational parameters and media physical properties are as follows:

Batch length or growth rate = 5 days
Bioreactor turnaround rate = 2 days
Number of operating days in a year = 300
Annual fermenter output = required production / recovery efficiency = 10,000 kg/yr.
Required batch size = (annual fermenter output / titer) (total batch duration / yearly operation) = 11,667 L = 12,000 L.
Number of bioreactors = 3 (assumed)
Size of each bioreactor = 4000 L

Broth properties of Newtonian fluid:
Effective viscosity = 5×10^{-3} Pa-s (glucose media Newtonian).
Density = 1000 kg/m^3
Interfacial tension (gas–liquid) = 7.2×10^{-2} N/m
O_2 broth diffusion coefficient = 4.3×10^{-9} m^2/s

Based on the above fermentation volume, important fermenter dimensions are initially estimated. The required mass transfer coefficient is calculated from the oxygen uptake rate (OUR). Superficial gas velocity is computed from the Sherwood number–Froude number correlation (equation 10.48). Final column dimensions are determined iteratively while ensuring that bubbly flow (as indicated by superficial gas velocity) is maintained within the riser. After finalizing column dimensions, the required volume is calculated, and the sparger is designed according to accepted rules of thumb. These and other details of the iterative solution procedure are tabulated in Table 10.12.

Figure 10.12 shows a schematic of the production scale concentric tube airlift fermenter. The detailed design specification for the triple-ring sparger is presented in Table 10.13.

TABLE 10.12
Airlift Fermenter Design Specification

Variable	Equation	Value
Column diameter [m]		1
Unaerated liquid height [m]		5
Ratio of column to draft-tube area		2
Draft-tube diameter [m]		0.707
Slenderness ratio		4.5
Draft-tube length [m]		4.5
Diameter of gas separator [m]		1.25
Disengagement ratio		0.566
Draft-tube bottom clearance ratio		0.25
Draft-tube top clearance ratio		1.25
Bubble separation group		0.313
Type of bottom head		ASME flange-dished head
Nominal unaerated media volume [m^3]		4.0
Aeration rate/media volume (VVM) [min^{-1}]		0.5
Actual aeration rate [m^3/min]		2.0

(Continued)

TABLE 10.12 ***(Continued)***
Airlift Fermenter Design Specification

Variable	Equation	Value
Superficial gas velocity [m/s]		0.043
Mass transfer efficiency		0.80
Required mass transfer coefficient [s^{-1}]	10.1	0.0028
Superficial gas velocity [m/s]		0.043
Pressure drop [Pa]		171
Non-dimensional parameters		
Sherwood number		6.5×10^5
Galileo number		3.9×10^{11}
Froude number	10.48	0.014
Schmidt number		1.2×10^3
Grashof number		4.4×10^4
Bond number		1.4×10^5
Euler number	10.52	93
Gas holdup		
In the riser	10.49	0.072
In the downcomer	10.50	0.055
In the airlift fermenter	10.51	0.051
Aerated height [m)]		5.4
Specific power input [W/m^3]	10.56	417
Shear rate in the riser [s^{-1}]	10.57	425

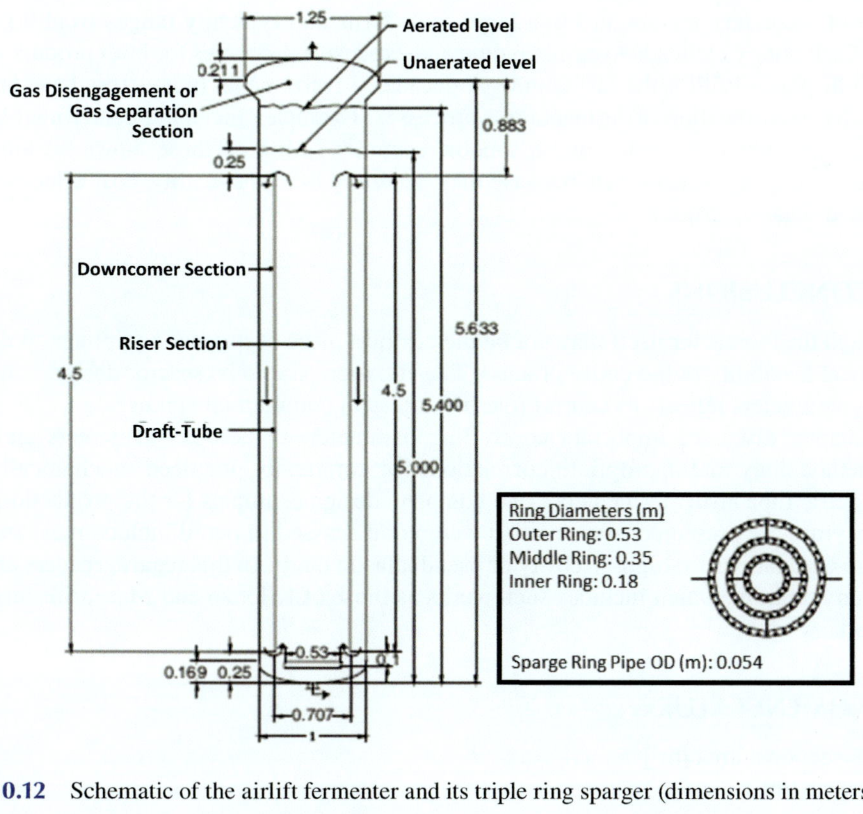

FIGURE 10.12 Schematic of the airlift fermenter and its triple ring sparger (dimensions in meters).

TABLE 10.13
Sparger Design Results

Variable	Equation	Value
Type: three-ring air sparger		
Actual volume of gas [m^3]		0.23
Ratio of column to draft-tube area		2
Superficial gas velocity in the riser [m/s]		0.085
Number of orifices		450
Orifice diameter [m]		2.54×10^{-3}
Velocity through orifice [m/s]		14.6
Orifice Weber number	10.59 (check)	9
Total cross-sectional area of the orifices [m^2]		2.28×10^{-3}
Minimum sparger pipe diameter [m]		0.054
Orifice velocity coefficient		0.8
Pressure drop across the orifice [Pa]	10.21	201
Bubble size (non-coalescing media) [m]	10.61	0.005
Orifice Reynolds number		2.5×10^3
Outer sparger diameter [m] and number of orifices		0.53, 225
Middle sparger diameter [m] and number of orifices		0.35, 150
Inner sparger diameter [m] and number of orifices		0.18, 75

10.7 EMERGING UNCONVENTIONAL FERMENTERS

The yield of secondary metabolites in a large-scale fermenter typically ranges from 0.1 to 12 g/L of broth. Such poor yield leads to cumbersome and expensive processes for both product separation and broth disposal. Within the last couple of decades, several novel bioreactors have been developed for the intensification of fermentation processes. Examples include a centrifugal bioreactor, a rotating packed bed fermenter, and a sonobioreactor.[26,27] Most of these, however, are yet to be implemented on a production scale because they generally lack practicality, cost-effectiveness, and well-defined scale-up criteria.

10.8 CONCLUSIONS

Even though the bioreactor itself may not be the costliest piece of process equipment, it determines the technical feasibility of the entire process. The attention paid to bioreactor design, scale-up, and operating parameters reflects its central role in successful commercial ventures.

This chapter discusses important controlling parameters of aerobic fermenters and provides design methodology and appropriate correlations for commonly practiced mechanically agitated and concentric tube airlift fermenters. Two complete design examples for the production of biocatalysis enzymes are presented. Aspects that were not discussed in detail include mass transfer and gas holdup correlations for high-viscosity or pseudoplastic fluids. In this regard, readers are referred to the open literature, which includes such works as those of Kilonzo and Margaritis, and Kawase and coworkers.[4,28–30]

10.9 NOMENCLATURE

a Cross-sectional area [m^2]
c Clearance [m]

d Diameter [m]
D Impeller diameter [m]
D Molecular diffusion coefficient [m^2/s]
f Wall heat transfer fouling resistance [m^2 K/W)
g Acceleration due to gravity [m/s^2]
h Pitch [m]
H Height [m]
K k-factor [–]
l Height/length [m]
L Length [m]
m Mass flowrate [kg/s]
n Number [–]
$\dot{n}$ Mole flowrate [kmol/s]
N Impeller speed [s^{-1}]
p Pressure [Pa]
P Power [W]
$\dot{Q}$ Heat flux rate [W]
s Impeller spacing [m]
t Temperature [°C]
T Tank diameter [m]
u Superficial gas velocity [m/s]
$\bar{u}$ Average superficial gas velocity [m/s]
U Overall heat transfer coefficient [W/m^2/K]
V Volume [m^3]
$\dot{V}$ Volumetric flowrate [m^3/s]
w Coolant velocity in the pipe [m/s]

Greek Letters

α Heat transfer coefficient [W/m^2/K]
$\dot{\gamma}$ Average shear rate [s^{-1}]
ε Power per mass [W/kg]
ε_g Gas power per volume [W/m^3]
ρ Density [kg/m^3]
Φ_g Gas holdup in the fermenter [–]
μ Dynamic viscosity [Pa s]
η Efficiency [–]
σ Interfacial tension [N/m]
ν Kinematic viscosity [m^2/s]
λ Thermal conductivity [W/m/K]
δ Tank wall thickness [m]
τ Shear stress [Pa]

Subscripts

A Axial impeller
b Bubble
btm Bottom
c Column
cool Coolant
g Gassed
gas Gas expansion
hc Horizontal coil
loc Local

L	Liquid
m	Motor
ug	Ungassed
r	Riser
R	Radial impeller
d	Downcomer
susp	Suspension (slurry)
t	Draft tube
t-t	Draft-tube top
t-b	Draft-tube bottom
tc	Turns per coil
s	Gas separator
to	Total
rs	Ring sparger
sp	Sparger pipe
rsi	Inner ring sparger
rsm	Middle ring sparger
rso	Outer ring sparger
o	Sparger orifice
p	Process side
j	Jacket side
w	Tank wall
i	Inner
in	Inlet
inst	Installation
eff	Effective

10.10 ACKNOWLEDGMENT

The authors would like to thank the publisher, Taylor & Francis, New York, for their permission to reproduce the entry "Fermenter Design" by Kar, K. K., Lueske, J., Cope, R., and Ed. S. Lee, In *Encyclopedia of Chemical Processing*, Vol. 1, 2005.

REFERENCES

1. Storhas, W., *Bioreaktoren und Periphere Einrichtungen*, Vieweg Verlag, Wiesbaden, p. 657, 1994.
2. Krishnan, M. S., Du, J. X., Cao, N. J., Gong, C. S., Tsao, G. T., Simultaneous production and recovery of fuel ethanol by gas stripping in a novel airlift loop fermenter improved biofuel production by corn fiber simultaneous saccharification and fermentation in an airlift-loop fermenter with side arm. In *213th National Meeting of the American Chemical Society, San Francisco, CA, Apr 13–17, 1997*, American Chemical Society, Washington, DC, 1997, BIOTECH-239.
3. Atkinson, B., Mavituna, F., Gas-liquid mass transfer. In *Biochemical Engineering and Biotechnology Handbook*, Macmillan Publishers Ltd, p. 801, 1983.
4. Kilonzo, P. M., Margaritis, A., The effect of non-Newtonian fermentation broth viscosity and small bubble segregation on oxygen mass transfer in gas-lift bioreactors: A critical review. *Biochem. Eng. J.*, 17, pp. 27–40, 2004.
5. Bartow, M. V., Supersizing the aerobic fermenter. *Chem. Eng.*, pp. 70–75, July 1990.
6. Bartow, M., Erin, S., Bioreactor design for mammalian cell cultures. *Chem. Eng.*, pp. 49–54, Jan. 2004.
7. Govier, G. W., Aziz, K., The flow properties of fluids. In *The Flow of Complex Mixtures in Pipes*, Robert E. Krieger Publishing Company, Malbar, p. 98, 1982.
8. Chisti, M. Y. Airlift reactor: Current technology. In *Airlift Bioreactors*, Elsevier, Amsterdam, pp. 74–79, 1989.

9. Moo-Young, M., Chisti, Y. Considerations for designing bioreactors for shear sensitive culture. *Biotechnology*, pp. 1291–1296, November 1988.
10. Ryu, D. Y., Oldshue, J. Y., A re-assessment of mixing cost in fermentation process. *Biotech. BioEng*, XIX, pp. 621–629, 1977.
11. Bakker, A., Smith, J. M., Myers, K. J., How to disperse gases in liquids. *Chem. Eng.*, 101, pp. 98–104, 1994.
12. Rehm, H. J., Reed, G., Puehler, A., Stadler, P., Design of aseptic, aerated fermenters. In *Biotechnology*, VCH Verlagsgesellschaft, Weinheim, Vol. 3, pp. 418–425, 1993.
13. Weiss, S., Stoffvereinigen in fluiden phasen. In *Verfahrenstechnische Berechnungsmethoden*, VCH Verlagsgesellschaft, Weinheim, p. 186, 1988.
14. Lee, S., Tsui, Y., Succeed at gas-liquid contacting. *Chem. Eng. Prog.*, 95, pp. 23–49, July 1999.
15. Strek, F., Heat transfer in liquid mixers—study of a turbine agitator with six flat blades. *Int. Chem. Eng.*, 3, pp. 533–556, 1963.
16. Kurpiers, P. A., Steiff, P. M., Weinspach: Zum wärmeübergang in gerührten mehrphasensystemen. *Chem. Ing. Techn.*, 56(3), pp. 234–235, 1984.
17. Stein, W. A., Schmidt, W., Heat transfer in agitated tank jackets with liquid media. *Ger. Chem. Eng.*, 9(6), pp. 362–371, 1986.
18. Merchuk, J. C., Ladwa, N., Cameron, A., Bulmer, M., Pickett, A., Concentric-tube airlift reactors: Effects of geometrical design on performance. *AIChE J.*, 40(7), pp. 1105–1117, 1994.
19. Nakanoh, M., Yoshida, F., Gas absorption by Newtonian and non-Newtonian liquids in a bubble column. *Ind. Eng. Chem. Process Des. Dev.*, 19, pp. 190–195, 1980.
20. Schumpe, A., Decker, W. D., Viscous media in tower bioreactors: Hydrodynamic characteristics and mass transfer properties. *Bioprocess. Eng.*, 2, pp. 79–94, 1987.
21. Moo-Young, M., Halard, B., Allen, D. G., Burrell, R., Kawase, Y., Oxygen transfer to mycelial fermentation broth in an airlift fermenter. *Biotechnol. Bioeng.*, 30, pp. 746–753, 1987.
22. Chisti, Y., Moo-Young, M. Communication to the editor on the calculation of shear rate and apparent viscosity in airlift and bubble column bioreactors. *Biotechnol. Bioeng.*, 34, pp. 1391–1392. 1989.
23. Kilonzo, P. M., Margaritis, A., The effects of non- Newtonian fermentation broth viscosity and small bubble segregation on oxygen mass transfer in gas-lift bioreactors: A critical review. *Biochem. Eng. J.*, 17, pp. 27–40, 2004.
24. Fair, J. R., Designing gas-sparged reactors. *Chem. Eng.*,74, pp. 67–74, July 1967.
25. Bhavaraju, S. M., Russel, T. W. F., Blanchard, H. W., Design of gas sparged devices for viscous liquid system. *AIChE J.*, 24(3), pp. 454–466, 1978.
26. Voit, H., Goetz, F., Mersmann, A. B., Overproduction of lipase with Staphylococcus carnosus (pLipPS1) under modified gravity in a centrifugal field bioreactor. *Chem. Eng. Technol.*, 12, pp. 364–373, 1989.
27. Chisti, Y., Sonobioreactors: Using ultrasound for enhanced microbial productivity. *Trends Biotechnol.*, 21(2), pp. 89–93. 2003.
28. Kawase, Y., Moo-Young, M., Volumetric mass transfer coefficients in aerated stirred tank reactors with Newtonian and non-Newtonian media. *Chem. Eng. Res. Des.*, 66, pp. 284–288, 1988.
29. Kawase, Y., Moo-Young, M., Theoretical prediction of gas hold-up in bubble columns with Newtonian and non-Newtonian fluids. *Ind. Eng. Chem. Res.*, 26, pp. 933–937, 1987.
30. Kawase, Y., Hallard, B., Moo-Young, M., Theoretical prediction of volumetric mass transfer coefficients in bubble columns for Newtonian and non-Newtonian fluids. *Chem. Eng. Sci.*, 42(7), pp. 1609–1617, 1987.

11 Jet Mixing Systems

11.1 INTRODUCTION

Jet mixers are widely used in industrial stirred tanks and railcars because they are a cost-effective means of homogenizing the macro-scale concentration, temperature and pH of the working media. Because of cost and practicality, installations of top-entry mechanical agitation systems are typically avoided in large-scale storage tanks. Also, jet mixers in many instances with small and mid-size tanks provide an adequate mixing solution without any submerged moving parts; hence, they are quite popular. A typically simple jet mixing system, as shown in Figure 11.1, consists of a plumbed recirculation and/or an addition loop, a pump and a nozzle. The latter could be an in-house tapered pipe (reducer) nozzle or a commercial nozzle that is submerged in the working media (see Figure 11.2). High-velocity flow through the nozzle creates the jet that entrains adjacent fluid. The relative velocity of the jet, with respect to the fluid in the tank, produces either laminar or turbulent mixing depending on the jet Reynolds number, Re_j.

The mass momentum of the fluid jet, i.e., mass flow times the velocity $\left(\rho_f . \nu_j^{\,2}\right)$ transferred to the vessel contents, is responsible for the mixing action depicted in the entrainment and impingement flow patterns of Figure 11.1. Use of multiple jets is common in large (> 20k-gal) storage tanks containing caustics, monomers, hydrocarbons, LNG, polymers, and products subjected to weather-related temperature nonuniformities leading to self-polymerization/stratification/striation. Fluid aeration jets are used with biomass in wastewater treatment ponds. Applications of multiple jets will be discussed in detail later, together with a few design examples.

Besides commercial nozzles, simple standard pipes can be used for jets within vessels. Table 11.1 lists some American (BS1600) and mainland European (DIN) standard steel pipe sizes with nominal and bore diameters for potential use as a relatively precise jet diameter.

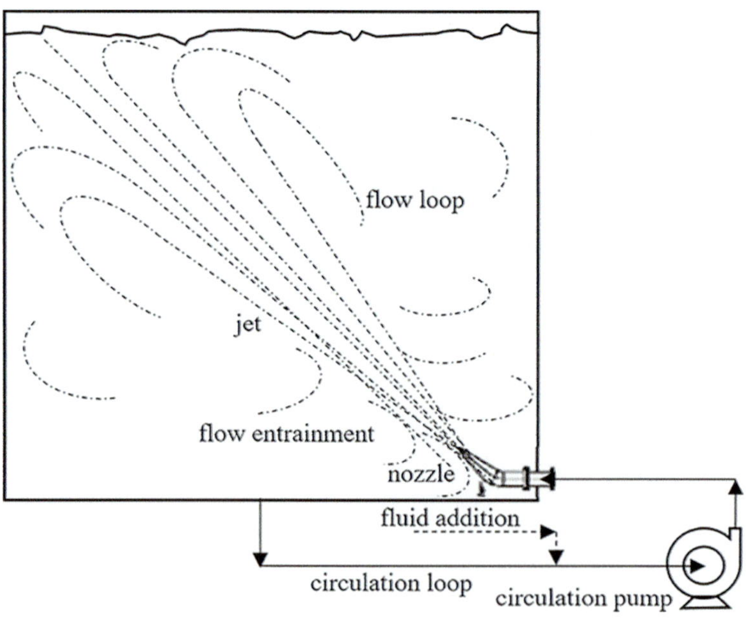

FIGURE 11.1 Schematic of a jet mixer with the expected flow pattern.

DOI: 10.1201/9781003606390-11

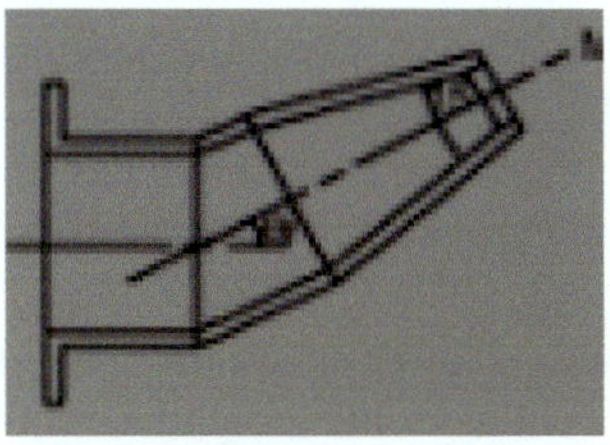

Commercial Nozzles

Pipe Reducer Nozzle

FIGURE 11.2 Cutaway schematics of a commercial nozzle[1] and a simple pipe reducer.[2]

TABLE 11.1
BS1600 and DIN Pipe Standards and Actual Bore Diameters[3]

Nominal Size Pipe (mm)		15	20	25	32	40	50	65	80	100	150
Bore (mm)	Schedule 40	15.8	21	26.6	35.1	40.9	52.5	62.7	77.9	102.3	154.1
	Schedule 80	13.8	18.9	24.3	32.5	38.1	49.2	59.0	73.7	97.2	146.4
	Schedule 160	11.7	15.6	20.7	29.5	34.0	42.8	53.9	66.6	87.3	131.8
	DIN 2448	17.3	22.3	28.5	37.2	43.1	60.3	70.3	82.5	107.1	159.3

11.2 DESIGN OF A COLDREY JET MIXING SYSTEM

The jet Reynolds number, Re_j, is defined as

$$Re_j = v_j \, d_j \, / \, \nu \tag{11.1}$$

where, v_j = jet velocity at the nozzle, d_j = jet diameter and ν = kinematic viscosity of the fluid.

Depending on the jet Reynolds number, jets are generally classified as laminar ($Re_j \leq 2000$) or turbulent ($Re_j > 2000$). For the typical Coldrey[4] jet design shown in Figure 11.3, an acceptable jet velocity, v_j, is $\gtrsim$ 10m/s. The jet decay velocity, v_Z, at the end of jet path length, Z, should be between 0.10 to 0.3 m/s to avoid splashing and surface aeration that could form foam and/or froth. Since entraining liquid into the jet is critical to mixing effectiveness, the nozzle exit must be located away from solid boundaries such as the tank wall and bottom to avoid choking of the entrainment flow. To avoid a wall-hugging jet (Coandă effect), the location of the jet discharge must not be flush with the wall. Researchers at BHRG-FMP[5] have established proprietary empirical correlations with physical rationale based on their extensive studies of the jet mixing systems while varying geometric scale and other relevant parameters. In the past, authors had the privilege of using them. However, there are abundant research publications[6–22] since 1940s that address applications of jet mixers, their design and expected mixing performance. Blend time being the most important parameter in designing a jet mixer, one of the latest literatures is discussed here in detail with a workout example.

The circulation and jet turbulence models for prediction of mixing time, θ, are the two accepted theories that evolve from the effects of shear flow at the jet boundary and jet kinetic energy dissipation at the end of the jet. These are used to establish mixing time correlations for free-jet configurations based on experimental data with varying jet and vessel parameters, such as jet-induced volumetric flowrate, $\dot{Q}$, jet path length, Z, and liquid height to tank diameter ratio, Z_s / T, etc. For discussion purpose, Figure 11.3 schematically shows a popular upward-pointing jet design in a cylindrical tank.

$$\dot{Q} = \dot{Q}_j + \dot{Q}_e \tag{11.2}$$

where, $\dot{Q}_j$ = flowrate through the nozzle and $\dot{Q}_e$ = entrainment flowrate.

$$\dot{Q}_j = \left(\pi/4\right) d_j^2 \, v_j \tag{11.3}$$

Momentum of a free jet remains constant along its path. However, as the jet radially spreads, velocity decreases; thus, the jet entrained flowrate, $\dot{Q}_e$, increases with the path length, as shown in Figure 11.4.

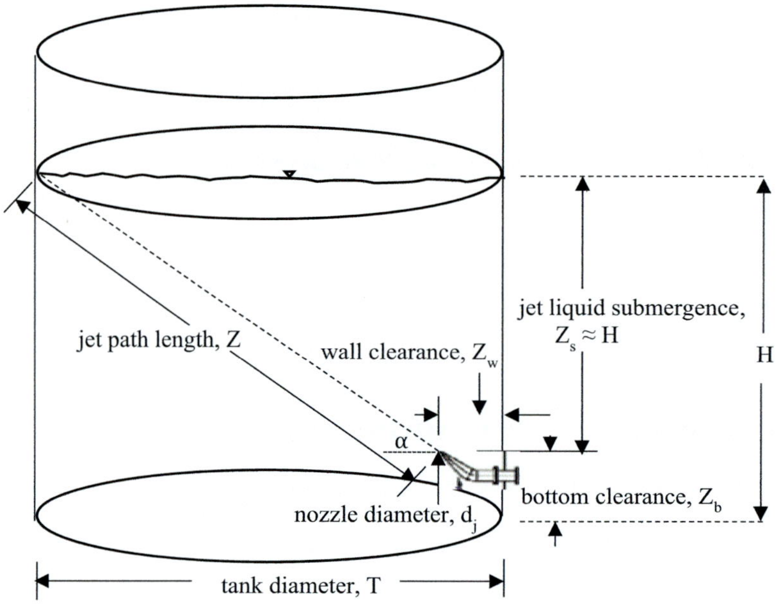

FIGURE 11.3 Schematic of a Coldrey type jet mixing system in a circular tank.

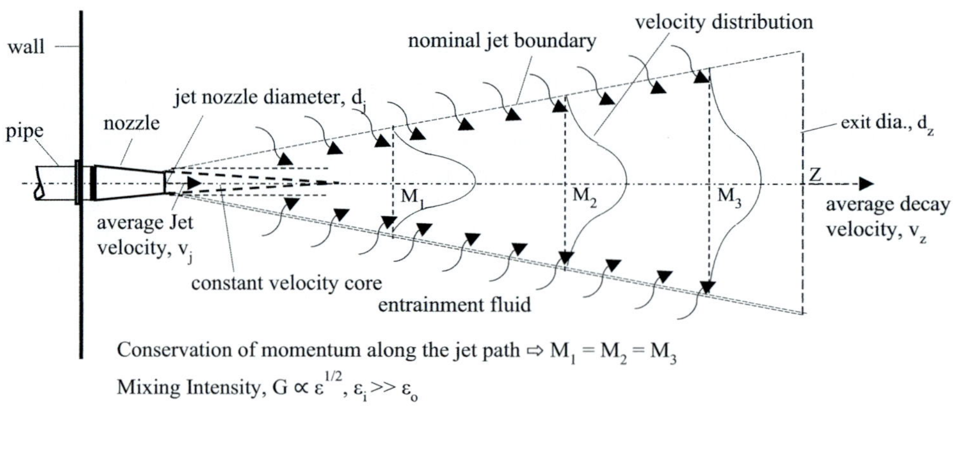

FIGURE 11.4 Spreading characteristics of a circular free jet including the jet boundary, entrainment flow and velocity distribution.

Based on momentum balance, total flowrate at the end of the jet path, Z, is expressed as[6]

$$\frac{\dot{Q}}{\dot{Q}_j} = k\frac{Z}{d_j} \tag{11.4}$$

where, k is an experimental coefficient.

For a free jet, the k value is experimentally determined to be 0.32.[7]

$$\dot{Q} = 0.32\dot{Q}_j \frac{Z}{d_j} \tag{11.5}$$

11.2.1 Circulation Model

Circulation time, θ_c, is defined as the ratio of volume of liquid in the tank, V, to the total jet volumetric flowrate, $\dot{Q}$, as follows:

$$\theta_c = \frac{V}{\dot{Q}} \tag{11.6}$$

Circulation model relates circulation time to mean residence time, θ_r, required to achieve homogeneity of contents in the tank. Since the latter is a function of the tank geometry, T and Z_s, and the jet flow velocity, v_j, mixing time is directly proportional to the circulation time, i.e., $\theta \propto \theta_c$.[8] For all practical purposes, in this case assume liquid submergence depth, Z_s, to be the liquid height, H, in the tank for volume calculations. Mixing time is expressed in equation 11.7.

$$\theta = \varphi \frac{T^2 Z_s}{v_j d_j Z} \tag{11.7}$$

Grenville and Tilton[9] concluded that φ varies as follows, depending on jet angle to the horizontal plane, α:

$$\varphi = 13.8 \text{ for } \alpha < 15^\circ\text{, and } \varphi = 9.34 \text{ for } \alpha > 15^o$$

where $\alpha \approx tan^{-1}\left[Z_s / \left(T - Z_w\right)\right]$, for $0.4 \le H/T \le 1.0$

11.2.2 Jet Turbulence Model

For isotropic turbulent mixers, Corsin[10] theoretically derived that the mixing time, θ, is a function of (a) integral scale of concentration fluctuations, l; (b) turbulent energy dissipation rate, ε; and (c) Schmidt numbers, S_c, as shown in equation 11.8.

$$\theta \approx \frac{1}{2}\left[4\left(\frac{l^2}{\varepsilon}\right)^{1/3} + \left(\frac{\mu}{\rho\varepsilon}\right)^{1/2} \log Sc\right] \tag{11.8}$$

In low-viscosity media, the second term in equation 11.8 is negligible,[11] thus, blend time can be considered as

$$\theta \approx 2\left(\frac{\varepsilon}{l^2}\right)^{-1/3} \tag{11.9}$$

Based on this theory, Grenville and Tilton[11] introduced the jet turbulence model and successfully correlated FMP experimental blend time data[5] with the kinetic energy dissipation rate at the end of the jet, ε_z, (power per mass) and the characteristic length as the jet path length, z.

$$\theta \propto \left(\frac{\varepsilon_z}{Z^2}\right)^{-0.345} \tag{11.10}$$

with correlation coefficient of 0.990 and standard error of exponent of ±0.0062.

After rounding off the exponent to $-1/3$, further regression analyses provided the following relationship:

$$\theta\left(\frac{\varepsilon_z}{Z^2}\right)^{1/3} = 32.37 \pm 8.73\% \tag{11.11}$$

By rearranging,

$$\theta = 32.37\left(\frac{Z^2}{\varepsilon_z}\right)^{1/3} \tag{11.12}$$

Turbulent kinetic energy dissipation rate at the end of jet path, z:

$$\varepsilon_z \propto \frac{v_z^{\,3}}{d_z} \tag{11.13}$$

where, v_z is the decay velocity of the jet at its end.

Substituting equation 11.13 in equation 11.12,

$$\theta = 32.37\left(\frac{d_z Z^2}{v_z^{\,3}}\right)^{1/3} \tag{11.14}$$

Free jet conservation of momentum principle implies

$$d_z = \frac{v_j d_j}{v_z} \tag{11.15}$$

Velocity at the end of a free jet, v_z, is related to the jet velocity, v_j, as follows:[6]

$$v_z = 6\frac{v_j d_j}{Z} \tag{11.16}$$

Substituting d_z and v_z from equations 11.15 and 11.16, respectively, into equation 11.14, blend time explicitly depends on the free jet parameters such as jet diameter, jet velocity (flowrate) and jet path length, as shown in equation 11.17.

$$\theta = 32.37\left(\frac{Z^2}{6^{4/3} v_j d_j}\right) = 2.97\left(\frac{Z^2}{v_j d_j}\right) \tag{11.17}$$

Or, in the familiar non-dimensional mixing (blend) time correlation form, expressed as,

$$\frac{v_j\theta}{d_j} = 2.97\left(\frac{Z}{d_j}\right)^2 \tag{11.18}$$

for $0.4 \le H/T \le 1.0$.

Later, further data analyses by Grenville and Tilton[11] included the blending data of Van de Vusse,[12] generated using a 36m diameter ($1200m^3$) vessel with a 0.5m diameter nozzle. The correlation constant turned out to be 3.00.

$$\frac{v_j\theta}{d_j} = 3.00\left(\frac{Z}{d_j}\right)^2 \tag{11.19}$$

with a relative standard deviation of ±11.83%.

Or if expressed explicitly in terms of mixing time for the tank aspect ratio, $0.2 \le H/T \le 3.0$, as

$$\theta = 3.00\left(\frac{Z}{d_j}\right)^2\left(\frac{d_j}{v_j}\right) \tag{11.20}$$

Equation 11.20 allows users to produce a less conservative design without extrapolating beyond the limits of the earlier experimental data,[11] which are

$$0.178 < V < 1200m^3, \qquad 1.32\times10^{-2} < {}^{v_j d_j}\!/_{Z} < 0.137ms^{-1}, \qquad 86 < {}^{Z}\!/_{d_j} < 753$$

Circulation time and jet turbulence models (equations 11.7 and 11.20) give very similar predictions over the range where most of the data were collected ($0.4 < \frac{H}{T} < 1, \alpha > 15°$). For example, with ${}^{H}\!/_{T} = 1$, blend time predicted by the circulation time model is 110% that predicted by the jet turbulence model.[13]

11.2.3 Pump Energy

Pressure loss across the nozzle is calculated based on the dynamic pressure drop equation and a design factor, κ, i.e., $1.5 \le \kappa \le 2.5$ is used for design purpose. Typically, $\kappa = 2$.

$$\Delta p = \kappa \rho v_j^2 / 2 \tag{11.21}$$

Pump energy, E_j, imparted to the jet only, is expressed as pressure drop times the flowrate.

$$E_j = \dot{Q}_j(\Delta p) \tag{11.22}$$

However, pump delivery pressure, P, must include (a) pressure drop across the nozzle, Δp; (b) frictional pressure loss in the delivery pipeline (including elbows, valves, etc.), p_f and (c) net hydrostatic pressure head, p_s. Latter includes the tank head pressure, p_{tank}, if any, i.e., $p_s = p\, g\, Z_s + p_{tank}$.

$$P = \Delta p + p_f + p_s \tag{11.23}$$

11.2.4 Example

Problem: Authors present a simple example of a single Coldrey jet system design as shown in Figure 11.3. A cylindrical flat-bottom tank containing calcium hydroxide solution having dynamic viscosity and density of 15cP and 1.4g/cc, respectively, needs to be neutralized with a minor quantity of pH adjuster. The tank diameter and its height are 3m and 6m, respectively. Design of the system is described as follows and in mixit-io SS11.1.

Solution: The approximate maximum volume of the liquid in the vessel should be ~34m^3 (9,000gal), nozzle inner diameter, d_j, is 30mm and the 99% turbulent mixing time, $\theta_{99\%}$, is 4 min.

Design steps:

1. Physical properties of the working media:

   Viscosity, $\mu = 15 \times 10^{-3}$ Pa-s
   Density, $\rho = 1400$kg-m^{-3}

2. Tank dimensions:

   Tank diameter, $T = 3$m
   Tank height, $H_T = 6$m

3. Maximum liquid height must not exceed 80% of the straight-side tank height, H_T:

   Liquid height, $H = 6 \times 0.8 = 4.8$m

4. Aspect ratio, $H/T = 1.6$ ⇨ use of equation 11.20
5. Volume of liquid, $V = \left(\pi/4\right) T^2 H = 33.91 m^3$
6. Considering the nozzle wall and bottom clearances, respectively as follows:

   $Z_w = 0.3m$ and $Z_b = 0.3m$

7. Submergence depth of the nozzle exit:

   $Z_s = H - Z_b = 4.5m$

8. Jet upward inclination angle with respect to horizontal surface:

   $$\alpha \approx tan^{-1}\left[Z_s / \left(T - Z_w\right)\right] \approx tan^{-1}\left[4.5 / \left(3 - 0.3\right)\right] = 59°$$

9. Jet path-length:

   $$Z = \left[Z_s^{\,2} + \left(T - Z_w\right)^2\right]^{1/2} = \sqrt{\left[(4.5)^2 + (3 - 0.3)^2\right]} = 5.25m$$

10. Considering the ratio of jet path length to nozzle diameter ratio, i.e., Z / d_j as 180

    $$d_j = Z / 180 = 5.25 / 180 = 0.029m$$

    In lieu of having a tapered nozzle, the designer may decide on using a pipe (reducer) with the closest bore diameter to 0.029m. From Table 11.1, use of a Schedule 80, nominal 32mm pipe provides a bore size of 32.5mm.

    $$d_j = 0.0325m$$
    $$Z / d_j = 161$$

    Designers must check the value of Z/d_j before using correlation equation 11.20, i.e., $86 < Z/d_j < 753$.

11. Maintain jet velocity at nozzle, $v_j = 10ms^{-1}$
    Jet flowrate at nozzle exit:

$$\dot{Q}_j = \left(\pi/4\right) d_j^2\, v_j = \left(\pi/4\right)(0.0325)^2 \times 10.0 = 0.0083 m^3 s^{-1}$$

12. Total jet flowrate at its end:

$$\dot{Q} = 0.32 \dot{Q}_j \frac{Z}{d_j} = 0.32 \times 0.0083 \times 161 = 0.428 m^3 s^{-1}$$

13. Jet flow Reynolds number:

$$Re_j = v_j\, d_j / \nu = 10 \times 0.0325 \times \left[1400 / \left(15 \times 10^{-3}\right)\right] \approx 3 \times 10^4 \Rightarrow \text{Turbulent flow}$$

14. Blend time using correlation equation 11.20 (jet turbulent model)

$$\theta = \left[3.00 \left(\frac{Z}{d_j}\right)^2 \left(\frac{d_j}{v_j}\right)\right] = 3.0 \times (161)^2 \times \left(\frac{0.0325}{10}\right) = 253s \approx 4.2min$$

15. Blend time using correlation equation 11.7 (circulation model*)

$$\theta = 9.34 \frac{T^2 Z_s}{v_j d_j Z} = 9.34 \times \left[\frac{(3)^2 \times (4.5)}{(10 \times 0.0325 \times 5.25)}\right] = 221s \approx 3.7min$$

    [* $H/T = 1.6$, whereas, the correlation condition is $0.4 \le H/T \le 1.0$.
    Discrepancy based on mid-value = ±13%]

16. Pressure loss in the nozzle:

$$\Delta p = \rho\, v_j^2 / 2 = \frac{1400 \times (10)^2}{2} = 7 \times 10^4\, Pa.$$

    Design pressure with a design factor of $\kappa = 2$:

$$\Delta p = \kappa\, \rho\, v_j^2 / 2 = 2 \times \frac{1400 \times (10)^2}{2} = 14 \times 10^4\, Pa$$

    Pump delivery pressure must account for additional losses such as pipe frictional loss (including elbows and valves etc.), liquid hydrostatic head and its overhead pressure, if any.

The nozzle wall and bottom clearances, Z_w and Z_b, are critical in the design of jet mixing systems, and depend on the jet flowrate, $\dot{Q}_j$, and vessel size. A practitioner's experience and intuition will often dictate these dimensions, but must be $\lesssim (10 \times d_j)$. Optimization of nozzle position to avoid a wall-hugging jet and improved mixing efficiency can be addressed by computational fluid dynamics (CFD) modeling.

11.3 JET MIXERS

There are multiple submerged jet configurations to introduce fluid into the bulk fluid in a tank. Figure 11.5 (a–d) depicts a few self-explanatory jet locations with the corresponding dead/slow-flow regions in the vessel. The desired outlet locations for fluid re-circulation should be farthest from the jet, or in the likely poor-flow regimes to avoid jet-flow bypass, and to stimulate pump suction flow in the dead region. Apart from the jet configurations shown as follows, special jets such as downward-pointing plunging jets in the headspace of railcar tanks, pulsating jets and multiple tangential jets near the bottom of large-scale vertical cylindrical tanks are not uncommon. This topic will be discussed later in the chapter. Atypical submerged moving jets such as rotary and reciprocating jets are also used for special mixing applications. However, in many instances, because of the mutual flow interactions and wall impingements, multiple jets are no longer considered as free jets. In large-scale tank applications, performance of the multi-jet systems is often critical but the mixing time is not; for example, effectively dislodging the biomass or sludge from the bottom of a tank, which may have an acceptable prolonged mixing time. If the blend time is important, practitioners may consider (a) taking a conservative approach in using the established free jet mixing correlations, (b) performing pilot-scale simulations in the lab and/or (c) seeking numerical solutions such as computational fluid dynamics (CFD).

Revill's recommendations[14] on ideal jet systems based on the liquid level, H (note: $Z_s \approx H$), and tank geometry, T, are:

Single axial jets for $0.75 \le H/T \le 3.0$
Single side-entry jets for $0.25 \le H/T \le 1.5$
Multiple side-entry jets $H/T \le 0.25$ (shallow tank) or $H/T > 3.0$

Jet length, Z, to be at a maximum as shown in Figure 11.5 (a–d) is one of the most important aspects to designing a good jet mixing system for liquid blending processes.[4]

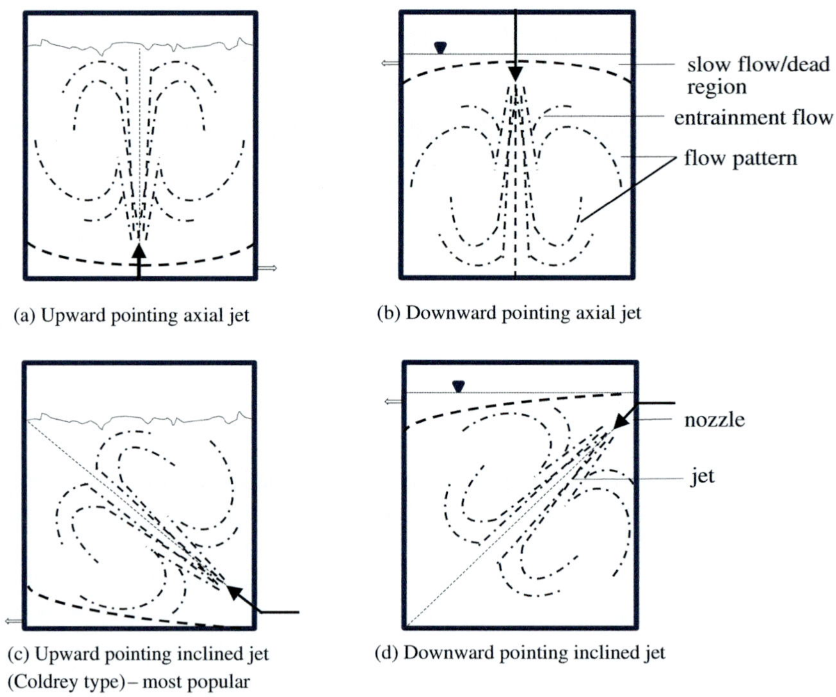

FIGURE 11.5 Schematics of axial (a–b) and side-entry (c–d) jets, and their idealized circulatory patterns.

11.4 JET MIXER MIXING TIME

In most applications, mixing time, θ, is the sought-after design criteria; it is especially important for system house/onsite formulations of chemicals, home, healthcare and ag products. Time required for the desired homogeneity of the components all throughout the vessel translates into the required residence time, θ_r, and consequently, dictates batch turnover. Typically, for the batch and continuous processes, θ_r should be greater than 5 and 10 times θ, respectively, to conservatively assess mixedness of the components prior to their downstream applications. This rule-of-thumb is also typical of stirred tanks.

11.4.1 Experimental Techniques

In the lab, a tracer injection technique using a pulse of electrolyte solution (NaCl, NaOH, I_2 etc.) or a colorant (dye) is mostly used to establish mixing (blend) time correlations. Respectively, concentration and intensity of color are measured using appropriate probes (pH, thermal conductivity and optical-light transmittance) at multiple locations in the tank, including in the intuitively predetermined poorly mixed regions. Although visual observation of color is another simple, least time-consuming approach with transparent vessels, it is quite subjective on the experimentalist's assessment/judgement. Furthermore, the mixing duration is often too small in the lab scale to visualize the time to achieve the predetermined (calibrated) steady-state color all throughout the tank. Details of these experimental techniques are addressed in Chapter 18, "Experimental Techniques in Mixing." As a standard practice, mixing time, θ_{95}, for 95% macro mixedness is typically measured. The desired mixing time for x% ($0.95 \le x < 1.0$) homogeneity is calculated from equation 11.24 (see Chapter 6).[15] The typical 99% blend time, θ_{99}, is $1.6 \times \theta_{95}$.

$$\theta_x = \theta_{95} \frac{\ln(1-x)}{\ln(0.05)} \tag{11.24}$$

11.4.2 Mixing Time Correlations

Following the mixing time guidelines on jet mixing presented by Revill,[14] the authors present a few empirical mixing time correlations and their limited applications in Table 11.2. Mixer designers must review respective references prior to their use because of the variations in jet configurations

TABLE 11.2
Experimental Correlations for Jet Mixing Time[13]
[$Re_j \lesssim 2000$ ⇨ laminar flow and $Re_j > 2000$ ⇨ turbulent flow]

References	Mixing Times	Experimental Conditions/Remarks
Okita and Oyama[8] (1963)	<u>Axial and Side entry Jets</u> $\theta = 9.0\, T^2/(v_j . d_j)$ when, $t_{inj} > t_{mix}/2$ for $4500 < Re_j < 8x10^4$	$T = 1.52m$, Fluid physical property ratios, μ_j/ρ_j were constant and Z_s, d_j, v_j were varied. This correlation is of limited quantitative use.[14] Non-ideal jet angle, α according to Coldrey.[4] Long tracer pulse injected into the re-circulation loop. 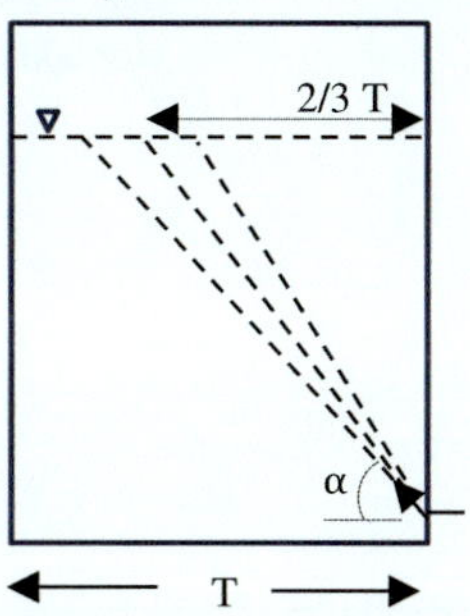

(Continued)

TABLE 11.2 *(Continued)*
Experimental Correlations for Jet Mixing Time[13]
[$Re_j \lesssim 2000$ ⇨ laminar flow and $Re_j > 2000$ ⇨ turbulent flow]

References	Mixing Times	Experimental Conditions/Remarks
Fossett[17] (1951)	$\theta = 4.5\, T^2 / (v_j . d_j)$ when, $t_{inj} < t_{mix} / 2$	-do- Short tracer pulse ⇨ more appropriate correlation.
Fox and Gex[18] (1956)	Side entry jet $\theta = \frac{A\, T\, Z_s}{v_j\, d_j}\left(d_j / Z_s\right)^{0.5}$ where $A = \frac{7.8 \times 10^5\, Fr_j^{0.17}}{Re_j^{1.33}}$ for $250 < Re_j < 2000$, $A = \frac{120\, Fr_j^{0.17}}{Re_j^{0.17}}$ for $2000 < Re_j < 1.6x10^5$ Froude number, $Fr_j = v_j^2 / \left(g\, d_j\right)$	$T = 0.29, 1.52$ (with visual observation) $T = 4.27m$ (with iodine concentration determination) Wide data scatter. Correlation data range ± 50%. Hence, further data analyses carried out. In case of laminar jet, mixing time is a strong function of Reynolds number. In case of turbulent jet, mixing time is a weak function of jet Reynolds number. 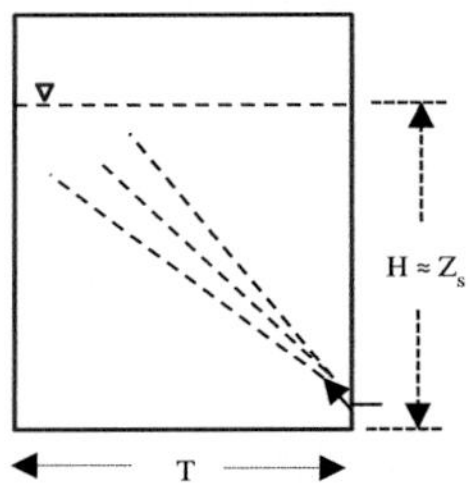
Okita and Oyama[8] (1963)	Axial and Side entry Jets $\theta = \frac{2.8x10^4}{Re_j} T^{1.5} . Z_s^{0.5} / (v_j . d_j)$ for $1000 < Re_j < 5000$	$T = 0.4, 1.0\, m$ μ_j / ρ_j = constant Tracer pulse at the tank center; near the liquid surface Mixing time is a function of jet Reynolds number. 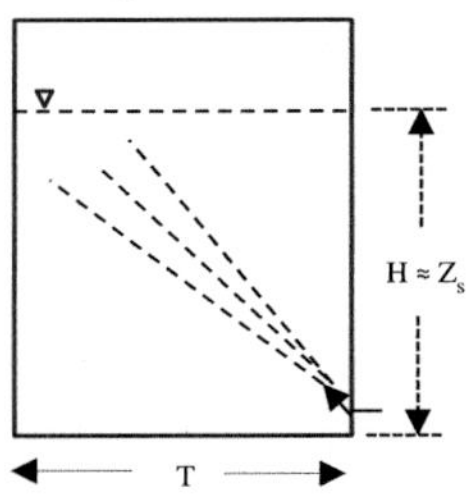
-do-	$\theta = 5.5\, T^{1.5} .\ Z_s^{0.5} / (v_j . d_j)$ for $5000 < Re_j < 8x10^4$	-do- Mixing time is independent of jet Reynolds number
Hiby and Modigell[19] (1978)	Axial Jet Systems $\theta_{99} = 3.2\, T^2 / (v_j . d_j)$ for $6000 < Re_j < 3.6x10^4$	Tracer technique to monitor 95% mixing. $Z_s, d_j, T, \mu_j / \rho_j$ are constant, only v_j varied. In case of turbulent Jet, mixing time is independent of jet Reynolds number 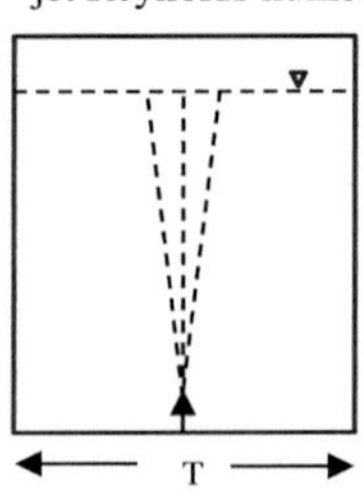

(Continued)

TABLE 11.2 *(Continued)*

Experimental Correlations for Jet Mixing Time[13]

[$Re_j \lesssim 2000 \Rightarrow$ laminar flow and $Re_j > 2000 \Rightarrow$ turbulent flow]

References	Mixing Times	Experimental Conditions/Remarks
Rácz and Wassink[20] (1974)	Axial downward Cases: $H/T = 3,2,1 \; and \; 0.5$ $\theta_{99} = 1.1\left(T/d_j\right)^{2.75}(d_j/v_j)$ for $H/T = 1$ $\theta_{99} = 2.6\left(T/d_j\right)^{2}(d_j/v_j)$ for $H/T = 1/2$ Overall proposed correlation, $\theta_{99} = 3.9\,T^2/(v_j.d_j)$	-do- 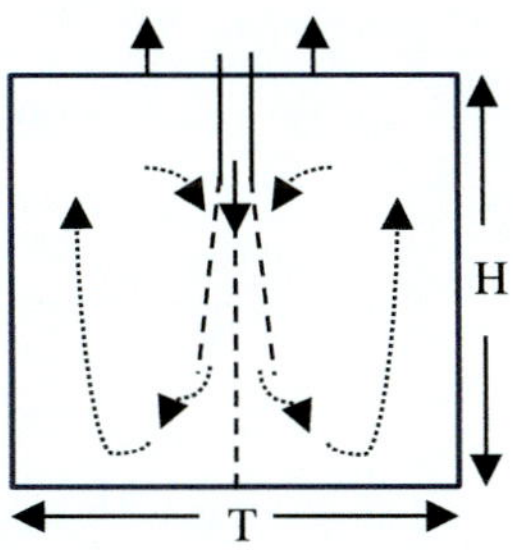
Lane and Rice[21] (1982)	Side entry jets $\theta = \dfrac{A\,T\,Z_s}{v_j\,d_j}\left(d_j/Z_s\right)^{0.5}$ $A = \dfrac{1.13\times10^6\,Fr_j^{0.17}}{Re_j^{1.34}}$ for $200 < Re_j < 2000$ $A = \dfrac{154\,Fr_j^{0.17}}{Re_j^{0.17}}$ for $2000 < Re_j < 6.0x10^4$	Tracer technique to monitor 95% mixing. Correlation range: $0.31 \le T \le 0.57m$ $0.90 \le Z_s/T \le 1.10m$ Jet wall clearance, $Z_w = 5d_j$ Jet bottom clearance, $Z_b = 5d_j$ 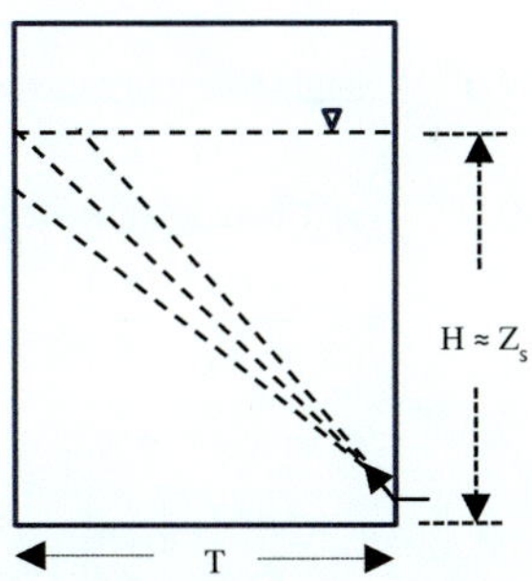
	Axial Jets $\theta = \dfrac{A\,T\,Z_s}{v_j\,d_j}\left(d_j/Z_s\right)^{0.5}$ $A = \dfrac{7.08\times10^5\,Fr_j^{0.17}}{Re_j^{1.30}}$ for $100 < Re_j < 2000$ $A = \dfrac{145\,Fr_j^{0.17}}{Re_j^{0.17}}$ for $2000 < Re_j < 10^5$ $A = 22.8\,Fr_j^{0.17}$ for $10^5 < Re_j < 1.4x10^5$	Upward pumping jet, flushed with tank bottom Correlation range: $0.31 \le T \le 0.57m$ $0.50 \le Z_s/T \le 3.0m$ 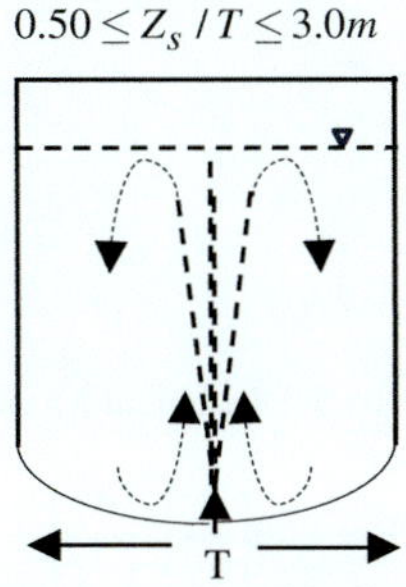

(Continued)

TABLE 11.2 ***(Continued)***

Experimental Correlations for Jet Mixing Time[13]

[$Re_j \lesssim 2000$ ⇨ laminar flow and $Re_j > 2000$ ⇨ turbulent flow]

References	Mixing Times	Experimental Conditions/Remarks
Fox and Gex[18] (1956)	$\theta = \frac{A\,T\,Z_s}{v_j\,d_j}\left(d_j / Z_s\right)^{0.5}$	Improved ± 20% correlation based on specific tank diameter, T. 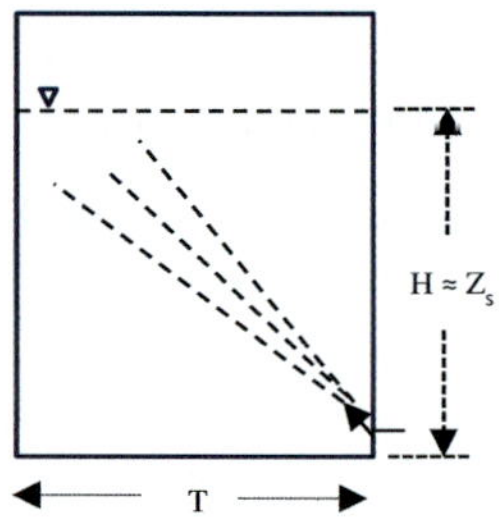
	for $250 < Re_j < 2000$,	
	$A = \frac{7.82 \times 10^5\, Fr_j^{0.17}}{Re_j^{1.33}}$	T = 0.29m
	$A = \frac{5.27 \times 10^5\, Fr_j^{0.17}}{Re_j^{1.33}}$	T = 1.52m
	for $2000 < Re_j < 1.6 \times 10^5$,	
	$A = \frac{120\, Fr_j^{0.17}}{Re_j^{0.17}}$	T = 0.29m
	$A = \frac{85.5\, Fr_j^{0.17}}{Re_j^{0.17}}$	T = 1.52m
	$A = \frac{182\, Fr_j^{0.17}}{Re_j^{0.17}}$	T = 4.27m
Grenville, Tilton[9, 11] (1997)	Single side-entry upward jet	Experimental conditions:[22]
		Cylindrical tank
	(a) Circulation Model:	NaCl soln. used as is as tracer fluid.
		$T = 0.61, 1.68, 3.98; 36$m (van de Vusse[12])
	$\theta = \frac{13.8\,T^2}{v_j\,d_j}\left(\frac{Z_s}{Z}\right)$	$\frac{Z_s}{T} = 0.2–1.0; 3$ (van de Vusse[12])
	for jet angle, $\alpha < 15°$	$d_j = 5.8–50$mm
	$\theta = \frac{9.34\,T^2}{v_j\,d_j}\left(\frac{Z_s}{Z}\right)$	$v_j = 2.2–24.8$m
	for jet angle, $\alpha > 15°$	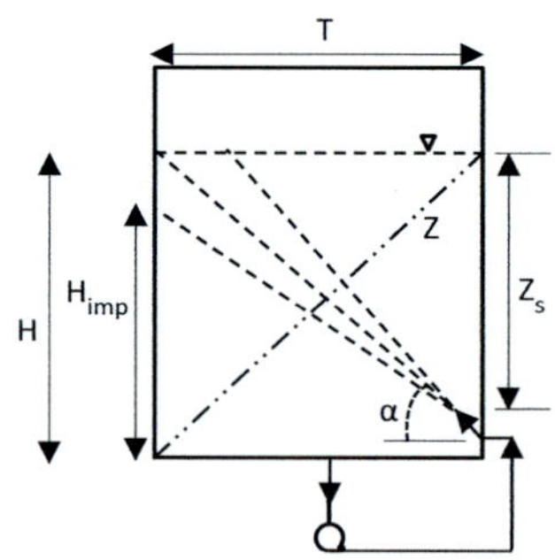
	(b) Jet Turbulent Model:	Correlation conditions:
	$\theta = 3.00\left(\frac{Z}{d_j}\right)^2\left(\frac{d_j}{v_j}\right)$	(a) $0.178 < V < 1200 m^3$
		(b) $1.32 \times 10^{-2} < {}^{v_j d_j}\!/_{Z} < 0.137 ms^{-1}$
		(c) $86 < {}^{Z}\!/_{d_j} < 753$

relative to the tank and multiple experimental constraints. In addition, the authors have added the Grenville-Tilton[9,11] jet mixing time correlations developed from the experimental data of FMP[5] and van de Vusse.[12]

A review article, "Design of Jet Mixing Tank,"[23] on jet mixer designs and well-established mixing time correlations, which may be pursued by practicing engineers. Also, reference 23 compiles the effects of various jet parameters on its blending performance. Although most correlations are case specific and related to a single jet, recent years have seen the emergence of multiple-jet system designs using advanced numerical analysis software. This topic is further discussed in Chapter 16, "Numerical Methods in Fluid Mixing," with an example. Although the jet mixers are more economically feasible for processes, from the blending performance vs. energy consumption view point, in most applications they are less efficient than the top-entry mechanical agitators.

11.5 JET MIXING SYSTEMS FOR LARGE TANKS

As mentioned earlier, multiple jets are ideal for mixing in large-scale tanks and storage vessels, due to lower capital and maintenance costs. Tangential jets are used to break up stratification and/or facilitate heat transfer from the bottom plate/coil heater, sometimes together with Coldrey angled jets for bulk mixing of the tank contents. The authors have successfully designed many such systems to maintain uniform temperature in cold climates and to neutralize working media. CFD modeling sometimes assists in optimizing jet positioning to achieve uniform thermophysical properties in a short duration. In many instances, jets operate continuously in large-scale uninsulated storage tanks to produce temperature uniformity and to compensate for heat loss to the surroundings. The jet configuration in Figure 11.6. produces chaotic mixing and prevents stagnated flow at the bottom.

To maintain efficient bulk mixing, practitioners must adjust powerful tangential jets positioned near the tank bottom so that strong swirl and solid body rotation is avoided. Typically, the tangential nozzle is angled both upwards (α) and inwards/outwards (β) depending on its radial (optimized) location in the tank. In shallow basins, Coldrey angled jet(s) may not be necessary. Given the available free space to produce unhindered jet spreading, one could apply the jet mixing time, $\theta_{95\%}$, equations 11.7 and 11.20, or from Table 11.2 depending their suitability. If necessary, however, a tank should be appropriately segmented into different compartments to maintain the stated application limits of

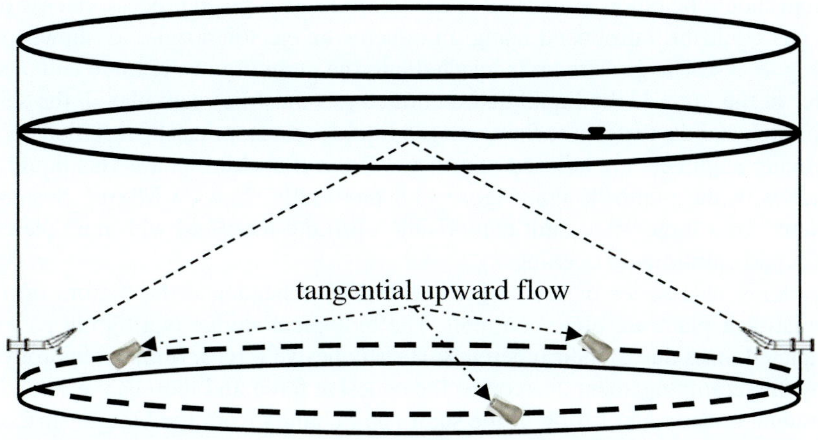

FIGURE 11.6 Positioning of three tangential upward-flow pipe reducers and two nozzles configured for Coldrey-type flow in a large storage vessel.

these empirical correlations. Engineering intuition often plays a pivotal role in configuring nozzle locations within a tank. Mixing process performance of the multi-jet systems, in terms of $\theta_{95\%}$, are assessed based on the total pump energy consumption of the system, $E_{J-system}$.

Figure 11.7 shows a dual nozzle configuration for jet mixing the contents of a horizontal cylindrical storage vessel, such as a tanker railcar or a monomer storage tank.

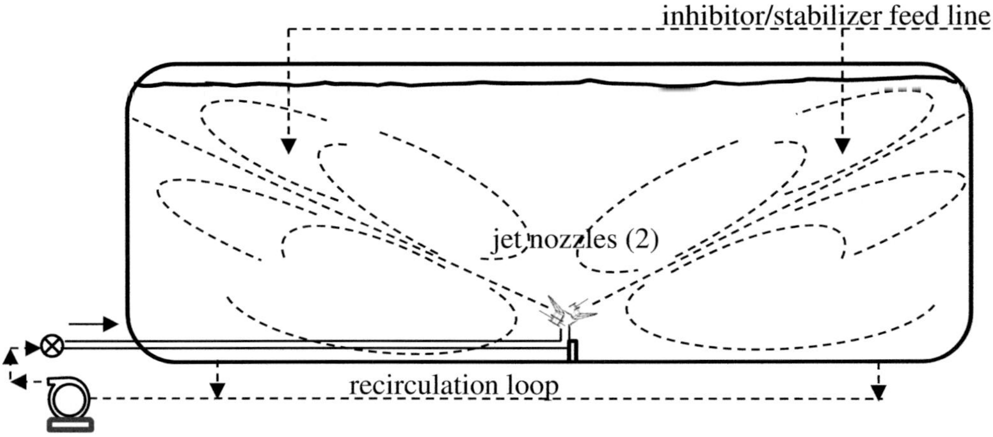

FIGURE 11.7 A dual jet mixing system in a horizontal cylindrical tank.

One common example is mixing minor quantities of such additives as polymerization inhibitors (e.g., phenothiazine (PTZ) and/or hydroquinone monomethyl ether (MeHQ)) into self-polymerizable acrylates (vinyl monomer) that are susceptible to polymerization in moderate temperatures. It's a simple, cost-effective approach to good mixing in a large-scale vessel that requires minimal in-tank metal hardware. The jet arrangement in the slender cylindrical tank of Figure 11.7 produced mild instabilities, resulting in jet oscillations that further enhanced mixing performance.[24]

Jet mixing systems, because of their design simplicity, are also used to suspend light solids, dislodge sludge from tank bottoms and to intimately contact gas and liquid at a low vvm (ratio of volumetric flow of gas per minute to the volume of liquid at $\lesssim 0.4$). The specific design of each jet mixing system should be tailor-configured to accommodate the given process. In case of gas–liquid contacting, gas could be introduced using an eductor or ejector nozzle, as shown previously in Chapter 3, Figure 3.4. The gas stream is injected into the main liquid stream to enhance gas–liquid mass transfer in the tank. If the liquid and entrained gas must be recirculated, the selected pump must appropriately handle two-phase flow, or a gas-liquid separator must be installed upstream of the pump. Practicing engineers are advised to review "Tutorial on Multiphase Gas-liquid Pumping"[25] for assistance with the pump design. Figure 11.8 pictorially shows a Mixing Systems, Inc.'s jet mixing system[26] in a large-scale tank comprising a mixing manifold with multiple eductors that incorporate liquid entrainment openings.

Implementation of a series of submerged jets fed by a header at the bottom of a wastewater biological-treatment plant are quite common. The biological sludge is a tightly packed particle–particle entangled mass due to their inherent surface cohesive forces. Initially, hydrodynamic force generated by the jet impingement overcomes the cohesive force and loosens the settled sludge, followed by gradual erosion and lifting of the semi-solid chunks from the bottom surface. Turbulent shear-force of the jet dislodges/disintegrates the particles while hydrodynamic force keeps them suspended in the continuous phase. Figure 11.9 shows an idealized process of uplifting the sludge from the bottom of a large tank. As opposed to a mechanical agitation system, such applications are

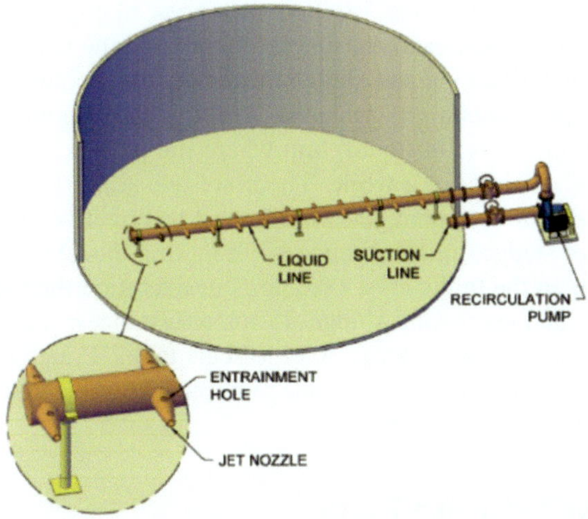

FIGURE 11.8 A pictorial representation of Mixing Systems Inc.'s mixing system for large-scale vessels.[26]

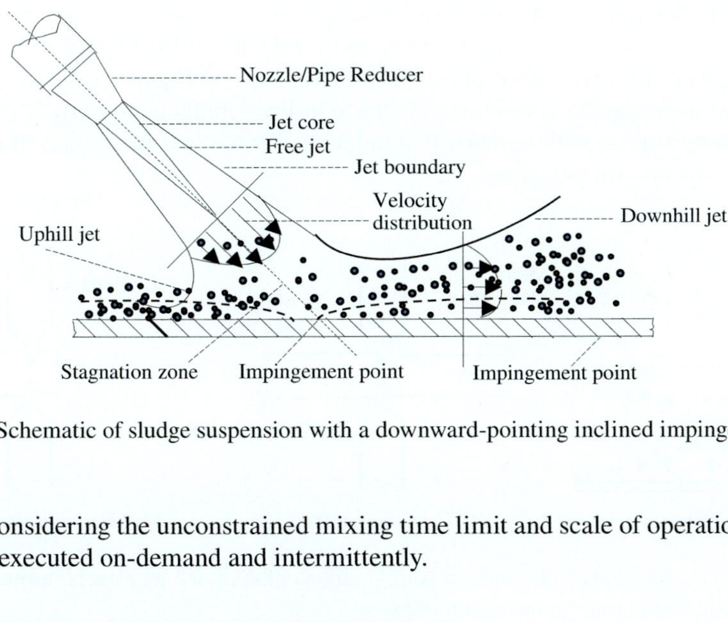

FIGURE 11.9 Schematic of sludge suspension with a downward-pointing inclined impingement jet.

cost effective considering the unconstrained mixing time limit and scale of operation. These operations are often executed on-demand and intermittently.

11.6 JET MIXING DESIGN CONSIDERATIONS

Configuration of the jet nozzle in the tank depends on the mixing process to be accomplished and the ease of maintenance without time-consuming and often unsafe vessel entry. For example, design of the Coldrey system's piping elbow radius should be such that the jet assembly can be easily removed through its corresponding wall nozzle. In the case of very low flow, i.e., stagnation or dead zones in the tank, where solids may accumulate due to laminar jets, fillets may be installed to redirect the flow. Flexibility of changing the nozzle angle to accommodate varying liquid heights is considered a sound design; otherwise, the nozzle should be designed for the expected average liquid level. In many instances, multiple angled nozzles are installed to address significant changes in the liquid height and to switch recirculation through the various different nozzles accordingly. Multiple directional jet mixers in large tanks are employed to dislodge and suspend solids. Spray nozzles are not meant to create jet mixing. However, in a few instances, high-pressure submerged sprays are simply used to uniformly distribute catalysts and inhibitors into the working media.

As mentioned earlier, stratification and sludge formation are encountered in large-scale storage tanks and basins. These phenomena are typical of working media containing components of different densities (liquid/liquid or solid/liquid) and where the jet mixing systems operate intermittently or intentionally remain idle for a while. There are two scenarios to circumvent, i.e., during blending and active biological treatment, jet mixing must be effective, whereas for L-L separation and physical extraction of sludge, in the absence of jet mixing, faster settling is necessary. Design and configurations of jets in the tanks must take into consideration the likelihood of plugging due to polymerized products and semi-solid sludge. In many instances, this is considered as a major operational drawback of jet mixing systems in large-scale tanks. Practicing engineers are advised to seek help from the manufacturers of large-scale multiple jet mixing systems (headers, manifolds and nozzles) for their specific applications.

11.7 CONFINED IMPINGEMENT JETS

For continuous chemical processes where the reaction is instantaneous and must not be micromixing limited, confined impingement jet reactors (CIJR) are a viable option in the pipes, preferably upstream of the motionless mixers. The impingement of monomer, initiator and catalyst jets with their short pathlengths create turbulent mixing with extremely high energy dissipation rate; thus, this enables the desired product selectivity and yield without the byproducts. The confined jet configurations are co-flow, cross-flow or radial flow, as shown in Figure 11.10 a–c. Typical applications range from on-site injection molding systems to in-line tubular nano-particle generators. There appears to be many studies with numerical simulations for the optimal design of such systems as opposed to experimental investigations.

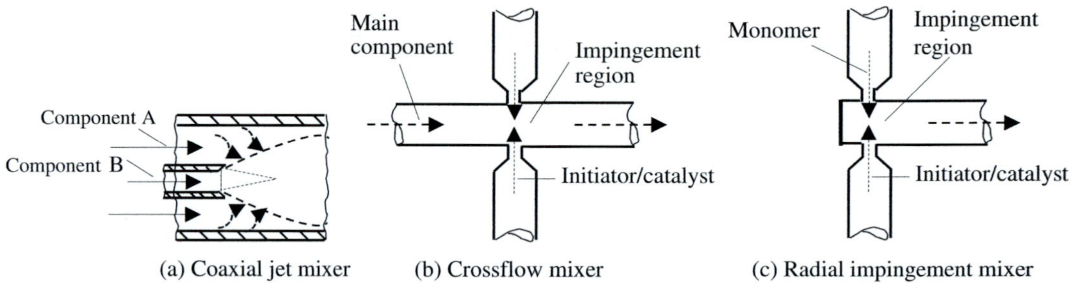

FIGURE 11.10 a–c. Schematics of common configurations of confined jet mixers in practice: (a) coaxial, (b) cross-flow and (c) radial impingement jet mixers.

REFERENCES

1. Dickey, D. S., *Mixing and Blending*, Wiley Online Library, 16 April 2010. https://doi.org/10.1002/0471238961.1309240908051318.a01.pub3
2. https://www.shutterstock.com/search/pipe+reducer
3. https://www.spiraxsarco.com/learn-about-steam/steam-distribution/pipes-and-pipe-sizing
4. Coldrey, P. W., Paper to I. Chem. E., *"Mixing in the Process Industries" Course*, Bradford University, UK, July 1978.
5. BHRG-FMP, Mixing time for jet mixing systems, *FMP-067*, June 2000.
6. Rajaratnam, N., *Turbulent Jets*, Elsevier Scientific Publishing, Co., Amsterdam, The Netherlands, 1976.
7. Ricou, F. P., Spalding, D. B., Measurement of entrainment by axisymmetric turbulent jets, *J. Fluid Mech.*, 11, pp. 21–32, 1961.

8. Okita, N., Oyama, Y., Mixing characteristics of jet mixing, *Chem. Eng.*, 27, pp. 252–260, 1963.
9. Grenville, R., Tilton, J. N., Turbulence for flow as a predicator of blend time in turbulent jet mixed vessels, in *Proceedings of the Nineth European Conference on Mixing*, Paris, France, pp. 67–74, 18–21 March 1997.
10. Corrsin, S., The isotropic turbulent mixer: Part II Arbitrary Schmidt number, *AIChE J.*, 10, pp. 870–877, 1964.
11. Grenville, R., Tilton, J. N., A new theory improves the correlation of blend time data from turbulent jet mixed vessels, *Trans. IChemE.*, 74(A), pp. 390–396, April 1996.
12. van de Vusse, J. G., Vergleischende Ruhrversuche zum Mischen loslicher Flussigkeiten in einem 1200m^3 Behalter, *Chem. Ing. Tech.*, 31(9), pp. 583–587, 1959.
13. Grenville, R. K., Tilton, J. N., Jet mixing in tall tanks: Comparison of methods for predicting blend times, *Chem. Eng. Res. Des.*, 89(12), pp. 2501–2506, 2011.
14. Revill, B. K., Jet Mixing, in *Mixing in Process Industries*, Eds. N. Harnby, M. F. Edwards and A. W. Nienow, p. 159, Butterworth-Heinemann, 1992.
15. Himmelsbach, W., Houlton, D., Keller, W., Lovallo, M., Mixing Systems: Design and Scale-up, *Chem. Eng.*, pp. 46–53, April 2006.
16. Fossett, H., Prosser, L. E., The Application of Free Jets to the Mixing of Fluids in Bulk, *Proc. Inst. Mech. Eng.*, 160, p. 224, 1949.
17. Fossett, H., The action of free jets in the mixing of fluids, *Trans. IChem. Eng.*, 29, p. 329, 1951.
18. Fox, E. A., Gex, V. E., Single phase blending of liquids, *AIChE J.*, 2, p. 539, 1956.
19. Hiby, J. W., Modigell, M., Experiments on jet agitation, in *Paper Presented at 6th CHISA Conf.*, Prague, 1978.
20. Rácz, I., Wassink, J. G., Strömungsverlauf und Mischzeiten in axialen Strahlmischern, *Chemie-Ingenieur-Technik*, 46, p. 261, 1974.
21. Lane, A. G. C., Rice, P., An experimental investigation of jet mixing employing an inclined side entry jet, *Trans. Inst. Chem. Eng.*, 60, p. 171, 1982.
22. Wasewar, K. L., A design of jet mixed tank, *Chem. Biochem. Eng. Q.*, 20(1), pp. 31–46, December 2005. https://www.researchgate.net/publication/27189305
23. Grenville, R. K., Mak, A. T. C., Ruszkowski, S. W., Blending of fluids in mixing vessels by turbulent re-circulating jets, in *The IChemE Research Event*, University of Manchester Institute of Science and Technology, UK, pp. 128–130, 1992.
24. Yu, Z., Kar, K., Storrs, T., Jian, C., Cope, R., Feng, J., Aglave, R., Oscillation of mixing jets in a long horizontal tank, in *Proceedings Paper FEDSM2014-22034, ASME 2014 4th Joint US-European Fluids Engineering Division Summer Meeting*, Chicago, July 2014.
25. Cooper, P., Schiavello, B., de Marolles, C., de Sallis, J., Prang, A. J., Broussard, D. H., *Tutorial on Multiphase Gas-Liquid Pumping*, Texas A&M University, OakTrust Library, 1996.
26. *Jet Mixing Systems Explained*, Mixing Systems Inc. https://www.mixing.com/jet-mixing

12 In-Line Motionless Mixers

12.1 INTRODUCTION

This chapter is dedicated to non-moving, or static, devices that are installed inside of pipes and tubes to mix multiple components. Such devices include well-known in-line static mixers such as those shown in Figure 12.1. These are widely used in polyethylene loop reactors for controlled polymerization of olefins and in the recirculation lines of stirred tank reactors for addition of dilute homogeneous and heterogenous catalyst slurry streams to the working media. Other major applications of static mixers include admix of initiator to epoxy resins, flocculants to wastewater, pigments to latex and additives to home and healthcare products.

(a)

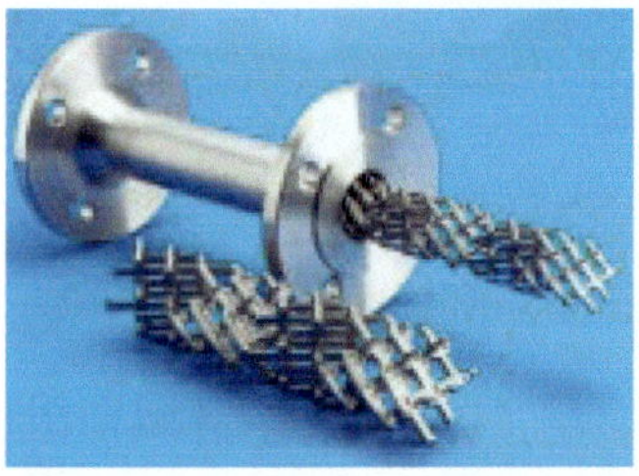

(b)

FIGURE 12.1 Static mixers by: (a) Kenics KM (NOV- Chemineer)[1] and (b) SMX (Sulzer Chemtech AG).[2]

12.2 STATIC MIXERS

12.2.1 Working Principles

In the case of laminar flow, lamellar fluid structures undergo multiple splits, rotations, flow reversals and recombination while passing through the tortuous path comprising well-designed static mixer elements with opposite chirality (see Figure 12.2). During this process, multidirectional interception of fluid streams produces chaotic advection phenomenon to enhance global transport processes such as bulk diffusion.

In the case of turbulent flows, static mixer elements promote additional turbulence to the feed streams and generate intense radial mixing, even near the wall. In the turbulent regime, eddy

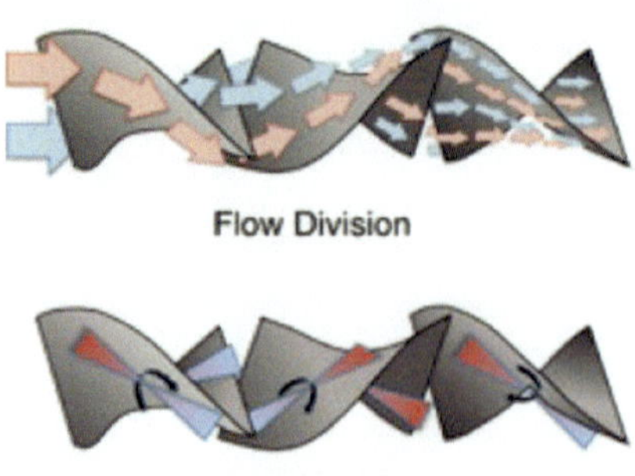

FIGURE 12.2 Illustration of flow division, rotation and radial mixing in a twisted tape static mixer.[3]

DOI: 10.1201/9781003606390-12

diffusion provides sufficient mixing for most industrial processes, i.e., blending, dispersion (liquid-liquid or gas-liquid), heat and mass transfer.

Figure 12.3 shows a simple pipe with flow obstructions such as tabs attached to the wall that alter the direction of turbulent flow streams to promote chaotic mixing. The HEV™ mixing element consists of special patented trapezoidal tabs mounted at an acute angle relative to the downstream surface of the mixer housing. The length, width and attack angle of the HEV mixing elements have been optimized to maximize mixing performance while limiting pressure drop. The alternating rotations of the tip vortices induce vigorous cross-stream mixing that results in rapid uniformity of the process components.[4]

FIGURE 12.3 Photographs of Kenics HEV high-efficiency static mixer (NOV-Chemineer Inc.).[4]

12.2.2 Static Mixer Elements

A long list of commercially available static mixers and their corresponding manufacturers is presented in the review article by Ghanema et al.[5] Table 12.1 presents photographs of a few common commercial static mixers and their typical applications.[6,7] Readers are advised to review the static mixer product bulletins of StaMixCo LLC, Charles Ross & Son Company (Ross), Komax Systems Inc. (Komax) and SPX Flow Systems (Lightnin), since each offers slightly differentiated products for various applications. Typical material of construction (MOC) of the static mixers are stainless steel, plastics, Teflon® and glass-lined carbon steel; housings are made of stainless steel or Fiberglas® depending on their applications.

TABLE 12.1
Several Commercially Available Static Mixers and Their Applications

Manufacturers	Models and Applications	Photographs
NOV Chemineer Inc. Kenics™ Series[6]	KM™ Laminar and turbulent blending, liquid–liquid and gas–liquid dispersions	
	KMX™ Laminar blending	
Vortex mixer	HEV™ Turbulent blending. Re > 10,000.	

(Continued)

TABLE 12.1 *(Continued)*
Several Commercially Available Static Mixers and Their Applications

Manufacturers	Models and Applications	Photographs
Sulzer Chemtech AG[7]	SMV™ Distributive and homogeneous mixing or blending in turbulent regime	
	SMX™ Homogenization and dispersing in a laminar regime, and dosing of a low-viscosity component into a high-viscosity media	
	SMI™, KVM™ Liquid–liquid mixing in the turbulent regime, and gas–liquid mixing	
	CompaX™ Admixing of additives in the turbulent flow regime	 A circular static mixer with a complicated interior structure is used to blend ingredients in the flow regime.

12.2.3 Reactors with Static Mixers

Helical ribbon or twisted tape static mixer elements, when brazed to the inner walls of the tubes in a shell-and-tube heat exchanger, significantly improve mixing and heat transfer (three to five times) compared to one with mixer-less (empty) tubes. Tubular reactors filled with static mixing elements are often used as plug flow reactors for both high- and low-viscosity chemical reaction systems.

Figure 12.4 shows the photograph of a compact commercial Kenics® static mixer heat exchanger/reactor, used as an in-line polymerization reactor.[8]

FIGURE 12.4 Shell-and-tube heat exchanger and/or plug flow reactor with Kenics® static mixers.[8]

Figure 12.5a shows the StaMixco company's tubular plug flow reactor with GX™ static mixers (see Figure 12.5b) for viscous media.[9]

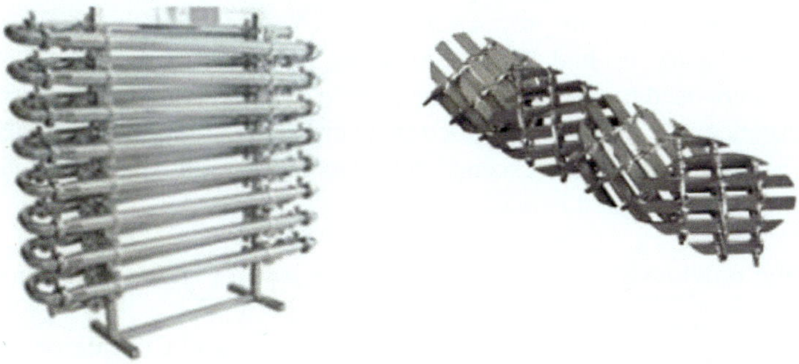

(a) (b)

FIGURE 12.5 (a) Layout of a skid-mounted, small-diameter, long residence time static mixer plug-flow reactor, (b) StaMixco GX™ static mixer for mixing of viscous media.[3]

In empty tubes or pipes, a laminar velocity profile is parabolic, with the centerline velocity being two times the cross-sectional average velocity. Adding static mixers, however, forces radial and rotational fluid transport to create a nearly flat velocity profile (see Figure 12.6). The advantage of this uniform velocity profile (plug flow) is that it produces trans-lamellar shear mixing and a shorter residence time distribution (RTD).

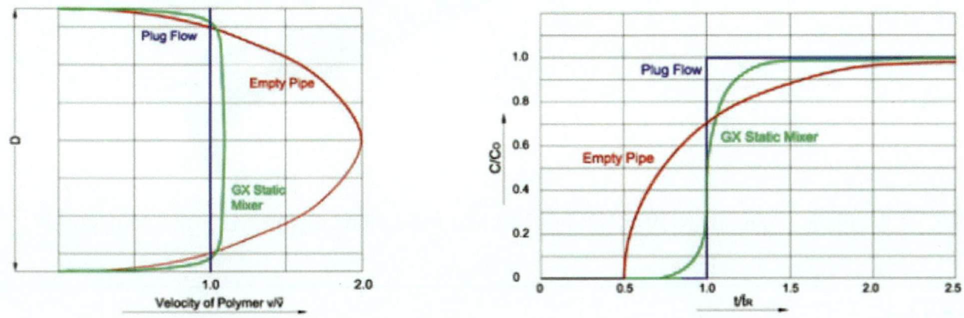

(a) (b)

FIGURE 12.6 (a) Velocity profiles and (b) Step Response Curves in a pipe that is either empty (red), containing a GX static mixer (green), or ideal plug flow (blue).[9]

12.2.4 CoV and Mixing Length

The coefficient of variation (CoV) is recognized as a simple parameter to quantify the quality of mixing delivered by a static mixer. CoV is defined in Equation 12.1.

CoV = Standard Deviation/Mean

$$C_v = \sigma / \bar{\mu} \tag{12.1}$$

A CoV of 0.05% corresponds with mixing to 95% homogeneity. To make such a determination, a tracer such as a salt solution is typically injected into the main stream. Then, concentration changes due to mixing are detected by conductivity probes mounted at various downstream locations. Multipoint detections at a single pipe cross-section can yield a concentration mean value and standard deviation, which can produce a local CoV. For example, at cross-section A-A', in Figure 12.7, the values of six residual chlorine readings, μ, are 3, 5, 3, 4, 6, 5 at a given point in time.

Other typical tracers used to characterize the degree of mixing are dyes, neutralizer (pH adjusters), iodine fluorescence, etc. Tracers are added in a very short duration (i.e., $t \approx 0s$) and monitored by optical probes, pH probes, or fluorescence detectors at different spans. The calculation of CoV for the above example is presented below.

The mean of the previously listed n = 6 readings is $\bar{\mu} = \left(\sum_{1}^{n} \mu / n\right)$. The standard deviation of the six readings is $\sigma = \sqrt{\sum_{1}^{n} \left(\mu - \bar{\mu}\right)^2 / \left(n-1\right)}$. Values of $\bar{\mu}$, σ, and CoV are computed as shown in Table 12.2.

The lower the value of CoV, the better the mixture quality. From Table 12.2, the calculated value of CoV is 0.28 or 28% variation. Such a value is rather high. In most instances, a value ≤5% variation is very acceptable and corresponds to ≥95% mixing homogeneity. However, in the case of colorant/pigment distribution in paints, ideally ≥99% homogeneity is required to provide uniform hiding and opacity characteristics.

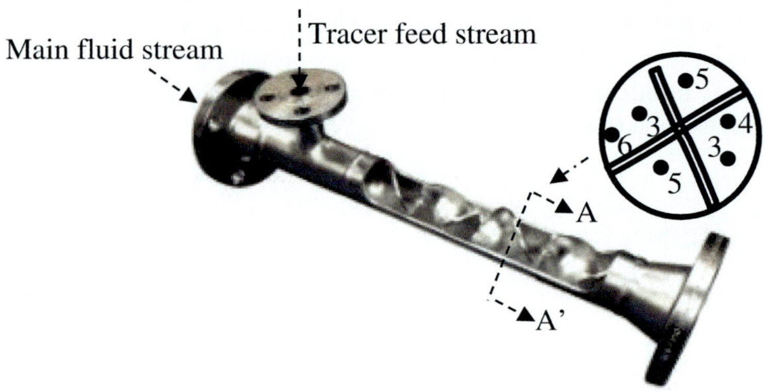

FIGURE 12.7 Photograph of a NOV–Chemineer Kenics mixer showing the tracer or admix port.[10]

TABLE 12.2
Calculation of the Coefficient of Variation, CoV

μ	$\bar{\mu}$	$(\mu-\bar{\mu})$	$(\mu-\bar{\mu})^2$	σ	CoV
3		−1.33	1.77		
5		0.66	0.44		
3		−1.33	1.77		
4		−0.33	0.11		
6		1.66	2.76		
5		0.66	0.44		
$\sum_{1}^{6} \mu = 26$	$\frac{1}{6}(26) = 4.33$		$\sum_{1}^{n} (\mu-\bar{\mu})^2 = 7.29$	$\sqrt{\frac{7.29}{(6-1)}} = 1.21$	$\frac{\sigma}{\bar{\mu}} = \frac{1.21}{4.33} = 0.28$

An alternate approach is to use computational fluid dynamics (CFD) to numerically monitor residual concentrations of a tracer at multiple grid points and several cross-sections, at a given time (or distance, since distance = time x velocity). Then, the location of the desired mixing level must be identified from a CoV vs. distance curve.

12.3 IN-LINE TEE MIXERS

12.3.1 Applications

For light-duty mixing and dispersion processes, in-line Tee mixers are quite popular because they adapt easily to pipe lines, using flanges, and are cost effective. Typical applications are safe dilution of chemicals such as sodium hypochlorite, sodium/ammonium hydroxide and sulfuric acid to the aqueous streams, and upstream injection of initiator into monomers feeding the plug flow reactor.

Figure 12.8a shows an Inyo Process' single Tee mixer.[11] Low-pressure feed of an admix (gas or liquid) into the main stream at the throat of a venturi is a unique approach, often practiced for rapid dispersion followed by diffusion. As shown in Figure 12.8b, aeration venturis are often used for gas–liquid contacting downstream of the contraction section. Accelerating fluid flow (velocity increases, pressure decreases) at the throat sucks in gas to create fine bubbles for enhanced mass transfer.

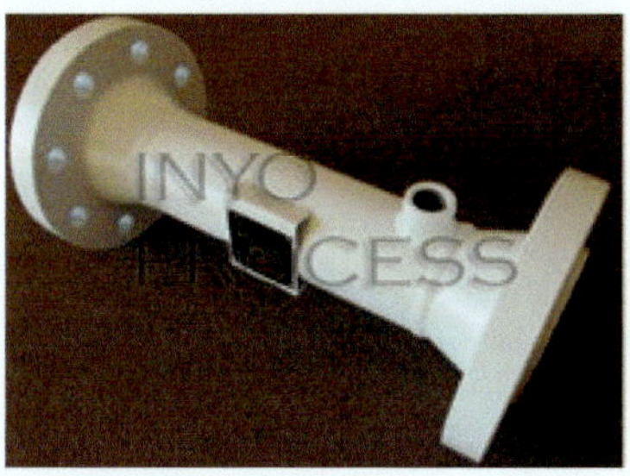

(a)

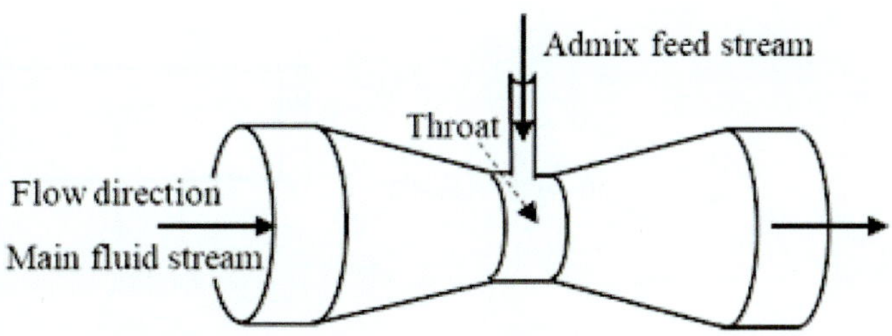

(b)

FIGURE 12.8 (a) Photograph of a simple pipe Tee mixer,[11] (b) schematic of a venturi as a Tee mixer.

12.3.2 Admix Feed System Design

As shown in Figure 12.9(a–b), to significantly shorten the mixing duration, a tee mixer injection tube should ideally extend up to the centerline (CL) of the main pipe to prevent the admix stream from merely hugging the wall. It is critical for the admix phase to be delivered into the highly turbulent region. The admix tube diameter, relative to the main pipe, is critical for several reasons. First, if possible, to maintain higher admix velocity with respect to the main flow velocity. Second, to create minimum obstruction to flow that may create secondary flow or a wake region behind the admix tube. In the industrial scale, where flowrates of both streams are equally prominent, in order to avoid entrainment of the admix stream into the main stream's secondary flow region, the admix tube needs to be elbowed and extended along the centerline of the pipe; furthermore, a discharge area that tapers on the downstream side, as shown in Figure 12.9b, is advised. Computational fluid dynamics modeling played a significant role in troubleshooting, redesigning and successfully implementing a large-scale pipeline Tee mixer (see Figure 12.10(a–b)), that added significant value to one world-scale manufacturing process.

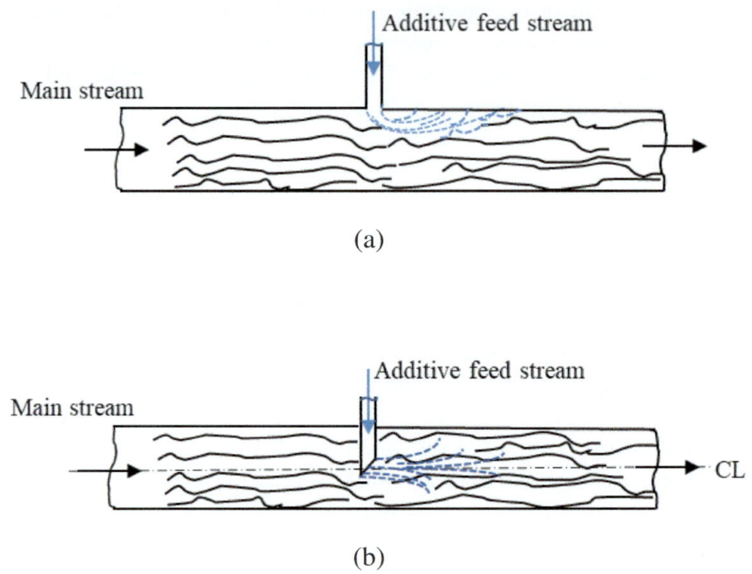

FIGURE 12.9 Schematics of Tee mixer admix tubes showing (a) wall hugging flow and (b) central flow near their discharge regions.

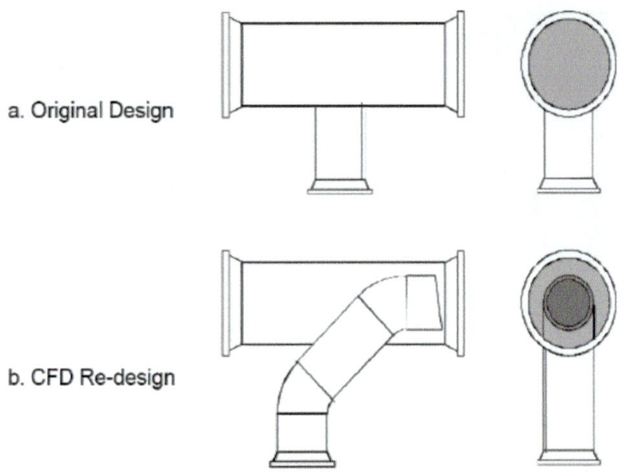

FIGURE 12.10 CAD drawings of a large-scale pipeline Tee mixer before and after the redesign.

12.3.3 Multi-Tee Mixers

In the case of fast-reacting components, Multi-Tee mixers are sometimes used to promote reaction and avoid byproducts. The objective is to uniformly deliver secondary streams into the main flow through multiple ports for faster mixing. Multi-Tee mixers are commercially available, but they are often designed and fabricated in-house for a user's specific applications. Typically, the secondary stream is delivered through a manifold to the multiple peripheral ports or admix tubes on the main pipe, as shown in Figure 12.11. The simple "rule of thumb" is that pressure drop across the port (orifice) or in the tube must be ≥ 10 times the pressure drop within the manifold to maintain uniform flow through each reactant injection port. The total cross-sectional area of all the injection tubes should not exceed the cross-sectional area of the manifold pipe.

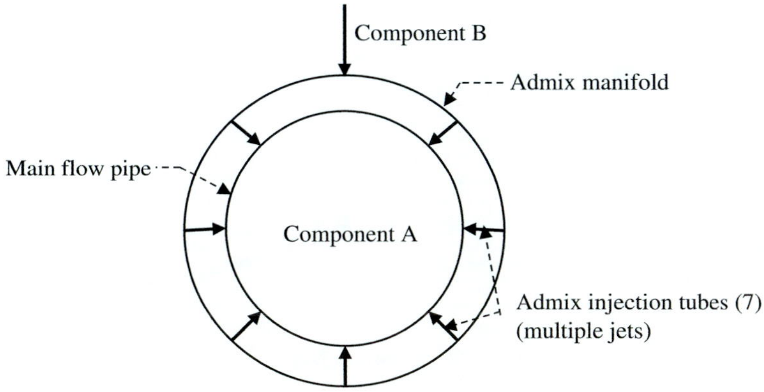

FIGURE 12.11 Schematic of a Multi-Tee mixer with injection manifold.

Figure 12.12 shows an Inyo multiport admix injection Tee mixer ring with a straight pipe manifold,[11] which is to be mounted between the flanges of the main pipe.

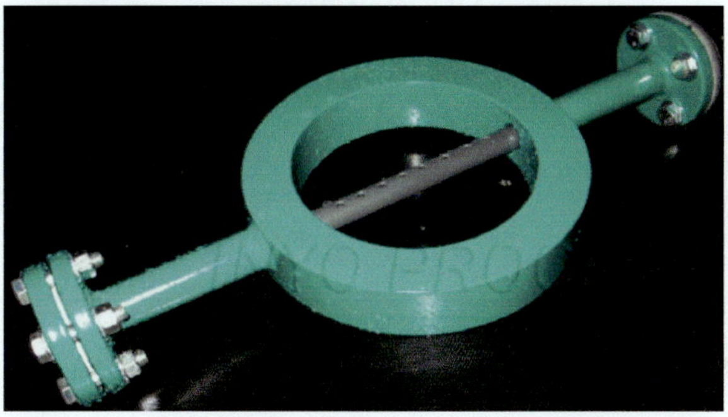

FIGURE 12.12 Inyo Process Tee mixer with multiple orifice injectors.[11]

12.3.4 Mixing in Empty Pipe

Orientation of the jet discharge angle, α, and its crossflow intensity with respect to the main flow, $\dot{q}/\dot{Q}$, dictate mixing efficacy of a Tee mixer. There are a few design correlations, for a limited number of jet configurations, that operate in the turbulent flow regime for pipe Tees based on experimental and numerical studies under reactive and non-reactive conditions.[12–15] A simple side jet in a pipe is shown in Figure 12.13. As expected, the non-dimensional mixing length, X_{95}/d, is a strong function of the ratio of jet (admix) velocity, u, to the main flow velocity, U (see Figure 12.14).[12]

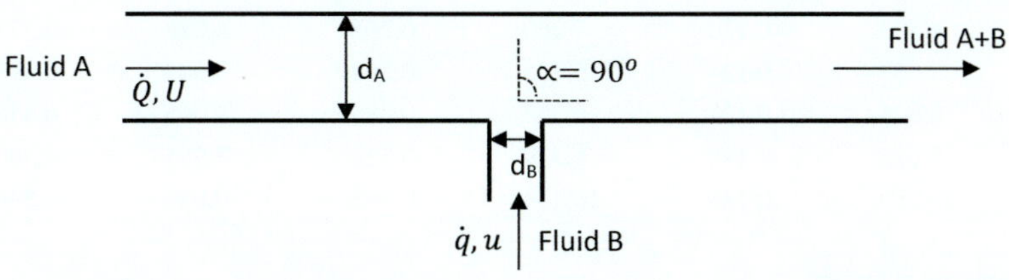

FIGURE 12.13 Schematic of a simple Tee mixer side pipe/jet (α = 90°) connected to a main pipe, respectively carrying fluids B and A.

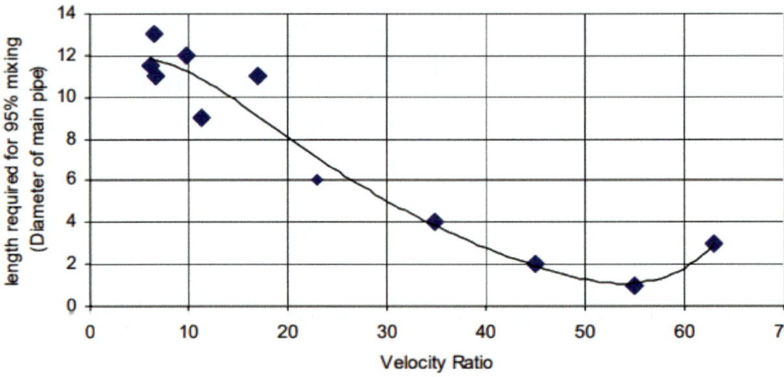

FIGURE 12.14 Graph of mixing length in diameters of main pipe versus *u/U*, for 95% mixedness in a ¼″ right-angled side tee.[12]

12.3.5 Static Mixer Pipes

The typical materials of construction (MOCs) of Tee mixers are stainless steel, Hastelloy, FRP, PVC, and PTFE (Teflon®)-, epoxy- or glass-lined carbon steel. Material of choice depends on the required chemical compatibility and erosion resistance in case of slurries. Table 12.3 presents the inner diameters and wall thicknesses of a few BS1600 Schedule steel pipes to assist practicing engineers in the sizing of Tee and static mixers.

TABLE 12.3
Physical Dimensions of BS1600 Schedule Steel Pipes

Nominal Pipe Size	Outside Diameter	Wall Thickness	Inside Diameter	Wall Thickness	Inside Diameter	Wall Thickness	Inside Diameter
		Schedule 10		Schedule 40		Schedule 80	
1/4	0.540	0.065	0.410	0.088	0.364	0.119	0.302
1/2	0.840	0.065	0.710	0.109	0.622	0.147	0.546
3/4	1.050	0.083	0.884	0.113	0.824	0.154	0.742
1.00	1.315	0.109	1.097	0.133	1.049	0.179	0.957
1-1/4	1.660	0.109	1.442	0.140	1.380	0.191	1.278
1-1/2	1.900	0.109	1.682	0.145	1.610	0.200	1.500
2.0	2.375	0.109	2.157	0.154	2.067	0.218	1.939
2.5	2.875	0.120	2.635	0.203	2.469	0.276	2.323
3.0	3.500	0.120	3.260	0.216	3.068	0.300	2.900
4.0	4.500	0.120	4.260	0.237	4.026	0.337	3.826
5.0	5.563	0.134	5.295	0.258	5.047	0.375	4.813
6.0	6.625	0.134	6.357	0.280	6.065	0.432	5.761
8.0	8.625	0.148	8.329	0.322	7.981	0.500	7.625
10	10.750	0.148	10.454	0.365	10.020	0.594	9.562

Note: All dimensions are in inches.

12.3.6 Design of In-Line Mixers

Pressure drop, Δp, for steady-state incompressible fluid flow through a constant diameter horizontal pipe is presented as follows by the well-known Darcy-Weisbach friction loss equation:

$$\Delta p = f\,\rho L v^2\,/\,2d \tag{12.2}$$

where, f = friction factor [-]
ρ = density of the fluid [kg/m³]
L = length of pipe [m]
v = velocity in the pipe [m/s]

$$v = \dot{Q}\,/\,A = \dot{Q}\,/\left(\frac{\pi d^2}{4}\right)$$

where, $\dot{Q}$ = flowrate [m³/s], A = pipe area of cross-section [m²],
d = inner diameter of the pipe [m], Δp= pressure drop [Pa]

Flow Reynolds number, Re, in the pipe is defined as $Re = \rho v d\,/\,\mu$.

As mentioned earlier, flow in the pipe is laminar if $Re \lesssim 2000$. The Darcy friction factor is $f = 64\,/\,Re$. For laminar flow, pressure drop, Δp_L, over a pipe length **of** L is determined from Poiseuille's law.

$$\Delta p_L = 128\dot{Q}\,\mu L\,/\,\pi d^4 \tag{12.3}$$

For fully turbulent flow, $Re \gtrsim 3000$, friction factor, f, is determined from the Moody diagram,[16] which is a function of Re and the pipe roughness factor, $\in/d$ (see Figure 12.15). A typical roughness value, $\in$, of drawn tubes is 0.0025mm. Turbulent flow pressure drop across the empty pipe is

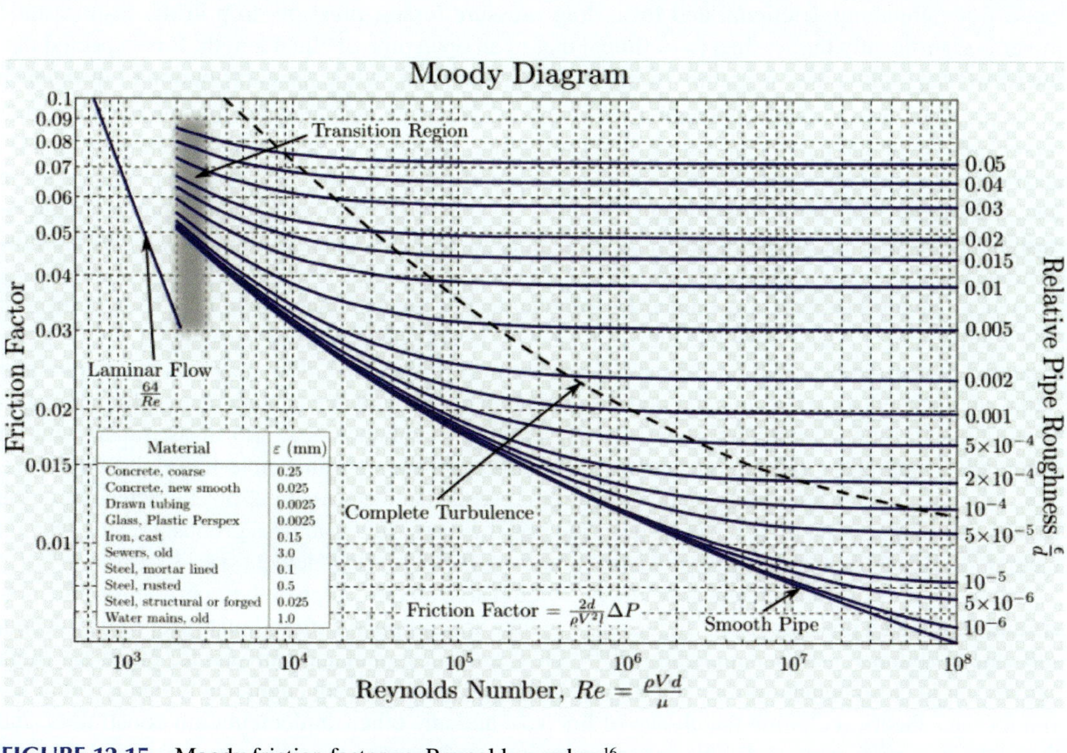

FIGURE 12.15 Moody friction factor vs. Reynolds number.[16]

determined using equation 12.2. An explicit correlation for the Moody friction factor is presented in equation 12.4.

$$f = \frac{1.325}{\left[\ln\left(\frac{\in}{3.7d} + \frac{5.74}{Re^{0.9}}\right)\right]^2} \text{ for } 5000 \le Re \le 10^8 \text{ and } 10^{-6} \le \frac{\in}{d} \le 10^{-2} \tag{12.4}$$

The implicit Colebrook equation[17] (equation 12.5) is used to solve the Darcy friction factor, f, for $Re \gtrsim 4000$.

$$\frac{1}{\sqrt{f}} = -2.0\,log\left(\frac{\in / d}{3.7} + \frac{2.51}{Re\sqrt{f}}\right) \tag{12.5}$$

If, instead of the Moody friction factor, f, the Fanning friction factor, f', vs. Re curve is used, the relationship $f = 4f'$ must be applied to the Moody pressure drop and friction factor equations, i.e.,

$$\Delta p = (4f')\rho L v^2 / 2d = (2f')\rho L v^2 / d \tag{12.6}$$

The Fanning friction factor, f', for laminar flow is

$$f' = 16 / Re \tag{12.7}$$

The length of open pipe required to blend two streams to 95% homogeneity is a strong function of flow turbulence. Although the well-designed Tee mixers shorten the pipe length, it is not to the same extent as that produced by wall tabs or static mixer elements. Because obstructions to free flow cause flow stretching, frictional and form drag pressure losses, pressure drop in the in-line static mixer is significantly higher than (>10 times) that in an open pipe of equal length. It is expected that the resistance to fluid flow is mainly from the mixing elements, while the contribution of turbulence to the pressure drop is relatively small. Energy loss, $\dot{Q}_s \times \Delta p$, associated with flow through the static mixer is the essential parameter responsible for mixing. Hence, the magnitude of pressure loss, Δp, across the static mixer is an indication of its mixing performance. Normally, the higher the pressure loss, the shorter will be the mixer length required to achieve the desired CoV. The pump should be designed to deliver the required pressure, plus more to accommodate additional losses in the flow system.

Helical ribbon elements are known for producing the least pressure drop of all static mixers. The static element's length-to-diameter ratio, L / D, is critical to mixing—usually, 1.0 to 2.5. As expected, when the aspect ratio decreases, pressure drop increases, flow becomes predominantly radial as opposed to axial, and the degree of turbulence is enhanced. Turbulence characteristics of the fluid in the static mixers are effectively applied to enhance heat and mass transfer and to accomplish multiphase mixing such as gas–liquid contacting, or immiscible liquid–liquid dispersion. Dimensional analysis based on the π theorem reveals that the pressure drop characteristic of the Kenics and Sulzer static mixers can be described by the three dimensionless groups, i.e., the friction factor (f'), Reynolds number (Re), and aspect ratio (AR) of a mixing element, i.e., L / D.[18]

Manufacturers of static mixers are quite knowledgeable on the design of their pertinent static mixers and anticipated pressure drops. In most cases, they are equipped with in-house proprietary design software to address customer applications.[19–25] Hence, the design of static mixers for plants are often carried out in close collaboration with their manufacturers. Practicing engineers typically fill in the vendor data sheets, including the choice of line size and any other implementation constraints, and then rely on manufacturers' design recommendations such as type of static mixer, required size and number of elements, and expected pressure loss. In cases of particular uncertainty, R&D efforts tend to validate the design in the pilot plant or to conduct scaled-down laboratory experiments with simulated fluids and non-intrusive colorimetry (DISMIT) or laser-induced fluorescence (LIF) techniques. See Chapter 18 for details of these experimental techniques. Recent numerical simulations of static mixers using commercial CFD packages have been an advantageous approach.

The BHR Group has, during their HILINE research program, done extensive research on various commercially available static mixers.[26] Their proprietary design guidelines are an outstanding resource for practicing engineers. The authors often used them for preliminary design of static mixers prior to the vendor selection and in-house pilot studies. However, here, based on published correlations, the mixit-io SS12.1 (Static Mixer Design) provides a blending workout guide to determine pressure drop in static mixers. At the end of this chapter, authors illustrate designs of a couple industrial-scale static mixers with examples.

It is always worth adding a few more elements to assure a better CoV and a reaction process that is not mixing controlled. Additional energy is always required, considering the typical consumption of energy in the production plant, unless it's a retrofit challenge facing site footprint or pump delivery pressure constraints.

Figure 12.16 represents a generic computer aided design (CAD) drawing of a static mixer tube with flanges and a secondary stream addition Tee pipe. Because of the continuation of lamellar ultra-thin fluid structures (laminar flow) and the trailing vortices (turbulent flow) downstream of the static mixers, mixing persists to some extent in the open pipe. Designers must allow required tailpipe length for such beneficial mixing. Furthermore, mixing improves with intermittent spacing between a set of mixer elements and injection of the admix, which is positioned after the first set of elements. Such designs are typical for laminar/transitional flows. The admix stream is generally the lower-viscosity component relative to the mainstream. In the case of high-viscosity ratio components (≥10,000:1), multiple injection ports are used for graduated addition of admix to avoid creation of viscous blobs that take much longer time to mix.

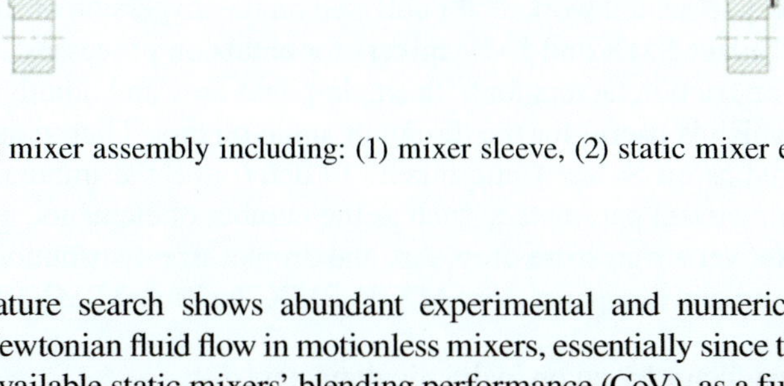

FIGURE 12.16 Static mixer assembly including: (1) mixer sleeve, (2) static mixer elements, and (3) flange gasket.[27]

The authors' literature search shows abundant experimental and numerical research involving Newtonian and non-Newtonian fluid flow in motionless mixers, essentially since their inception. In most cases, commercially available static mixers' blending performance (CoV), as a function of pressure loss (Δp), has been compared with that of the simple open pipe. Here, for completeness, a few academic and industrial experimental studies that developed empirical correlations for the friction factor of static mixers are included. Li et al.[28] pursued experimental research with different numbers ($n_e = 6,12,18$) of Sulzer SMX mixers (tube id, $d = 1.6cm$, aspect ratio, $AR = 0.8$, porosity, $\varnothing = 0.84$) in a stainless-steel tube using CMC, glycol-water solutions and water at varying flowrates (2.78–111.2 m^3/s). Following the Fanning friction factor, f', correlations were developed for SMX mixers from their 397 experiments.

Laminar flow regime ($0.08 \leq Re_g \leq 15$):

$$f' = 368 / Re_g \tag{12.8}$$

Transition flow regime ($15 < Re_g \leq 1000$):

$$f' = \frac{220}{\left(Re_g\right)^{0.8}} + 0.8 \tag{12.9}$$

Turbulent flow regime ($Re_g > 1000$):

$$f' = 12 / Re_g^{\,0.25} \tag{12.10}$$

For laminar flow in an empty tube ($Re_g \leq 2100$), the Fanning friction factor vs. Reynolds number relationship was verified with water and glycol-water solutions, i.e.,

$$f' = 16 / Re_g \tag{12.11}$$

where, Re_g is the generic Reynolds number based on the apparent viscosity of the fluid near the tube wall.

In the laminar flow regime, for an SMX static mixer, pressure drop increases by a factor of 23 compared to flow through an open pipe, all other conditions being equal.[28] As expected, desired homogeneity comes with a pressure drop penalty.

Later, Morancais et al.,[29] using Newtonian fluids of different viscosities, developed a universal experimental equation for SMX mixers for the laminar and inertia regimes. The static mixer was considered as a porous media. Pressure drop was analyzed using a capillary model for the determination of such geometric parameters as pore diameter, d_p, and tortuosity. Fanning friction factor, f', as a function of pore Reynolds number, Re_p, is represented in equation 12.12.[29] The transition from a laminar flow to an intermediate flow occurred at a pore Reynolds number of about 200. The turbulent flow regime was observed at a pore Reynolds number of between 1500 and 3000.

$$f' = 16 / Re_p + 0.194 \tag{12.12}$$

where, $Re_p = \rho v_p d_p / \mu$ and pore or interstitial velocity, $v_p = \dot{Q} / \varnothing A = \dot{Q} / \left(\frac{\pi d_p^{\,2}}{4}\right)$

There are several experimental works[30–36] published on the dispersion of immiscible liquid using Kenics KM and also Sulzer SMX and SMV mixers for emulsion processes. These articles provide multiple correlations on friction factors for both single-phase flow and liquid–liquid flows in the turbulent flow regime, which is useful for the design of static mixers. Theron and Le Sauze[34] focused their experimental studies on Sulzer static mixers to determine the influence of dispersed-phase concentration and geometrical parameters, such as the number of elements, n_e, and on the structure of the motionless mixer versus pressure drop, Δp, and droplet size distribution (dsd) characteristics. The three motionless mixers compared are SMX™, SMXPlus™ and SMV™.

(a) Fanning friction factor based on hydraulic diameter (d_h):

$$f_h' = \frac{\Delta p \varnothing^2 d_h}{2 \rho v_o^{\,2} L} \tag{12.13}$$

(b) Reynolds number based on hydraulic diameter:

$$Re_h = \frac{\rho v_o d_h}{\varnothing \mu} \tag{12.14}$$

where, v_o is velocity of the working media through the open pipe, and $\varnothing$ is the mixer porosity, i.e., void fraction in the presence of the static mixer.

In the turbulent regime, using a linear profile through a logarithmic representation, experimental correlation established a Blasius type of equation.

$$f_h' = \frac{2C_1}{{Re_h}^{0.25}} \tag{12.15}$$

where, $C_1 = 2, 1, or\, 1$, for SMX, SMV or SMXPlus static mixers, respectively.

Pressure drops of SMV and SMXPlus is half of what's expected in the SMX. This is also valid for dilute emulsions up to 25% dispersed phase. Continuous-phase physical properties are used for the determination of the above non-dimensional parameters, f_h' and Re_h.

In one research effort,[34] experimental results with a water plus Tween80™/cyclohexane emulsion system correlate with Sauter mean diameter, d_{32}, as a function of the mean energy dissipation per unit mass of the fluid, ε_m. The continuous-phase density, ρ_c, was held constant in all cases. See Table 12.4 for geometric characteristics of the three different mixers used.

$$d_{32} = C_2 {\varepsilon_m}^k \tag{12.16}$$

Predicted k values are −0.36, −0.52 and −0.62, respectively, for the SMX, SMXPlus and SMV static mixers. Thus, with the similar emulsion system, SMV produces smaller d_{32} compared to SMX and SMXPlus at equal mean specific energy, ε_m.

With the advancement of numerical techniques, pressure drop data are generated from computational fluid dynamics (CFD) calculations, thus avoiding typical experimental limitations in obtaining data that is comprehensive enough to develop a reliable pressure drop correlation.

Finally, the authors present research with Kenics static mixers wherein the aspect ratio, AR $\left(L/D = 1, 1.5, 2.0, 2.5\right)$, is included in the pressure drop correlation covering the whole range of Reynolds number, Re, from laminar to turbulent.[35] The reliability of the proposed correlation was validated by comparison with various pressure drop data reported in the literature.

$$f' \,.\, \left(AR\right)^{2.04} = k\left[\frac{Re}{\left(AR\right)^{2.15}}\right]^n \tag{12.17}$$

In the three distinct regimes, k and n values are presented in Table 12.5.

TABLE 12.4
Geometric Characteristics of the Three Mixers Used by Theron and Sauze[34]

Mixer's Design	SMX	SMX+	SMV
D (mm)	10.15	10.30	9.45
L/D	≈1	≈1	≈1
D_h (mm)	2.45	2.1	3.5
Φ	0.67	0.84	0.83
Crossbars number (for SMX and SMX+) or corrugated plates number (SMV)	6	6	5
Crossbars (for SMX and SMX+) or corrugated plates (SMV) thickness (mm)	0.99	1.10	0.14

TABLE 12.5
***k* and *n* Values for Different Regimes**[35]

Regime	$0 < Re/(AR)^{2.15} < 100$	$100 < \frac{Re}{(AR)^{2.15}} < 1000$	$1000 < Re/(AR)^{2.15}$
k	320	32	2.66
n	–0.86	–0.36	0

As mentioned earlier, performance of the static mixer is a strong function of the pressure drop in its elements due to fluid friction and/or form drag that facilitates stretching and directional change of the mixing media. The structural geometries of the elements are optimized to address laminar and/or turbulent fluid flow (see Table 12.1). Furthermore, for a particular type of static element, a couple of its predominant fabrication features dictate the pressure drop: the aspect ratio of the element, *AR*, and the thickness of the ribbons/plates/bars. These significantly alter the three parameters such as the pitch/twist angle with respect to the pipe axis, packing void fraction, and consequently the hydraulic diameter. Although there are numerous pressure-drop correlations in literature, these are pertinent to a specific type(s) of static mixer(s) and its limited variations in geometry; thus, they lack the usefulness for industrial-scale applications. In addition, published correlations related to homogeneity, in terms of CoV as a function of mixer length, are rare. As stated before, designers most often rely on vendor recommendations and lean towards overdesigning of their mixing systems. Therefore, it is imperative to add a couple more additional static mixer elements to the design and incorporate enough flexibility for assembly/disassembly and routine maintenance of the mixer.

Streiff et al. (1999)[36] have pursued extensive research and critical review of published literatures on various static mixers. Here, the authors briefly describe their empirical correlations related to pressure drops, blend performance (CoV) and shear rates of a few selected static mixers from the published articles.

(a) Respectively, the laminar and turbulent flow pressure drops in the mixers, i.e., Δp_{mL} and Δp_{mT}, with respect to the pressure drop in an empty pipe, Δp_p, of equal diameter and length are expressed in equations 12.18 and 12.19. The constants K_L and K_T are presented in Table 12.6.

$$\Delta p_{mL} = K_L \times \Delta p_p \tag{12.18}$$

$$\Delta p_{mT} = K_T \times \Delta p_p \tag{12.19}$$

TABLE 12.6
Values of Constants Used in Equations 12.18–12.19 and 12.22–12.23 for Different Type Mixers[36]

	Laminar Flow		Turbulent Flow		
Mixers	K_L	K_{iL}	$2f'$	K_T	K_{iT}
Empty pipe	1	–	0.01	1	0.95
KMS—Twisted type	6.9	0.87	1.5	150	0.50
SMV—Corrugated	–	0.79	1–2	100–200	0.21–0.46
SMX—Rod structured (45 deg)	37.5	0.63	5	500	0.46
SMXL—Rod structured (30 deg)	7.8	0.85	1	100	0.87

where, $\Delta p_p = \left(2\rho_m \times v_m^{\ 2} \times \frac{L_m}{D}\right) f'$

L_m = length of the mixer, D = diameter of pipe,
ρ_m = density of mixture fluid, v_m = velocity of mixture fluid in the pipe

The Fanning friction factors, f', for laminar and turbulent flows in the empty pipe are a function of the pipe flow Reynolds numbers as shown:

$$\text{Laminar flow, } Re < 2000 \Rightarrow f' = 16 / Re \tag{12.20}$$

$$\text{Turbulent flow, } 4000 < Re < 100{,}000 \Rightarrow f' = \frac{0.074}{Re^{0.25}} \tag{12.21}$$

(b) The degree of homogeneity is a function of the type of mixer element, its diameter, D, and length of the mixer, L_m, as shown below in the exponential form. The homogeneity constants, K_{im}, of these mixers are presented in Table 12.6.

$$\text{Laminar flow, } CoV_r = \left(K_{iL}\right)^{L_m/D} \tag{12.22}$$

$$\text{Turbulent flow, } CoV_r = \left(K_{iT}\right)^{L_m/D} \tag{12.23}$$

where, CoV_r is a reduced coefficient of variation from the initial unmixed state, CoV_0, defined as,

$$CoV_r = \frac{CoV}{CoV_0} \tag{12.24}$$

(c) In the case of laminar flow, the shear rate, $\dot{\gamma}$, experienced by the fluid in an empty pipe and in presence of the mixers are given by the following equation:

$$\dot{\gamma} = \kappa \frac{v_{ave}}{D} \tag{12.25}$$

where, κ = shear rate constant (see Table 12.7), v_{ave} = average fluid velocity through the pipe/mixer and D = diameter of pipe.

A publication by Myers et al.[37] addresses functional aspects of static mixers versus mechanical agitation systems in the flowchart shown in Figure 12.17. There are several guidelines and recommendations for static mixer selection and design pertinent to NOV–Chemineer's Kenics and HEV mixers. For example, Table 12.8 provides a priori information to practitioners on the number of helical elements needed when designing the mixer based on flow Reynolds number.

Many static mixer types are offered commercially. The authors are familiar with the performance of only a few, as discussed in this book, without bias towards any. Table 12.9 gives rough

TABLE 12.7
Approximate Shear Rate Constants, κ for Laminar Flow through Mixers[36]

Mixers	Empty Pipe	Kenics (helical/twisted)	SMX (rod-structured)	SMXL (rod-structured)	SMV (corrugated plate)
Shear rate constant, κ	8	28	64	30	50

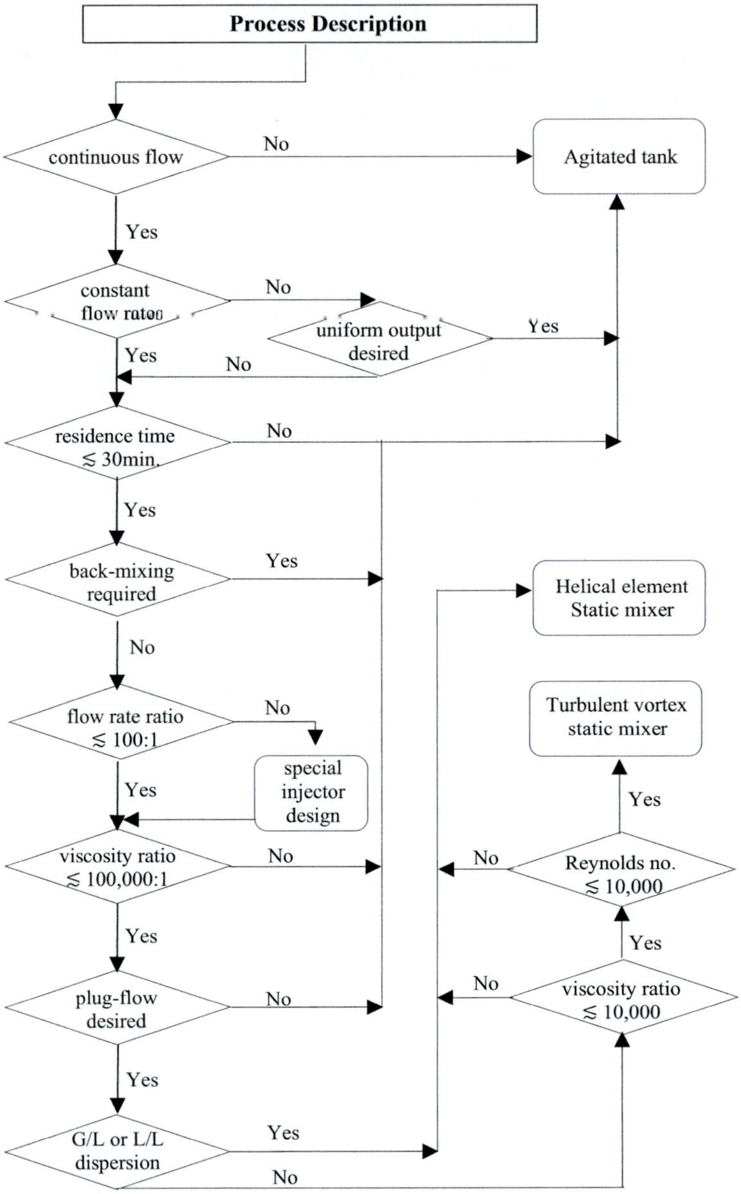

FIGURE 12.17 Flowchart of the main criteria in deciding whether to use a static mixer or agitated tank.[37]

TABLE 12.8
Helical Element Selection Guidelines for Blend Homogeneity[37]

Reynolds Number	Recommended Number of Helical Elements
< 10	18
10–100	12
100–1,000	6
1,000–5,000	4
> 5,000	2

TABLE 12.9
Rough Guidelines for Applying Static Mixers in the Laminar and Turbulent Flow Regimes[38]

	Static Mixer Designs									
Flow Regime	**KMS**	**KMX**	**HEV**	**SMV**	**SMX**	**SMXL**	**SMR™**	**KVM**	**SMF™**	**ISG™**
Laminar										
Mixing/blending	c	a			c	c			a	a
High-low viscosity		a			c	a				a
Dispersion	a	a			c	a				a
Heat transfer	c				b	c	c			
Plug flow	b				c	b	c*			
Turbulent										
Mixing/blending										
High turbulence	a		c	c'				c		
Low turbulence	c			c	a	a			a	
Dispersion										
Liquid-liquid	c			c	a	a	c*		a	
Gas in liquid	c			c	a	a	a*		a	
Liquid in gas	a			c	a					
Fluidized beds					c''					

Notes: 1. a, applicable; b, typically applied; c, best design choice. *where temperature control is required; "especially for very large diameters and non-round cross-sections." gas fluidized solid particles specialized design.

2. For details of (a) SMR[39] (b) SMF[40] (c) ISG,[41] see their manufacturers product brochures.

guidelines for selecting static mixers in laminar and turbulent flow applications, as suggested by Etchells and Meyer (2004).[38]

12.4 DESIGN EXAMPLES

The authors now provide a couple of examples of industrial-scale designs of static mixing systems for dilution of propylene glycol and corn syrup with water. These examples illustrate the use of published correlations relevant to the choice of static elements. Table 12.10 presents the two cases with the flowrates of the components and their physical properties. The desired blend homogeneity is ≥95%.

TABLE 12.10
Flow Rates and Physical Properties of Miscible Fluid Components

Example #	**1. 10% dilution of polyethylene glycol (PEG-400) with water**			**2. 10% dilution of high-fructose corn syrup (Iso-glucose) with water**		
	main stream	**addition stream**	**mixture properties***	**main stream**	**addition stream**	**mixture properties***
Flowrate (gpm)	500	50	550	300	30	330
Viscosity (cP)	90	1.0	45.6	3,000	1.0	803
Density (kg/m³)	1128	1000	1116	1310	1000	1284

* Mixture density and viscosity are determined from equations 12.27–12.28 and 12.29–12.30, respectively, for examples 1 and 2.

12.4.1 Example 1—Low-Viscosity Dilution

Design Procedure

(a) Fluid flowrates:
Main liquid flowrate, $\dot{Q}_1 = 3.2\text{x}10^{-2}$ m^3-s^{-1}
Admix liquid flowrate, $\dot{Q}_2 = 3.2\text{x}10^{-3}$ m^3-s^{-1}
Total flowrate downstream of injector pipe,

$$\dot{Q}_m = \dot{Q}_1 + \dot{Q}_2 = 3.2\times10^{-2} + 3.2\times10^{-3} = 3.52\times10^{-2}\ \text{m}^3\text{-s}^{-1}$$

Flowrate volume fraction of admix: $\phi_2 = \dot{Q}_2 / \dot{Q}_m = 0.091$
Flowrate volume fraction of main flow: $\phi_1 = 1 - \phi_2 = 0.909$
Initial unmixed fluids' coefficient of variation,

$$CoV_0 = \left(\frac{1-\phi_2}{\phi_2}\right)^{1/2} = \left(\frac{1-0.091}{0.091}\right)^{1/2} = 3.16 \tag{12.26}$$

(b) Physical properties:
Main liquid density, $\rho_1 = 1128$kg-m^{-3}
Admix liquid density, $\rho_2 = 1000$kg-m^{-3}
Main liquid mass flowrate, $\dot{m}_1 = \rho_1 \dot{Q}_1 = 1128\times\left(3.2\times10^{-2}\right) = 36.1$ kg-s^{-1}

Admix liquid mass flowrate, $\dot{m}_2 = \rho_2 \dot{Q}_2 = 1000\,x\left(3.2x10^{-3}\right) = 3.2$ kg-s^{-1}
Total mass flowrate downstream of injection pipe:

$$\dot{m}_m = \dot{m}_1 + \dot{m}_2 = 36.1 + 3.2 = 39.3\ \text{kg-s}^{-1}$$

Flowrate mass fraction of admix: $\chi_2 = \dot{m}_2 / \dot{m}_m = 3.2/39.3 = 0.081$
Flowrate mass fraction of main liquid: $\chi_1 = 1 - \chi_2 = 0.919$
Mixture density, $\rho_m = \dot{m}_m / \dot{Q}_m = 39.3/\left(3.52x10^{-2}\right) = 1116$ kg-m^{-3} (12.27)
Main liquid viscosity, $\mu_1 = 9\text{x}10^{-2}$ Pa-s; $\nu_1 = \mu_1/\rho_1 = 80$cSt
Admix liquid viscosity, $\mu_2 = 1\text{x}10^{-3}$ Pa-s; $\nu_2 = \mu_2/\rho_2 = 1.0$cSt
Mixture viscosity, μ_m, is calculated using Refutas' blend viscosity[42,43] equation 12.28.

$$\nu_{12} = \exp\left[\exp\left(\frac{A12 - 10.975}{14.534}\right)\right] - 0.8 \tag{12.28}$$

where, $A_i = 14.534\ \ln\left[\ln\left(\nu_i + 0.8\right)\right] + 10.975\ (i = 1,2)$
$A_{12} = \chi_1 A_1 + \chi_2 A_2$

TABLE 12.11
Calculated Values of $A_1, A_{2,}, A_{12}$ and ν_{12} and μ_m

ν_1	A_1	ν_2	$A_{2,}$	χ_1	χ_2	A_{12}	ν_{12} [cSt]	ρ_m[kg-m^{-3}]	μ_m [Pa-s]
80	32.48	1.0	3.25	0.919	0.081	30.11	40.9	1116	45.6x10^{-3}

(c) Design steps:

(i) Typical pipe velocity of single-phase water-like liquid in plants: $1 \lesssim$ pipe velocity $\lesssim$ 2m-s^{-1}

Consider a pipe diameter, D, to achieve a desired main-flow velocity, v_1 of $\approx$1.5m-s^{-1}.

$$v_1 = \frac{\dot{Q}_1}{\frac{\pi}{4}D^2} \Rightarrow D = \left(\frac{\dot{Q}_1}{\frac{\pi}{4}v_1}\right)^{1/2} = \left(\frac{3.2\times10^{-2}}{\frac{\pi}{4}\times1.5}\right)^{1/2} = 0.165m\ \left[\sim 6.5''\right]$$

Choose a Sch.-40, 6″ nominal SS-316 pipe from Table 12.3. Main pipe id = 6.065″ (0.154m).

Revised pipe diameter size, $D = 0.154m \Rightarrow v_1 = \frac{3.2\times10^{-2}}{\left(\frac{\pi}{4}\right)D^2} = 1.71\text{m.s}^{-1}$.

Diameter of the static element, $D = 0.154m$

Consider admix pipe diameter, $d = \frac{D}{4} = \frac{0.154}{4} = 0.038m$ [1.50”]

Admix velocity in the injection pipe, $v_2 = \frac{\dot{Q}_2}{\frac{\pi}{4}d^2} = \frac{3.2\times10^{-3}}{\left(\frac{\pi}{4}\right)d^2} = 2.82ms^{-1}$.

Choose a Sch-10, 1-1/2” nominal SS-316 pipe in Table 12.3. Admix pipe id = 1.682” (0.043m).

Revised pipe diameter size, $d = 0.043m \Rightarrow v_2 = \frac{3.2\times10^{-3}}{\left(\frac{\pi}{4}\right)d^2} = 2.20\text{m.s}^{-1}$. $v_2 > v_1$.

Mixture velocity in the pipe downstream of the injector,

$$v_m = \frac{\dot{Q}_m}{\frac{\pi}{4}D^2} = \frac{3.52x10^{-2}}{\left(\frac{\pi}{4}\right)D^2} = 1.89m.s^{-1} \text{ [preferably, should be} \lesssim 2\text{m-s}^{-1}]$$

(ii) Preliminary mixer selection based on empty pipe flow Reynolds number:

Determine pipe-flow Reynolds number, *Re*, prior to the mixer elements. Since the admix stream is rarely or inadequately mixed with the mainstream, designers may assume main stream density and viscosity, but they must use mixture velocity for Reynolds number, *Re*, calculation.

$$Re = \frac{v_m D \rho_1}{\mu_1} = \frac{1.89x0.154x1128}{90x10^{-3}} = 3.65x10^3 \Rightarrow \text{Turbulent flow (see Figure 12.15)}$$

Based on the low-turbulent-flow Reynolds number, consider Sulzer SMV™ (corrugated plate type) static mixer elements. According to Sulzer's Technical Specification, available SMV diameter sizes range from 8mm to 250mm in SS-316.[44]

Typical element aspect ratio (AR) values are 1, 1.5, 2, 2.5. Consider element AR to be 1.5.

$$\text{Consider, } AR = L / D = 1.5$$

$$\Rightarrow L = 1.5\ x\ 0.154m = 0.231m$$

Consider a set of static mixers inside the pipe, consisting of six SMV elements each of which has length, L, as shown in Figure 12.18.

$$\text{Length of the mixer, } L_s = 6 \times L = 6 \times 0.231 = 1.386m$$

(iii) Determine mixer pipe flow Reynolds number, Re_h, based on mixture velocity and hydraulic diameter, D_h. Since mixing of the streams are in evolution, mixture physical properties are used for a conservative pressure drop calculation. Using equation 12.14:

$$Re_h = \frac{v_m D_h \rho_m}{\varnothing \mu_m} = \frac{1.89 \times (0.154 \times 0.37) \times 1116}{0.83 \times (45.6 \times 10^{-3})} = 3.17 \times 10^3$$

where, $\varnothing$ = mixing element porosity or void fraction;
for SMV, $\varnothing = 0.83$.[34]
D_h = hydraulic diameter of the pipe in presence of SMV mixer
$D_h = 0.37 x D$ ($\xi_h = \frac{D_h}{D} \approx 0.37$ is calculated from Table 12.4.[34])
See Table 12.12 and its notes for $\varnothing$ and ξ_h values for a few common types of mixers.
Using equation 12.15, Fanning friction factor, f_h', is calculated as follows:

$$f_h' = \frac{2C_1}{Re_h^{0.25}} = \frac{2 \times 1.0}{(3.17 \times 10^3)^{0.25}} = 0.266$$

where, constant $C_1 = 1$ for SMV.[34]

(iv) Determine pressure-drop in the mixer.
- Using equation 12.13:

$$\Delta p_m = \left[2\rho_m \left(\frac{v_m}{\varnothing} \right)^2 \frac{L_s}{D_h} \right] f_h' = \left[2 \times 1116 \times \left(\frac{1.89}{0.83} \right)^2 \left(\frac{1.386}{0.154 \times 0.37} \right) \right] x 0.266$$

$$= 74.88 \times 10^3 \, Pa (10.86 psi)$$

- Pressure drop based on Sulzer Chemtech's SMV design correlation:[45]

$$\Delta p_m = 2 \times \rho_m x v_m^2 \times {}^{L_s}\!/_{D} = 2 \times 1116 \times (1.89)^2 \times \left({}^{1.386}\!/_{0.154} \right)$$

$$= 71.76 \times 10^3 \, Pa (10.41 psi)$$

- Discrepancy between the above pressure drops = $\dfrac{|10.86 - 10.41|}{\left(\dfrac{10.86 + 10.41}{2} \right)} = 0.04 \, or \, 4\%$

⇨ Results from both the SMV correlations[34,45] seem to be remarkably close.

(v) Average shear rate in the mixer is calculated from equation 12.25 and Table 12.7:

$$\dot{\gamma} \approx 50 \times \left(\frac{1.62}{0.128} \right) \approx 632 s^{-1}$$

(vi) Determine tail-pipe pressure-drop.
Since the vertices shed from the mixer continue to travel in the empty pipe until their decay, a tail-pipe is attached to accomplish additional mixing. Typically, its length, L_t, ≥ mixer length, L_s, depending on turbulent intensity.

$$L_t = 1 \times L_s = 1.386m$$

$$Re_p = \frac{v_m D \rho_m}{\mu_m} = \frac{1.89 \times 0.154 \times 1116}{45.6 \times 10^{-3}} = 7.12 \times 10^3$$

From Moody's diagram (Figure 12.15), the Moody friction factor, f, for a smooth SS 316 pipe ≈ 0.034 at $Re_p \approx 7 \times 10^3$. The Fanning friction factor, $f' = \frac{f}{4} = 0.0085$.

$$\Delta p_t = \left(2 \rho_m v_m^2 \frac{L_t}{D}\right) f' = \left(\frac{2 \times 1116 \times (1.89)^2 \times 1.386}{0.154}\right) 0.0085 = 610\, Pa \text{ (0.09 psi)}$$

As expected, $\Delta p_m >> \Delta p_t$ with $\frac{\Delta p_m}{\Delta p_t} \approx 118$.

(vii) Mixing performance (using equation 12.23 and Table 12.6):

$$CoV = CoV_0 \times CoV_r = 3.16 \times \left(\frac{0.21 + 0.46}{2}\right)^{\frac{1.386}{0.154}} = 0.003$$

⇨ Blend homogeneity ≈ 100%

(viii) Power requirements of the main and admix pumps for energy loss in the static mixer:

- Total energy loss in the mixer:

$$E_m = \dot{Q}_m \times \Delta p_m = (3.52 \times 10^{-2}) \times (71.76 \times 10^3) = 2.5kW$$

- Main pump power:

$$P_1 = \dot{Q}_1 \times \Delta p_m = (3.2 \times 10^{-2}) \times (71.76 \times 10^3) = 2.3kW$$

- Admix pump power:

$$P_2 = \dot{Q}_2 \times \Delta p_m = (3.2 \times 10^{-3}) \times (71.76 \times 10^3) = 0.23kW$$

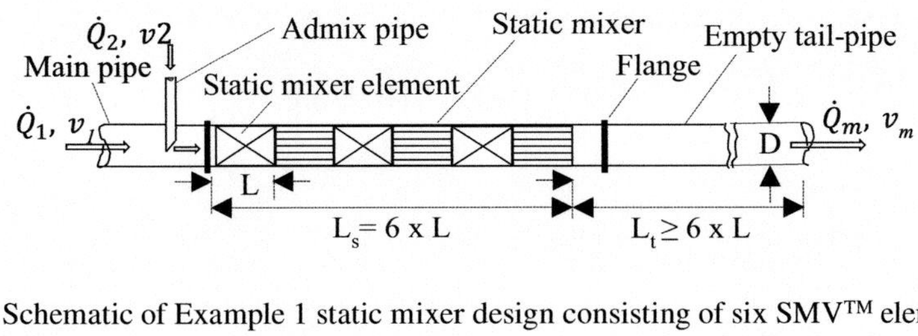

FIGURE 12.18 Schematic of Example 1 static mixer design consisting of six SMV™ elements.

TABLE 12.12
Approximate Void Fraction and Hydraulic Diameter Values* for Four Static Mixer Elements Types[34+]

Mixer Elements	KMS	SMV	SMX	SMXL
Void fraction or porosity, Φ	0.9	0.83	0.67	0.84
Hydraulic diameter ratio, ξ_h	0.5	0.37	0.24	0.20

* 1. These values depend on mixer sizes and their fabrication.
2. Design engineers are strongly suggested to consult with the mixer manufacturers.

12.4.2 Example 2—High-Viscosity Dilution

Design Procedures A and B

(a) Fluid flowrates:

Main liquid flowrate, $\dot{Q}_i$ = 1.89×10^{-2} m^3-s^{-1}
Admix liquid flowrate, $\dot{Q}_2$ = 1.89×10^{-3} m^3-s^{-1}
Total flowrate downstream of injector pipe:

$$\dot{Q}_m = \dot{Q}_1 + \dot{Q}_2 = 1.89\times10^{-2} + 1.89\times10^{-3} = 2.08\times10^{-2}\ \mathrm{m^3 s^{-1}}$$

Flowrate volume fraction of admix: $\phi_2 = \dot{Q}_{.2} / \dot{Q}_m = 0.091$
Flowrate volume fraction of main flow: $\phi_1 = 1 - \phi_2 = 0.909$
Initial unmixed fluids' coefficient of variation, $CoV_0 = \left(\frac{1-\phi_2}{\phi_2}\right)^{1/2} = \left(\frac{1-0.091}{0.091}\right)^{1/2} = 3.16$

(b) Physical properties:
Main liquid density, ρ_1 = 1310kg-m^{-3}
Admix liquid density, ρ_2 = 1000kg-m^{-3}
Main liquid mass flowrate, $\dot{m}_1 = \rho_1\dot{Q}_1 = 1310\times\left(1.89x10^{-2}\right) = 24.76\ \text{kg-s}^{-1}$

Admix liquid mass flowrate, $\dot{m}_2 = \rho_2\dot{Q}_2 = 1000\times\left(1.89\times10^{-3}\right) = 1.89\ \text{kg-s}^{-1}$
Total mass flowrate downstream of injection pipe:

$$\dot{m}_m = \dot{m}_1 + \dot{m}_2 = 24.76 + 1.89 = 26.65\ \text{kg-s}^{-1}$$

Flowrate mass fraction of admix: $\chi_2 = \dot{m}_2 / \dot{m}_m = 0.071$
Flowrate mass fraction of main liquid: $\chi_1 = 1 - \chi_2 = 0.929$

Mixture density, $\rho_m = \dot{m}_m / \dot{Q}_m = 26.65/\left(2.08\times10^{-2}\right) = 1281\ \text{kg-m}^{-3}$ (12.29)
Main liquid viscosity, μ_1 = 3Pa-s; ν_1 = 2290cSt
Admix liquid viscosity, μ_2 = 1×10^{-3} Pa-s; ν_2 = 1.0cSt
Mixture viscosity, μ_m, is calculated using Refutas[42,43] blend viscosity equation (12.30).

$$\nu_{12} = \exp\left[\exp\left(\frac{A12 - 10.975}{14.534}\right)\right] - 0.8 \tag{12.30}$$

where, $A_i = 14.534\ \ln\left[\ln\left(\nu_i + 0.8\right)\right] + 10.975$ (i = 1,2)
$A_{12} = \chi_1 A_1 + \chi_2 A_2$

TABLE 12.13
Calculated Values of A_1, A_2, A_{12} and ν_{12} and μ_m

ν_1	A_1	ν_2	A_2,	χ_1	χ_2	A_{12}	ν_{12} [cSt]	ρ_m [kg-m^{-3}]	μ_m [Pa-s]
2290	40.71	1.0	3.25	0.929	0.071	38.05	627	1281	803×10^{-3}

(c) Design steps:

(i) Typical pipe velocity of single-phase water-like liquid in plants: $1 \lesssim$ pipe velocity $\lesssim 2$m-s^{-1}

Consider a pipe diameter, D, to achieve a desired main-flow velocity, v_1 of $\approx$1.2m-s^{-1}.

$$v_1 = \frac{\dot{Q}_1}{\frac{\pi}{4}D^2} \Rightarrow D = \left(\frac{\dot{Q}_1}{\frac{\pi}{4}v_1}\right)^{1/2} = \left(\frac{1.89\times10^{-2}}{\frac{\pi}{4}\times1.2}\right)^{1/2} = 0.142m\ [\sim 5.6'']$$

Choose a Sch.-40, 5″-nominal SS pipe from Table 12.3. Main pipe id = 5.047″ (0.128m)

Revised pipe diameter size, $D = 0.128m \Rightarrow v_1 = \dfrac{1.89\times10^{-2}}{\left(\frac{\pi}{4}\right)D^2} = 1.47\text{m.s}^{-1}$.

Diameter of the static element, $D = 0.128m$

Consider admix pipe diameter, $d = \dfrac{D}{4} = \dfrac{0.128}{4} = 0.032m$ [1.26″]

Admix velocity in the injection pipe, $v_2 = \dfrac{\dot{Q}_2}{\frac{\pi}{4}d^2} = \dfrac{1.89\times10^{-3}}{\left(\frac{\pi}{4}\right)d^2} = 2.35ms^{-1}$.

Choose a Sch-10, 1-1/4″ nominal SS pipe in Table 12.3. Admix pipe id = 1.442″ (0.036m).

Revised pipe diameter size, $d = 0.036m \Rightarrow v_2 = \dfrac{1.89\times10^{-3}}{\left(\frac{\pi}{4}\right)d^2} = 1.86\text{m.s}^{-1}$.

Mixture velocity in the pipe downstream of injector,

$$v_m = \frac{\dot{Q}_m}{\frac{\pi}{4}D^2} = \frac{2.08\times10^{-2}}{\left(\frac{\pi}{4}\right)D^2} = 1.62m.s^{-1}\ \text{[preferably, should be} \lesssim 2\text{m-s}^{-1}]$$

(ii) Preliminary mixer selection based on empty pipe flow Reynolds number:

Determine flow Reynolds number, Re, prior to the mixer elements. Since the admix stream is rarely or inadequately mixed with the mainstream, designers may assume mainstream density and viscosity, but they must use mixture velocity for Reynolds number calculation.

$$Re = \frac{v_m D \rho_1}{\mu_1} = \frac{1.62\times0.128\times1310}{3000\times10^{-3}} = 90 \Rightarrow \text{Laminar flow (see Figure 12.15)}$$

Based on the laminar flow Reynolds number ($Re \lesssim 10^2$), consider Kenics™ KMS (helical—right twist–left twist; angle of blade twist 180°) static mixer elements. According to NOV–Chemineer's Technical Specification, available KMS diameter sizes range from 8mm to 148mm in SS-316.

Typical *AR* values of the KMS elements are 1, 1.5, 2 and 2.5. The standard design length of the helical element for pipe diameter less than 0.3m or less is 1.5 times its diameter.[37]

Consider aspect ratio (AR) of the element to be 1.5.

$$AR = L / D = 1.5$$

$$\Rightarrow L - 1.5 \times 0.128m - 0.192m$$

(iii) Determine Moody friction factor, f, from Reynolds number, $Re_p = Re$ using equation 12.15.[35]

$$f = K\left[\frac{Re}{(AR)^{2.15}}\right]^{n} \times (AR)^{-2.04}$$

$$\frac{Re}{(AR)^{2.15}} = {}^{90}\!\Big/_{(1.5)^{2.15}} = 37.6 \Rightarrow K = 320 \text{ and } n = -0.86 \text{ (from Table 12.5)}$$

$$f = 320\left[\frac{Re}{(AR)^{2.15}}\right]^{-0.86} (AR)^{-2.04} = 320(37.6)^{-0.86}(1.5)^{-2.04} = 6.18$$

(iv) Average shear rate in the mixer is calculated from equation 12.25 and Table 12.7:

$$\dot{\gamma} \approx 28 \times \left(\frac{1.62}{0.128}\right) \approx 354 s^{-1}$$

A. Mixer Design:

Myers et al.[37] for 10 < Re < 100, recommend use of 12 KMS elements for blending homogeneity (see Table 12.8). First, the authors considered two sets of static mixers inside the pipe, consisting of six KMS elements, each having length, L, as shown in Figure 12.19. Total number of static mixers used is 12. An empty space of length, $L_e = L$, is provided for (a) disruption of regular striations, (b) flange connections between the sets of elements for ease of maintenance, (c) a typical installation feature of industrial-scale static mixers.

$$\text{Length of the mixer: } L_m = 2L_s = 12 \times L = 12 \times 0.192 = 2.304m$$

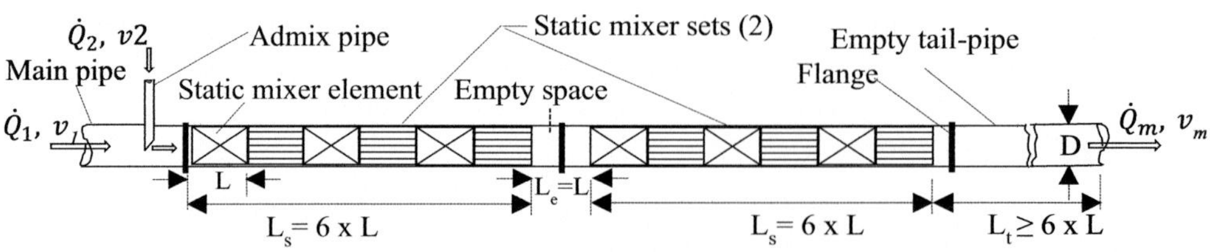

FIGURE 12.19 Schematic of Example 2 Case A static mixer design with two sets of KMS™ element.

(i) Determine pressure drop in the mixer

$$\Delta p_m = \left[\frac{\rho_m (v_m)^2}{2}\frac{L_m}{D}\right] f = \left[1281 \times (1.62)^2 \left(\frac{2.304}{0.128}\right)\right] \times 6.18$$

$$= 3.74 \times 10^5 \, Pa = 54.2 \text{ psi}$$

(ii) Mixing performance

$$CoV = CoV_0 \times CoV_r = 3.16 \times (0.87)^{\frac{2.304}{0.128}} = 0.26$$

⇨ Blend homogeneity ≈ 74% ⇨ unacceptable homogeneity!

B. Mixer Design:

Using equations 12.22 and 12.24, required mixture length, L_m, for ~95% of homogeneity is derived from the following relationship. The covariation reduction (CoV_r) constant for laminar flow regime, K_{iL}, is determined from Table 12.6.

$$L_m = \frac{D \times \ln\left(\frac{CoV}{CoV_r}\right)}{ln(K_{iL})} = \frac{0.128 \times ln\left(\frac{0.05}{3.16}\right)}{ln(0.87)} = 3.8m$$

⇨ Number of elements required, $n_e = \frac{L_m}{L} = \frac{3.8}{0.192} = 19.8 \approx 20.$

The authors considered three sets of static mixers inside the pipe, each consisting of six KMS elements having length, L, as shown in Figure 12.20. Total number of static mixers used is 18. Additionally, two empty spaces each of length, $L_e = L$, is provided for disruption of continuous lamellar striations and flange connections between the sets of elements for ease of their installations and maintenance.

Length of the mixer: $L_m = 3L_s = 18 \times L = 18 \times 0.192 = 3.46m$

(i) Determine pressure drop in the mixer

$$\Delta p_m = \left[\frac{\rho_m (v_m)^2}{2}\frac{L_m}{D}\right] f = \left[\frac{1281 \times (1.62)^2}{2}\left(\frac{3.46}{0.128}\right)\right] \times 6.18$$

$$= 2.81 \times 10^5 \, Pa = 40.75 psi$$

(ii) Mixing performance

$$CoV = CoV_0 \times CoV_r = 3.16 \times (0.87)^{\frac{3.456}{0.128}} = 0.07$$

⇨ Blend homogeneity ≈ 93%

(iii) Determine tailpipe pressure drop.

Since the ultra-thin lamellar flow structures shed from the mixer continue in the empty pipe until they are damped by the viscosity, normally a tail-pipe is attached to accomplish additional mixing. Typically, its length, L_t, ≥ mixer length, L_s, depending on flow Reynolds number.

Tail-pipe pressure-drop:

$$L_t = 1 \times L_s = 1.15m$$

$$Re_p = \frac{v_m D \rho_m}{\mu_m} = \frac{1.62 \times 0.128 \times 1281}{803 \times 10^{-3}} = 331$$

From the f vs. Re correlation for laminar flow $\left(Re_p \lesssim 2000\right)$ in a pipe, i.e., $f = \frac{64}{Re_p}$

$$f = \frac{64}{331} = 0.19$$

$$\Delta p_t = \left(\frac{\rho_m v_m^2}{2}\frac{L_t}{D}\right) f = \left(\frac{1281 \times (1.62)^2 \times 1.15}{2 \times 0.128}\right) 0.19 = 2.87 \times 10^3\, Pa \approx 0.42 \text{ psi.}$$

As expected, in the case of laminar flow, pressure drop, Δp_m, in the static mixer is much greater than that of the tailpipe, Δp_t, i.e.,

$$\frac{\Delta p_m}{\Delta p_t} = \frac{2.81 \times 10^5}{2.87 \times 10^3} = 98$$

(iv) Power requirements of the main and admix pumps for energy loss in the static mixer:

- Total energy loss in the mixer:

$$E_m = \dot{Q}_m \times \Delta p_m = \left(2.08 \times 10^{-2}\right) \times \left(2.81 \times 10^5\right) = 5.8 kW$$

- Main pump power:

$$P_1 = \dot{Q}_1 \times \Delta p_m = \left(1.89 \times 10^{-2}\right) \times \left(2.81 \times 10^5\right) = 5.3 kW$$

- Admix pump power:

$$P_2 = \dot{Q}_2 \times \Delta p_m = \left(1.89 \times 10^{-3}\right) \times \left(2.81 \times 10^5\right) = 0.53 kW$$

Although the examples cited the use of Sulzer's SMV and NOV–Chemineer's KMS, many other equivalent static mixers would equally serve the purpose. Manufacturers' websites provide guidance to selection of static mixers for different applications based on flow regime, viscosity ratio and admix to bulk fluid flow ratio. Applications range from simple blending to liquid–liquid dispersion (emulsion), gas–liquid mass transfer, and heat transfer. Practicing engineers are suggested to read a recent article by Meijer et al.[46] on the blending performance and quantitative pressure-drop comparison of a few commercial static mixers, and the static mixer selection, sizing and troubleshooting guide presented in the *Kolmez Handbook of Process Equipment Design*.[47]

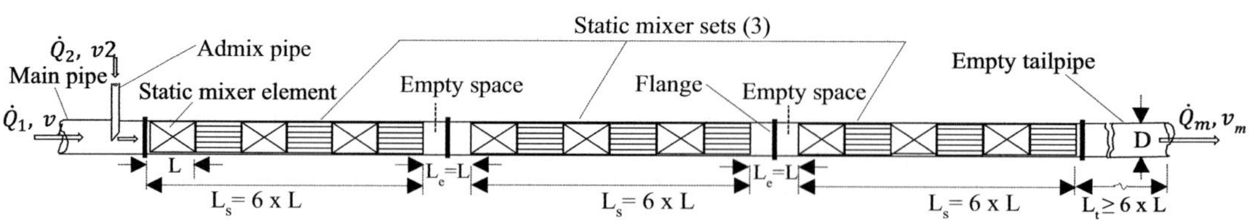

FIGURE 12.20 Schematic of Example 2, Case B static mixer design with three sets of KMS™ elements.

In many instances, in-line mixing systems with static mixers are considered cost-effective implementations in the production scale. As mentioned earlier, their continuous motionless-process characteristics with easy installation, minimum maintenance, smaller footprints, shorter residence times, narrow residence time distributions (RTDs), and enhanced heat transfer are all attractive features. The advantages are application specific, however. Static mixers do not always meet required overall mixing and/or shear intensity for subsequent reaction and dispersion (mass transfer) as compared to mechanically agitated batch/continuous processes. Thus, these mixers have limited applications targeted toward fast-reacting, controlled feed-flow processes. Their usage is widely accepted in the side-arm/loop-around pipes of stirred tanks for on-line proportional controlled dosing of additives and catalysts, dissolution of sparingly soluble gasses into liquids and for simple liquid–liquid blending or dilution tasks in commercial-scale formulation system houses.

REFERENCES

1. https://www.chemineer.com/products/kenics.html
2. https://www.sulzer.com/-/media/files/products/static-mixers/flyer_smx_plus_d.ashx?la=en
3. https://www.stamixco-usa.com
4. https://www.flexachem.com/wp-content/uploads/2013/10/HEV-Static-Mixer-bulletin_811.pdf
5. Ghanema, A., Lemenanda, T., Della Valle, D., Peerhossaini, H., Static mixers: Mechanisms, applications, and characterization methods—A review, *Chem. Eng. Res. Des.*, 92, p. 205, 2014.
6. https://www.chemineer.com/images/chemineer/documents/mixing-product-catalog.pdf
7. https://www.sulzer.com/en/products/static-mixers/general-purpose-mixers
8. https://www.chemineer.com/products/kenics/heat-exchangers.html
9. Stamixco Plug Flow Reactor Static Mixers, *Product Bulletin PF-1.0*, http://www.stamixco-usa.com/files/3-Content-PDF-Files/PDF-14.0)%20Plug%20Flow%20Reactor%20Technical%20Bulletin.pdf
10. https://www.nov.com/-/media/nov/files/products/caps/process-and-flow-technologies/ultratab-mixer/kenics-static-mixer-technology-brochure-en.pdf
11. Inyo Process, http://www.inyoprocess.com/mixing-tees.html
12. Cozewith, C., Busko Jr., M., Design correlations for mixing tees, *Ind. Eng. Chem. Res.*, 28, p. 1521, 1989.
13. Cozewith, C., Strate, G. V., Dalton, J. T., Frederick, J. W., Ponzi, R. P., Computer simulation of tee mixers for nonreactive and reactive flows, *Ind. Eng. Chem. Res.*, 30, p. 270, 1991.
14. Forney, L. J., Gray, G. E., Optimum design of a Tee mixer for fast reaction, *AIChE J.*, 36(11), p. 1773, November 1990.
15. Zughbi, H. D., Siddiqui, S. W., Fatehi, A. I., Reactive mixing in a pipeline with a side-tee, *Arab. J. Sci. Eng.*, 31(1B), p. 13, April 2006.
16. https://en.wikipedia.org/wiki/Moody_chart
17. Colebrook, C. F., White, C. M., Experiments with fluid friction in roughened pipes, proceedings of the royal society of London, Series A, *Math. Phys. Sci.*, 161(906), p. 367, 1937.
18. Zlokamik, M., *Scale-up in Chemical Engineering*, Wiley-VCH, New York, 2002.
19. http://www.staticmixers.com/motionless-howtoselect.asp
20. https://www.fluitec.ch/download.php?f=3fe420f8be822be48f8d6fe85fdfb79f
21. https://www.sulzer.com/-/media/files/products/static-mixers/technicalarticles/cr8047_sulzer_compax_short_report1.ashx?la=en
22. Stec, M., Synowiec, P. M., Analysis of the pressure drop calculation method impact on the accuracy of the experimental results in the Koflo® static mixer, *Inż. Ap. Chem.*, 54(4), p. 201, 2015.
23. *Stamixco: Static Mixing, Reaction, Heat Transfer & Fluid Dynamics Technology*, Technical Bulletin GMB-1.1-(7/18/07). stamixco-usa.com/files/3-Content-PDF-Files/PDF-11.0)%20General%20Static%20Mixing%20Technical%20Bulletin.pdf
24. Kenics Static Mixing Technology, *Bulletin 813, 2013*, Chemineer, Inc. static_mixer_bulletin_19 (6).pdf
25. Storm Mixer, https://stormmix.com/index.php/selecting
26. Dawson, M. K., BHRG, *Design Guide for Motionless (Static) Mixers FMP-086: Static Mixer Blending Design*, BHR Group (now Framatome BHR Limited), Cranfield, Bedfordshire, UK, December 2005.
27. Fluidyne Static Mixers, https://www.indiamart.com/proddetail/design-principle-for-static-mixer-19141219530.html

28. Li, H. Z., Fasol, C., Choplin, L., Pressure-drop of Newtonian and non-Newtonian fluids across a Sulzer SMX static mixer, *Trans. IChemE*, 75(A), p. 792, November 1997.
29. Morancais, P., Hirech, K., Carnelle, G., Legrand, J., Friction factor in static mixer and determination of geometric parameters of SMX sulzer mixers, *Chem. Eng. Comm.*, 171(1), pp. 77–93, 1999.
30. Middleman, S., Drop size distributions produced by turbulent pipe flow of immiscible fluids through static mixer, *Ind. Eng. Chem. 1, Proc. Des. Dev.*, 13, p. 78, 1974.
31. Streiff, F. A., In-line dispersion and mass transfer using static mixing equipment, *Sulzer Tech. Rev.*, 108, p. 113, 1977.
32. Pahl, M. H., Muschkelnautz, E., Static Mixer and their applications, *Int. J. Chem. Eng.*, 22, p. 197, 1982.
33. Berkman, P. D., Calabrese, R. V., Dispersion of viscous liquids by turbulent flow in a static mixer, *AIChE J.*, 34, p. 602, 1988.
34. Theron, F., Le Sauze, N., Comparison between three static mixers for emulsification in turbulent flow, *Int. J. Multiph. Flow*, 37(5), p. 488, 2011.
35. Song, H.-S., Han, S. P., A general correlation for pressure drop in a Kenics static mixer, *Chem. Eng. Sci.*, 60, p. 5696, 2005.
36. Streiff, F. A., Jaffer, S., Schneider, G., The design and application of static mixer technology. In: *3rd International Symposium on Mixing in Industrial Processes*, Osaka, Japan, pp. 107–114, 1999.
37. Myers, K. J., Bakker, A., Ryan, D., Avoid agitation by selecting static mixers, *Chem. Eng. Prog.*, 93(6), pp. 28–38, June 1997.
38. Etchells, A. W., Meyer, C. F., Mixing in pipelines, Chapter 7 in *Handbook of Industrial Mixing*, Paul, E. L. et al., eds., Wiley, Hoboken, NJ, pp. 391–477, 2004.
39. StaMixco-SMF™, https://www.stamixco-usa.com/products
40. Sulzer-SMR™, https://www.sulzer.com/en/shared/products/heat-exchanger-smr
41. Ross-ISG™, https://www.mixers.com/products/static-mixers/isg-motionless-mixer/
42. Maples, R. E., *Petroleum Refinery Process Economics*, 2nd ed., Pennwell Books, 2000. ISBN 0-87814-779-9
43. Baired, C. T., *Guide to Petroleum Product Blending*, HPI Consultants, Inc., 1989. HPI Website.
44. Static Mixer SMV™, Mix efficiently with low energy consumption, *Technical Specification*, https://www.sulzer.com/en/shared/products/static-mixer-smv
45. Mixing and Reaction Technology Pace Setting Technology, *Sulzer*, www.sulzer.com/-/media/files/products/static-mixers/mixing_and_reaction_technology_e_23270640.ashx?la=en (for correlation)
46. Meijer, Han E. H., Singh, M. K., Anderson, P. D., Performance of static mixers: A quantitative comparison, *Prog. Polym. Sci.*, 37, pp. 1333–1349, 2012.
47. Kolmetz, K., Dwijayanti, A., *Static Mixer Selection, Sizing and Troubleshooting, Kolmez Handbook of Process Equipment Design*, KLM Technology Group, November 2012.

13 In-Line Dynamic Mixers

13.1 INTRODUCTION

Chapter 8, "Immiscible Liquid-Liquid Mixing/Dispersion," describes formation of coarse droplets and their dispersions in a continuous media using reasonable shear strains in mechanically agitated tanks. Typical droplet sizes are greater than 10μm, and the working media often demonstrate heterogeneous flow behavior.

This chapter focuses on high-shear mixers (HSMs) known as rotor-stators. They consist of a rotor (typically blades or a toothed ring) spinning at high speeds inside a motionless stator (commonly a mesh, or a ring of slots/teeth/holes) to generate micron-sized droplets (emulsions) or submicron-sized droplets (micro-emulsions). Shear rates of 10^4 to 10^5 s^{-1} are commonly produced from HSM rotor-tip speeds of 10 to 50 m-s^{-1}. The highly localized dissipation of shear energy, adjacent to the rotor head, produces droplet sizes on the order of 0.01–10μm and a relatively homogeneous product media. Production-scale rotor-stators (RS) in a process line sometimes operate independently and sometimes combined with a mechanically agitated tank, as in the bottom-mounted S-Jet homogenizer[2] in Ekato's Unimix® SRT.[3] R&D laboratories often use such systems as batch mixers for process evaluation. Rotor-stators are relatively mature emulsification and homogenization technology in manufacturing industries including food, pharmaceuticals, cosmetics, agricultural, home-care, health-care, and fine chemicals. Besides making emulsions, HSMs can also disperse fine particles or droplets in a liquid, promoting fast reactions that might otherwise be micro-mixing controlled in a STR. This chapter also includes specialized rotor-stators, called colloid mills, that are designed for high-shear breakage and dispersion of solids within a continuous media.

13.2 ROTOR-STATORS

Based on their designs, rotor-stators are primarily classified into either batch mixers or in-line/continuous mixers. Figures 13.1–13.4 represent a few typical HSMs manufactured by Silverson, Ika, Ross and Greerco.

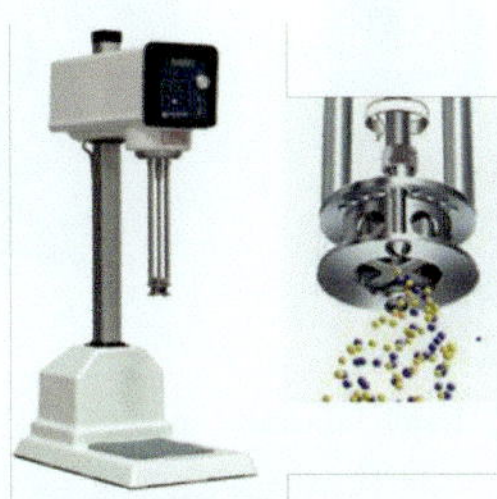

(a) Silverson L5M-A batch laboratory-scale mixer

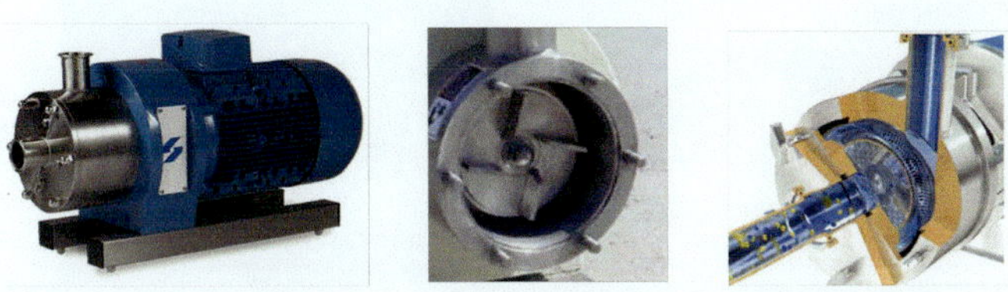

(b) Silverson industrial-scale in-line (continuous) mixer

FIGURE 13.1 High-shear lab- and industrial-scale dispersers manufactured by Silverson Machines, Inc.[4]

DOI: 10.1201/9781003606390-13

The manufacturers represented in Figures 13.1–13.4 introduced a wide variety of rotor and stator disperser configurations to provide versatility in emulsion/homogenization applications for a diverse range of working media. The list of disperser types, and specific applications for each, are disclosed in their respective websites and product brochures.[4–7] While there are many other manufacturers of unique HSMs (e.g., the Oakes Homogenizer, Tetra Pack Scanima, Quadro® HV Emulsifier, Quadro-Ytron®-Z/ZC, and VMI etc.), they are excluded from discussion not because of bias, but due to the authors' limited knowledge and personal experience with them.

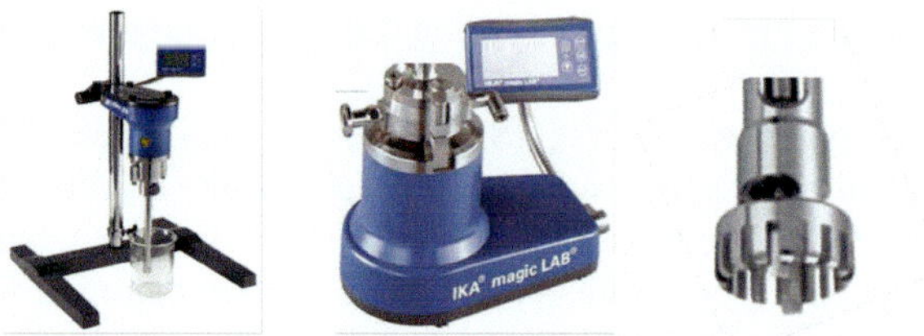

(a) Magic LAB® high-shear mixer in batch (left) and continuous (center) configurations, with UTC batch disperser (right)

(b) Labor Pilot (continuous operation)

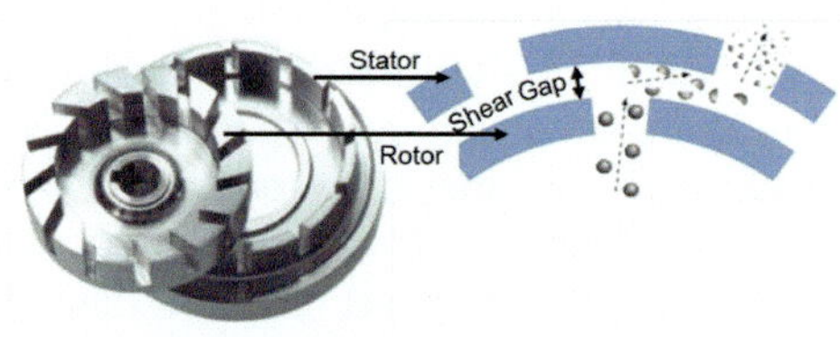

(c) UTL 2000 single stage disperser, with graphic of the dispersion process

FIGURE 13.2 High-shear lab- and industrial-scale dispersers manufactured by Ika Works, Inc.[5]

(a) Ross model LCI-T batch lab-scale mixer (left) with HSM head (right)

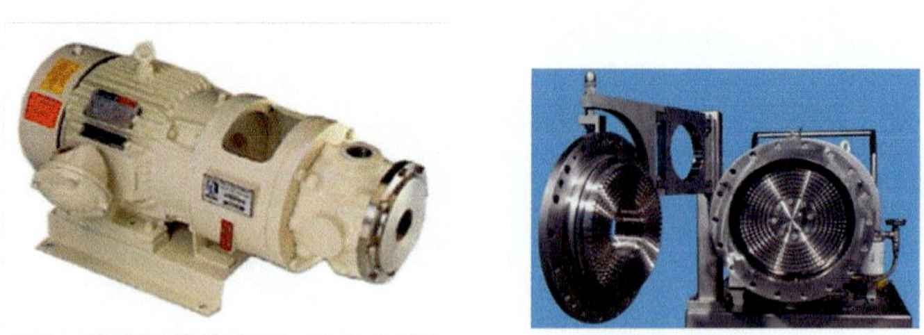

(b) Ross model 712XS-50 high-shear (left), and X-series ultra-high-shear (right) in-line continuous mixers

FIGURE 13.3 High-shear lab- and industrial-scale dispersers manufactured by Charles Ross & Son Co.[6]

(a) NOV Greerco batch high-shear mixer (right) and heads (center and right)

(b) NOV Greerco industrial-scale in-line (continuous) high-shear mixer

FIGURE 13.4 Greerco high-shear dispersers manufactured by National Oilwell Varco, Inc.[7]

13.2.1 Batch High-Shear Mixers

Batch high-shear mixers consist of a rotor, whose blades or teeth rapidly rotate across stationary openings in the stator. The size and number of stator openings (i.e., holes or slits) vary significantly. Process fluid is drawn axially into the mixing zone, then expelled (either radially or axially) at high speeds through the stator openings. The fluid experiences high mechanical and hydraulic shear[8] within the small (on the order of 100–1000 microns) gap between the tips of the rotor and the inner edge of the stator wall. Thus, this region is called the shear gap, and it is a key dimensional parameter in determining a high-shear mixer's hydraulic shear rate (velocity gradient). Under laminar flow conditions, fluid shear stresses are due to velocity gradients. Under turbulent flow conditions, shear stresses come from velocity gradients and inertial forces. Although rotor tip speeds range from 15–30 $m{\cdot}s^{-1}$ (~3000–6000 ft/min), interchangeable stators broaden the range of mechanical shear rates experienced by the process fluid as it exits the HSM (see Figure 13.5). There is a broad similarity between design features of the batch HSM rotors (paddles/blades) and interchangeable stators (holes/slits) manufactured by manufacturers such as those listed previously. Although plant-scale HSM devices are typically purchased from the manufacturers or their local representatives, laboratory-scale HSM devices are also available through lab equipment suppliers such as Fisher Scientific, Cole-Parmer, etc.

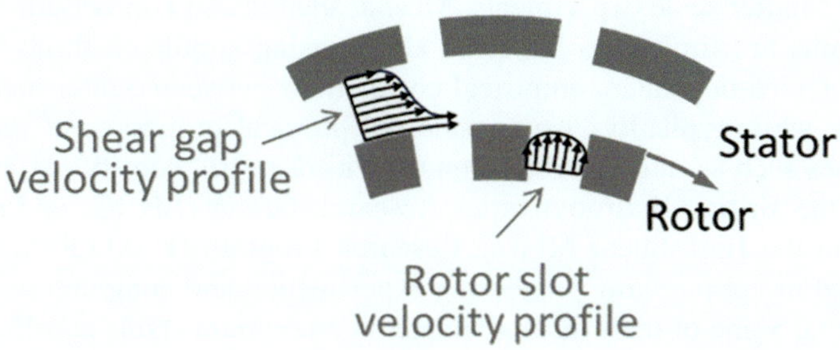

FIGURE 13.5 Velocity profiles in a stator slot and in the shear gap between the rotor and stator.

13.2.2 In-Line Continuous High-Shear Mixers

Ultra-shear in-line rotor-stator HSMs are commonly continuous and, in industrial emulsification and homogenization processes, often single-pass and followed by a stirred tank for coarser droplet dispersion. Recirculating the stirred tank contents through an in-line rotor-stator, as shown in Figure 13.6a, can create a multi-pass process to achieve a smaller desired droplet size.

Figure 13.6b shows the multi-stage head of an in-line HSM that has an intermeshing inner rotor (four rotating slots and teeth) and inner stator (30–40 stationary slots and teeth), followed by an intermeshing outer rotor (12–15 rotating slots and teeth) and outer stator (40–50 stationary slots and teeth). Multi-stage in-line rotor-stator heads with more than two intermeshing rotor-stator pairs are available to possibly produce even smaller droplets per processing pass.

Maximum hydraulic shear occurs in the radial gap between the rotor teeth's outer surface and corresponding stator teeth's inner surface, while the highest mechanical shear occurs at mutual projections (sharp edges) of the teeth. Coupling a series of multiple rotor-stator heads, or of multiple single-pass HSMs, interspersed with sequential feed streams, produces sequential dispersion processes for different species or additives.

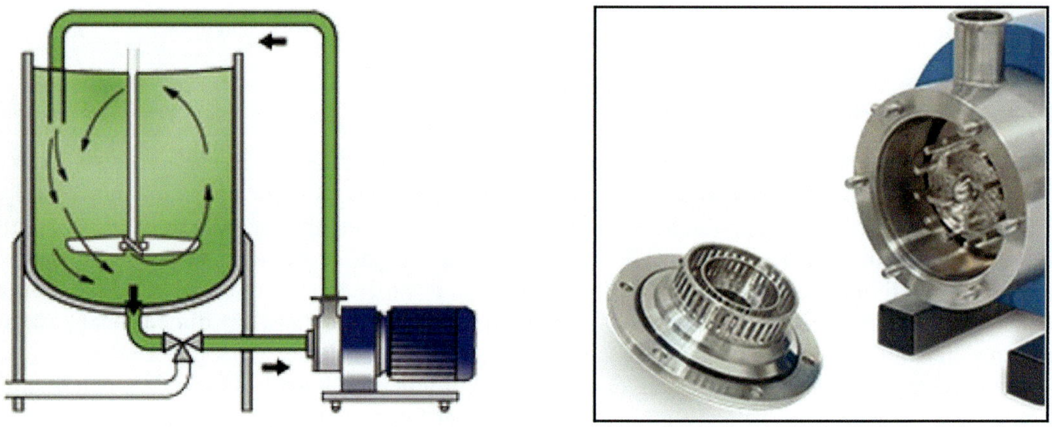

FIGURE 13.6 (a) A stirred tank with multi-pass recirculation through an in-line rotor-stator,[4] and (b) an in-line multi-stage rotor-stator head with dual concentric, intermeshing rotors and stators.[4]

13.2.3 Flow, Shear and Drop Size

Before a given high-shear process is implemented at production scale, the performance of the rotor-stator mixer (batch or in-line) is typically assessed while processing the specific process fluids of interest in smaller scale experiments. Most academic and consortium-based experimental research programs are limited to lab-scale HSMs using simulated fluids. They commonly establish flow and device-dependent empirical correlations between Sauter mean drop size (d_{32}) and energy density, while typically also considering physical properties of the continuous and discontinuous phases such as interfacial tension (σ), viscosity (μ), density (ρ), and phase fraction (Φ). For example, the British Hydrodynamics Research Group (BHRG, or Framatome BHRG more recently) [9] and the High Shear Mixing Research Program (HSMRP) at the University of Maryland[10] have and/or are pursuing pioneering experimental and computational fluid dynamics research in this arena. Some of their results[11] relate the maximum stable equilibrium drop size to the energy dissipation rate (per swept rotor volume) and physical properties of the working media, as shown in Table 13.1.

TABLE 13.1
Experimental Correlations Relating Sauter Mean Diameter, d_{32}, to Energy Dissipation Rate in a High Shear Mixer (HSM) and to Pertinent Physical Properties of the Immiscible Phases[11]

Experimental Correlations	Correlation Conditions (d_p = general term for droplet diameter [m])	Ref.
$d_{32} \approx (\sigma / \rho_c)^{3/5} \varepsilon^{-2/5}$	For $d_p \geq \lambda_k$ and dilute, low-viscosity dispersions. The drops are inviscid and the interfacial surface tension contributes to stability.	12
$d_{32} \approx (\rho_c \rho_d)^{-3/8} \mu_d^{3/4} \varepsilon^{-1/4}$	For $d_p \geq \lambda_k$ and high-viscosity dispersions. Viscous stress is more important than surface forces for resisting droplet deformation.	
$d_{32} \approx \sigma^{1/3} \rho_c^{-2/3} \mu_c^{1/3} \varepsilon^{-1/3}$	For $d_p \leq \lambda_k$, where drop breakage is from applied inertial stresses.	13
$d_{32} \approx \sigma^{1/2} \mu_c^{-1/2} \varepsilon^{-1/2}$	For $d_p \leq \lambda_k$, where drop breakage is from applied viscous stresses	

Nomenclature: d_{32} = Sauter mean diameter [m], d_p = drop size [m], σ = liquid-liquid interfacial tension [N.m^{-1}], ρ_c and ρ_d = respective densities of continuous and dispersed phases [kg.m^{-3}], μ_c and μ_d = respective dynamic viscosities of continuous and dispersed phases [Pa-s], ε = energy per unit mass of swept volume (specific energy) [W.kg^{-1}]

Notes: 1. Since d_{32} has been shown to be proportional to d_{max}, and has a greater design value than d_{max}, d_{32} can be substituted into the correlations for d_{max}.

2. Kolmogorov length scale, $\lambda_k = (\nu^3/\varepsilon)^{1/4}$ [m], where, ν = dispersion kinematic viscosity [m^2.s^{-1}]

For all practical applications, d_{32} is used instead of the maximum drop size, $d_{max,}$ due to its direct relationship to the dispersed-phase content, Φ, and the interfacial area per unit volume, "a", i.e., $d_{32} = (6\Phi / a)$. The Sauter mean diameter is important in mass transfer and reaction because it is related to interfacial area per unit volume. Furthermore, d_{32} is proportional to d_{max}. In addition to its design value, d_{32} gives a better representation of small and large droplets by simultaneously accounting for both volume and surface area.

As pointed out in Chapter 5, "Hydrodynamics," the physical nature of a fluid flow can be classified as laminar or turbulent depending on whether it is viscous or inertia dominated, and characterized in terms of Reynolds number (*Re*). In the case of large Reynolds number flow ($Re \geq 10^4$), the kinetic energy is transferred initially through energetic eddies without dissipation (inertial sub-range), until finally it is dissipated in the molecular viscosity scale (viscous sub-range), known as the Kolmogorov length scale. The properties of these isotropic eddies are a function of the energy dissipation rate per unit mass (ε). However, at Kolmogorov scale (λ_k), their properties also depend on viscosity. The cascading of turbulent eddies from large scale (order of the source, λ_l) to the small scale (order of the viscous dissipation scale, λ_k) occurs in HSMs to produce submicron droplets. The Kolmogorov eddies are created in a continuum cascading process, and yet their length scale is typically much greater than the molecular length scale of the working media, i.e., $\lambda_k \gg \lambda_m$. The eddy Reynolds number, Re_k, is quantified based on Kolmogorov length scale as

$$Re_k = \frac{v_k \lambda_k}{\nu} \tag{13.1}$$

where, v_k = characteristic eddy velocity scale [m·s^{-1}]; ν = kinematic viscosity of the fluid [m^2·s^{-1}]

Conceptually, at the Kolmogorov scale, the inertial and viscous forces balance,[14] i.e., $Re_k = 1$, and from dimensional reasoning, the Kolmogorov scale is defined as

$$\lambda_k = \left(\frac{\nu^3}{\varepsilon}\right)^{1/4} \tag{13.2}$$

where, ε = specific energy [W·kg^{-1}]

and, the velocity scale (v_k) and the time scale (τ_k) associated with it are respectively

$$v_k = (\nu\varepsilon)^{1/4} \tag{13.3}$$

$$\tau_k = \left(\frac{\nu}{\varepsilon}\right)^{1/2} \tag{13.4}$$

Padron's research at HSMRP included use of a batch Silverson L4RT rotor-stator with a slotted-head.[15] He used a mechanistic model to develop empirical dimensionless correlations between the Sauter mean diameter, d_{32}, and the Weber number, (*We*), Reynolds number (*Re*), and viscosity group, (V_i). As seen in Table 13.2, the inertial subrange correlation fits droplet breakup due to predominantly inertial stresses. The viscous subrange correlation is for droplet breakup due to predominantly viscous stresses. Mean drop size and drop size distributions (DSDs) were determined for multiple dilute (≤ 0.1v/v%) liquid formulations that dispersed immiscible silicone oils, of various viscosities (10–1000cSt), in aqueous surfactant and aqueous methanol solutions. The non-ionic surfactants used were Tergitol® TMN-6, Triton® X-100, and Triton® X-165. For clean systems with no surfactant, water and aqueous methanol were the continuous phase. DSDs were measured by video microscopy, followed by an automated image analysis.

When d_{32} data from the above experimental batch rotor-stator device are classified into two categories relative to the estimated Kolmogorov length scale (λ_k), i.e., $d_{32} > \lambda_k$ or $d_{32} \le \lambda_k$, the Table 13.2 correlation constants C_1 and C_2 are 0.055, 2.06 and 0.093, 24.44, respectively,[15] as explicitly expressed in Table 13.3.

TABLE 13.2

Non-Dimensional Correlations for d_{32} as a Function of We, Re, and V_i (experiments used a Silverson L4RT RSM with slotted cage stator[15])

Experimental Correlations	Flow Regions	Drop Size
$\frac{d_{32}}{D} = C_1 (We)^{-3/5}\left[1 + C_2 V_i \left(\frac{d_{32}}{D}\right)^{1/3}\right]^{3/5}$	Inertial subrange	$l_T \gg d \gg \lambda_k$
$\frac{d_{32}}{D} = C_1 (We\,Re)^{-1/3}\left[1 + C_2 V_i Re^{1/2}\left(\frac{d_{32}}{D}\right)\right]^{1/3}$	Viscous subrange – Drop breakup criteria: inertial stresses	$d < \lambda_k$
$\frac{d_{32}}{D} = C_1 We^{-1} Re^{1/2}\left(1 + C_2 V_i Re^{-1/4}\right)$	Viscous subrange – Drop breakup criteria: viscous stresses	$d \ll \lambda_k$

Nomenclature: D = rotor diameter [m], d = drop size [m], l_T = turbulent macro-eddy size [m], We = Weber no. $(\mu_c N^2 D^3 / \sigma)$, Re = Reynolds no. $(\mu_c N D^2 / \mu_c)$, V_i = Viscosity group $[(\mu_d N D / \sigma)(\rho_c / \rho_d)^{1/2}]$, C_1 and C_2 are empirical constants pertinent to the type of RSM.

TABLE 13.3
Non-Dimensional Experimental Correlations Relating d_{32} to *We* and *Re* for Silverson L4RT Rotor-Stator (with slotted cage) Mixer[15]

Experimental Correlations	Data RMSD	Drop Size
$\frac{d_{32}}{D} = 0.055(We)^{-3/5}\left[1 + 2.06V_i\left(\frac{d_{32}}{D}\right)^{1/3}\right]^{3/5}$	28.5%	$d_{32} > \lambda_k$
$\frac{d_{32}}{D} = 0.093(We\,Re)^{-1/3}\left[1 + 24.44\,V_i Re^{1/2}\left(\frac{d_{32}}{D}\right)\right]^{1/3}$	21.5%	$d_{32} \leq \lambda_k$

13.2.4 Experimental and Computational Research

At Cranfield University, UK, Thapar pursued in-line high-shear immiscible liquid–liquid dispersion research with the industrial-scale, single-stage Silverson 425LS in-line rotor-stator[16] shown in Figure 13.7. The stator in most of 13 tests was either of the two disperser screens (b and c), but limited testing used the slotted (d) and general-purpose (e) disintegrating heads instead to assess stator geometry effects on droplet size distribution (dsd).

Thapar dispersed kerosene in water, at dispersed-phase concentrations (φ) ranging from 5–30vol%. The dispersions were subjected to multiple passes through the in-line HSM at different flowrates and rotor speeds. The two immiscible phases were forced through the narrow rotor-stator gap, in a tangential direction primarily and a radial direction secondarily. They experienced intense hydraulic and mechanical shear while passing through the tiny circular (b) and square (c) stator perforations, at very high velocities. After some noted operating time (τ), the multi-pass process produced an equilibrium size distribution of kerosene droplets that were non-coalescing, even with no surfactants or stabilizers added. Drop size distributions were measured with a Malvern Mastersizer™ 2000 and also a conventional video probe technique.

Under turbulent conditions, the working media experienced a complex shear and extensional flow during consecutive passes through the in-line RSM. According to Thapar,[16] it is reasonable to assume that drops likely break due to (a) inertial stresses in the mixer inlet, (b) inertial and viscous stresses in the rotor region, (c) inertial sub-range eddies in the shear gap, (d) viscous stresses in the stator holes, and (e) inertial stresses in the volute. On the other hand, these regions can be put into

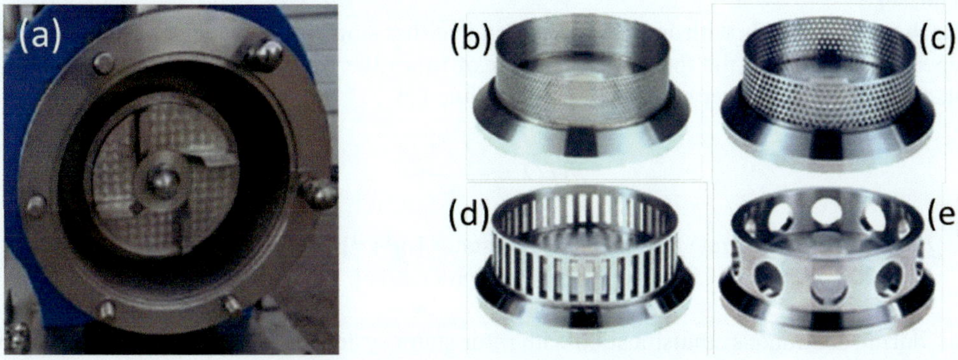

FIGURE 13.7 (a) Silverson LS425 in-line rotor and associated optional stators, (b) emulsor screen, (c) square-hole high-shear screen™, (d) slotted disintegrating head and (e) general-purpose disintegrating head.[4]

the following order of residence times: shear gap < stator < inlet < rotor < volute. He postulates that the very short residence times in the shear gap, where regions of high localized energy dissipation are very small, would indicate that a large percentage of the drops bypassed this region. Thus, Thapar[16] concludes that the relatively long residence times and the prevailing stress magnitudes would indicate that droplet size distribution is predominantly determined by viscous and inertial stresses in the rotor region. The Chen and Middleman correlation for liquid–liquid dispersion in stirred tanks[17] was found to be the most appropriate model for correlating d_{32}. This suggests that inertial stresses in viscous sub-range eddies are the dominant drop break-up mechanism. The drop size data could be correlated for a phase fraction (φ) of up to 20% by the dimensionless correlation in equation 13.5.[16] It is noted, however, that this mechanistic model is based on a single kerosene/water system, produced at only two different flowrates.

$$\frac{d_{32}}{D} = 2\times10^{9}\left(1+20\varphi\right)\left(We.Re^{4}\right)^{-1/7} \tag{13.5}$$

where, We = Weber number $\left(\dfrac{\rho_c N^2 D^3}{\sigma}\right)$; Re = Reynolds number $\left(\dfrac{\rho_c N D^2}{\mu_c}\right)$; d_{32} = Sauter mean diameter [μm]; and D = rotor diameter [m]

The limited Thapar[16] kerosene/water emulsion experiments with the HSM produced broad DSDs due to different turbulence and shear levels in the mixer, and also the observation that even above an identifiable critical operating speed, drop size could not be reduced further.

Droplet formation while emulsifying an immiscible working media within a mechanically agitated vessel depends on the input power per volume (ε_v) and the duration (τ) of exposure. Droplet size and stability depend on the liquid–liquid interfacial tension, which is often modified by surfactants and stabilizing agents added to the continuous or dispersed phases. A simple empirical correlation for the Sauter mean diameter (d_{32}) of a non-coalescing system, as a function of power density (P_v, which is power per unit volume in W·m^{-3}), is presented in equation 13.6.

$$d_{32} = C\left(\frac{P}{V}\right)^{-b} \tag{13.6}$$

where, P = power [W]; V= media volume [m^3]; C = proportionality constant; b = exponent constant

For non-recoalescing media, exponent b is −0.4 (low viscosity) to −0.25 (high-viscosity dispersed phase).[18]

In batch mixing vessels with an HSM operating in either single-pass or recirculation mode, drop size depends critically on the theoretical number of times the working media passes through the rotor-stator. A batch turn-over number (BTO) is defined as

$$BTO = \frac{\dot{Q}\,.\,\tau_p}{V} \tag{13.7}$$

where, $\dot{Q}$ = volumetric flowrate through the rotor-stator [m^3·s^{-1}]; V = batch media volume [m^3]; and τ_p = duration of the emulsion/homogenization process[es].

Thus, during continuous emulsification with rotor-stator systems, the influence of emulsion duration, in terms of mean residence time ($\overline{\tau}_r$) in the dispersing gap, must be considered along with the mean power density ($\overline{\varepsilon}_v$).[19] The recommended[18] modified form of volumetric specific energy is

$\varepsilon_v = \overline{\varepsilon}_v.\overline{\tau}_r$. If the conditions for droplet breakup are suitably fulfilled, and residence time influences the degree of breakup, then the Sauter mean diameter can be determined as a function of the energy density, ε_v[20–21], as in equation 13.8.

$$d_{32} = C\,\varepsilon_v^{-b} \tag{13.8}$$

The simple empirical correlation in equation 13.8 is an extension of predicted droplet size for non-coalescing immiscible media contained within a mechanically agitated vessel. It is widely used as an alternative to the mechanistic models generally pertinent to specific applications. Constants *C* and *b* are determined in the batch experiments and depend on the physical properties of the working media.

13.2.5 Phase Inversion

Phase inversion in liquid–liquid dispersion processes refers to the transformation of a disperse phase into a continuous phase and vice versa. This phenomenon occurs due to a change in the thermodynamic condition or imposition of ultra-high flow-induced deformation (extension) of the dispersed phase. Chapter 8, "Immiscible Liquid–Liquid Mixing/Dispersion," concisely describes the theory behind these occurrences. The shear and extensional flow induce high internal phase inversion characteristics that are typical of high-shear mixers such as rotor-stators. "Catastrophic" phase inversions (CPI) in HSMs that produce very high (≥0.74) dispersed-phase fractions are often intentional, and leveraged in the synthesis/manufacture of food, pharma and cosmetics products that meet required consistency, stability and solubility. Examples include butter preparation from milk/cream, nano-emulsion (droplet size ≤ 200nm) formulation for ocular drug delivery, and active nanoparticles formulation for sunscreens. HSMs are also used for both flow-induced phase inversions and for dispersions of such polymers as bitumen,[22] polyethylene, epoxides, polysiloxanes[23] and latex monomers[24] in surfactant-laden aqueous media subsequent to their solidification for special application in processes that are more advantageous than STRs. CPI of latex monomers produces high solid content dispersions and increases emulsification efficiency; concentrated latexes can be diluted and/or polymerized later.[24] CPI is a well-accepted industrially relevant technique used to produce paints and construction materials.

Basic research on flow-induced phase inversion (FIPI) was carried out by Akay[25] using a model multiple expansion contraction static mixer (MECSM), which is essentially a series of short capillaries with flow dividers to provide high rates of shear and extensional flow in oil/water working media. Mechanisms of phase inversion, and subsequent emulsion droplet size reduction, are proposed based on the behavior of micro-structured fluids under converging and diverging flows. The degree of phase inversion (o/w to w/o) is dependent on the rate of deformation and number of mixer units. The dependence of phase inversion on temperature and surfactant properties (e.g., concentration, composition, and mean hydrophile-lipophile balance) was investigated. The type of surfactant used in the experiments is discussed in facilitating the flow-induced phase inversion and subsequent emulsion droplet size reduction, as well as emulsion stability.

When an oil-in-water (o/w) immiscible media is in the presence of a stabilizer, and then subjected to high shear and extensional rates in an HSM, viscosity (η) and thermal conductivity (κ) generally increase with increasing dispersed-phase volume fraction (Φ). These increases continue until a critical volume fraction, Φ_{cr}, is reached. Thereafter, the emulsion inverts into a water-in-oil (w/o) emulsion, with oil as the continuous phase. The inverted w/o emulsion has a much lower aqueous phase volume fraction and lower viscosity.[26] The critical fraction (Φ_{cr}) is the maximum identified packing fraction, i.e., ~ 0.74. Unintended phase inversion due to extraneous shear and extensional flowrates commonly produces a soft semi-solid mass inside the HSM. This undesired coagulation phenomenon is detrimental to equipment and is an operational nuisance that should be avoided. Unfortunately, unforeseen incidents are common, and they mandate that the mixing equipment be dismantled and cleaned.

Either batch or in-line rotor-stators can be used for a high internal phase emulsion (HIPE). The batch process adds oil at a slow, steady flowrate (preferably below the rotor pumping rate) to an operating rotor-stator mixer (RSM) submerged in the aqueous phase (which incorporates an emulsifier). As the dispersed-phase volume fraction gradually increases, emulsion viscosity and rotor-stator power consumption increase until the flow-induced (catastrophic) interconversion of phases (from o/w to w/o) takes place. This is identified by the sudden drop in measured power consumption and/or viscosity (η) and emulsion thermal conductivity (κ).

In-line operations classically consist of a mixing vessel initially containing the aqueous phase and an external rotor-stator in the recirculation configuration, shown schematically in Figure 13.6a. Oil phase is slowly added to the bulk mixing chamber or the rotor through a streamline. The immiscible phases (oil and aqueous media) undergo multiple passes through the in-line rotor-stator to produce the desired HIPE. Since the rotor of an RSM typically lacks appreciable pumping capacity at the increased viscosity, a pump in the recirculation line handles the high-viscosity HIPE.

Figure 13.8 shows an Ekato SRT® mixer with submerged S-Jet rotor-stator[27] that the authors used in batch/in-line high-shear dispersions.

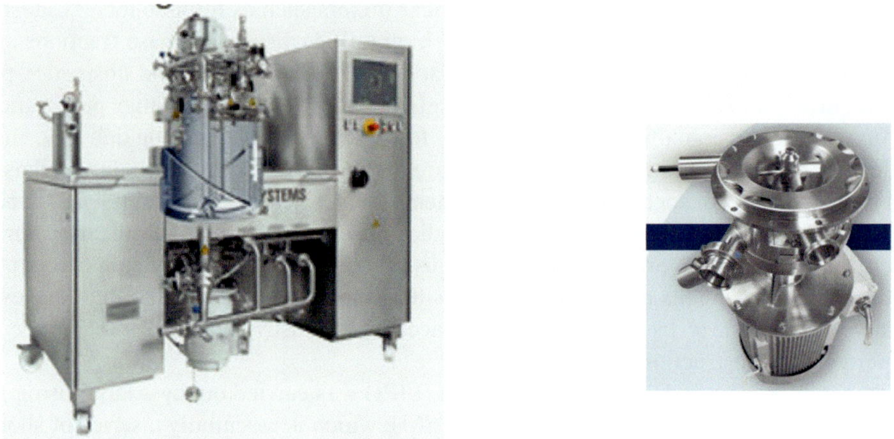

FIGURE 13.8 (a) Ekato Unimix SRT with (b) the S-Jet internal/external rotor-stator homogenizer.[27]

13.2.6 High Internal Phase Ratio (HIPR) Emulsion

In conventional emulsions, dispersed-phase droplets typically have a volume fraction of less than 0.74, the theoretical volume fraction of the most compact arrangement of mono-spheres. However, in high internal phase ratio (HIPR) emulsions, the dispersed (internal) oil phase has a volume fraction as high as 0.99, exceeding the aforementioned closed packing limit, resulting in deformation of the dispersed-phase droplets into polyhedral. The continuous phase forms a thin liquid film that separates polyhedral cells, and the HIPR emulsion is stabilized by the absorption of surfactants from the continuous phase, forming the cell surfaces.[28] Such emulsions are both kinetically and thermodynamically unstable; nevertheless, it is possible to prepare metastable systems that show no changes in properties or appearances over long periods of time.[29] The HIPR emulsions are useful precursors to high-molecular-weight polymers, composites and membranes for separation systems[30] that are widely used in the food, pharmaceutical and cosmetic industries.

HIPR emulsions can be prepared without phase inversion by continuously merging a stream of internal phase with the external phase inside a high-shear rotor-stator. One of the streams contains an appropriate quantity of surfactant(s) that serve(s) as an emulsifying and stabilizing agent. In US patent 5688842, Pate et al.[28] describe a method of preparing latex from HIPR emulsion using an IKA UTL 25 TURRAX™ in-line rotor-stator mixer operating at 7400 rpm. The dispersed phase, an ethylene-octene/toluene solution in the presence of dissolved sodium oleate as the surfactant, was continuously fed at a rate of 30g/min through one of the ports of a tee, located ahead of the RSM. Concurrently, the continuous water phase is pumped through the other port of the tee. Both streams were passed through the IKA RSM and subjected to sufficient shear to produce the targeted HIPR emulsion with dispersed-to-continuous phase fraction ranging from 10:1 to ~60:1 (i.e., with a constant water flowrate in the range of 0.5 to 3g/min). The toluene is removed from the resultant latex by rotary evaporation, and the psd is measured using a Coulter LS 130 analyzer. Respective latex volume-average mean particle size (d_{vm}) and polydispersity did not exceed 1.9 and 3.0, respectively.

Rotor-stator mixer (RSM) heads are equipped with rotors typically either of fixed blade or teeth designs; corresponding stators are interchangeable cages (teeth and slots) of various styles to fit different process requirements. Similar to the impeller in a batch agitated vessel, the shaft power draw of rotor blades can be expressed in terms of the dimensionless quantity Power number (Np), as shown in equation 13.9. Shaft power is measured using torque or a calorimetry method described in Chapter 17, "Experimental Techniques in Mixing," or from the energy dissipation rate derived from computational fluid dynamics (CFD).

$$N_p = P_s / \rho N^3 D^5 \tag{13.9}$$

where, P_s = power [W]; ρ = emulsion density [kg·m^{-3}]; N= shaft speed [s^{-1}]; and D = rotor OD [m]

Based on the blade Reynolds number, N_{Re} (defined in equation 13.10), there are three distinct flow regimes in the RSM head like the ones pointed out in Chapter 5, "Hydrodynamics," for STRs (see Figure 5.8). These are laminar flow $(N_{Re} \leq 10)$, transitional flow $(10 < N_{Re} < 10^4)$ and turbulent flow, $(N_{Re} \geq 10^4)$:

$$N_{Re} = \rho D^2 N / \mu \tag{13.10}$$

where, μ = emulsion dynamic viscosity [Pa·s]; and N_{Re} is based on the rotor tip velocity (~ND)

Batch experimental and computational investigations[1, 31–34] with blade-style rotors have revealed that for the fully turbulent flow condition, Power number is independent of Reynolds number, and varies between 1.0 and 3.0, depending on the RSM head's stator (cage) geometry and the vessel design.

Utmo et al.[32] investigated CFD simulations of the flow pattern and energy dissipation rates of a Silverson rotor-stator (RS) using the sliding mesh method with the k-ε model in turbulent flow $(N_{Re} = 26{,}000 - 52{,}000)$. The RS comprises a four-bladed rotor inside a cage-style stator like those seen in Figure 13.9 and Figure 13.10. The stator geometries (disintegrating head, slotted head and square hole head) were varied to understand their influence on flow pattern and power draw.

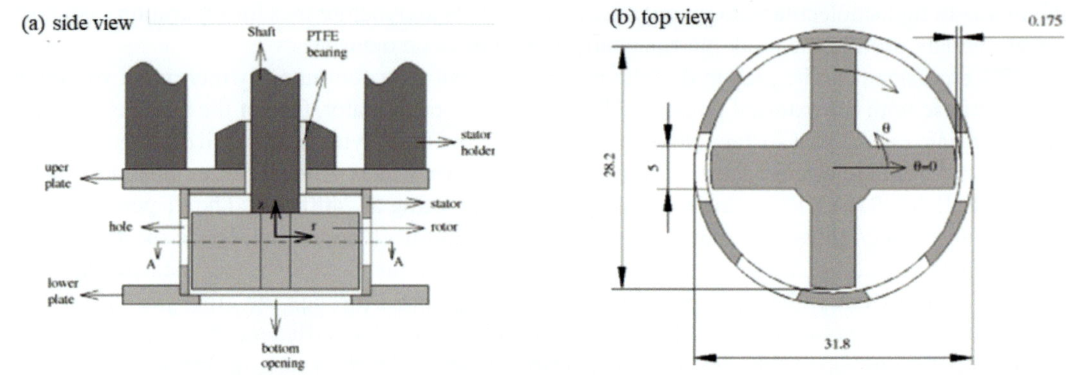

FIGURE 13.9 (a) Side view and (b) top view of Silverson RS mixer (rotor and disintegrating stator head). Not showing pins connecting upper and lower plates. All listed dimensions are in mm.[32]

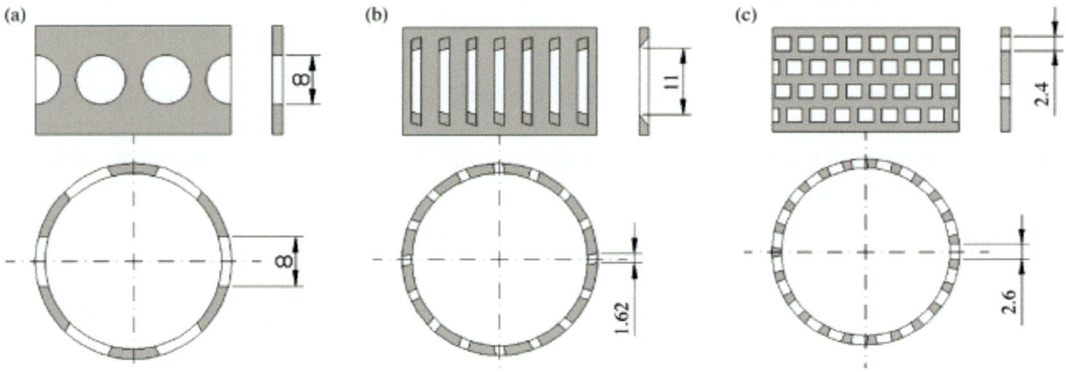

FIGURE 13.10 Tested Silverson stators including (a) disintegrating head, (b) slotted head, and (c) square head. All dimensions in mm.[32]

In the shear gap, radially discharged fluid passing through the stator experiences significant resistance to flow through holes in the above enclosures. Furthermore, the emerging fluid jets, depending on their flow pattern, likely strike the vessel wall. Head and tank design influence flowrate and torque, and thus, shaft power draw and power number. Because most pressure drop occurs in the stator, influence of vessel geometry on power draw will be much less than expected in a stirred tank.

Figure 13.11 *and* Figure 13.12 show the flow patterns and contours of energy dissipation rate for different stators determined from CFD simulation with Fluent® 6.3. The numerical simulations showed power number (N_p) at constant rotor speed to be proportional to flowrate through the holes; thus, N_p can be correlated to total hole area of the stator cage.[32]

In the case of a rotor-stator mixer operating in the continuous mode, required flow through the device is generally assisted by an external pump. Flow is process controlled and independent of the rotor speed. Several experimental and numerical studies indicate that power number (N_p) increases with inlet flowrate ($\dot{Q}_i$) at constant rotor speed.[35] The relationship between the rotor speed, flowrate and energy dissipated in an in-line HSM cannot be adequately described by a single power number.[36] Flow through the rotor has two components: the first resulting from pumping action of the rotors ($\dot{Q}_r = N_q N D^3$) and the second representing the inlet flowrate ($\dot{Q}_i$). Rotors expend the power

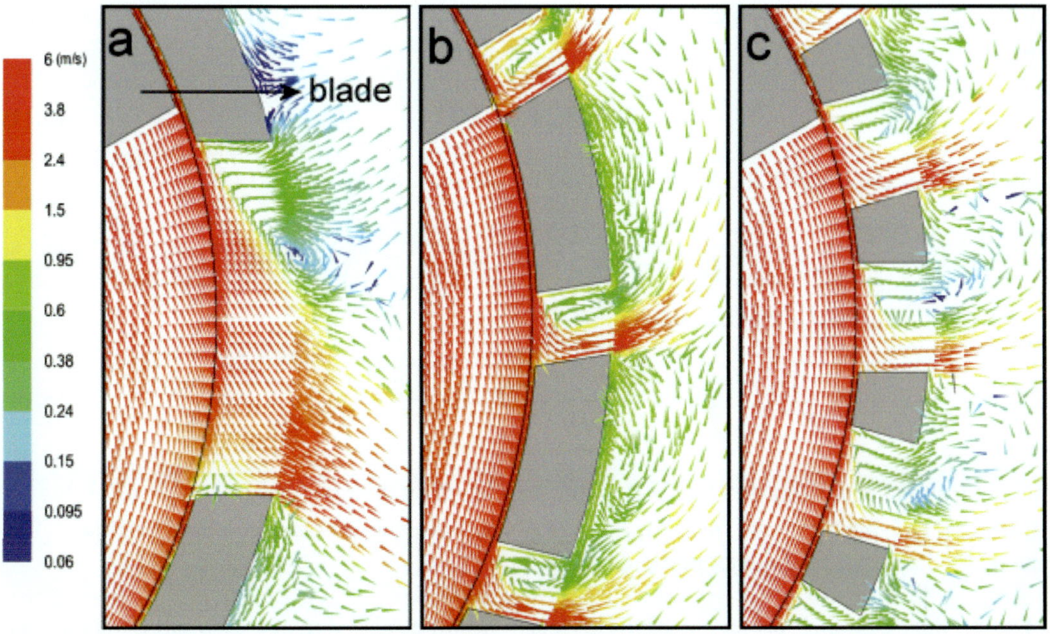

FIGURE 13.11 Radial and tangential jet velocities emerging from different Silverson stator holes in: (a) disintegrating head, (b) slotted head and (c) square hole head, recorded at planes (a) z = −0.8 mm, (b) z = 0 mm and (c) z = 1.3 mm. Clockwise (CW) rotor rotation is top to bottom.[32]

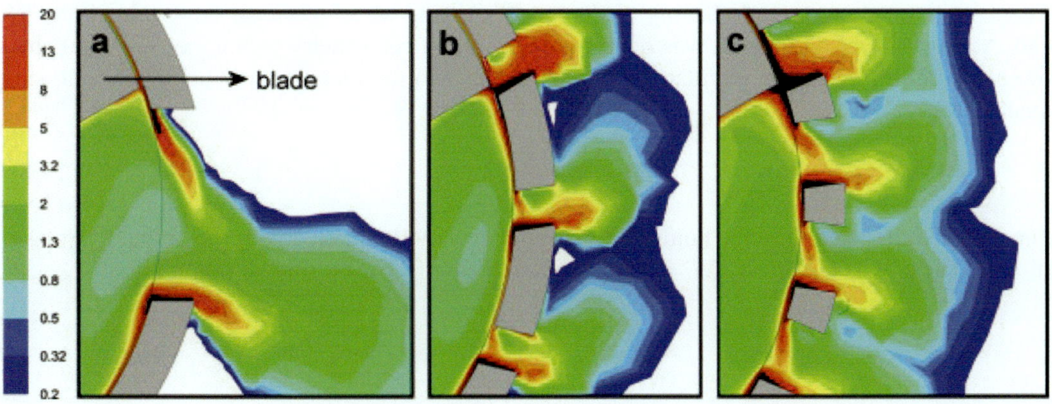

FIGURE 13.12 Contours of Silverson RSM energy dissipation rate (normalized by N^3D^2) at 4000 rpm in: (a) disintegrating, (b) slotted and (c) square hole heads. Black areas where normalized energy dissipation rate > 20. Planes as noted in Figure 13.11. CW rotor rotation is top to bottom.[32]

to accelerate the fluid so that kinetic energy increases to $C\rho\left(ND\right)^2$ where C is a proportionality constant. Thus, the power draw (P) of an RSM, as expressed in equation 13.11, is the sum of these two distinct operating parameters, Π_r and Π_v, and additionally, the expected frictional and vibrational power losses, Π_l, in the rotating device. This three-component power draw model for in-line RSM has been proposed by several investigators.[34–37] Terms Π_r and Π_v are, respectively, indicative of the power draw of a batch RS head to overcome the fluid resistance, and the centrifugal energy imparted to the fluid as flow convects it through the device.[38]

$$P = \Pi_r + \Pi_i + \Pi_l \tag{13.11}$$

Hydralic power imparted to the fluid, as defined by shaft power, $P_s = P - \Pi_l$, is

$$P_s = C\rho(ND)^2\left(\dot{Q}_r + \dot{Q}_i\right) = N_{p1}\rho N^3 D^5 + N_{p2}\rho N^2 D^2\,\dot{Q}_i \tag{13.12}$$

The power number (N_p) is expressed as

$$N_p = \frac{P_s}{\rho N^3 D^5} = N_{p1} + \frac{N_{p2}\,\dot{Q}_i}{ND^3} \tag{13.13}$$

where, ρ = fluid density, [kg-m^{-3}]; $\dot{Q}_r$ = flowrate developed by the rotor [m^3-s^{-1}]; N_q = rotor flow number, [-]; $\dot{Q}_i$ = inlet flowrate [m^3-s^{-1}]; N_{p1} and N_{p2} are respective power numbers for flow induced by the rotor and the inlet pumping rate.

During the "Domino" project at BHRG, Ozcan-Taskin et al.[37] pursued both experimental and computational research with three in-line commercial rotor-stators. Two were custom-built, two-stage RS designs from Silverson (150/250MS), and the third was a multi-stage Ytron Z open-teeth RS from Ytron-Quadro, as seen in Figure 13.13 and Figure 13.14. The Silverson RS comprises a four-bladed rotor with two different types of stator screens, one being two concentric emulsor screens (EMSC), and the other being the general-purpose disintegrating head (GPDH) as the inner screen with an outer square-hole high-shear screen (SQHS). The Ytron-Z RS consisted of three rows of concentric rotors and stators having teeth, and a pumping wheel to create flow through the RS head. Average power dissipation was determined using the calorimetry technique. The total shaft power draw of the system is calculated from the increase in fluid temperature using the following equation:

$$P_s = \rho V c_p \Delta T / \Delta t \tag{13.14}$$

where, P_s = shaft power into the fluid [W]; ρ = fluid density [kg·m^{-3}]; V = fluid volume [m^3]; c_p = fluid heat capacity [J·kg^{-1}·K^{-1}]; ΔT = temperature increase [K]; Δt = time duration [s]

FIGURE 13.13 Silverson RSM stators (a) EMSC and (b) GPDH-SQHS, and (c) the rotor.[37]

Mechanical and disc friction losses were estimated and subtracted from the total power, as explained in the PhD dissertation of Sparks.[39] Operating Reynolds numbers, based on rotor diameter as a characteristic length, ranged from 2.0×10^5 to 7.4×10^5.

CFD simulations were also conducted using Fluent® 6.3. Rotation of rotor was simulated with the multiple reference frame (MRF) technique. Turbulence was modelled using the realizable k-ε model.[37] Flow and power characteristics were determined for the three rotor-stator heads, i.e., Silverson GPDH-SQHS, EMSC, and Ytron-Quadro Ytron Z. Power draws for these devices were derived from the CFD-predicted torque values (T) and integrated energy dissipation rate (ε), then compared with their respective experimental values. See equations[35] 13.15–13.16. For the two Silverson heads, theoretical power as calculated from torque compared well with experimental values. CFD predicted powers based on energy dissipation rate, however, were significantly lower than the experimental result.

According to Ozcan-Taskin et al.,[37] the MRF method used in the simulation was unable to properly capture rotor-stator interaction and transient flow behavior. Thus, the authors used the sliding mesh technique to possibly overcome the deficit. For the Ytron Z head, predicted power based on CFD torque was higher than the experimentally measured value. The investigators attributed the discrepancy to two possible sources: (1) the complexity of the Ytron head geometry (i.e., concentric multi-stage tooth-style rotor and stator), as opposed to the simpler Silverson RS single rotor and stator, and (2) the CFD's poorly predicted flow through the Ytron Z unit. Different approaches for calculating a characterstic power number were evaluated before it was decided to adopt equation 13.12 proposed by Baldyga et al.[35] A summary of power numbers N_{p1} and N_{p2} for the three geometries is presented in Table 13.4.[37]

$$P_s = 2\pi NT \tag{13.15}$$

$$P_s = \rho h \int \varepsilon \, ds \text{ (2D surface integral of energy dissipation rate)} \tag{13.16}$$

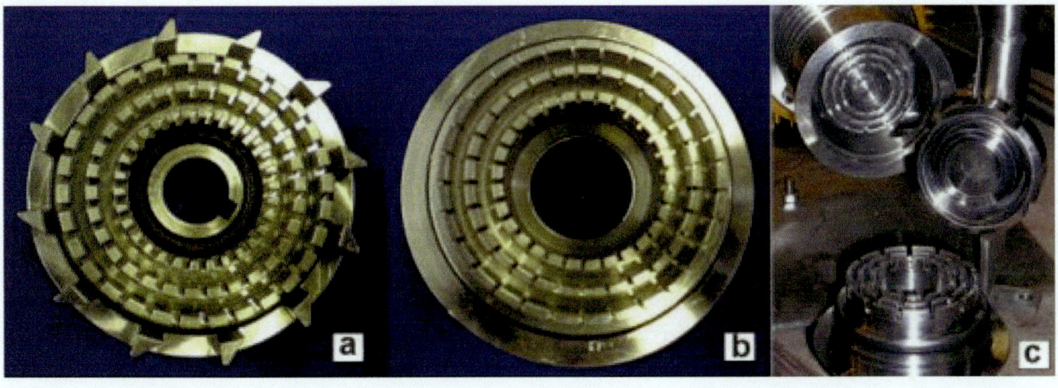

FIGURE 13.14 Ytron RSM Z geometry: (a) rotor + pumping wheel, (b) stator, and (c) mixing head.[37]

TABLE 13.4
Power Numbers for the Three Specific Inline Rotor-Stator Geometries[35]

Rotor-Stator Head	N_{p1}	N_{p2}
Silverson—GPDH-SQHS	0.13	9.1
Silverson—EMSC	0.11	10.5
Quadro—Ytron Z	0.18	10.6

where P_s = power [W]; N = rotor speed [s^{-1}]; T = torque [N-m]; ρ = density of fluid [kg-m^{-3}]; ε = energy dissipation rate [m^2-s^{-3}]; h = active height of the outer stator [m]

Shi et al.[40] pursued single-pass emulsification experiments with two different pilot-scale in-line HSMs shown in Figure 13.15. The dispersed and continuous phases in these tests were kerosene/water, kerosene/aqueous-glycerine, and silicone oil/water. In the experimental rig, a centrifugal pump provided continuous-phase pumping while dispered phase was introduced at a "tee" junction (upstream of the rotor-stator) using a peristaltic pump. Tween® 80 surfactant was used as an emulsifier to prevent coalescence of the dispersed phase. Drop size distribution (dsd) was measured with a Malvern Mastersizer S laser diffraction technique. Respective rotor tip speed and jet velocity through the slots were determined on the basis of outer swept rotor diameter (D) and open area of the outer stator. Power (P_f) consumed by the fluid is calculated by subtracting bearing frictional losses from the mixer power, as shown:

$$P_f = 2\pi NT - 2\pi NT_o \text{ [W]} \tag{13.17}$$

where T = torque [N-m]; N = speed, [s^{-1}]; T_o = tare torque [N-m] measured in water with no rotor

A systematic experimental parametric study was carried out while varying rotor speed, dispersed-phase volume fraction, continuous-phase flowrate, and continuous-phase viscosity. Of particular interest was fluid bypassing the high-shear and high-energy dissipation zone in the rotor-stator, an obvious issue with any single-pass in-line continuous flow system. This phenomena broadens the dsd, and in the worst-case scenario, produces bimodal drop size distributions. Shi et al.[40] found that dsd from the toothed HSM generally produced smaller droplets and a narrower dsd than did the blade-screen unit, suggesting that in-line HSMs with multiple rows of rotor-stator teeth might more effectively overcome bypassing and short-circuiting than single-row devices can.[40] Dispersed-phase Sauter mean diameter (d_{32}) decreased with increased rotor speed by increasing energy dissipation rate and shear rate. Increased continuous-phase viscosity also decreased d_{32} by enhancing effective transmission of shear stress to the drop interface. On the other hand, increased oil phase volume fraction (φ) and continuous-phase flowrate ($\dot{Q}$) only weakly affected d_{32}. Sauter mean diameter correlations, based on these experimental conditions, are summarized in Table 13.5.[40]

Mortensen et al.[41] used measured shaft torque and angle-resolved particle image velocimetry (PIV) to study power and flow charactertsics in a full-scale Tetra Pack Scanima (TPS) batch rotor-stator. Figure 13.16 shows spatially varying mean velocity magnitudes (normalized by rotor tip velocity) at four rotor positions.

The working media was water and light-scattering polyamide particles with a μm diameter (average) and 1.03 specific gravity. The rotor-stator was eccentrically located at the vessel bottom. Glass windows in the bottom head and a horizontal side nozzle gave the laser and camera optical access to

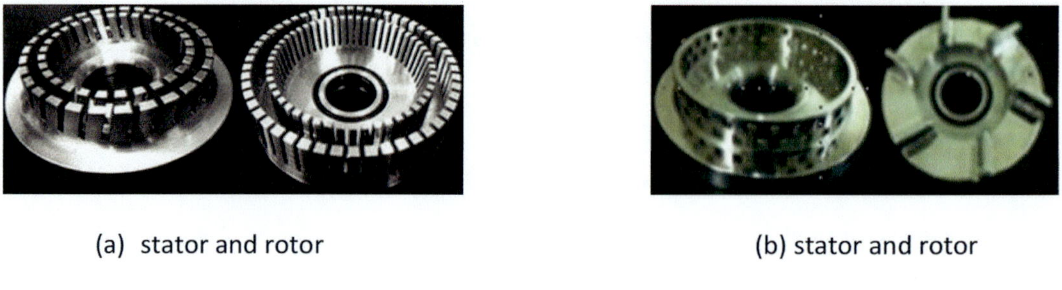

(a) stator and rotor (b) stator and rotor

FIGURE 13.15 (a) Tooth-tooth and (b) blade-screen rotor-stator pairs for in-line single-pass flow tests.[40]

TABLE 13.5
Effects of Various Parameters on Dispersed-Phase Drop Size, d_{32}

Influencing Parameters	Emperical Correlations	Experimental Conditions
Dispersed-phase volume fraction	$d_{32} / D = 0.043(1+2.32\varphi) We^{-0.53}$	$\varphi \leq 0.1$, $N \leq 3500$ rpm
Continuous-phase flowrate	$d_{32} / D = 0.029(1+3.46\varphi)(P_f / \dot{Q})^{-0.23}$	$\varphi \leq 0.1$, $N \leq 3500$ rpm, $0.2 \leq \dot{Q} \leq 1.5 m^3 h^{-1}$
Continuous-phase viscosity:		
Toothed in-line HSM	$d_{32} / D = 0.048(\mu_d / \mu_c)^{0.0074} We^{-0.55}$	$0.1 \leq \mu_d / \mu_c \leq 1.0$, $N \leq 3500$ rpm
Blade-screen in-line HSM	$d_{32} / D = 0.11(\mu_d / \mu_c)^{0.021} We^{-0.63}$	$0.1 \leq \mu_d / \mu_c \leq 1.0$, $N \leq 3500$ rpm

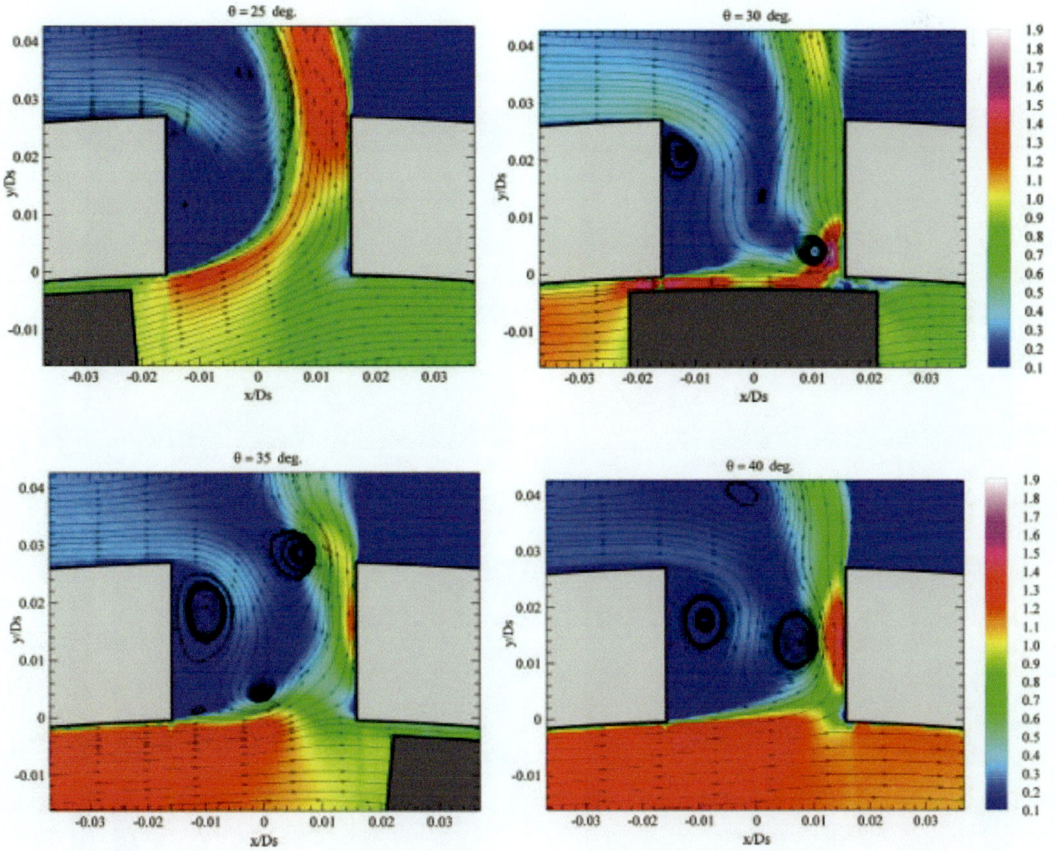

FIGURE 13.16 PIV-deduced topology: mean velocity magnitude, normalized by rotor tip speed, for rotor positions of 25°–40° in a region near a stator slot. Measurements obtained at 1033 rpm.[41]

the vessel interior. Additionally, a transparent (Plexiglas®) stator, with vertical slots, permitted the interior illumination needed for the camera to capture raw 2D PIV images inside the swept interior. Flow in the rotor region is predominantly tangential, except near the stator slots, where centrifugal forces and pressure differences across the slots[41] created radial flow. Instantaneous (periodic) changes in flow direction, as rotor blades pass the stator openings, created a few noticeable vortices in the slots.

The following excerpt (figure and discussion) is taken from reference 41. Figure 13.17 shows distinct jets emanating from a stator slot before incorporating slower-moving surrounding fluid. It also shows jet transience with respect to the rotor positions. The velocity of the emanating jet is reportedly greater than the rotor tip velocity (~1.3x), while that of volute fluid downstream from the stator is smaller than rotor tip velocity (< 0.8x).[41] According to Mortensen et al., the strong vortices that form in the stator slots produce cavitation bubbles in their low-pressure cores. As these vortices are carried out of a slot by radial jets, vortex strength decreases and core pressure increases until the cavitation bubbles implode. This produces pressure waves and velocity gradients that likely impact emulsification and dispersion performance of the mixer.[41]

Power dissipated into the fluid (P_s) and power number (N_p) obtained from the full-scale prototyped TPP rotor-stator mixer are plotted as functions of speed (N) in Figure 13.18. The power dissipation closely follows theoretical power curves of conventional stirrers operating in the turbulent flow regime, consistent with the use of standard dimensionless power numbers and flow numbers. The constant power number was determined to be 1.2. Based on their experimental investigation with the TPP,

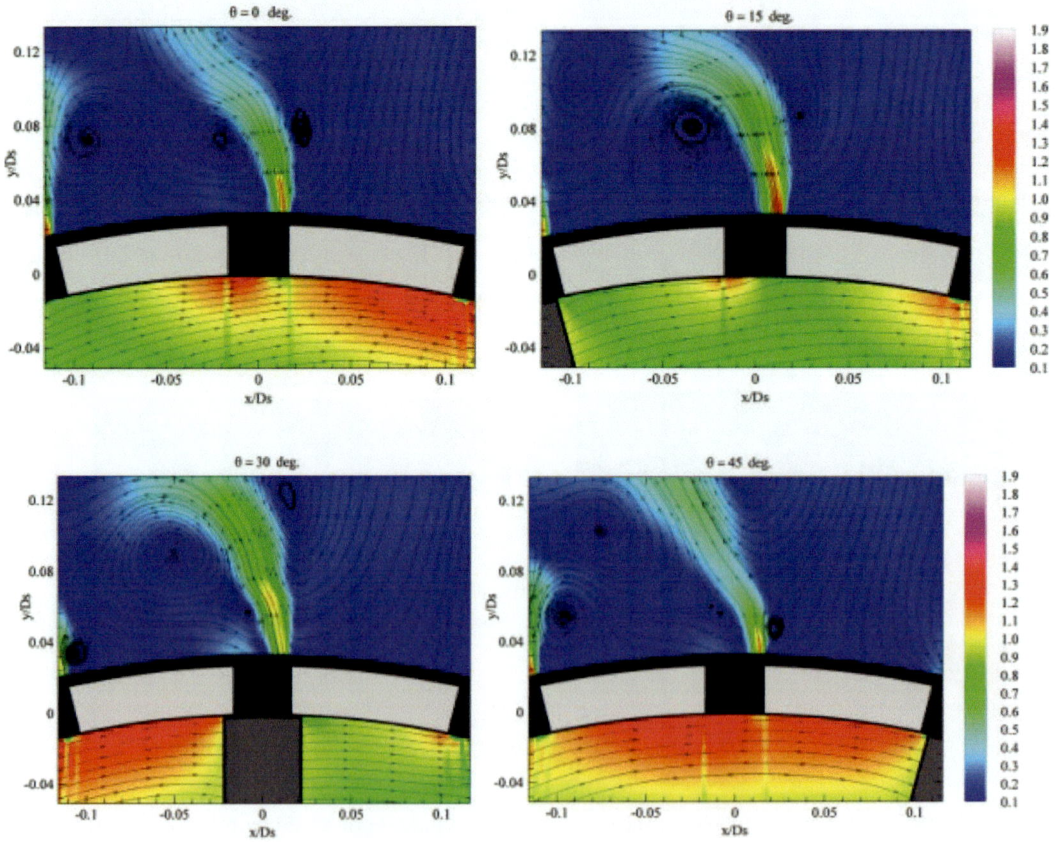

FIGURE 13.17 PIV-deduced topology: mean velocity magnitude, normalized by rotor tip speed, for rotor positions of 0°–45° in a region near the stator. Measurements obtained at 1033 rpm.[41]

Mortensen et al. conclude: "if geometric similarity is maintained, the scale-up of a batch rotor-stator mixing process that depends on macro-scale quantities should be based on constant rotor speed, since this leads to self similar flow topology, mean velocity, strain rate and turbulence intensity fields." [41]

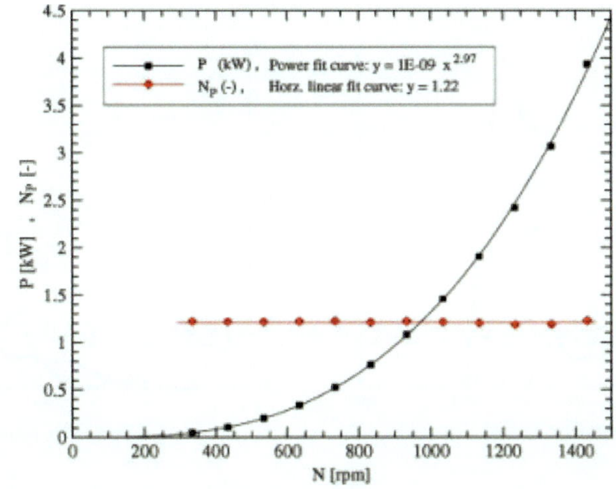

FIGURE 13.18 Power dissipated in the fluid (P_s) and power number versus rotor speed (N_p vs. N).[41]

Gallassi et al.[42] combined numerical and experimental approaches to understand the droplet breakage of oil-water emulsions using a batch IKA Ultra-Turrax® rotor-stator with a S25N-18G dispersing element. Experiments in a glass 1-liter vessel (see Figure 13.19a) used an emulsion of mineral oil (viscosity ≈ 19cP) and surfactant in a synthetic sea water solution. A 0.4-liter working volume and an oil phase concentration of 10% by volume were used to simulate conditions experienced by a centrifugal pump operating in an oil production setting. Drop size distribution (dsd) was measured by laser diffraction with a Malvern Mastersizer® 2000 PSA coupled with a Malvern Hydro 2000MU dispersing unit. Breakage of individual drops was analyzed using Photron's FASTCAM Mini UX100 high-speed camera. DSD experiments examined the effects of different mixing times, at rotor speeds of 6600, 9600 and 13,600 rpm. As suggested for relevancy in reference 37,

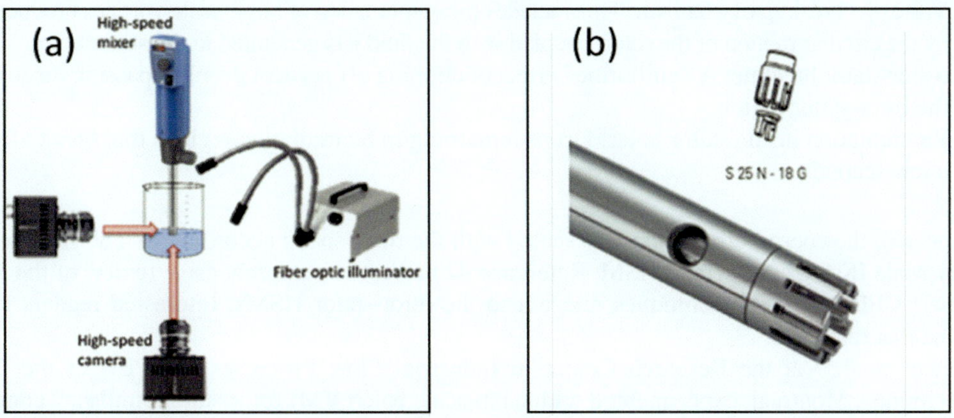

FIGURE 13.19 (a) IKA RSM experimental setup to visualize drop breakup mechanisms, from two different angles, and (b) IKA Ultra-Turrax® batch dispersing element in the RSM.[42]

experiments ran for at least 60 minutes, or until the fluid temperature increased by 10°C. Results in Figure 13.20[42] show DSDs shifting from a bimodal towards a monomodal distribution, with simultaneous decreases in drop size, as energy dissipation (speed) and/or mixing duration increase.

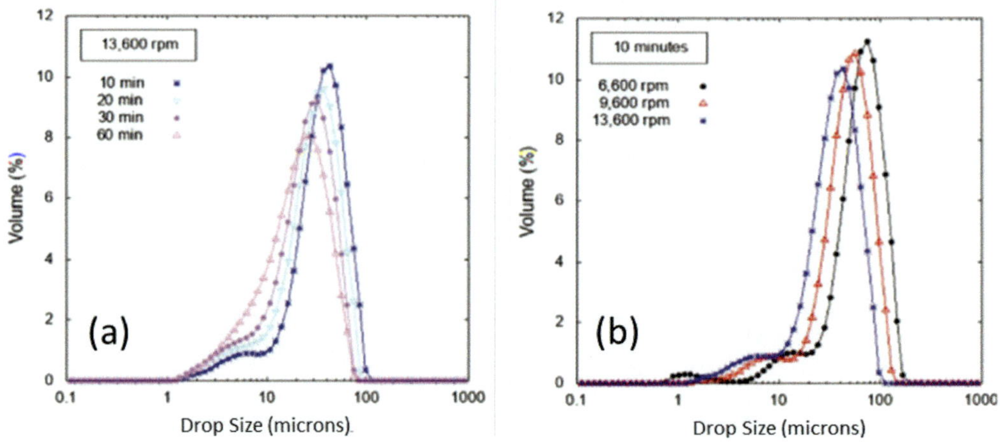

FIGURE 13.20 Drop size distributions for (a) different mixing time using the same rotational rate (13,600 rpm), (b) all rotational rates studied using the same mixing time (10 min.).[42]

Transient numerical simulations were performed using Fluent 16.0, with a standard $k-\varepsilon$ turbulence model with enhanced wall treatment. All of the simulations at 6600, 9600 and 13,600 rpm were initialized with steady-state results using a moving reference frame (MFF) model, followed by the transient sliding mesh model (SMM). A second-order upwind scheme was selected for spatial discretization. The modeled fluid was considered to be sea water at 20°C.

The complementary approaches of experimental fluid dynamics (EFD) with visualization and computational fluid dynamics (CFD) coupled with a population balance model (PBM) using an appropriate ad-hoc breakage kernel[43] revealed both qualitative and quantitative descriptions of the key phenomena. These include flow interactions, streamlines and energy dissipation rates involved in drop breakage. Like the previously mentioned investigations,[32,37,41] Gallassi et al. confirmed the following three main mechanisms relevant to drop breakup.[42]

1. Vortex-jet breakup mechanism: This includes the combination of large-scale vortices produced by the circular motion of the rotor, coupled with the fluid jets generated in the stator holes.
2. Rotor-stator breakup: A "guillotine" effect of clipping off parts of droplets passing through the rotor-stator region.
3. Recirculation inside stator holes: Drops remaining in recirculation regions that break after a few seconds.

As expected, the energy dissipation rate scaled with the rotor speed according to εN^3 for the laboratory-scale IKA Ultra-Turrax® HSM. Reference 42 presents a comprehensive review of the state-of-the-art CFD and EFD techniques relevant to the rotor-stator HSMs. Interested readers might examine that reference.

Doucet et al.,[44] at the Research Center in Industrial Flow Processes (URPEI) of the Ecole Polytechnique, Montreal, experimented with a lab-scale batch VMI rotor-stator emulsor[45] comprising a four-bladed rotor inside a slotted stator (see Figure 13.19). Experimental fluid contained in a transparent polycarbonate tank was drawn axially into circular holes at the top of the stator cage and exited radially through the stator slots.

FIGURE 13.21 VMI rotor-stator impeller.[45]

High-viscosity Newtonian (glucose solution) and non-Newtonian (Xanthan gum and CMC solutions) fluids were used in the experiments to visualize hydrodynamics internal and external to the rotor-stator volute and determine power draw of VMI RSM in the laminar, transition and turbulent regimes. Viscosities of the glucose solutions ranged from 0.5Pa-s to 29Pa-s. The rheological properties of the pseudoplastic fluids followed a power law model, as shown in equation 13.18. Their consistency coefficient, m, and the flow behavior index, n, are presented in Table 13.6.

$$\mu_a = m\dot{\gamma}_{av}^{\,n-1} \tag{13.18}$$

where, μ_a = apparent viscosity of the fluid [Pa] and $\dot{\gamma}_{av}$ = average shear rate [s^{-1}].

TABLE 13.6
Rheological Properties of the Non-Newtonian Test Fluids[44]

Solutions	wt.%	τ_y [Pa]	m [Pa.s^n]	n [-]
Xanthan	0.20	–	0.65	0.34
	0.30	–	0.90	0.3
	0.35	–	1.2	0.3
	0.50	4	2.2	0.26
	1.0	9	4	0.26
CMC	1.0	10	5	0.4
	2.0	20	55	0.24

Rotor speed was varied up to 1500 rpm to maintain both laminar and transitional flow fields adjacent to the rotor. To understand how the stator cage influenced power draw in the Newtonian fluid, power was calculated using equation 13.17. Shaft torque values were "tared" by measuring torque from the rotating impeller, with and without the stator. The without-stator values were then subtracted from the with-stator values.

Figure 13.22 shows the classical Newtonian Power number (P_o) vs. Reynolds number (Re) curve for the VMI rotor-stator configuration. In the laminar region, the curve slope is close to the theoretical value of –1, i.e., Power number is inversely proportional to Reynolds number. From these particular rotor-stator experiments, the power constant $K_p = P_o \cdot Re = 314$, while the Power number

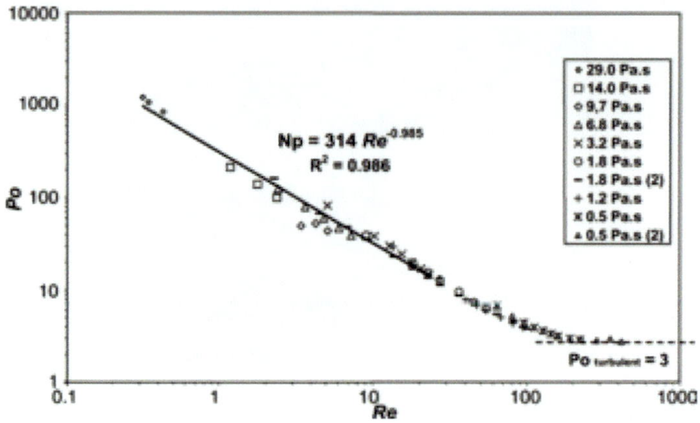

FIGURE 13.22 Newtonian power curve for an experimental rotor-stator.[44]

in the turbulent-flow regime is constant at 3. These values are in the same range as those reported by Padron,15 in the case of the Silverson LS425 rotor-stator with a slotted-stator head.

The Power number vs. Reynolds number curve, for the "without-stator" case, is presented in Figure 13.23 In the laminar flow region, the slope of the P_o vs. Re is again found to be −1. However, the power constant $K_p = P_o \cdot Re$ is 92.7. The power number in the turbulent regime remains unchanged at 3. These plots also show the transition regime starting at Re =10 for the rotor alone and at Re = 100 for the rotor-stator pair. The following inferences by Doucet et al.[44] confirm the previous postulations by Calabrese et al:[46]

(a) The power constant, K_p, in the presence of the stator, is three times the value observed in the absence of the stator. Thus, the stator cage significantly increases power draw in the laminar regime, likely by increasing the resistance to laminar (viscous-dominated) flow coming off from the rotor.
(b) In turbulent (inertia-dominated) flow, constant power (P_o) with and without the stator indicates that power draw is controlled by jets discharged from the rotor.

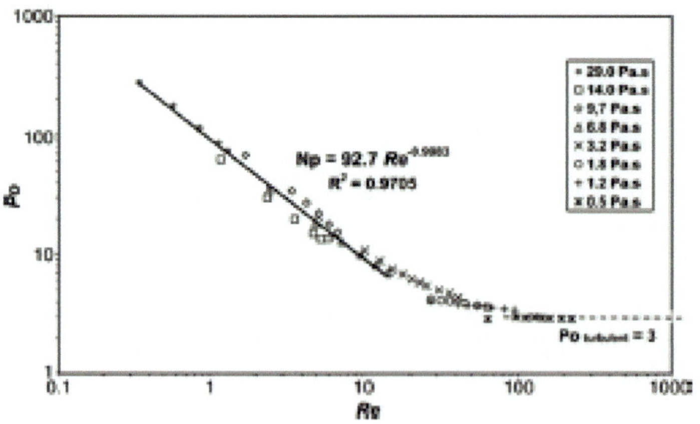

FIGURE 13.23 Newtonian power curve for an experimental rotor only.[44]

Experimental visualization studies using acid-base colorimetry (see Chapter 15, "Experimental Techniques in Mixing) revealed that the formation of pseudo-caverns in the vicinity of the rotor are almost an unavoidable phenomenon when mixing viscous and/or non-Newtonian shear-thinning fluids with rotor-stator impellers. In fact, fluid in the rotor-stator does not have to be a yield-plastic media to form such caverns in the RS volute and beyond. The size of pseudo-caverns formed in the rotor-stator, while using Newtonian and non-Newtonian fluids, can be scaled with the classical Reynolds number (Re) and modified Reynolds number (Re_y), respectively, as defined:

$$Re = \rho\, D^2 N / \mu \tag{13.19}$$

$$Re_y = \rho\, D^2 N^2 / \tau_y \tag{13.20}$$

where, Re_y is the ratio of inertial force due to fluid motion/fluid yield stress (τ_y) [Pa], as defined in the Herschel-Bukley model, i.e., $\tau = \tau_y + m\,\dot{\gamma}_{av}^{n}$

Later, Barailler et al.,[47] at URPEI, used the virtual finite element method in laminar 3D transient CFD simulations of the same VMI mixer as was used by Doucet et al.[41] Calculated shaft torque values agreed well with the experimental measurements of Doucet et al. Pseudo-caverns in the previous experimental study were highlighted by analyzing the shear stress field. Further analysis of the stress field showed maximum stress to occur in the clearance between the rotor blade tip and the stator.[47]

13.3 COLLOID MILLS

Colloid mills are specially designed high-shear rotor-stator mixers intended to reduce particle size in slurries and colloids by means of grinding and/or deagglomeration. Because they perform the conventional task of uniformly dispersing (homogenizing) finely granulated/disintegrated particles in a continuous media, rotor-stators and colloid mills are synonymous HSM devices. Their designations are often expressed interchangeably in the mixing community.

In a typical in-line application, a powder is added to a liquid stream ahead of the colloid mill. The slurry is subjected to very high hydraulic shear (10^4–10^6 s^{-1}) and mechanical grinding in the mix-head, with the latter often dominating in the process of further granulating and homogenizing solids. One such example is the Silverson—Flashmix® powder/liquid RSM[48] (see Figure 13.24a), which draws down solids and mixes them into a continuous liquid stream immediately prior to their entry into the rotor-stator head. Prior to storage, coarse-grain particles (≥75μm) and fibers are normally comminuted, pre-milled, or disintegrated by means of either dry grinding or wet milling with ball or hammer mills. Besides traditional RSMs, multiple proprietary colloid-mill designs have also been patented for functional flexibility and effectiveness in producing a desired product-specific psd and polydispersity in the continuous phase.[49] In most cases, however, their fundamental operating principles are the same as that of rotor-stators.

Figure 13.24b shows a simple cone-type colloid mill that uses the frustum of a right circular cone as the high-speed rotor inside a stator that is a circular stationary block, whose internal surface is conformal to the outer periphery of the rotor. This design produces a closed-controlled uniform annular shear gap through which the working media passes. Gap width can be adjusted by vertically translating the rotor position to fine tune the desired shear stress and resulting psd. Slurry is typically introduced through a chute or hopper that is sometimes an integral part of the mix-head. Charging and discharging of the slurry occurs through the top and bottom annular zones seen in Figure 13.24b. In a batch process, the discharged material is recirculated back to the hopper for continued processing until the desired product quality is achieved.

The average particle size in slurry discharged from a colloid mill is typically between 1μm and 25μm. The processing capacities of colloid mills vary from 0.5 to 120gpm, depending on the mix-head size. These capabilities make colloid mills appropriate for a variety of applications, including colloid and oil/water emulsion processing for cosmetics, mayonnaise, silicone/silver amalgams, and roofing tars.[49]

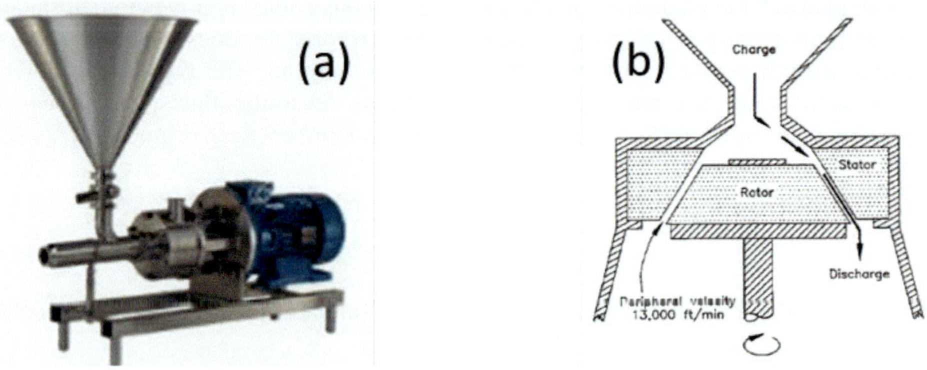

FIGURE 13.24 (a) Silverson Flashmix® powder-liquid mixer[48] and (b) schematic of a typical colloid mill.[50]

Since the gap in a colloid mill or other rotor-stator device may be as small as 25μm, a significant amount of heat can be generated by particle grinding and viscous energy dissipation. Consequently, stator heads commonly incorporate a cooling jacket to minimize thermal expansion and contraction of mix-head components. Besides reducing slurry temperatures, such cooling also helps maintain close clearances in the mix head. Rotors and stators are typically constructed from materials that resist wear and have a low coefficient of thermal expansion. Such materials include toughened stainless steel (e.g., 420 stainless steel), corundum, Invar etc.

Important media properties and operating parameters that influence the size of final product particle are (a) viscosity of the continuous phase, (b) volume fraction of the dispersed/solid phase, (c) shear intensity, (d) gap size and (e) residence time/duration. Most major RSM manufacturers, such as Silverson, Ika, Ross, Sina, Ekato and NOV-Greerco, make colloid mills with toothed and serrated rotors for efficient particle sizing. Unfortunately, there is too little research reported on dispersing powders rather than liquids (i.e., emulsions). Different mechanisms for particle breakage have been developed, but work is still needed to establish applicable correlations.[36]

13.4 OTHER IN-LINE DYNAMIC MIXERS

Apart from the rotor-stator and colloid mills, there are multiple other commercially available high-shear in-line dynamic mixers to accomplish liquid–liquid or solid–liquid dispersions, emulsions and their homogenization. Chapter 15, "Emerging and Unconventional Mixing Systems," addresses a few such in-line mixers, namely Higee—rotating packed bed (RPB), ResonantSonics® and Hydrodynamics devices. The high shear and elongational rates imparted to the immiscible media such as the gas–liquid or liquid–liquid systems are derived from maintaining high velocity in the confined space or by harnessing cavitation. The authors have deliberately excluded their further discussion in the chapter.

In-line agitators are used to continuously homogenize miscible liquids flowing through pipes. SPX Flow Lightnin's two-stage Line Blender,™ as shown in Figure 13.25, is one such device used to produce a uniform effluent from various feed streams.[51] Its design features two separate mixing chambers where high-speed, high-power impellers in confined domains create reasonably high shear to produce uniformity of fluid components at flowrates up to 20,000gpm. A separator plate divides the mixer into two separate zones and directs all flow through the impeller zone, thus preventing short-circuiting and back-mixing. Three stirred tank-like baffles are installed to ensure an optimum flow pattern and reliable power consumption. The result is a far more intense blending and dispersing action than a single-stage mixer could provide. Typical application areas include: chemical and food processing, petroleum refining, and pulp and paper service.[51]

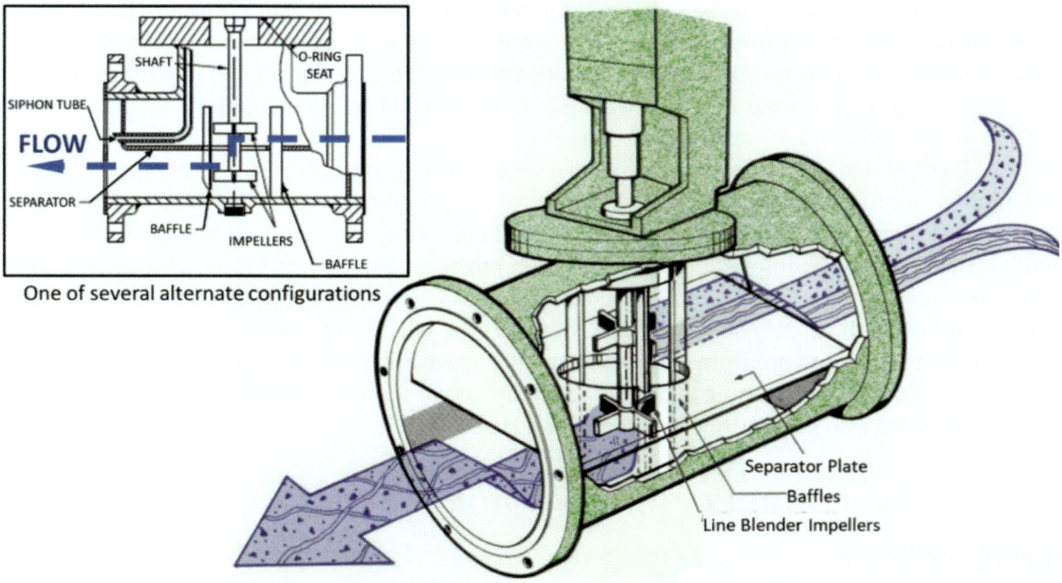

FIGURE 13.25 SPX Flow Lightnin in-line dynamic mixer.[51]

13.5 BATCH VS. CONTINUOUS HIGH-SHEAR MIXERS (HSMs)

In general, a batch-mode RSM converts a much higher portion of the shaft energy into high-intensity turbulence, in the region where mixing and emulsification takes place, than does a continuous-mode RSM, which uses much of this energy for pumping (i.e., increasing the head of the flow). Several different theories have been suggested, but there is still no clear literature consensus on how continuous-mode RSMs should be operated to give the same emulsion as does a batch-mode RSM.[52]

Continuous operations are often preferred over batch operations for a variety of reasons, including lower capital intensive and higher reproducibility than batch processing. In-line mixers are recognized for their process intensification benefits, such as smaller footprints with their environmental and cost-effectiveness benefits. Batch processes, however, are still commonly employed in many production proceeses.[52] In most instances, minimum required residence time, uncertainty in scale-up from laboratory batch process, and control over local tempearture (viscous and frictional heat dissipational) make batch process genuinely a more risk-averse approach.

Apart from the discussed lack of specific scale-up/-down criteria, practicing process engineers must be aware of some inherent drawbacks in high-shear mixers (HSMs). These are energy-intensive devices that generate heat that must be instantaneously dissipated to the coolant. Continuous HSM processes could be prone to high-speed vibration, fluid bypass and foam generation. Furthermore, the colloid mill's rotor and stator components are susceptible to attrition, wear and tear.

13.6 CURRENT CAPABILITIES IN PREDICTING ROTOR-STATOR DROP SIZE

Design of the stator element surrounding the high-speed rotor significantly influences the hydrodynamics inside the head that are responsible for drop breakup and coalescence. In turbulent flow, evidently, the dispersion in the mixer-head is subjected to hydraulic shear and elongational rates in the shear gap, mechanical shear in the orifices/slots/teeth, local turbulent energy dissipation rates along the flow path and in radially emanating jets. As already mentioned, some report evidence of imploding cavitation bubbles at the stator outlets. As mentioned in Chapter 8, "Immiscible Liquid-Liquid Dispersions," for droplet sizing of most emulsions, suitable emulsifiers (surfactants and/or stabilizers) are used to minimize organic/aqueous phase interfacial tension and maximize interfacial resistance to droplet coalescence. See droplet formation and stabilization mechanisms.

Though the steady-state mean power consumption rate of an HSM is reasonably predictable, complex 3D flow characteristics of the mix-head make *a priori* prediction of emulsion dsd, volume median diameter (d_{vm}) and Sauter mean diameter (d_{32}) currently unlikely. This chapter cites a few familiar drop size correlations from experimental and numerical investigations using selected commercial batch and in-line rotor-stator mixers, which represent a mere fraction of a broad and ever-expanding literature base. Although the cited correlations have limited application to emulsion and homogenization applications using specific RSMs, they are useful to process designers in providing a basic understanding and preliminary assessment of some available RSMs, based on their drop size distribution efficacies. Developing energy-efficient processes requires designers to consider dispersion capability but also power draw in selecting an HSM.[36]

13.7 CURRENT CONSIDERATIONS IN ROTOR-STATOR SCALE-UP

As briefly outlined, a fair number of recent experimental and computational studies of high-shear mixers are reported in the literature. There are scale-up suggestions based on geometric similarity,[15] constant rotor tip velocity,[41,44,53] power per volume[54] and processing time.[36] Since these works are very specific to their pertinent head designs, and are limited in the scope of their applications, their recommendations are not universal.

For example, Hall et al.[53] performed silicon oil/water emulsion scale-up experiments using Silverson MS 150/250 (pilot scale) and 088/150 (lab scale) RSMs. The effects of rotor speed, total flowrate and dispersed-phase viscosity on drop size distributions and average drop size were investigated. The Sauter mean diameter was correlated with tip speed, mechanical energy dissipation rate, total energy dissipation and Weber number. Investigators found that tip speed is the most appropriate scaling parameter for the specific in-line rotor-stator mixers for low to intermediate dispersed-phase viscosities.[53] The rotor tip speed is varied to provide the required shear rate that plays a major role for drop breakage in laminar and transition regimes. Note that irrespective of the RSM size, in order to maintain the predetermined confined flow for drop breakup,[55] manufacturers often maintain constant shear gap depending on the design of their stators. This constant confinement poses a major deviation from scale-up based on geometric similarity, and hydraulic shear and extensional rates turn out to be proportional to the rotor tip speed.

Industrial-scale RSM heads often incorporate multiple concentric rotors and stators of varying designs. Such complexity in design and inherent flow features makes the *a priori* prediction of generalized performance characteristics of HSMs, such as specific power requirement (ε), drop size distribution (dsd), and Sauter mean diameter (d_{32}), seem justifiably unattenable. According to Atiemo-Obeng et al.[1] (2004), "None of the research clearly addresses the scale-up rules on a fundamental basis." Similar opinions have been published in later review articles by Zhang et al.[36] (2012), and more recently, by Vashisth et al.[56] (2021). Practicing engineers normally adhere to a judicious scale-up based on a demonstrated laboratory HSM, then follow up with a geometrically identical commercial unit selected with much-needed help from its manufacturer regarding required tip speed, power per volume, and residence time or batch turn-over number (BTO). They then pursue a trial-and-error approach using their full-scale demonstration units.

In many instances, the authors of this book collaborated with chemists to create desirable o/w dispersions (dsd and d_{vm}) with laboratory-scale RSMs and real recipes, prior to pursuing scale-up with a similar device. In collaboation with the manufacturers, minor mix-head design alterations and operating parameters tweaks (such as rotor speed and residence time) led to multiple successful commercial products. Generally, this is the most prudent, cost effective, and time-conscious approach to intellectual property rights for a specific production process.

REFERENCES

1. Atiemo-Obeng, V. A., Calabrese, R. V., Rotor-stator mixing devices, in: E. I. Paul, V. A. Atiemo-Obeng, S. M. Kresta (Eds.), *Handbook of Industrial Mixing: Science and Practice*, John Wiley & Sons Inc., NJ, pp. 479–505, 2004.
2. https://img.sigmaequipment.com/equipment-docs/47912/UMX_S-JET_E_921-01-059_04.10.pdf
3. www.ekato.com/wp-content/uploads/2020/12/EKATO_UNIMIX_SRT_E.pdf
4. www.silverson.com/images/uploads/documents/Silverson_Brochure_US.pdf
5. www.ika.com/ika/pdf/flyer-catalog/Disperser_Brochure_EN_USD_IWW_screen.pdf
6. https://www.mixers.com/products/high-sher-mixers/
7. https://pdf.agriexpo.online/pdf/chemineer/greerco-high-shear-mixers/184431-15777-_2.html
8. https://www.mixers.com/resources/mixing-technology-reports/how-to-increase-shear-without-sacrificing-throughput/
9. https://framatomebhr.com/
10. https://chbe.umd.edu/research/calabrese-group-high-shear-mixing-research-program
11. Hall, S., Cooke, M., El-Hamouz, A., Kowalski, A. J., Droplet break-up by in-line Silverson rotor-stator mixer, *Chem. Eng. Sci.*, Vol. 66, pp. 2068–2079, 2011.
12. Leng, D. E., Calabrese, R. V., Immiscible liquid-liquid systems, Chapter 12, in: Paul, E. L., Atiemo-Obeng, V. A., Kresta, S. M., (Eds.), *Handbook of Industrial Mixing*, John Wiley & Sons, Inc., NJ, 2004.
13. Shinner, R., On the behavior of liquid dispersions in mixing vessels, *J. Fluid Mech.*, Vol. 10, pp. 259–279, 1961.
14. Harnby, N., Edwards, M. F., Nienow, A. W., *Mixing in the Process Industries*, Second edition, Butterworth-Hienemann Ltd., Oxford, 1992.
15. Padron, G. A., *Effect of Surfactants on Drop Size Distributions in a Batch Rotor-Stator Mixer*, Ph.D. Dissertation, University of Maryland, College Park, MD, 2005.
16. Thapar, N., *Liquid-Liquid Dispersions from In-Line Rotor-Stator Mixers*, Ph.D. Thesis, Cranfield University, UK, July 2004.
17. Chen, H. T., Middleman, S., Drop size distributions in agitated liquid-liquid systems, *AIChE J.*, Vol. 13, pp. 989–995, 1967.
18. Karbstein, H., *Untersuchungen zum Herstellen und Stabilisieren von Oil-in-Wasser-Emulsionen*, Dissertation, Universitat Karlsruhe, 1994.
19. Koglin, B., Paweloski, J., Schonring, H., Kontinuierliches emulgieren mit rotor-stator-maschinen: Einfluss der volumenbezogenen dispergierleistung und der verweilzeit auf die emulsionsfeinheit, *Chem. Ing. Rech.*, Vol. 53, pp. 641–647, 1981.
20. Schubert, H., Karbstein, H., Mechanical emulsification, in: T. Yano, R. Matsuno, K. Nakamura (Eds.), *Developments in Food Engineering, Part I, Proc. of 6th. International Congress on Engineering and Food, May 23–27, Chiba, Japan*, Blackie Académie and Professional, London, 1994.
21. Ekato. *The Book*, Ed. Himmelsbach, W., Ekato Holding GmbH, Freiburg, Germany, 2012.
22. Arenas-Calderon, E., Sadtler, V., Marchal, P., Choplin, L., Delfosse, F., Maze, M., Preparation of highly concentrated bitumen emulsions by catastrophic phase inversion: Follow-up of the emulsification process, *Colloids Surf. A: Physicochem. Eng. Asp.*, Vol. 458, pp. 25–31, 2014.
23. Astruc, M., Navard, P., A flow-induced phase inversion in immiscible polymer blends containing a liquid-crystalline polymer studied by *in situ* optical microscopy, *J. Rheol.*, Vol. 44, pp. 693–712, 2000.
24. Campbell, S. B., Larson, T., Smeets, N. M. B., El-Jaby, U., McKenna, T. F. L., Mini-emulsification by catastrophic phase inversion, *Chem. Eng. J.*, Vol. 183(15), pp. 534–541, 2012.
25. Akay, G., Flow-induced phase inversion in the intensive processing of concentrated emulsions, *Chem. Eng. Sci.*, Vol. 53(2), pp. 203–223, 1998.
26. Tadros, T., Phase inversion, in: *Encyclopedia of Colloid and Interface Science*, Springer Link, p. 930, 2013.
27. https://www.ekato.com/products/process-plants-and-units/vacuum-processing-units-unimix/
28. Pate III, J. E., Latensky, E., Peters, J., Pelletier, R. R., Process for preparing high internal phase ratio emulsions and latexes derived thereof, US Pat. 5,688,842, United States Patent Office, 1997. https://www.uspto.gov/patents/search
29. Cameron, N. R., Sherrington, D. C., High internal phase emulsions (HIPEs)—Structure, properties and use in polymer preparation, in: *Biopolymers Liquid Crystalline Polymers Phase Emulsion. Advances in Polymer Science*, Vol. 12, Springer, Berlin, Heidelberg, 1996.

30. Ruckenstein, E., Ebert, G., Platz, G., Phase behavior and stability of concentrated emulsions, *J. Colloid Interface Sci.*, Vol. 133(2), pp. 432–441, 1989.
31. Håkansson, A., Rotor-stator mixers: From batch to continuous mode of operation—A review, *Processes*, Vol. 6, p. 32, 2018. https://doi.org/10.3390/pr6040032
32. Utomo, A., Baker, M., Pacek, A. W., The effect of stator geometry on the flow pattern and energy dissipation rate in a rotor-stator mixer, *Chem. Eng. Res. Des.*, Vol. 87, pp. 533–542, 2009.
33. Padron, G. A., *Measurement and Comparison of Power Draw in Batch Rotor-Stator Mixers*, Master's thesis, University of Maryland, College Park, MD, 2001.
34. Cheng, Q., Xu, S., Shi, J., Li, W., Zhang, J., Pump capacity and power consumption of two commercial in-line high shear mixers, *Ind. Eng. Chem. Res.*, Vol. 52, pp. 525–537, 2012.
35. Baldyga, J. A., Kowalski, M., Cooke, M., Jasinska, M., Investigation of micro mixing in rotor-stator mixer, *Chem. Process Eng.*, Vol. 28, pp. 667–877, 2007.
36. Zhang, J., Xu, S., Li, W., High shear mixers: A review of typical applicationsand studies on power draw, slot pattern, energy dissipation and transfer propeties, *Chem. Eng. Process.*, Vol. 57–58, pp. 25–41, 2012.
37. Ozcan-Taskin, G., Kubicki, D., Padron, G., Power and flow characteristics of three rotor-stator heads, *Can. J. Chem. Eng.*, Vol. 89, pp. 1005–1017, 2011.
38. Cooke, M., Naughton, J., Kowalski, A. J., A simple measurement method for determining the constants for the prediction of turbulent power in a Silverson MS 150/250 in-line rotor stator mixer, *Presented at the ISMIP VI*, Niagara Falls, Canada, August 17–21, 2008.
39. Sparks, T., *Fluid Mixing in Rotor/Stator Mixers*, PhD thesis, Cranfield University, 1995.
40. Shi, J., Xu, S., Qin, H., Liu, Y., Li, W., Zhang, J., Single-pass emulsification process in two different inline high shear mixers, *Ind. Eng. Chem. Res.*, Vol. 52, pp. 14463–14471, 2013.
41. Mortensen, H. H., Calabrese, R. V., Innings, F., Rosendahl, L., Characteristics of batch rotor-stator mixer performance elucidated by shaft torque and angle resolved PIV measurements, *Can. J. Chem. Eng.*, Vol. 89, pp. 1076–1095, 2011.
42. Gallassi, M., Cöncalves, G. F. N., Botti, T. C., Moura. M. J. B., Carneiro, J. N. E., Carvalho, M. S., Numerical and experimental evaluation of droplet breakage of o/w emulsions in rotor-stator mixers, *Chem. Eng. Sci.*, Vol. 204, pp. 270–286, 2019.
43. Gao, Z., Li, D., Buffo, A., Podgórska, W., Marchisio, D. L., Simulation of droplet breakage in turbulent liquid-liquid dispersions with CFD-PBM: Comparison of breakage kernels, *Chem. Eng. Sci.*, Vol. 142, pp. 277–288, 2016.
44. Doucet, L., Ascanio, G., Tanguy, P. A., Hydrodynamics characterization of rotor-stator mixer with viscous fluids, *Chem. Eng. Res. Des.*, Vol. 83(A10), pp. 1186–1195, 2005.
45. https://www.vmimixing.com/healthcare-specialty/mylabmixer-bench-top-mixers/
46. Calabrese, R. V., Francis, M. K., Kevala, K. R., Mishra, V. P., Padron, G. A., Phongikaroon, S., Fluid dynamics and emulsification in high shear mixers, *3rd World Congress on Emulsion, Theme II, Conf. 03*, Lyon, France, 2002.
47. Barailler, F., Heniche, M., Tanguy, P. A., CFD analysis of a rotor-stator mixer with viscous fluids, *Chem. Eng. Sci.*, Vol. 61, pp. 2888–2894, 2006.
48. https://www.findtheneedle.co.uk/companies/silverson-machines-ltd/products/flashmix-powder-liquid-mixers
49. Colloid Mill, Korstvedt, H. O., US Pat. 6745961B2, 2004.
50. https://en.wikipedia.org/wiki/Colloid_mill
51. Mechanical Line Blenders, *Brochure B-528C-9M-02-94*. SPX Flow Systems.
52. Vananroye, A., Puyvelde, P. V., Moldenaers, P., Effect of confinement on droplet breakup in sheared emulsions, *Langmuir*, Vol. 22(9), pp. 3972–3974, 2006.
53. Hall, S., Pacek, A. W., Kowalski, A. J., Cooke, M., Rothman, D., Scaling up of Silverson rotor-stator mixers, *North American Mixing Forum, Mixing XXII*, Victoria, BC Canada, June 20–25, 2010.
54. Thapar, N., *Scale-up of Rotor-Stator Equipment*, Tech Trends, The Pharmaceutical Solutions Update. http://vertassets.blob.core.windows.net/download/f4e83bc9/f4e83bc9-1644-418f-a62d-ab728d4ec27a/springpharm-silverson-techtrends.pdf
55. Häkansson, A., Rotor-stator mixers: From batch to continuous mode of operation—a review, *Processes*, Vol. 6(32), 2018.
56. Vashishth, V., Nigam, K. D. P., Kumar, V., Design and development of high shear mixers: Fundamentals, applications and recent progress, *Chem. Eng. Sci.*, Vol. 232(15), 116296, March 2021.

14 High-Viscosity Media Mixing

14.1 INTRODUCTION

This chapter is devoted to the special blending and solid dispersion systems used to produce working media with viscosities of 50,000cP to 8 million cP. Typically, such high-viscosity products are silicone, epoxy and polyurethane-based industrial products, automotive adhesives, sealants, fillers, stiffeners and damping materials. The solid pigment and filler material dispersions are often carbon fibers, carbon black, fumed silica such as CAB-O-SIL® and calcium carbonate. Certain foods (doughs, peanut butter, jam and jelly) and pharma ingredients (dental fillers and whiteners, transdermic drugs) are also considered high-viscosity media. Depending on the process fluid viscosity, multi-shaft, planetary and kneader mixers are used to achieve desired product consistency and particle dispersion uniformity. There are numerous types of laboratory- and production-scale equipment to address homogenization of high-viscosity materials that pose extraordinary mixing challenges.[1–4]

14.2 WORKING PRINCIPLE

Working media is subjected to both mechanical and hydraulic shear forces for distributive mixing followed by intimate contact for homogeneous blending and dispersion. As the viscosity increases, mixing operation becomes complex, and energy input significantly increases. For ultrahigh-viscosity ($\gtrsim 1x10^6$cP) suspensions, typically vessel configuration is horizontal. Multiple plows in a row are mounted onto a central shaft having single or double helical ribbons. These plows are used to lift and throw chunky fluid elements to the core region of the vessel. Thus, the elements are exposed to mechanical and hydraulic shear provided by the rotating ribbons. The designs of the ribbon blades are as such that the working media in the core region is forced to a synergistic axial motion. Usually, solid suspensions are shear thinning, which facilitates in particle migration in the continuous phase and consequently creates uniform dispersion. A high-viscosity positive displacement pump is used in a bottom central pit to discharge the product into drums.

As mentioned in Chapter 7, the addition sequence of the components is critical to the homogenization process. Particles are slowly added to the liquid phase to uniformly incorporate those and maintain non-Newtonian pseudoplastic rheology. Because ultrahigh-viscosity media are sticky and adhere to the wall, the ploughs and helical ribbons are normally equipped with sacrificial Teflon® or rubber scrapers to maintain cleanliness of the vessel wall. As expected, the wall friction due to scrapers significantly add to the energy consumption. However, this is essential to maintain the required process-side heat transfer by minimizing film formations that lead to wall fouling. According to Oldshue, in presence of the scrapers, heat transfer coefficient, h, essentially doubles, as well as the power consumption, P. So, there is a tradeoff between the increased heat transfer and the additional heat input from increased power.[5]

The rate of energy generation per unit volume, ε_{visc}, as a result of viscous dissipation is proportional to the square of the shear rate, $\dot{\gamma}$, represented by

$$\varepsilon_{visc} = \mu \dot{\gamma}^2 \text{ [W/m}^3\text{]} \tag{14.1}$$

where, μ = dynamic viscosity [Pa-s]
$\dot{\gamma}$ = shear rate (velocity gradient) [1/s]

DOI: 10.1201/9781003606390-14

The rate of media temperature increase per unit mass, ΔT, due to viscous dissipation of heat is given by the following formula:

$$\Delta T = \varepsilon_{visc} / \rho c_p \text{ [C/s]} \tag{14.2}$$

where, ρ = density of the working media [kg/m^3]
c_p = specific heat [J/kg-C]

In order to minimize temperature rise due to viscous heating, most high-viscosity mixers run at lower speeds and produce low hydraulic shear. There are instances during the solid dispersion process, however, when high-shear dispersers (HSDs) are used to achieve particle deagglomeration in a relatively flowable media.

14.3 MIXER SELECTION GUIDE

In R&D and manufacturing, the authors have dealt with multiple commercial agitation systems to achieve particle and fiber dispersions into high-viscosity fluids, followed by blending of such solid–liquid media at scales ranging from 1 to 2,000 gallons. In this section, some experiences with different agitation configurations are discussed briefly without bias towards any specific ones. Table 14.1 may be used by practicing engineers to identify an ideal agitation system, depending on the expected media viscosity during processing.

TABLE 14.1
High-Viscosity Mixer Selection Guide

Working Media	Viscosity Range (cP)	Agitation Systems
Single-phase systems	20×10^3—50×10^3	Closely spaced open impellers
	20×10^3—100×10^3	Anchor, Gate impeller, Paravisc®
	20×10^3–1×10^6	Helical ribbon w/screw
	20×10^3—5×10^6	Double helical ribbons
Multiphase systems	20×10^3—0.5×10^6	HSD + Anchor
(solid–liquid dispersions)	20×10^3—0.75×10^6	HSD + Helical ribbons
	20×10^3—2×10^6	HSD + Single orbiting planetary
	50×10^3—3×10^6	HSD + Double orbiting planetary
	50×10^3—10×10^6	Kneader (Sigma blade, Extruder)
	$>5 \times 10^6$	Plows w/Helical ribbon

14.4 SINGLE-PHASE HIGH-VISCOSITY MEDIA

Open impellers presented in Chapter 4 have limited use with high-viscosity media (>50Pa·s) because of their poor mixing effectiveness throughout the tank. In slender cylindrical shot tanks $\left(\frac{Z}{T} \geq 3\right)$, as shown in Figure 14.1, multiple closely spaced hydrofoils on a shaft may efficiently blend single-phase high-viscosity working media of 20–50Pa-s. The tank is unbaffled and the impeller diameter to the tank diameter ratio, $\frac{D}{T}$, is atypical, i.e., $\gtrsim 0.6$. In high-viscosity fluids, an ideal flow pattern is similar to a draft-tube-impeller agitation system, with downward flow in the impeller domain that turns back upwards in the impeller free annular region near the wall, as shown in Figure 14.1. In the absence of hydrofoils such as the A315, A320, Maxflo-W® and Viscoprop® impellers, alternatively,

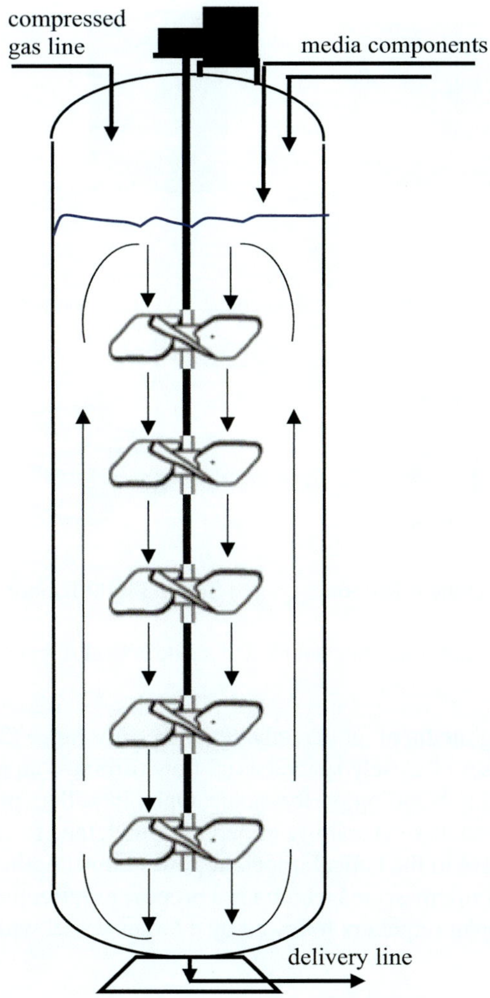

FIGURE 14.1 SPX Flow-Lightnin A320 impellers mixing a viscous fluid in a slender chemical feed tank.

sets of propellors or pitch-blade turbines (PBTs) may be used to achieve blending in viscosities as high as 50Pa-s. In high-viscosity media, a mixed flow impeller such as PBT behaves as an axial flow impeller.

Proximity impellers such as Anchor and Gate[6] agitators, as shown in Figure 14.2a–b, are used in media having a viscosity of up to approximately 100Pa-s. There are multiple modified versions of these impellers. One such predominantly used is Ekato's Paravisc,[7] presented in Figure 14.2c. Because of its unique blade geometry, Paravisc creates a combination of axial and radial flow. The superior blending performance of the Paravisc relative to the anchor agitator is clearly observed in a laboratory study conducted with similar viscosity media (~50Pa-s). In order to minimize power consumption and heat dissipation, typically in high-viscosity fluids, these proximity impellers rotate at a reasonably low speed. The hydraulic shear experienced by the media in the annular region between the impeller blade and the vessel wall causes laminar mixing.

When appropriately designed and installed, helical ribbons will blend almost freely non-flowable ultrahigh-viscosity fluids up to 5 million centi-Poise. As shown in Figure 14.3, dual ribbons built onto the shaft, with their optimized blade width, pitch and slope, create synergistic down-pumping and up-pumping (or vice versa) flow that promote laminar mixing.

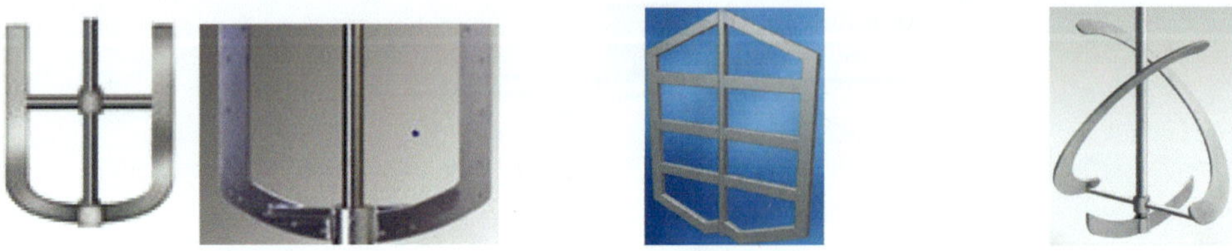

(a) (b) (c)

FIGURE 14.2 Proximity impellers including: (a) Anchors, (b) Gate Impeller[6] and (c) Paravisc.[7]

(a) Double Helical Ribbon (b) Single Flight Ribbon w/screw

FIGURE 14.3 Proximity impellers including double and single helical ribbons.[8]

In a single-phase working media of moderately high-viscosity range ($20x10^3$–$50x10^3$cP), the agitation system options are a set of closely spaced axial flow turbines, an anchor, or a helical ribbon in the order of their decreasing blending performance. Open impellers provide intense micro-scale mixing with a higher shear rate than proximity impellers, which tend to create macro-scale mixing. At constant blend time relative to the helical impeller, power consumption of the turbines would be 4x, but cost would be x/4.[5] If micro-scale blending is a process requirement, a smaller vessel such as a shot tank equipped with open impellers followed by a larger vessel with a helical impeller would be a very practical choice.

14.5 MULTIPHASE HIGH-VISCOSITY MEDIA

Multiphase viscous media are fine particles, fibers and chips dispersed in a viscous fluid. Rheology of the product media depends on the solid content. Table 14.2 illustrates a few freely non-flowable high- to ultrahigh-viscosity ($5x10^4$ to $8x10^6$cP) products.

TABLE 14.2
A Few Multiphase High-Viscosity Products and Their Components

Products	Solid Dispersers and Fillers	Liquid—Continuous Phase
CeraVe™ moisturizing cream	Ceremide particles	Glycerin-water etc.
Bread dough	Flour	Water
BETASEAL™ glass adhesive[9]	Carbon black, Calcium Carbonate	Urethane prepolymer
Silicone sealant[10]	Fumed silica (Cabosil®)	Silicone-rubber resin
Methocel® intermediates	Wood chips/cellulose fibers	Sulfuric acid
Cement	Cement powders/sand	Water
Asphalt	Gravel chips	Tar

14.6 MULTIPHASE MIXING

Because quality of the multiphase high-viscosity product depends solely on the uniformity of particle dispersion in the liquid media, special attention is given to its formulation. As the solids are added to the liquid, suspension viscosity increases; rheology becomes pseudoplastic or Bigham plastic depending on particle concentration. For ease of uniform dispersion, it's intuitive that solids should be added when and where the media viscosity is low, i.e., in the beginning of the mixing process or in a shear thinning zone. High-shear dispersers (HSDs) similar to the ones shown in Figure 14.4 are used to create local shear zones where solid particles could be added. The rate of addition must be slower than the local fluid circulation rate to avoid particle accumulation, and consequently non-uniform dispersion in the working media. Since in most cases, the desired working media properties are affected by excessive viscous heating, local heat in the HSD domain is

(a) (b) (c)

FIGURE 14.4 High-shear dispersers: (a) Morehouse Cowles blade,[11] (b) SPX Flow-Lightnin R500 and (c) Ekato Dissolver.[12]

commonly dissipated by bulk circulation from proximity impellers such as anchors, gates, helical ribbons, Paravisc® etc. As mentioned earlier, in addition to mixing, these proximity impellers also generally transfer materials away from the wall and thus help maintain the required process-side heat transfer.

A planetary mixer is as common as a household kitchen appliance that mixes bread doughs. Figure 14.5 shows the Cusimax® dough mixer and its attachments.[13] In this configuration, the attachment rotates about its centerline axis, which is offset from the simultaneously rotating centerline axis of the mixer to create the indicated orbitals. The various attachments are designed to mix products of different consistencies.

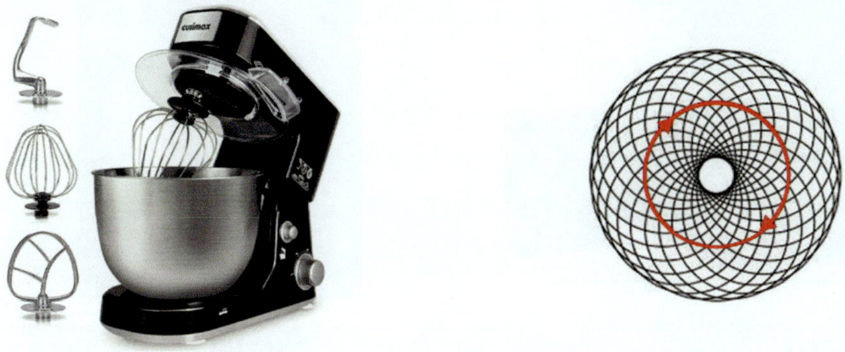

(a) Mixer and attachments (b) Orbit of attachment center (red) and blade edge (black)

FIGURE 14.5 Cusimax® dough mixer with dough hook, whisk and mixing beater attachments.[13]

Planetary mixing with the dispersion principle is widely used in the industrial scale with or without the HSDs. Equipment designs differ as to the shape of blades, number of shafts and vessel capacity. Figure 14.6 and Figure 14.7 show multiple designs[14–16] offered by high-viscosity mixer manufacturers. As mentioned earlier, for uniform particle distribution, solids are typically added to the high-shear domain created by the HSD. Planetary blades transport the slurry from the high-shear domain to the rest of the vessel through bulk media mixing. During this transport process, the media *is* also subjected to relatively low-shear zones located between the rotating blades and adjacent to the wall. Working media undergo very complex laminar mixing that further disperses the particles to produce the desired viscous homogeneous suspension. These devices are used to make polyurethane and silicone-based sealants for automotive gaskets and glass bonding.

FIGURE 14.6 Ross double planetary mixer with kneader attachment.[14]

(a) (b)

FIGURE 14.7 (a) Ross single-blade planetary mixer with dual-blade high-shear disperser[15] and (b) Oliver+Batlle Hydrobat multishift planetary mixer with a central high-shear disperser.[16]

Industrial kneaders are commonly used to mix ultrahigh-viscosity media such as putties, pastes and doughs. A list of familiar products made using the well-known Sigma blade mixers is outlined in reference.[17] Sigma Kneader consists of two sets of Z-shaped blades that rotate in opposite directions in a W-shaped or double U-shaped trough (see Figure 14.8). As these blades rotate, working media is subjected to cross mixing in the high-shear close clearance zones between the counter rotating blades. Also, the media experiences additional shear mixing in the region adjacent to the wall of the trough enclosure. Practicing engineers may view the animation of a Ross kneader-extruder, which clearly demonstrates the working principle of the sigma blade mixer.[18] Sigma mixer enclosures are horizontally oriented. Product mass is discharged through the cover plate simply by tilting or inverting the tank (see Figure 14.9). Because dislodging sticky media is problematic, the Ross sigma kneader extruder uses a screw at the bottom of the enclosure for product transfer. Also, the same screw when rotated in the reverse direction helps in the solid addition process.[18] In one instance, authors used a pressurized cover plate to constrain free upward flow of the working media in a sigma blade mixer. Such a technique helped in folding and compressing of the dough and significantly improved its homogeneity with minimum energy penalty.

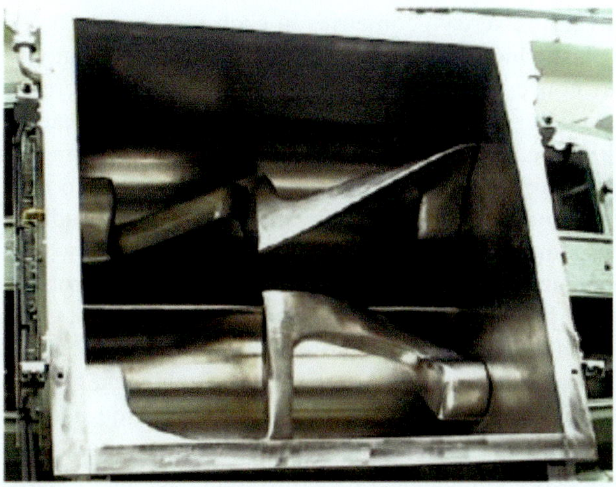

FIGURE 14.8 Photograph of Jaygo double arm Z-blade sigma mixer manufactured by Jaygo Inc.[19]

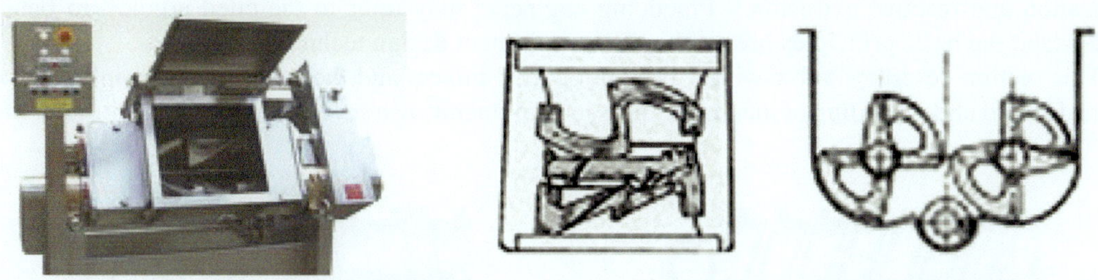

FIGURE 14.9 Photograph of a Ross sigma kneader-extruder and CAD drawings showing its blades.[18]

Twin-screw extruders (TSEs) are intermeshing mixing screws in a barrel; they are widely used in the plastic industries to continuously mix the solid additives and pigments into the molten plastic to produce a homogeneous pasty media. Figure 14.10 shows a photograph of B&P Littleford's twin-screw compounder (TSC) elements.

FIGURE 14.10 Photograph of B&P Littleford's twin-screw compounder elements.[20]

Typically, the extrusion process consists of three zones, i.e., melting, compounding and transfer sections, with varying screw geometries in each section to efficiently accomplish these required functions. In plastic industries, small plastic pellets are fed into the hopper of the screw extruder, where they are conveyed through the barrel by the rotating screws into a die. Because it is a specialized subject matter in manufacturing, full insight on extrusion technology is beyond the scope of this book. Here, the authors present the twin-screw compounders that are extensively used as continuous mixers to produce a wide variety of high- to ultrahigh-viscosity products in the chemical, food and healthcare industries, such as sealants, adhesives, putties, noodles, pastas and creams. Compounding extruders are used to mix together two or more materials into a homogeneous mass in a continuous process. Working media components are continuously added to the barrel through the hopper(s) and secondary port(s). Initially, the constituents undergo distributive mixing in the solids conveying section of the barrel; subsequently, dispersive mixing takes place as the media passes through the intermeshing high-shear zones of the rotating flights. Significant energy is consumed during particle de-agglomeration and the uniform dispersion process.

There are several twin-screw compounder manufacturers with their proprietary designs of flights on the screws. The TSEs with co-rotating intermeshing models, as shown in Figure 14.11a, dominate the compounding market, having captured over 90% of current installations. Counter-rotating designs (see Figure 14.11b) are primarily used for specialty applications, such as high-level devolatilization and reactive extrusion.[21] Practicing engineers may refer to the cited article[21] to better understand the basic principles involved in TSE equipment design technology.

This section describes a few widely used horizontal mixers and their working principles. For a more detailed understanding of their usefulness in manufacturing a specific product, practitioners are

(a) (b)

FIGURE 14.11 Commercially available screw configurations for TSEs: (a) intermeshing co-rotating and (b) intermeshing counter-rotating.[21]

advised to review the equipment brochures and/or contact the manufacturers directly. Photographs in Figure 14.12 show the Hockmeyer paste mixer with its "whale tail" plow blades and specially designed end blades that are contoured to precisely follow the wall and avoid dead spots. This equipment is used for uniform wetting and fast mixing of a large volume of highly viscous freely non-flowable media such as sealants, caulking materials and joint compounds. Its basic working principles are (a) tumbling and distributive mixing of the fluid chunks due to the plows, (b) wetting and dispersive shear mixing along the walls due to close clearance of the blades and (c) double rolling action of the blades that kneads for additional wetting and homogeneity.[22] An animation[23] of the mixing inside a Hockmeyer Model 500 series mixer may be viewed.

FIGURE 14.12 Photographs of a Hockmeyer mixer and its internals (two "whale tail" blades in the middle of a tri-blade blender-scraper on each end).[22]

14.7 POWDER MIXERS

The Ross ribbon mixer mounts specially designed double helical blades on a shaft positioned within a U-shaped trough, as shown in Figure 14.13.[24] The outer ribbon blades move the solid particles toward the center of the trough and middle of the mixer, whereas the inner blades force them outwards towards the mixer's end and side walls. This creates significant lateral and axial mixing and

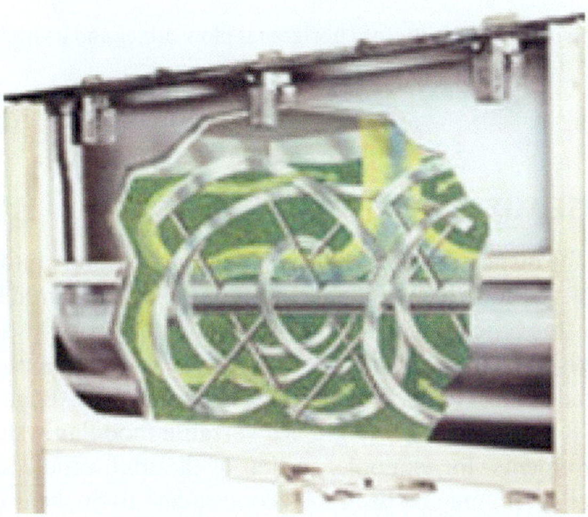

FIGURE 14.13 The Ross horizontal ribbon mixer with its double helical blades.[24]

a continuous flow of the working media. The mixer is commonly used for solids blending and spray coating. The cited animation[25] shows how the Ross horizontal ribbon blender works.

B&P Littleford's horizontal plow mixer shown in Figure 14.14[26] is used where high-volume production is a pre-requisite in the processing of dry products, dry products with liquid additions, and product granulations. The plows create a three-dimensional mixing environment as the solids pass through the mixer to a discharge weir at the outlet end. A control valve at the mixer outlet maintains the required residence time. Sticky and cohesive materials are typically discharged through a specially designed central port.[26] A series of spray nozzles on a manifold may be used to coat or moisten the solid particles and fibers as they mix.

FIGURE 14.14 Photographs of a B&P Littleford horizontal plow mixer and its multiple staggered 3D blades.[26]

14.8 MULTIPURPOSE MIXERS

Coaxial mixers are used for dispersion and intensive blending of solids into low to extremely high-viscosity media. Figure 14.15 shows a coaxial mixing system manufactured by Ekato Corporation,[27] which is fairly typical for this type of mixer. It consists of dual Viscoprops® (open impellers) on the central shaft and a Paravisc® (proximity impeller) on the peripheral shaft (sleeve). There are two interdependent drives[28] and control units, which can be driven with different shaft speeds and directions of rotation. Ekato's Koaxial mixing system is a versatile modular system, which easily adapts to varying process requirements, for example, in lieu of the high-efficiency Viscoprop, an Ekato dissolver, shown in Figure 14.4c, may be used to accommodate high-shear dispersion of solids into a viscous liquid. The Koax system has the option of additional features such as flexible or rigid scrapers and a bottom anchor or Z-anchor.[27]

FIGURE 14.15 Paravisc Koaxial agitator[27] and Ekato Coaxial agitation system.[28]

Figure 14.16 schematically presents the idealized flow patterns of a Paravisc and that of the Koax agitation system comprising a Paravisc and dual Viscoprops.[29] The addition of Viscoprops to the Paravisc creates collaborative flow and significantly improves the mixing performance.

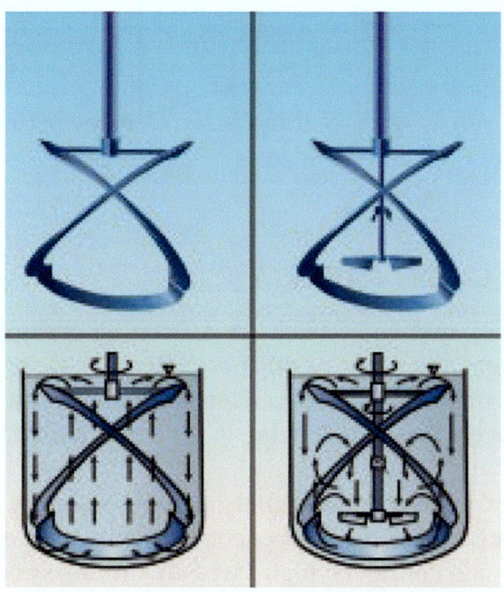

FIGURE 14.16 Flow patterns of a Paravisc vs. a Koax agitation system.[29]

Coaxial mixers are efficient in high-viscosity mixing because of the synergistic fluid dynamic effects of two independent rotating impellers. Proximity impellers rotating at a slower speed serve as baffles for the flow generated by the higher-speed open impeller. The review article cited here[29] is an excellent source for practicing engineers.

Figure 14.17 shows details of the Ekato UNIMIX SRT® mixing system used by the authors for pilot-scale manufacture of multiple single-phase and multiphase high-viscosity specialty products ranging from healthcare lotions to industrial sealants. Some distinguishing features of this unit are the top-mounted Paravisc depicted in the tank and a bottom-mounted Ekato Unimix S-Jet rotor-stator homogenizer (shown separately on the right) in the bottom discharge trough for internal and external homogenization. Besides clean-in-place (CIP) capability, further features include vacuum and heating/cooling systems, as well as an optional-use recirculation loop. All of the features integrated into this unit enable it to accommodate the manufacture of health and homecare products: solid–liquid dispersions and liquid–liquid emulsions followed by reaction. An online animation[32] clearly demonstrates the working principles of the Unimix agitation system. Although the process is in progress, particle deagglomeration and/or dispersion homogeneity of a slurry sample can be determined using a Hegman gauge[33] and/or an online particle size analyzer.

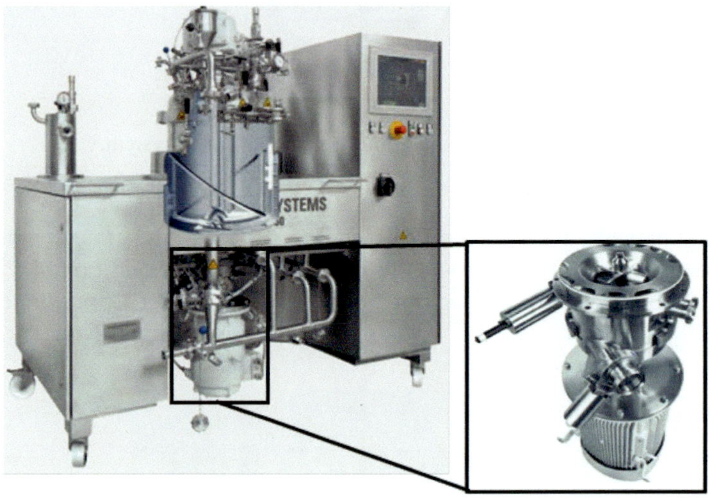

FIGURE 14.17 Ekato UNIMIX SRT pilot system[30] and its S-Jet homonizer[31] assembly.

14.9 EQUIPMENT SELECTION CRITERIA

Various types of high-viscosity mixing systems are offered by multiple manufacturers, and their detailed descriptions are beyond the scope of this book. Because these heavy-duty drives and agitation systems are typically quite expensive, practicing engineers must be prudent in selecting the equipment based on their process needs. Following is a list of criteria to consider:

1. If the working media experiences significantly varying viscosity changes during the process, as happens, for example, in homogeneous systems undergoing bulk and solution polymerization, one should consider a high-viscosity mixing system that is acquiescent to the highest achievable viscosity.
2. In order to accommodate viscosity changes during a process, or adapt to multiple products with differing viscosities, consider equipping systems with variable-speed drives.
3. Equipment must have the option to alter the agitators to optimize its mixing efficacy.
4. If sanitization is of paramount interest, as in the production of pharma, home and healthcare products, the equipment must include easy cleaning-in-place (CIP) features such as sprays, heating systems, vacuum systems, and surface finishes such as glass-lining or a dairy-finish.

5. Because high-viscosity media are typically sticky, proximity impellers should have flexible, easily replaceable wall scrapers, and the vessel itself must have appropriately located spray devices for fast cleaning and minimal batch time.
6. To prevent material bypass, especially in continuous mixing processes, locate the impeller between the media addition and discharge ports. This forces added material to pass through the impeller zone before it can reach the discharge port.
7. Because most working media have a preferred processing temperature range to control the consistency and performance of the product, excessive temperatures that may degrade the product and/or its desired physical properties must be avoided. This is typical of sealants and caulking materials.
8. The materials from which wetted internals are constructed must be compatible with the process fluids and relatively resistant to solids abrasion in multiphase applications.
9. Mechanical and electrical safety aspects of the equipment must meet OSHA regulations.

Practicing engineers should diligently explore each of the following options prior to deciding on a particular type of high-viscosity mixing device:

1. Consider using equipment based on in-house experience with a pilot- or production-scale unit that efficiently (batch time) produces material with a similar viscosity (rheology).
2. Discuss the project scope and seek recommendations from technical sales representatives of multiple high-viscosity mixing equipment manufacturers. Technical competencies of the manufacturers' R&D should be leveraged.
3. Pursue in-house pilot-scale R&D and/or follow-up at the manufacturers' sites with actual materials under confidentiality agreement or with appropriately designed cold-flow simulations.
4. Cost effectiveness based on capital costs (equipment, peripherals and installation) and running costs (utilities and maintenance) is critical to the equipment choice. However, manufacturers' long-term commitment to provide technical assistance must not be overlooked.
5. A risk-averse approach to make the product "first time right" is a key goal in most capital projects. Although cost cutting is often a goal, because in most instances plant operating costs outweigh the one-time equipment purchase and installation expenses, manufacturing must have the precise mixer to achieve desired product quality according to the planned schedule.
6. Practicing engineers must remember to carefully review the effectiveness of the entire mixing operation, including times and techniques of material loading, mixing, cleaning and product discharge.

14.10 MIXER POWER CHARACTERISTICS

When mixing a highly viscous Newtonian fluid, as expected, the flow regime within the vessel is laminar. For a single impeller, the power number, N_P, is inversely proportional to the Reynolds number, Re, and the constant of proportionality is K_P.

$$N_P \cdot Re = K_P \text{ for } Re = \frac{(\rho N D^2)}{\mu} \leq 10 \tag{14.3}$$

Table 14.3 briefly presents experimentally determined K_P values for a few commonly used proximity impellers, such as the anchor, double helical ribbon and Paravisc®.[34]

Variations in measured K_P values for similar impellers are attributed to differences in (1) blade width, W, and pitch, P_r, and (2) impeller location in the tank, i.e., off-bottom clearance, C, and proximity to the vessel wall, D/T ratio. Equation 14.4 may be used to calculate the expected power draw of the impeller.

TABLE 14.3
K_P Values of a Few Viscous Media Mixing Impellers[34]

Impellers	K_P	References	Details of the Impeller
Anchor	260	Oldshue[5]	
	253	Thaibault and Tanguy[35]	
	185	Kuriyama et al.[36]	$C/T = 0.05$
	203	Takahasi et al.[37]	$C/T = 0.025, W/T = 0.102$
	194	Novak and Rieger[38]	$C/D = 0.05$, $D/T = 0.9$
Double helical ribbon	277	Rieger et al.[39]	$P_r/D = 1.5$, $C/D = 0.03$, $W/T = 0.1$
	351	Takahashi et al.[37]	$C/T = 0.031$, $P_r/T = 0.98$
	330	Nagata et al.[40]	$P_r/D = 1.0$, $D/T = 0.95$
	296	Novak and Rieger[38]	$P_r/D = 1.0$, $D/T = 0.95$, $C/D = 0.025$
Paravisc®	315	Delaplace et al.[41]	
	329	Barar Pour et al.[42]	
	290	Stein[43]	

$$P = N_P \rho N^3 D^5 \tag{14.4}$$

Most high-viscosity fluids tend to be non-Newtonian pseudo-plastic in nature. In 1957, Mezner and Otto proposed that in the laminar flow regime, the effective shear rate, $\dot{\gamma}$, experienced by the media is proportional to the rotational speed, N, of the impeller,[44] as shown in equation 14.5.

$$\dot{\gamma} = K_s.N \ \text{ for } Re \le 10 \tag{14.5}$$

The proportionality constant, K_s, is known as the mixer shear rate constant or the well-known Metzner-Otto constant. For a few single impellers, K_s values are presented in Table 14.4. As mentioned earlier, variations in the K_s values for the anchor and double helical ribbon are due to differences in impeller geometry and location within the vessel.

TABLE 14.4
Published K_s Values of a Few Viscous Media Impellers[34]

Impellers	K_s	References
Anchor	25	Nagata et. al.[40]
	22	Foucault[45]
Double helical ribbon	28.2	Takahashi[46]
	30.6	Chowdhury and Tiwari[47]
	30	Nagata et al.[40]
Paravisc	33.7	Delaplace et al.[41]

The Metzner-Otto concept is widely used to design impellers for non-Newtonian shear-thinning working media. Initially, a theoretical engineering design of an impeller is targeted for the media, with its diameter, D, and rotational speed, N. Following equation 14.5, the effective shear rate, $\dot{\gamma}$, is determined based on the speed, N, and Metzner-Otto constant, K_s. Shear rate is used to calculate the apparent viscosity of the fluid using the measured function of viscosity vs. shear rate generated for the specific media. Then the viscosity is used to determine the Reynolds number, *Re*. Subsequently, the power number, N_p, is obtained from the power number vs. Reynolds number curve. Finally, the shaft power is determined from equation 14.4.

Experimental and numerical studies on power consumption and mixing performance of coaxial impellers using moderately viscous Newtonian and non-Newtonian single-phase fluids in laminar regime are reported in a few literature references.[48–51] Coaxial agitation systems are configured with one or two turbines (PBT, Rushton, Scaba®) as central open impeller(s) and an anchor as the proximity impeller. Based on the power consumption and homogenization time as the criteria for mixing efficiency, performance of the coaxial system in the co-rotating mode is found to be superior to its operation in counter-rotating mode. It is concluded that the power consumption of the proximity impeller depends on the tip velocity ratio of the coaxial impellers. However, power draw of the open impeller is independent of the proximity impeller speed.[50]

Flow phenomena due to multi-shaft and coaxial mixing systems in ultrahigh-viscosity media are chaotic and quite complex, perhaps even initially non-continuum. The flow pattern and mixing intensity depend on impeller type (blade geometries) dimensions, rotational speeds and locations within the mixing vessel. Variations in speed and direction of rotation of the individual impellers significantly affect their dispersion and blending performance. Because determining the blend time needed to achieve a desired homogeneity is so critical, as mentioned earlier, pilot-scale testing with actual material or an appropriately designed simulant is indispensable. Multiple articles on ultrahigh-viscosity media mixing, system design, operating parameters and process optimization have been published by the equipment manufacturers. Readers are advised to review articles[52–54] by Charles Ross & Son Company and Hockmeyer Equipment Corporation for further understanding on the subject.

REFERENCES

1. Meyers Mixers Company. https://myersmixers.com/type/multi-shaft-dual-tri-and-quad-shaft-designs/
2. Hockmeyer Equipment Company. https://hockmeyer.com/products/categories/high-viscosity-mixers/
3. Charles Ross & Son Company. https://www.mixers.com/results.asp
4. B&P Littleford. https://www.bplittleford.com/
5. Oldshue, J. Y., *Fluid Mixing Technology*, McGraw-Hill Publishing Co., New York, 1983.
6. https://www.directindustry.com/prod/jongia-nv/product-104055-1694320.html
7. https://www.youtube.com/watch?v=YNwiSWI9HPo
8. https://proquipinc.com/industrial-mixing-basics-high-viscosity-mixing-impellers/
9. BetaSeal®. https://patents.justia.com/patent/9102854
10. Silicone Sealant. https://patents.google.com/patent/US8075862B2/en
11. https://fusionfluid.com/products/mixing-impellers/cowles-disperser-multi-blade-stx
12. https://www.ekato.com/en-us/products/agitator-components/impellers/
13. https://www.amazon.com/Cusimax-Stand-Mixer-5-Quart-Tilt-head/dp/B07CXDV9MH
14. Planetary Mixer with Dual HSD. http://www.planetarymixers.com/
15. http://www.planetarymixers.com/
16. https://oliverbatlle.com/en/prod/mixing-equipment-planetary-hidrobat/
17. https://www.karvil.com/application.html
18. https://www.mixers.com/products/kneader-extruder/
19. https://jaygoinc.com/products/double-arm-mixer-sigma-blade-mixers/
20. https://www.bplittleford.com/twin-screw.html
21. Martin, C., Twin Screw Extruders as Continuous Mixers for Thermal Processing: A Technical and Historical Perspective, *AAPS Pharma Sci. Tech.*, Vol. 17, p. 3, 2016.
22. https://hockmeyer.com/products/horizontal-mixer/
23. https://myersmixers.com/product/agitators-model-500-series/600-500-horizontal-mixer/
24. https://www.mixers.com/proddetails.asp?ProdId=71

25. https://www.youtube.com/watch?v=tpt5VerLSsg
26. https://www.bplittleford.com/continuous-plow-mixer.html
27. https://www.ekato.com/products/agitators/coaxial-agitators-ekato-koax/
28. https://www.induchemgroup.com/products/coaxial-agitators-ekato-koax-/
29. Himmelsbach, W., Houlton, D., Keller, W., Lovallo, M., Mixing Systems: Design and Scale Up, Ekato Mixing Technology, *Chem. Eng.*, p. 46, April 2006.
30. https://www.ekato.com/wp-content/uploads/2020/12/EKATO_UNIMIX_SRT_E.pdf
31. https://www.directindustry.com/prod/ekato-group/product-36897-541326.html
32. https://www.youtube.com/watch?v=xzWau0AyCaA
33. ASTM International, Standard Test Method for Fineness of Dispersion of Pigment-Vehicle Systems by Hegman Type Gage, *ASTM D1210-05*, 2014. https://www.astm.org/d1210-05r14.html
34. Rudolph, L., *Characterization of Co-axial Mixing Systems to Blend Highly Viscous and Non-Newtonian Fluids*, Dr.-Ing Dissertation, The Technical University of Berlin, 2009.
35. Thibault, F., Tanguy, P. A., Power draw analysis of a coaxial mixer with Newtonian and non-Newtonian fluids in the laminar regime, *Chem. Eng. Sci.*, 57, p. 3861, 2002.
36. Kuriyama, M., Inomata, H., Arai, K., Saito, S., Numerical solution for the flow of highly viscous fluid in agitated vessel with anchor impeller. *AIChE J.*, 28, 3, p. 385, 1982.
37. Takahashi, K., Arai, K., Saito, S., Power correlation for anchor and helical ribbon impellers in highly viscous liquids. *J. Chem. Eng. Jpn*, 13, 2, p. 147, 1980.
38. Novàk, V., Rieger, F., Homogenization efficiency of helical ribbon and anchor agitators. *Chem. Eng. J.*, 9, p. 63, 1975.
39. Rieger, F., Novàk, V., Havelková, D., Homogenization efficiency of helical ribbon agitators. *Chem. Eng. J.*, 33, p. 143, 1985.
40. Nagata, S., Nishikawa, M., Katsube, T., Takahashi, K., Mixing of highly viscous non-Newtonian liquids. *Int. Chem. Eng.*, 12, 1, p. 175, 1972.
41. Delaplace, G., Leuliet, J. C., Relandeau, V., Circulation and mixing times for helical ribbon impellers: Review and experiments. *Exp. Fluid*, 28, p. 170, 2000.
42. Pour, S. B., Fradette, L., Tanguy, P. A., Laminar and slurry blending characteristics of a dual shaft impeller system, *Chem. Eng. Res. Des.*, 85, A9, p. 1305, 2007.
43. Stein, W. A., Rührleistung und Wärmeübergang auf der Innenseite eines Rührbehälters mit verschiedenen Wendel-Rührern. *Forsch. Ingenieurwesen—Eng. Res. Bd*, 59, 9, p. 165, 1993.
44. Metzner, A. B., Otto, R. E., Agitation of non-Newtonian fluids, *AIChE J.*, 3, 1, p. 3, 1957.
45. Foucault, S., Ascanio, G., Tanguy, P. A., Power characteristics in coaxial mixing: Newtonian and non-Newtonian fluids, *Ind. Eng. Chem. Res.*, 44, p. 5036, 2005.
46. Takahashi, K., Yokota, T., Konno, H., Power consumption of helical ribbon agitators in highly viscous pseudoplastic liquids, *J. Chem. Eng. Jpn.*, 17, p. 657, 1984.
47. Chowdhury, R., Tiwari, K. K., Power consumption studies of helical ribbon screw mixers, *Ind. Eng. Chem. Process Des. Dev.*, 18, 2, p. 227, 1979.
48. Foucault, S., Ascanio, G., Tanguy, P. A., Coaxial mixer hydrodynamics with Newtonian and non-Newtonian fluids, *Chem. Eng. Tech.*, 27, 3, p. 324, 2004.
49. Rivera, C., Foucault, S., Heniche, M., Espinosa-Soleres, T., Tanguy, P. A., Mixing analysis in a coaxial mixer, *Chem. Eng. Sci.*, 61, 9, p. 2895, May 2006.
50. Rudolph, L., Schaefer, M., Atiemo-Obeng, V., Kraume, M., Experimental and numerical analysis of power consumption for mixing high viscosity fluids with a co-axial mixer, *Trans. IChemE, Chem. Eng. Res. Des.*, 85, A5, p. 568, 2007.
51. Kazemzadeh, A., *Mixing of Complex Fluids with the Coaxial Mixers Composed of Two Central Impellers and an Anchor*, Doctoral Theses, Ryerson University, Toronto, 2016.
52. High Viscosity Mixer Design and Applications, *A White Paper Prepared* by Charles Ross & Son Company. http://www.mixers.com/whitepapers/hvm_designs_whitepaper.pdf
53. Langhorn, K., Digiannurio, T., Lu, G., Appreciate the art of high viscosity mixing: Optimization requires considering crucial issues during specification and scale-up, *Chem. Process.*, January 2016. https://www.chemicalprocessing.com/processing-equipment/fluid-handling/article/11326162/process-engineering-appreciate-the-art-of-high-viscosity-mixing-chemical-processing
54. https://hockmeyer.com/blog/articles/understanding-high-viscosity-mixing/

15 Emerging and Unconventional Mixing Systems

15.1 INTRODUCTION

This chapter provides an overview of a few novel and unconventional mixing technologies that are both practiced and familiar to the authors. These technologies have been embraced by various industries, including food, pharma, specialty chemical, and wastewater, to solve unmet mixing needs. A concise list of technologies that will be discussed, and their key applications, is presented in Table 15.1. Related references and website addresses are included in the chapter for those interested in a detailed understanding of these mixers and applications. The size and mixing characteristics of some of these unique technologies are ideally attractive for chemical process intensification (PI). To illustrate, a couple of recent innovative emerging technologies (Hydro Dynamics, which harnesses cavitation, and ResonantSonics®, which utilize low-frequency resonance) are comprehensively addressed by their respective technology pioneers (Mancosky and Farrar, respectively) as contributing authors to subchapters 15.6-I and 15.6-II.

TABLE 15.1
Emerging and Unconventional Mixing Technologies

Technologies	Key Applications
Oscillatory-Flow Baffled Reactor (OBR)	Blending, gas–liquid and liquid–liquid dispersions
Pulsair® Mixer	Blending, solid dispersions with large-scale bubbles
Sanitaire® Membrane Diffuser	Gas–liquid mass transfer
Higee Devices (RPB, RZB, RPBF, CFBR)	Gas–liquid contacting and mass transfer
Hydrodynamics	Emulsions, gas–liquid contacting, solid dispersions
ResonantSonics® (low frequency)	Solid–liquid (pastes), gas–liquid, and emulsions

15.2 OSCILLATORY-FLOW BAFFLED REACTOR (OBR)

OBR is a novel continuous plug-flow laminar reactor with significantly enhanced radial mixing. It consists of multiple columns or tubes, which have periodically spaced orifice-plate baffles, as shown in Figure 15.1.[1] Oscillatory flow is applied to the working media by means of various pulsating

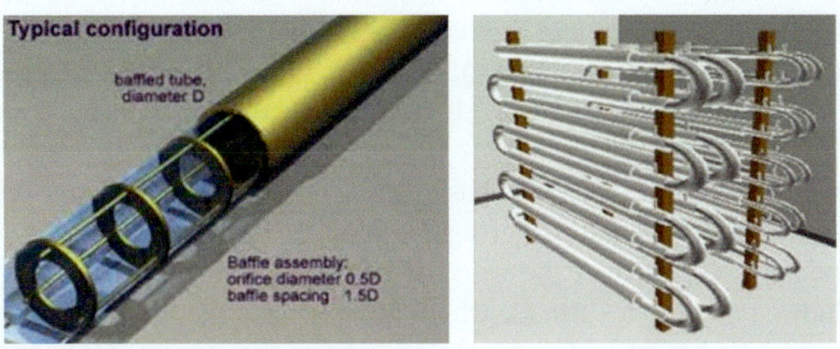

FIGURE 15.1 Oscillatory Baffled Reactor (OBR) utilizing an assembly of orifice plate baffles in each tube and multiple sequential tubes in a typical parallel-series arrangement.[1]

DOI: 10.1201/9781003606390-15

devices such as a piston-, bellow-, diaphragm-, syringe- or peristaltic pump. Flow through the orifices generates eddies, as depicted in Figure 15.2. Enhanced uniform mixing is produced by continuous generation and cessation of eddies within each mixing chamber.

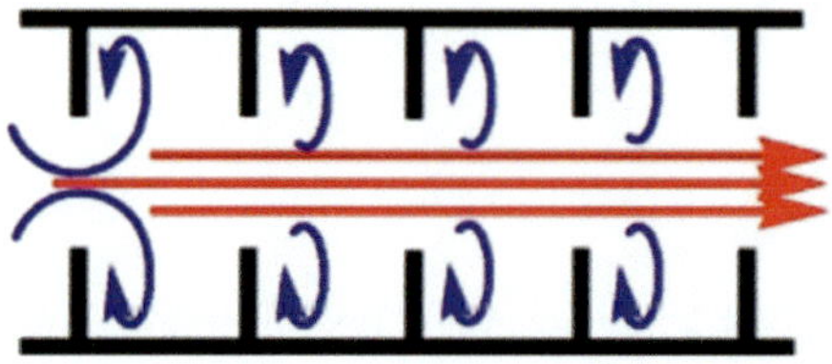

FIGURE 15.2 Eddies generated by flow passing through the multiple OBR orifices.

Continuous oscillatory baffled reactors have been proven to globally intensify chemical processes, leading to sustainable methods that use less energy and produce fewer wastes,[2,3] potentially with higher product yields and greater process control.[2] Long residence times are achievable in an OBR, despite its greatly reduced length-to-diameter ratio[4] and smaller working volume. Thus, the numerous issues facing large-volume tubular flow reactors, such as high pressure drop, large footprint, and poor residence time distribution (RTD), can be obviated.[5] An OBR can increase reaction rates for specific bioprocesses owing to its potential for lower shear, better global mixing, and enhanced mass transfer, when compared with conventional reactors.[6] Such features also make it suitable for continuous bioprocessing.

OBR superimposes an oscillatory flow component over the net flow through a conventional plug flow reactor. Because energy dissipated by flow through the OBR induces radial mixing and dispersion, it is often considered the most critical OBR performance parameter. Power per volume, ε_v, or energy density ($\overline{W} / V$), is determined as the ratio of pressure drop, Δp, through the OBR to mixing chamber volume, V. A recent article[7] discusses non-dimensional analysis of the experimental pressure drop and energy dissipation measurements in oscillatory baffled reactors. A summary of the key dimensionless numbers used to characterize oscillatory flow reactors, and their interpretations, is presented in Table 15.2.[5,7]

Table 15.3 presents two empirical power density models, quasi-steady[8] (QSM) and eddy-enhancement[9] (EEM), available for OBRs. Their multiple improved variations help determine energy dissipation in OBRs, depending on the amplitude, x_o, and frequency, f, ranges of the superimposed pulsating or oscillatory flow. These two models (QSM and EEM) are developed from laboratory experiments using single-orifice baffles in pulsed batch columns, and in oscillatory flow in tubes with no net flow, at varying amplitudes (Strouhal numbers). Additionally, these models assume high oscillatory velocities and a turbulent flow regime.

The quasi-steady flow model (QSM) postulates that the instantaneous pressure drop in an oscillation period is the same as the pressure drop that would be achieved in steady-state flow with the same velocity. It assumes no pressure recovery in the mixing chamber, due to the short distance between orifice baffles. The QSM is based on the standard pressure drop across a simple orifice.[10] Its use is appropriate for high oscillation amplitudes, x_o, and low oscillation frequency, f.

The eddy-enhancement model (EEM) is based on acoustic principles and the concept of eddy viscosity. The model was developed considering the acoustic resistance of a single orifice plate and by replacing the kinematic viscosity of the working media with its eddy viscosity at high Reynolds number. It also includes a mixing length, l_m, which is an adjustable parameter corresponding to the average travel distance of turbulent eddies (expected to be of the same order as the reactor diameter, D). Because l_m depends on reactor geometry and amplitude, x_o, of the oscillatory flow,[10] EEM has limited applications.

TABLE 15.2
Dimensionless Numbers for Characterizing Oscillatory Flow Reactors[5,7]

Dimensionless Groups	Formula	Interpretation
Orifice discharge coefficient	$C_o = \sqrt{\frac{\rho v^2}{2\Delta p}}\left(1/s^2 - 1\right)$	Flow obstructions due to sudden change in cross-sectional area
Net flow Fanning friction factor	$f_n = \frac{\Delta p}{2\rho v^2}\frac{D}{L}$	Ratio of net flow shear stress and local kinetic energy density
Oscillatory-flow Fanning friction factor	$f_o = \frac{\Delta p_{max}}{2\rho\left(2\pi f x_o\right)^2}\frac{D}{L}$	Ratio of oscillatory flow shear stress and local kinetic energy density
Power number	$P_o = \frac{\overline{W_o}}{n_b\,\rho\left(2\pi f x_o\right)^3 D^2}$	Ratio of resistance force to the inertia force
Net flow Reynolds number	$Re_n = \frac{\rho v D}{\mu}$	Measure of the mixing intensity in net flow
Oscillatory flow Reynolds number	$Re_o = \frac{x_o 2\pi f \rho D}{\mu}$	Measure of the mixing intensity in oscillatory flow
Velocity ratio	$\Psi = \frac{Re_o}{Re_n}$	Measure of the dominance of the oscillating flow over the net flow. Recommendation: $\Psi > 1$
Local shear stress Strouhal number	$St_r = \frac{D}{4\pi x_o}$	Degree of eddy propagation in an OBR
Peclet number	$Pe = \frac{vL}{D_a}$	Degree of axial back-mixing

ρ = fluid density (kg.m^{-3}), v = superficial fluid velocity (m s^{-1}), Δp = pressure drop (Pa), s = open cross-sectional area (m^2), D = tube inner diameter (m), L = length between pressure ports or tube length (m), Δp_{max} = maximum oscillatory pressure drop (Pa), f = oscillation frequency (Hz), x_o = center to peak oscillation amplitude (m), $\overline{W_o}$ = average power density over one cycle (W), n_b = number of baffles, μ = dynamic viscosity (Pa-s), D_a = axial dispersion coefficient (Wm^{-2}K^{-1}).

TABLE 15.3
OBR Models for the Calculation of the Energy Dissipation and Power Consumption[7]

Model	Quasi-Steady[11]	Eddy-Enhancement[9]
Power density	$\varepsilon_{v-qs} = \frac{2n_b\rho}{3\pi L C_o^2}\left(1/s^2 - 1\right)\left(2\pi f x_o\right)^3$	$\varepsilon_{v-ee} = 1.5\frac{\rho\omega^3 x_o^2 l_m}{1-s}$
Power consumption	$\overline{W_{qs}} = \frac{n_b\rho\left(2\pi f x_o\right)^3\left(1/s^2 - 1\right)}{1.5C_o^2}\frac{D^2}{4}$	$\overline{W_{ee}} = 1.5\frac{\rho n_b\omega^3 x_o^2 l_m}{s}\frac{\pi D^2}{4}$
Characteristic parameters	Discharge coefficient, C_o	Mixing length, l_m
Reference values	C_o = 0.6[8,11]–0.7[12] (simple orifice w/sharp edges)	l_m = 7mm[9] or l_m = 9mm[13]
Recommended application	x_o = 5–30mm, f = 0.5–2Hz	x_o = 1–3mm, f = 5–14Hz

Note: Angular frequency, $\omega = 2\pi f$ (rad-s^{-1})

To clarify, OBR technology is not intended to replace or supersede conventional technologies such as large-scale continuous stirred tanks. However, it is a real and viable alternative where appropriate, it is easily accessible, and it can successfully be applied to niche processes or products, especially for pilot-scale screening and manufacturing of specialty and high value-added products.[1] It could be a finishing reactor for achieving maximum conversion, within a small footprint, with low heat transfer requirements. There are multiple types of inserts and baffles that could be used, in tubes or columns with pulsating or oscillatory flow, to maximize mixing relative to that of empty tubes (see Figure 15.3).

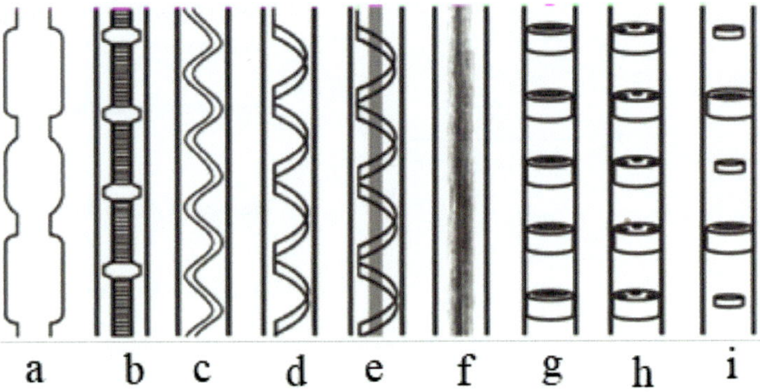

FIGURE 15.3 Configurations of various oscillatory baffled reactors: (a) integral baffles with periodic sharp or smooth constrictions, (b) central axial baffles, (c) round-edged helical baffles, (d) sharp-edged helical baffles, (e) sharp-edged helical baffles with central rod, (f) wire wool baffles, (g) single-orifice baffles, (h) multi-orifice baffles, and (i) disc and doughnut baffles.[5]

It is difficult to clearly quantify how these various insert and baffle geometries impact reactor performance, since it depends on multiple parameters.[5] In the presence of oscillatory flow, these internals generate intermittent spiral flows and vortices, which enhance transverse mixing. Practicing engineers are advised to pursue suitability of these baffle configurations over well-defined single- or multi-orifice-plate baffles.

Recently, investigations of novel meso-OBRs have begun for rapid high-throughput screening applications. They exhibit strong plug flow behavior at low and moderate net flowrates (e.g., 0.3–8 mL/min). For example, good plug flow is seen when net flow Re_n = 4.3–34 for St_r = 0.4–0.8,[14] and ψ = 4–8 in integral baffle configurations (Figure 15.3 a), or ψ = 5–10 in central baffle configurations (Figure 15.3 b). Reference 14 has a state-of-the art review of meso-OBRs.

15.3 PULSED-AIR MIXER

Pulsair® is a patented[15] pneumatic technology widely accepted for effective macro-mixing of working media at various scales, ranging from a small 300-gal tote tank to a large 30,000-gal railcar tanker. A flow controller intermittently pulses compressed air, or other gas, into an accumulator located just above the vessel floor (see Figure 15.4). Gas released beneath the round accumulator plate forms multiple streaming bubbles that instantaneously coalesce to form a disc-shaped bubble above the accumulator. As the bubble rises, due to buoyancy, it draws in surrounding fluids and forms a large mushroom-shaped bubble. This creates the well-established, up-pumping, vertical flow pattern shown in Figure 15.4. Fluid displacement ahead of the bubble induces this flow-suction phenomena. Finally, the large bubble that reaches the working-media surface instantaneously bursts and displaces adjacent liquid radially outward. Sequentially forming and bursting bubbles generate a distinct synergistic, up-pumping, circular-toroidal flow pattern that helps mix and blend the vessel contents.

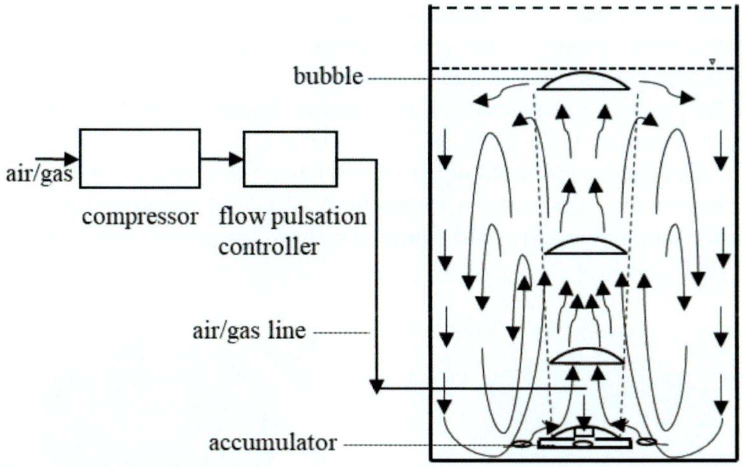

FIGURE 15.4 A single injection, pulsed-air system with rising bubbles and an ideal liquid flow pattern.

Pulsair technology is used in open or vented tanks. Theoretically, the mixing efficacy of a single Pulsair mixer is limited to a cylindrical domain whose radius is influenced by various parameters such as pulsation frequency and quantity of gas, working-media density and viscosity, etc. Depending on the vessel size, multiple injection openings are used to fully cover the fluid domain. Implementations often require design assistance from technical representatives of the manufacturer,[16] Pulsair® Systems, Inc. Control equipment and gas-pulsing valves control pulse frequency and duration, gas pressure, and plate sequencing to create optimal mixing conditions within the tank.[17] Generally, the time, τ, a bubble needs to reach the liquid surface determines the pulsation frequency, $f = 1/\tau$. The desirable injection time is a fraction of a second, and the preferred accumulator plate diameter is 1/16th of the tank diameter.[15]

Pulsair technology has no submerged moving parts. Because it lacks the extraneous capital and maintenance-overhead costs typically associated with mechanically agitated systems, it is a cost-effective approach to blending and dislodging solids from the bottom of open tanks. Pulsair offers a wide variety of tank mixing solutions for all types of liquids, including agriculture (fertilizer and seed treatment), food and beverage, wastewater, lubricants and fuels, etc.[16] Wastewater treatment research at the US Department of Energy (DOE)[17] has pointed out a few tank-specific issues that may arise from the use of pulsed-air technology. When the accumulator plates are operated at a pressure sufficient to generate bubbles (e.g., up to 100psig), a shock wave can be produced within the waste. The effect of this shock on the vessel integrity must be examined. A fine mist of waste slurry is formed when large bubbles break at the waste surface. This aerosol generation is primarily an operating-cost issue, because a higher aerosol-generation rate requires that the tank ventilation system be designed with greater aerosol-handling capability. The large number of plates required to cover the floor of a large-diameter wastewater treatment tank can present a challenge for pulsed-air mixing applications. In sludge media, like other gas-sparged slurry systems, accumulators are also susceptible to clogging during extended periods of mixer inactivity.[17]

15.4 SANITAIRE® DIFFUSED-BUBBLE AERATORS

Different types of gas spargers used in mechanically agitated gas-liquid contactors are represented in Figure 9.8. Because these are susceptible to particulate clogging, their applications are limited to sludge-free working media. Even distribution of the gas bubbles within the vessel is accomplished with a mechanical stirrer. The novelty of the patented,[18] subsurface Sanitaire aeration system is its self-creation and dispersion of uniform, fine bubbles in a wastewater basin, by means of an innovative flexible diffuser assembly that is not exposed to sludge or particle plugging. This makes application of the Sanitaire aerators unique to municipality and industrial sewage and wastewater treatment plants. Xylem, Inc. manufactures Sanitaire fine-bubble diffusers for biological aeration, pre-discharge aeration and sludge holding, and odor control, as well as coarse-bubble diffusers for grit chambers, retention, and sludge holding tanks.[19] A few types of diffuser designs and their applications are described on the manufacturer's website. A couple of widely used fine-bubble aerators (flexible and ceramic membrane) extracted from the website are represented in Figure 15.5.

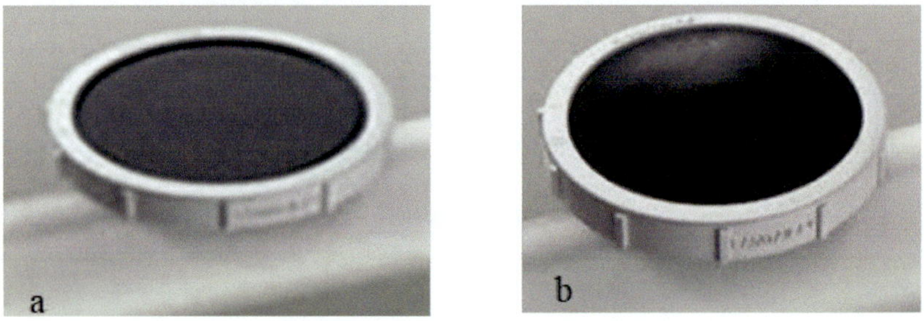

FIGURE 15.5 Sanitaire® (a) ceramic disc and (b) flexible membrane fine-bubble diffusers.[20]

The membrane disc aerator, shown in Figure 15.6, has a uniformly perforated flexible (EPDM elastomer) membrane mounted on top of a circular base plate consisting of a central hole that's connected to the air flow control orifice of the air distributor pipe. The membrane is attached to the disc circumference by means of an O-ring and a retainer ring. During the aeration process, air pressure inflates the membrane and uncovers the holes (see Figure 15.5 of Reference 18). Air flows through these holes and generates fine bubbles in the sewage media. In the absence of aeration, the membrane collapses on the base plate and prevents particles and sludge from plugging the sparger assembly. Ideally, the membrane should provide the minimum required pressure drop to maintain uniform flow through its perforations, relative to its location in the manifold grid. According to Sanitaire, experience has shown that the orifice head loss at each diffuser should be at least 2.5 times greater than the distribution header friction head loss with a minimum of 1.0" water column (actual compressed conditions). Furthermore, for a fine-bubble aeration system, the commonly accepted minimum airflow rate (defined as mixing criteria) to impart turbulence and keep solids in suspension is 0.12 scfm/ft^2 (2 Nm^3/m^2).[23]

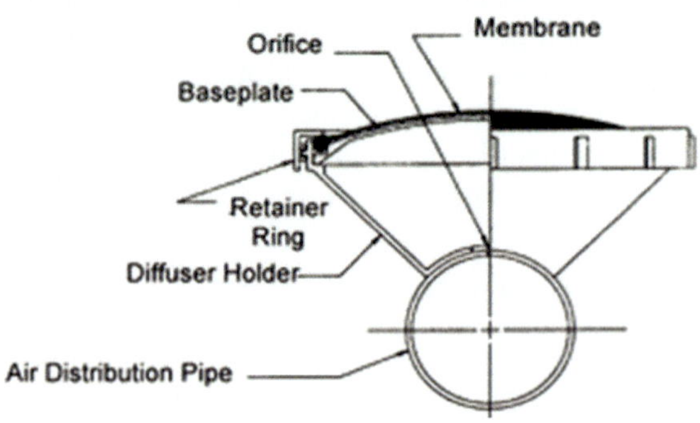

FIGURE 15.6 Cross-sectional view of a Sanitaire® flexible membrane fine-bubble aerator.[21,22]

Normally, wastewater treatment plant (WWTP) basins are longer than they are wide. In continuous discharge systems, required oxygen demand for the metabolism of microorganisms decreases as the flow proceeds. Hence, from the process optimization viewpoint, supplying oxygen and mixing, as performed by an aerator, must be tailored by an appropriate aeration system design (required airflow rate, manifold/grid design, and number of diffusors) to maximize aeration efficiency. For a client's specific application, Xylem provides process optimization, diffuser and distributor (manifold) designs, and turnkey help, including the supply of peripherals such as blowers and controllers, etc.

Figure 15.7 shows the major components of a Sanitaire aeration grid system.[24] Practicing engineers are advised to seek design assistance from the manufacturer. Fine-pore diffusers are prone to chemical and biological fouling, which may impair oxygen transfer efficiency and generate high head loss.[21] As a result, diffusers need routine cleaning to maintain uniform flow and distribution of dissolved oxygen.

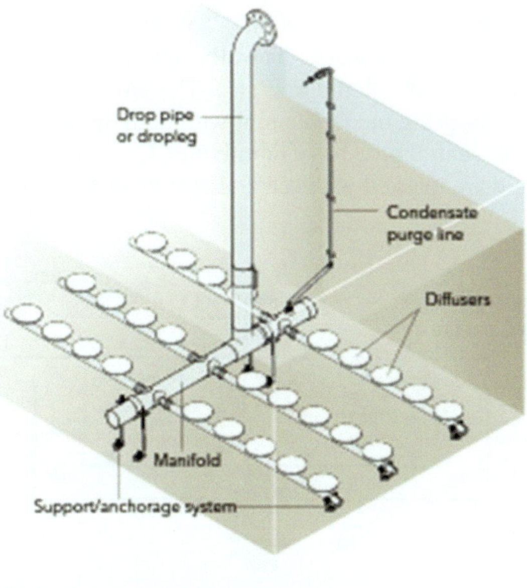

FIGURE 15.7 Sanitaire® diffused aeration grid components.[24]

15.5 HIGEE ROTATING PACKED BED (RPB)

Gas–liquid contactors such as stirred tanks, bubble columns, and pneumatically or jet-agitated airlift reactors are discussed in Chapters 9–10. RPB is an unconventional gas–liquid contactor ideally suitable for process intensification (PI). This device exploits the centrifugal force field in a cylindrical two-phase fluid contactor that rotates at very high speed to produce accelerations well above that of the earth's gravity (i.e., 1.0g). This excessive acceleration produces intense micro-mixing and enhanced mass transfer between the phases.

Since the early 1990s, RPB use has drawn special attention in distillation, absorption, scrubbing/stripping, heterogeneous catalytic reactions, and aerobic fermentation applications. After being patented by Pilo and Dahlbeck[25] in 1960, it lay dormant for several decades, before researchers[26] in the chemical industries produced multiple variants of the RPB rotor and/or its packing, to enhance resulting volumetric gas-liquid mass transfer coefficients, $k_g a$ or $k_l a$, with respect to specific energy consumption, ε. Industrial applications of these variants first appeared in the 1970s, when Imperial Chemical Industries (ICI) worked on the technology derived from NASA research projects. In 1981, the patent of Ramshaw and Mallinson[27] further promoted ongoing RPB technology. Nearly a decade later came PI activities at academic institutions and consortiums such as Delft University of Technology (the Netherlands), New Castle University Tyne (UK), Beijing University of Chemical Technology, BUCT—HIGRAVITEC Center (China), and the BHR Group (UK).

Recent emerging industrial and municipal applications of RPB reinforce the continuing exploration of this cost-effective PI novelty. The authors of this work know of a successful RPB application in a Dow Chemical Company HOCL plant,[28] which increased yield 10%, reduced gas stripping 50%, and reduced volume 40-fold relative to more conventional and capital-intensive spray towers. The major mass-transfer limitations of liquid-side chlorine absorption and gas-side chlorine desorption are minimized by increased liquid flow turbulence and gas loading in the rotating packed bed.[29]

A basic drawing (Figure 15.8) enhances the description of conventional rotating packed bed, RPB, workings. The packed bed, P, is an annular, cylindrical element attached to a high-speed, bottom-mounted rotating shaft, S. The rotating bed contains a flow-through porous media, M, that is typically innocuous to the working media. A variety of porous media, e.g., rigid wire mesh, wire gauge, metal-foil ribbons, glass beads and glass wools, etc., are packed inside P to a density (10–60 vol.%)[30] suitable to the specific application. The stationary, annular, liquid pipe, L, is centrally mounted to serve as a liquid distributor. Its walls and those of the packed bed are perforated, or slotted, to allow transfer of fluids (liquid and gas). The packed bed, P, and the central pipe, L, are enclosed in a stationary working-media plenum with both an inlet, I_g, for the gas/vapor and an outlet, O_l, for the liquid. The central pipe has a liquid inlet, I_l, and a gas/vapor outlet, O_g. The directional flow of gas and liquid are marked with red and blue arrows in Figure 15.8.

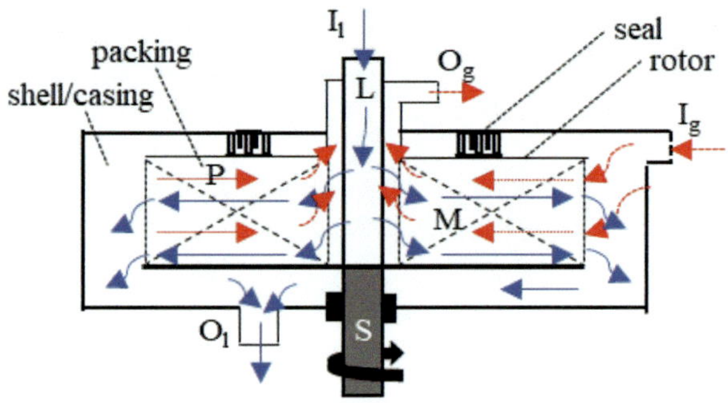

FIGURE 15.8 RPB schematic showing counter-current gas (red arrows) and liquid (blue arrows) flows.

In this RPB setup, the shaft is driven by a bottom-mounted motor. During operation, the very high speed (200–1500 rpm)[31] at which the packed bed rotates creates a centrifugal field of 100–1000 times the acceleration of gravity.[30] Liquid from the central pipe is forced radially outwards, through the packing material. Gas, being too light to have its flow direction altered by the rotation, moves radially inward through the packing material. The countercurrent gas and liquid flows, passing through the packing media, make intimate contact with each other and produce intense mass transfer between the two phases.

15.5.1 Rotor Designs

Many different high-gravity technology (Higee) RPB designs address inherent performance deficiencies by improving gas-liquid-flow slip velocity, lowering pressure drop and fouling in the packing material, preventing flooding (liquid hold-up), etc. The referenced article by Wang et al.[32] presents several such designs, with their pros and cons relative to a conventional RPB. For the sake of brevity, only a couple of simple rotor designs are described here.

Figure 15.9 is the schematic of a rotating zigzag bed (RZB) that eliminates the use of packing media. It consists of an upper stationary disk, D_s, and a lower disk, D_r, mounted on a rotating shaft, S. Respectively, these disks have a series of cylindrical baffles, B_s and B_r. Similar to the RPB, the rotor and stator units are enclosed in a stationary casing (plenum) with built-in gas and liquid inlets, I_g and I_l, and outlets, O_g and O_l. When disk D_r rotates at high speed, centrifugal force imposes orderly gas and liquid flow directions in the confined chambers (see the red and blue arrows, respectively). Intimate contact between the gas and liquid phases occurs in the annular domain, D, resulting in vigorous mass transfer. This design accommodates multiple feed and side streams essential for continuous distillation and catalyst addition for heterogeneous reaction.

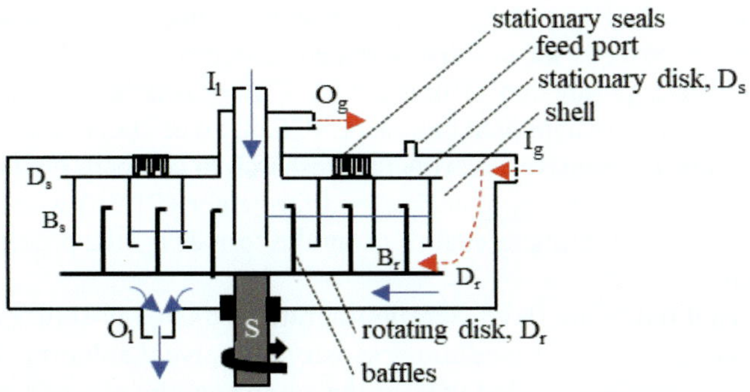

FIGURE 15.9 Schematic of a rotating zigzag bed (RZB).

[Design advantages over the RPB: stationary seals, feed/injection ports and no liquid distributor]

Figure 15.10 schematically illustrates the operations of a simple split-packing RPB (SP-RPB). The latter consists of two rotating disks, i.e., the upper disk, D_{ru}, and the lower disc, D_{rl}, with a series of vertical packing rings, P_u and P_l, that create an interweaving annular assembly, as shown.

Respectively, the disks are mounted on top and bottom shafts driven by two separate motors and enclosed in a stationary casing (plenum). The plenum entails gas and liquid inlets (I_g, I_l,) and outlets (O_g, O_l). During the operation of SP-RPB, liquid enters into the central distributor through the annular space and due to centrifugal force, flows outwards past the rotating packings, whereas the gas enters the chamber and flows inwards, i.e., follows a counter-current flow path, as shown with red and blue arrows.

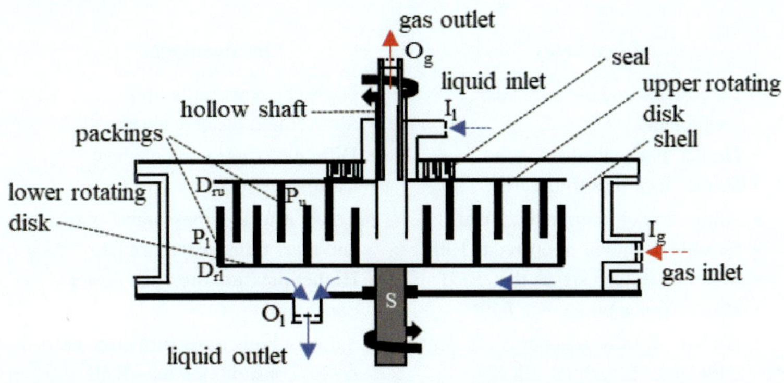

FIGURE 15.10 Schematic of a split rotating packed bed (Sp-RPB).[32]

[Shafts rotate in opposite directions for mass transfer performance advantage]

15.5.2 Applications

The RPB and RZB are devices that intensify gas–liquid contacting and significantly enhance gas/vapor–liquid micro-mixing and mass transfer. They have found niche industrial applications in absorption, desorption, and heterogeneous catalytic reaction. Table 15.4 summarizes a few industrial applications that surpass the efficacies of conventional packed columns.[32] The intensification achieved thus far falls short of the desired two to three times volume reduction, however, as compared to that required with classical columns.[30]

The inherent mass transfer benefits of Higee technology have been further leveraged to multiple mass-transfer-limited applications such as in industrial wastewater treatment and bioreactors. The aerobic activated sludge process (AAS) is widely used in municipal wastewater treatment plants. Results showed that the Higee-AAS aeration process significantly enhances oxygen transfer and microbial activity, improving wastewater treatment efficiency.[33] At the Technical University of Munich, growth and over-production of lipase with *S. carnosus* have been investigated under increased acceleration in a centrifugal field bioreactor (CFB), an offshoot design of the Higee technology.[34] The authors pursed extensive experimental research to compare the flow and mass transfer characteristics of a 30L-scale rotating packed bed fermenter (RPBF) with that of an identical stirred tank fermenter, with the ultimate goal of using the former for the process intensification of α-Amylase production.

All tests were carried out using the deaeration-aeration, unsteady-state mass transfer determination technique using water and medium viscosity polyglycol solution (170cP) to simulate fermentation broth. Test results revealed that, at the same specific power consumption and gas flowrates, the volumetric mass transfer coefficient ($k_L a$) of the RPB reactor was two to three times lower than that of the stirred tank reactor (STR), and as opposed to the STR, showed a very constant behavior in the $k_L a$ vs. power curve. The subpar energy-specific mass transfer coefficient, $(k_L a) / \varepsilon$, advantages of RPBF over the conventional gas–liquid STR may be attributed to the specific design of the RPBF. It was designed for co-current flow of gas and liquid through a standard wire mesh packing having relatively low volumetric surface area, i.e. ~300m²/m³. It maintained a low-velocity gradient of liquid introduced into the packing by means of an auger.

TABLE 15.4
Advantages and Disadvantages of RPB, RZB, and SP-RPB and Their Applications[32]

Higee RPB Type (main parts)	Advantages	Disadvantages	Applications
RPB—counter flow (Rotor)	• Largest volume mass transfer coefficient • Higher mechanical strength • Higher specific surface area	• Highest gas pressure drop • Short liquid residence time • Difficult to apply middle feed • Complex structure	• CO_2 capture • Distillation • Reaction
RZB (Rotor and Stator)	• Admission of intermediate feeds • Flexible operation compared to RPB • More theoretical stages than RPB • Higher contact time than RPB	• Lower effective mass transfer area and mass transfer rate (MTR) • Higher pressure drop and power consumption	• Distillation • Stripping
Sp-RPB (Two split packings)	• Higher gas mass transfer coefficient with larger tangential velocity • Higher effective area and mass transfer coefficient	• Same or higher pressure drop and power consumption than RZB • Difficulty in middle feed	• CO_2 adsorption

15.5.3 Advantage and Disadvantages

The tortuous and zigzag flow path of the working media in the Higee rotating bed induces turbulent fluid flow, which creates micro-mixing and fine gas bubbles with ultra-thin liquid films. Consequently, the high gas–liquid interfacial area, a, increases the liquid-side mass transfer coefficient by three- to seven-fold,[30] and similarly increases yield compared to traditional packed columns, all of course at the cost of higher-pressure drop. In addition, residence time is shortened significantly due to the high centrifugal gravitational forces, which also minimizes byproducts and maximizes selectivity.[35] The inherent characteristics of short residence time and high mass transfer rate, within the RPB, offer a good fit for the hypochlorous acid (HOCl) production process,[29] depicted in Figure 15.11: reactive gas–liquid chemical system for the production of hypochlorous acid,[29,31] which includes a very fast formation reaction followed by a fast decomposition reaction.

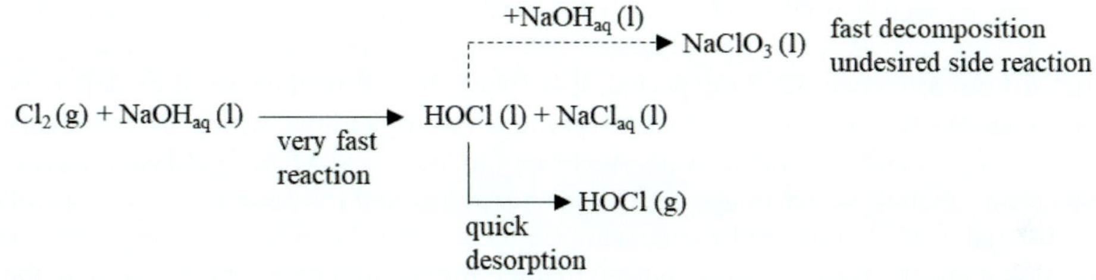

FIGURE 15.11 Reactive gas–liquid chemical system for the production of hypochlorous acid.[29,31]

The RPB and RZB installations are less capital intensive than tall packed columns[35] because of their smaller footprints, shorter heights, and greater flexibility in materials of construction. Furthermore, freedom to change rotor speed is an added benefit for optimizing process operations.

Despite voluminous research and numerous case studies[31] claiming significant performance benefits of the Higee technologies over conventional packed columns, for separation and distillation, industrial applications of RPBs in traditional multiphase fluid contacting applications are still rather scarce. The following drawbacks overwhelm the positive aspects of this technology, and many practicing industrial engineers are still too risk-averse to consider RPBs during process design. The rotor and the flow distributor of the Higee equipment tend to be quite complex in design, and the cumbersome rotating bed in the long run could be mechanically unstable, troublesome, and expensive to maintain. The RPB packings are prone to liquid flooding, particulate plugging, and may create ultra-shear zones that are detrimental to biological applications. Furthermore, the Higee devices generally lack reliable scale-up rules or well-established models essential for industrial applications.

15.5.4 Scale-Up

In general, Higee contactors are application specific, encompassing many variations in the design of major components such as the liquid distributor, rotating packed bed, and zigzag flow channels. As pointed out earlier, scale-up from R&D to production scale poses a significant challenge and often leads to a trial-and- error approach. In recent years, dimensional analysis and computational fluid dynamics (CFD) involving computed tomography (CT) reconstruction with volume graphical approaches, have been quite beneficial in the *a priori* study of pressure drop, flow, and mass transfer characteristics in full-scale Higee designs. The realizable k-ε turbulence CFD model, and the two-phase volume fraction (VOF) model, are amenable to handle various gas–liquid zones with varying cross-sectional areas (casing, porous packing bed and eye) internal to conventional RPB designs. In the case of countercurrent gas–liquid flows in the annular area of the rotor, the superficial inward radial velocity of gas increases as the outward radial velocity of the liquid also increases. By means of the distributor, liquid is typically injected as jets or sprays into the inner periphery of the packing.[30] The visual observations of liquid flow by Burns and Ramshaw[36] and Guo et al.[37] have been comprehensibly discussed in reference 30. As mentioned earlier, the contacting gas–liquid flow in the packing exhibits multidimensional transport of gas bubbles with thin liquid films. It creates intractable mathematical issues even with the VOF-based CFD models, and consequently mass transfer coefficient prediction is often unresolvable without simplifying assumptions.

Three common mathematical models (i.e., penetration theory, two-film theory, and surface renewal theory) have been extensively studied to understand their suitability to predict overall gas–liquid mass transfer coefficients, under assumptions of varied liquid patterns in high-gravity separation devices and their prototypes such as rotating blades and disks.[38–41] Complementary absorption experiments in the laboratory were carried out to systematically validate model predictions. See Table 15.5, which was extracted from reference 32. As expected, it is found that all mathematical models mentioned have their limitations based on assumptions of film flow or droplet flow, which are essentially functions of wetted region and angular velocity. According to Tung and Mah,[38] predicted mass transfer coefficients from penetration theory agreed with their experimental results in a certain range of effective areas. Munjal et al.[39] compared the penetration theory and the complete convection-diffusion model on mass transfer, and the results showed that the former was more reasonable than the latter at lower rotational speed. Assuming that the liquid consisted of liquid droplet and film flow, Sun et al.[40] published the two-film theory to estimate the overall volumetric mass transfer coefficient of the CO_2 and NH_3 reaction. It is considered to be a reliable mass transfer model under some appropriate assumptions in RPB.[32] Luo et al.[41] developed a surface renewal theory model to calculate the liquid-side volumetric mass transfer coefficient in an internal rotor, where significant spray existed. Experimental mass transfer results were close to those of model predictions, within ± 15%.[32]

TABLE 15.5
Overview of Different Mathematical Models and Their Applications[32]

Models	Liquid Patterns	Experimental Studies
Penetration theory	Film flow	Dissolved oxygen[38]
Penetration theory	Film flow	Chemical absorption of CO_2 in NaOH[39]
Two-film theory	Film and droplet flow	Absorption of CO_2 and NH_3 into water[37]
Surface renewal theory	Film, pore, and droplet flows	Chemisorption of CO_2 into a NaOH solution[41]

Table 15.6 summarizes formulae for three major contributing pressure drops experienced in the device. Because pressure losses from flow expansion and contraction (as the gas enters and exits the rotor) are small compared to the other pressure drops,[42] these are not noted in Table 15.6. Furthermore, since the pressure loss from momentum gain, Δp_m, is typically only ~5% of the total pressure loss, Δp_t, it is sometimes ignored. Thus, the sum of frictional and centrifugal pressure losses in the rotor account for almost all expected gas pressure drop in the device,[30] i.e., $\Delta p_t = \Delta p_f + \Delta p_c$.

The total power utilized in an RPB device overcomes frictional power loss and hydraulic (shaft) power consumption. The shaft energy expended in an operating RPB is a strong function of rotational speed and torque experienced by the shaft due to (a) tangential frictional forces on the rotor surfaces (windage drag) and (b) radial to tangential variation of liquid momentum in the packing

TABLE 15.6
Various Gas Flow Pressure Drops in a Conventional RPB[30]

Gas Pressure Drops	Remarks	Formulae
Pressure loss due to momentum gain[42]	Decreases in flow area of the packing, increase gas velocity and pressure loss.	$\Delta p_m = \rho_g \frac{1}{8}\left(\frac{Q_g}{\pi h}\right)^2\left(\frac{1}{r_o^2} - \frac{1}{r_i^2}\right)$
Frictional pressure loss[42,43]	Flow is predominantly radial. ΔP_f is predicted from the Ergun correlation. Pressure drop in the rotor is much greater than those in the rotor eye and casing.	$\int \Delta p_f = \int f\left[\frac{(1-\varepsilon)G^2}{\varepsilon^3 d_p \rho_g}\right] \mathrm{dr}$ where $f = fn.(Re)$ Approximated as: $\Delta p_f = \frac{(1-\varepsilon)}{\varepsilon^3}\frac{Q_g}{2\pi h d_p}\left[\frac{150(1-\varepsilon)\mu_g}{d_p} ln\frac{r_o}{r_i} + 1.75\frac{Q_g \rho_g}{2\pi h}\left(\frac{1}{r_i} - \frac{1}{r_o}\right)\right]$
Centrifugal force pressure loss[44–46]	Deduced from differences in centrifugal forces at the inner and outer rotor radius	$\Delta p_c = A\frac{1}{2}\rho_g \omega^2\left(r_o^2 - r_i^2\right)$ Constant A =1 for gas in pure rigid-body rotation. A = 0.5 to 2, otherwise. Rotor is not neutral, but acts as a blower or an expander.

Nomenclature: ρ_g = gas density [kg m^{-3}], μ_g = gas viscosity [kg m^{-1} s^{-1}], ε = porosity of the packing [-], Q_g = gas volumetric flowrate [m^3 s^{-1}], h = height of the bed [m], r_i = rotor inner radius [m], r_o = rotor outer radius [m], f = local friction factor [-], Re = Reynolds number [-], G = gas mass flowrate [kg s^{-1}], $d_p = \frac{6(1-\varepsilon)}{\alpha_p}$ is equivalent particle diameter [m] of the packing, where α_p = specific packing surface area [m^2. m^{-3}], r = radius [m], ω = rotational speed [rad. s^{-1}], Δp = pressure loss [Pa]

(liquid acceleration). Thus, $P_s = 2\pi NT$ [W], where, N and T are the rotational speed [s^{-1}] and shaft torque [N-m], respectively. The liquid or gas volumetric mass transfer performance of an RPB relative to power consumption (i.e., $k_l a / P_s$ or $k_g a / P_s$) is considered the standard to compare with mechanically agitated systems. Literature contains many experimental correlations for estimating gas and liquid volumetric mass transfer coefficients (i.e., $k_g a$ and $k_l a$). These are case study specific and summarized in Tables 8 and 10 of reference 32.

Similar to traditional packed columns, RPBs are also susceptible to flooding or separation of liquid and gas phases, which ultimately leads to formation of liquid droplets and/or striations, followed by ejection from the rotating device (see Figure 15.7). Such a scenario hinders intimate contact between the two phases and leads to inconsequential mass transfer. Hence, avoiding the separation point is critical to the design of conventional countercurrent rotating packed beds.

When the angular velocity decreases, centrifugal pressure drop, Δp_c, diminishes. The lack of centrifugal force tends to accumulate water in the packing, thus increasing the frictional pressure loss, Δp_f. Total pressure loss, Δp_t, is marginally altered. When the angular velocity is reduced beyond a critical value, however, separation of unrepelled water significantly increases frictional pressure drop. Water gradually collects at the rotor bottom, as shown in Figure 15.12.

At this point, total pressure loss increases until the liquid is ejected from the rotating packed bed. Design of an RPB must account for the operational flexibility of angular velocity ideally suited for the gas and liquid flowrates. In other words, flooding sets an upper limit of allowable throughputs in a specific Higee device. The point at which the pressure drop increases sharply with increasing superficial liquid velocity was first taken as a new Higee flooding criterion.[47,48]

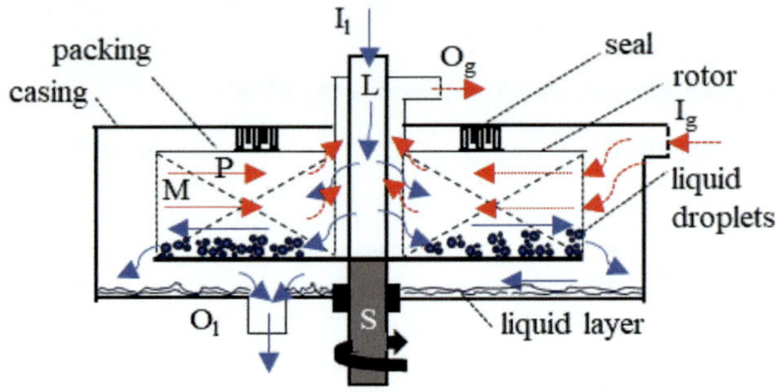

FIGURE 15.12 Schematic of a countercurrent gas–liquid flow RPB under flooding condition.

Flooding correlations for RPBs, based on their RSB experimental data as proposed by Lockett[48] and Rajan et al.,[49] in terms of maximum gas and liquid capacities (C_g and C_l), are presented in the equations of Table 15.7. Additional design information concerning the minimum height and inner radius of the RPB, to avoid flooding, are also presented.[50]

A recent (2021) review article[51] on modeling of an RPB is an excellent source for research and design engineers to understand the various parameters such as liquid distributors, packing media, rotor designs, flow characteristics, etc., that influence gas-liquid mass transfer rate, pressure drop and flooding performance (liquid hold-up), and power requirement of high-gravity rotating devices.

TABLE 15.7
Design Flooding Conditions for RPB[51]

Flooding Correlations	Associated Parameters and Nomenclatures	Ref.
$C_g^{0.5} + \dot{m}C_l^{0.5} = 1.57\beta^{0.25}a_p^{-0.25}\left(\frac{\mu_l}{\mu_w}\right)^{-0.03}$	$C_g = u_g\left(\frac{\rho_g}{\rho_l-\rho_g}\right)^{0.5}$ and $C_l = u_L\left(\frac{\rho_l}{\rho_l-\rho_g}\right)^{0.5}$, m-s^{-1}	48
$C_g^{0.5} + 1.51C_l^{0.5} = 130\beta^{0.434}\alpha_p^{-0.9343}\left(\frac{\mu_l}{\mu_w}\right)^{-0.03}$	$u_g = \frac{\dot{m}_g}{\rho_g A}$ and $u_l = \frac{\dot{m}_l}{\rho_l A}$ are superficial gas and liquid velocities, m-s^{-1} $\dot{m}$ = mass flowrate, kg-s^{-1} ρ_g, ρ_l = gas and liquid densities, kg-m^3 A = rotor cross-sectional area, m^2 β = liquid hold-up, volume of liquid sustained per unit volume of packing a_p = surface area of the packing per unit volume of the bed, m^2/m^3 μ_l, μ_w = viscosities of liquid and water [Pa-s)	49
Minimum RPB height without flooding, $h_f[m]$ $h_f = \frac{\dot{Q}_g}{2\pi R_i u_g}$	$\dot{Q}_g$ = gas flowrate, m^3-s^{-1} R_i = Length of rotor inner radius, m u_g = superficial gas velocity, m-s^{-1}	50
Minimum inner radius of RPB without flooding, r_{i-min} $r_{i-min} = \left[\frac{\dot{Q}_g}{\pi u_{jet}(1-f_d)}\right]^{1/2}\left(\frac{\rho_g \xi}{\rho_l}\right)^{1/4}$	f_d = fraction of volume the liquid distributer occupies in the inner radius of RPB ξ = Liquid maldistribution of injector, % u_{jet} = Liquid jet velocity, m-s^{-1}	50

Current mathematical models and dimensional analysis techniques based on *Re*, *Gr*, *Sc*, *Fr*, *We*, and shape factors used for prototype scale-up and benchmarking purposes are extensively discussed. Multiple process intensification case studies and applications such as stripping, distillation, chemisorption, bioremediation, and water deaeration are addressed.

15.6-I HYDRO DYNAMIC CAVITATION FOR MIXING, INCREASED MASS TRANSFER, AND PROCESS INTENSIFICATION

Douglas Mancosky

15.6-I.1 Introduction

Cavitation is the sudden formation and collapse of bubbles within a liquid by means of mechanical force. Unlike ultrasound, which is initiated acoustically, cavitation is initiated mechanically. Very rapid formation and collapse of these bubbles release pressurized shockwaves that can produce tremendous mixing. These shockwaves can provide breakthrough benefits when mixing a liquid with other liquids, gases, or solids at a microscopic level, thus increasing their reaction efficiency. Hydrodynamic cavitation is typically defined as mechanically induced cavitation in fluids that pressurize and depressurize while flowing around or through an obstacle in the flow field.

One typically adverse effect of hydrodynamic cavitation, the severe erosion of surfaces in contact with the cavitating fluid, has been documented extensively by many including Young.[52] Although hydrodynamic cavitation has been employed in some variation of a venturi or orifice plate in applications such as wastewater treatment, these devices are typically prone to erosion from the cavitation.[53–56] To facilitate practical commercial use, the equipment that induces fluid cavitation must harness or control the destructive forces of cavitation to eliminate equipment erosion and maximize hydrodynamic cavitation in the product. One example of controlled cavitation, the ShockWave Power Reactor (SPR) by Hydro Dynamics, Inc. of Rome, Georgia, USA, is the focus of this chapter section.

The SPR is a rotor-stator design, which employs concentric cylinders, one rotating and one stationary, with fluid flowing through the annular space. The surface of the inner rotating cylinder contains multiple cavities. The flowing fluid is repeatedly drawn into these cavities, where it is continually pressurized and depressurized. In the low-pressure region at the bottom of these cavities, low-pressure bubbles form, and because they are too unstable to maintain themselves, collapse. This releases a shockwave containing all the mechanical energy required for bubble creation. Thus, the spinning action of the rotor generates this hydrodynamic cavitation within the cavities, away from other metal surfaces.

This novel design allows a continuous flow of fluid to replenish multiple cavities as the rotor spins, facilitating cavitation and product flow through the system. Centrifugal forces from the inner rotating cylinder are countered with centripetal forces to avoid traveling cavitation, or cavitation that spills out of the cavity and travels through the system. Pressurizing the system during operation ensures continuous product contact with the inner rotor, and produces controlled cavitation in the fluid. This is depicted in Figure 15.13.

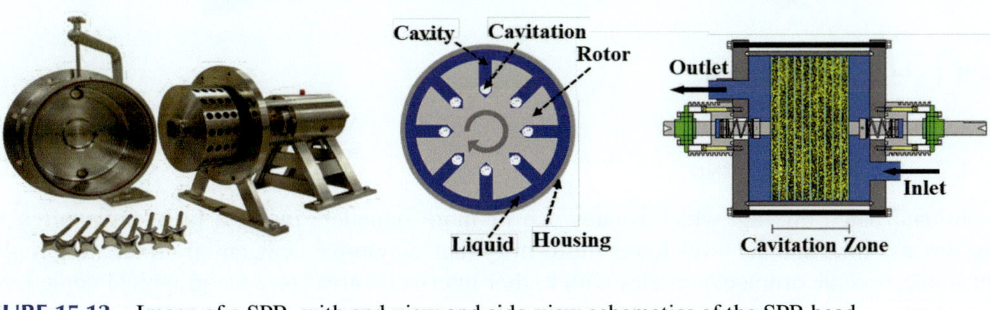

FIGURE 15.13 Image of a SPR, with end-view and side-view schematics of the SPR head.

Besides intense pressure waves, collapse of a vapor-filled cavity is also accompanied by vigorous shear forces, localized heating events, and in some cases, limited formation of free radicals.[52,57,58] Uncontrolled cavitation can damage surfaces and even the structural integrity of pipes, turbines, and pumps. In the SPR, controlled cavitation collapses occur within the bulk fluid, such that only a fraction of the initial cavitation energy strikes the metal surfaces. These mild cavitation shockwaves are akin to those of ultrasonic cleaners.

Cavitation energy is an exponential function of rotor rotation rate (tip speed) and enters the liquid as heat. In fact, ($T_{liq\text{-}out} - T_{liq\text{-}in}$) is a proxy for the cavitation energy. At a specified rotor size, rotor geometry, and operational speed (RPM), the liquid properties (tensile strength, specific weight, etc.) and the rate of liquid flow into the SPR annular space dictate the level of applied cavitation energy per unit volume.

The relationship between imparted cavitation heat (energy) and each individual operating parameter (all other individual parameters remaining constant) is shown below. Liquid flowrate is the only operating parameter inversely related to heat (energy) input.

TABLE 15.8
Impact of Changing Various Operating Parameters on Imparted Cavitation Energy (heat)

Changing Operating Parameter	Imparted Heat Increases	Imparted Heat Decreases
rotor speed	if increased	if decreased
rotor diameter	if increased	if decreased
liquid tensile strength	if increased	if decreased
liquid specific weight	if increased	if decreased
liquid flowrate	if decreased	if increased

While generally coupled with an electric motor on a skid, as shown in Figure 15.14, the SPR can also be coupled with other spin sources such as a diesel motor.

FIGURE 15.14 SPR skid system.

15.6-1.2 Cavitation Mixing

Conventional batch mixing, which rotates one or more impellers inside a large tank, often has longer process time and achieves lower uniformity than single-pass cavitation mixers that rapidly produce microscopic droplets/particles with higher interfacial areas (see the graphical depiction in Figure 15.15).

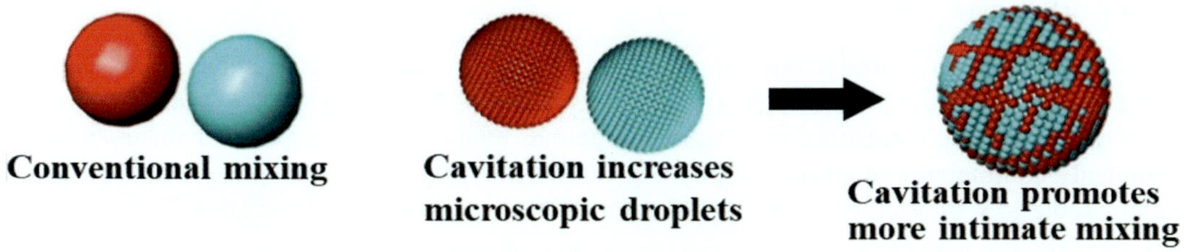

FIGURE 15.15 Comparison of conventional and cavitation mixing technology.

Cavitation is sometimes more effective at mixing shear-sensitive compounds and has been observed to produce superior results over conventional mixers in multi-phase chemical processes.[58]

15.6-I.3 Gum/Gel Hydration

Because cavitation shockwaves rapidly breakup the dry powder agglomerates, they tend to quickly hydrate gums, gels, and polymers, to produce higher viscosities and uniformities. Also, as they help straighten polymeric chains, without scission, they increase effective length and corresponding viscosity (see Figure 15.16a).

Figure 15.16b shows carrageenan gum and methylene blue mixtures processed by a conventional mixer (on left) and a cavitation mixer (on right). The latter mixture contains less partially hydrated gum, as evident by greater uniformity and a lower fraction of darker-blue regions. Cavitation can also reduce needed gum (raw material) and processing time.

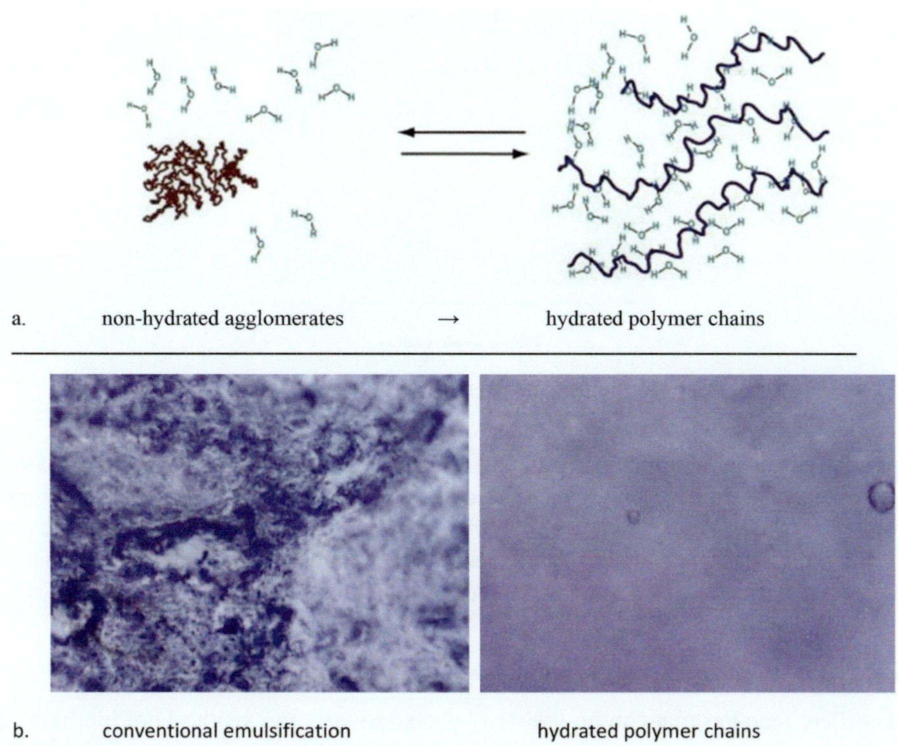

FIGURE 15.16 (a) Cavitation chain straightening and (b) gum hydration comparison.

15.6-I.4 Emulsification

Cavitation emulsifies immiscible liquids in seconds, whether in a batch or continuous mode. Despite its small footprint, the SPR easily produces emulsion droplets smaller than 1 micron. When it creates uniformly fine emulsions, even from compounds containing particulates or shear-sensitive components, continuous and instant control of cavitation intensity controls drop size without entrapping air. Table 15.9 shows the impact of SPR rotor diameter on resulting shear rates.

TABLE 15.9
Shear Rates of Standard Shockwave Power Reactors (SPRs)

Rotor Size (in)	Rotor Speed (min^{-1})	Tip Speed (in/s)	Clearance (in)	Shear Rate (1/s)
8	3600	1508	0.25	6032
10	3600	1885	0.25	7540
12	3600	2262	0.25	9048
14	3600	2639	0.25	10556
16	3600	3016	0.25	12064

Emulsification studies were conducted in the SPR with water, diesel fuel, and a surfactant, because a small water fraction could decrease combustion temperatures and lower NOx emissions.[59] Droplet size decreased as cavitation intensity was increased by changing SPR rotation rate. A micrograph from this study can be seen in Figure 15.17.

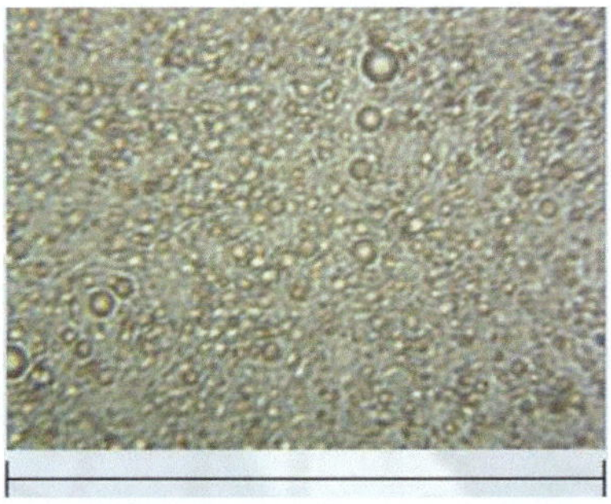

FIGURE 15.17 70:30 Diesel-water emulsion, formed by cavitation at 60 Hz.

Multiple passes did not dramatically reduce mean droplet size, but they did increase droplet uniformity, perhaps by emulsifying a small fraction of droplets that missed maximum cavitation on their first pass. Such "fly-through" can be greatly mitigated by minimizing void spaces within the SPR. Although surfactant can often be reduced with cavitation, some baseline amount is always required to coat all droplets and prevent coalescence/re-agglomeration.

15.6-I.5 Process Intensification of a Chemical Reaction Emulsion

Another excellent reaction that can be assisted by cavitation is the synthesis of biodiesel through transesterification. Biodiesel is created as a glycerin and alcohol react, in the presence of a catalyst, to form an ester. In its most common form, vegetable oil or animal fat are reacted with methanol and caustic to form a methyl ester, which is similar in character to diesel fuel. Successful reaction creates immiscible methyl esters and glycerin, which can then be separated by settling, centrifugation, or other methods. The reaction is depicted in Figure 15.18.

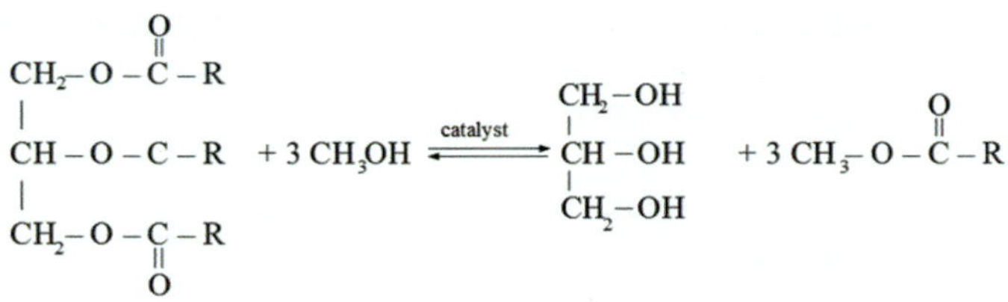

FIGURE 15.18 Transesterification reaction.

To react, the immiscible fat and methanol must be emulsified. Because cavitation creates extremely small droplets of both methanol and oil, the resulting high surface area drastically accelerates mass transfer, and thus, the overall reaction rate. With this cavitation enhancement, reaction can be completed almost instantaneously, rather than in approximately an hour, as commonly observed in more conventional commercial processes. This promotes continuous processing, and significant reductions in process footprint and upfront capital investment.

Typical transesterification processes require two stages, as equilibrium and chemical reversibility require either a large reactant excess, or nearly immediate product removal to drive reaction to near completion. Figure 15.19 depicts a two-stage continuous transesterification process using cavitation.

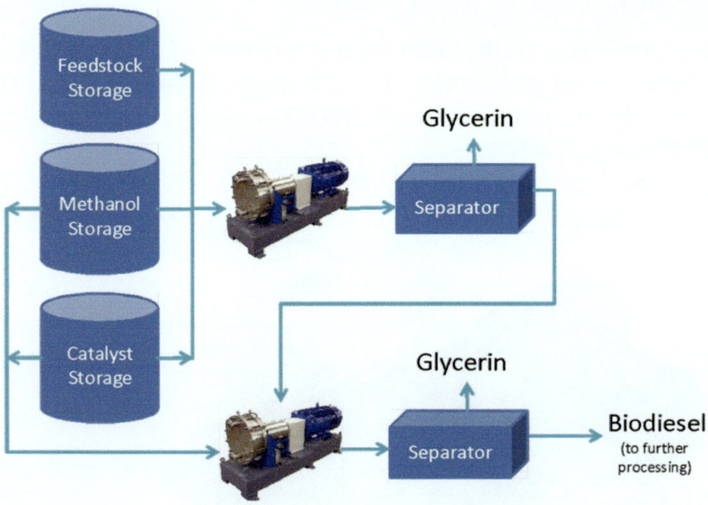

FIGURE 15.19 Two-stage transesterification.

Through more rapid and complete reaction, cavitation can essentially eliminate the monoglycerides "rag" layer that commonly slows processing and significantly reduces both yield and product quality in conventional transesterification processes.

15.6-1.6 Gas and Liquid Mixing

In gas–liquid mixing, the force of collapsing cavitation bubbles divides process gases and liquids into microscopic bubbles and droplets, thus increasing contact surface area and maximizing process efficiency within a small footprint. Enough liquid must be present to maintain the dominant liquid stream required to induce cavitation.

The volumetric gas–liquid mass transfer coefficient, k_La, is a widely used measurement that can help assess a system's mixing capability. The Gigas and Post k_La measurement method[60] uses sodium sulfite to deplete oxygen dissolved in a given water batch, then a timer and an oxygen sensor to help measure the rate at which a given mixer replenishes the dissolved oxygen. Gigas and Post measured a k_La value of $0.22 s^{-1}$ in the agitated tank in their work, while the k_La measured in the SPR was ~10 times that, as seen in Table 15.10.

TABLE 15.10
k_La Chart

Contactor	$k_L a$ (1/s)
Mechanically agitated tank	0.02–0.2
SPR (large lab size)	2

Hart[61] and Kazem[62] used cavitation in pulp and paper processing, including black liquor oxidation, and oxygen delignification of the pulp. Waste lignin, sulfur, and caustic in the black liquor are oxidized for energy recovery and recycling. In pulp delignification, oxygen helps degrade and remove colored lignin as part of a bleaching step. In each case, a high k_La (efficient oxygen use) and high conversion are important.

15.6-I.7 Enhancements Coupled with Cavitation

Tremendous surface refresh rates on the SPR rotor and housing surfaces produce high mass transfer rates during any surface-based process (e.g., UV, chemistry, electrochemistry, etc.). SPR performance can be enhanced by incorporating ultraviolet (UV) radiation (see Figure 15.20) for disinfection[63] or to catalyze chemical reactions.

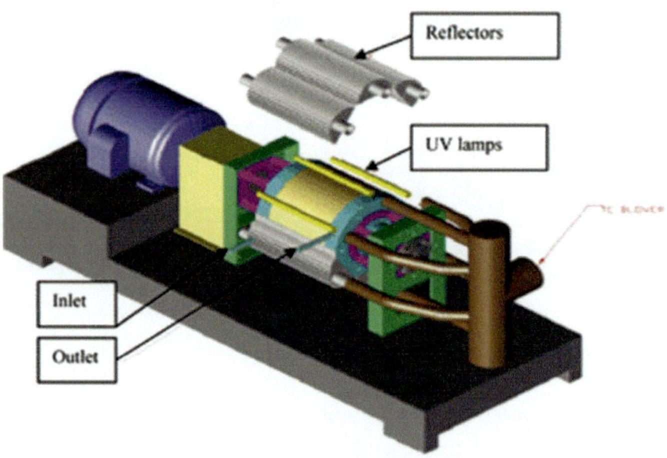

FIGURE 15.20 UV-SPR, a SPR coupled with UV system enhancement.

15.6-I.8 Conclusions

Cavitation can have many benefits over conventional technology. It can be used to intensely mix multiple liquids, liquids with solids, or liquids with gases. Cavitation could provide a paradigm shift for many mixing applications.

15.6-II RESONANTSONICS® (LOW-FREQUENCY) ACOUSTIC REACTOR

Lawrence Farrar

15.6-II.1 Introduction

Resodyn Acoustic Mixer (RAM) technology produces effective mixing without using blades, impellers, or other hardware typically inserted into the mixing chamber. The RAM mixing process creates an acoustic field that propagates high-intensity, low-frequency sound energy into an associated vessel. The highly uniform and intense mixing experienced by materials within that vessel is unmatched by other state-of-the-art mixing technologies. At the time of this writing, demonstrated mixing capabilities of the standard ResonantAcoustics® resonator platform had been commercially utilized in mixing a broad range of challenging materials of all phases, including solids, high- and low-viscosity pastes and liquids, and gases. Available RAM mixers range in scale from laboratory benchtops to commercial production plants and have evolved to include both batch and continuous mixing applications. The ResonantAcoustics mixing technology is unique, with advanced attributes and benefits such as:

- Rapid mixing (10 to 100 times faster)
- Reproducible mixing
- Excellent distribution of all constituents within the mixed matrix
- Uniform product properties
- Minimized waste and reduced cleanup
- Improved safety while processing hazardous materials
- Higher rates of heat and mass transport, independent of Reynolds number
- Viable production of materials that were previously unmanufacturable in any other way
- Highly preferred embodiment for many patented[64] methods
- Energy-efficient performance

The device design and operation, as well as its inherent mixing process, are unique from all other systems on the market. Working principles of Resodyn technology, and images illustrating its mixing capabilities, are presented for a variety of processes in the following discussion.

15.6-II.2 ResonantAcoustics® Working Principles

The key enabling feature of ResonantAcoustics technology is a plate and spring system that oscillates up and down to impose a maximum acceleration of 100 g-per-cycle onto the mix vessel and its contents.

Figure 15.21 shows a typical plate and spring arrangement in which the ResonantAcoustics kinetic energy of springs exchanges potential energy of the plates in a sinusoidal motion at mechanical resonance.

Equation 15.1 balances kinetic energy, potential energy, and a damping function with the forcing function being controlled to initiate a resonant condition.

$$m\frac{d^2}{dt^2}x(t) + c\frac{d}{dt}x(t) + kx(t) = F\,sin(\omega t) \tag{15.1}$$

where, m = total equivalent mass

$x(t)$ = spring position as a function of time, t

c = damping force

k = spring constant

F = driving force

ω = angular frequency of the spring, or $2\pi f$, where f is the linear frequency

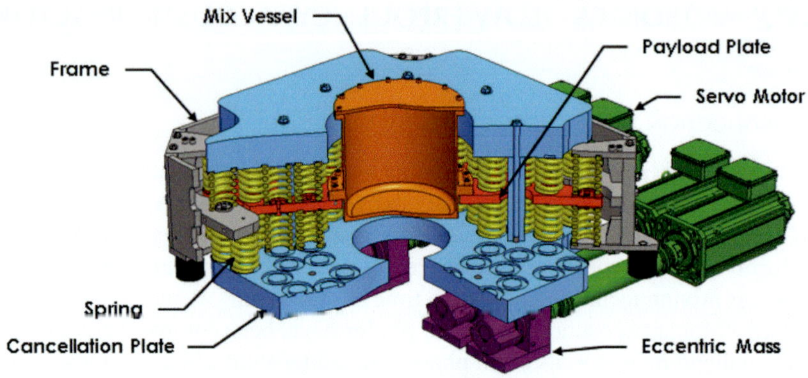

FIGURE 15.21 Rendering of plate, springs, damper, and excitation system in a mechanical-resonant system.

The plates and springs operating at resonance lose essentially no energy in each resonant cycle. When exciting resonant systems by an outside force, such as a motor, damping is essential to prevent uncontrolled system motion from overstressing and damaging mechanics and other system components. The working media being mixed in ResonantAcoustics technology becomes the damping force by modulating the system resonant excitation. Thus, virtually all energy entering the ResonantAcoustics resonator is imparted into the media, while essentially none is expended on movement of the plates and springs comprising the resonator. The total equivalent mass (dead mass + live mass) mounted to the payload plate must be ≤2.38kg for a 100 g-per-cycle operation. This is multiplied by a coupling coefficient (usually near 0.5 for paste) that depends on such conditions as vessel geometry, vacuum, headspace, etc.

ResonantAcoustics technology mounts the mixing vessel atop the resonator payload plate, then generates a frequency and displacement of the mixing-enabling ResonantAcoustics resonator. Behavior of the ResonantAcoustics resonator lies between those of an ultrasonic transducer (frequency ≤ 20 kHz; displacement ≤ 0.1 mm) and a paint shaker (frequency ≤ 2–11 Hz; linear and orbital displacement ≤ 25 mm). The forces that either of these devices induces in a vessel and its contents, however, are far below those from a ResonantAcoustics mixer.

Another key capability of ResonantAcoustics mixing is easy scalability from benchtop, to pilot, to production scales (in both batch and continuous processes), because frequency and displacement are the basis of oscillatory mixing methods. A single ResonantAcoustics mixing platform can produce the same mechanical/acoustic forces to mix all forms and combinations of materials, even though the mixing physics differ for each specific mixed media combination, described as follows.

15.6-II.3 Resodyn ResonantAcoustics® Mixers

The lab-scale embodiment of ResonantAcoustics technology is the compact and powerful LabRAM I mixer, shown in Figure 15.22. It delivers up to 100g of acceleration to developmental and repetitive batch mixes as large as 1.1 lb. (500 gram) in payload capacity. It provides intense but gentle acoustic mixing for rapid, repeatable, and ultrahigh-quality results. Besides introducing significant power to bench mixing, LabRAM I also scales easily to larger ResonantAcoustics mixers. The LabRAM I is capable of mixing/processing any combination of solids and powders (from nano-scale and larger); liquids and gases; liquids and solids (including slurries of potentially any viscosity); and materials to be coated, milled, or ground.[65]

A key feature of the LabRAM I mixer is an on-board computer operating system with an integrated touch screen. It not only provides real-time monitoring and control of key mixing parameters, but can

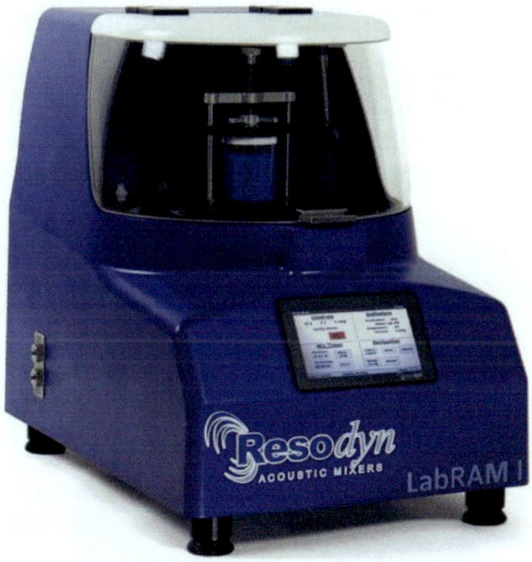

FIGURE 15.22 LabRAM I bench-scale mixer.[65]

also develop, store, and secure up to 100 batch recipes. It is programmable for optional control of the process and of such accessories as an RTD temperature probe and an inherent vacuum system (automatic or manual). Processing data can also be downloaded to an external computer for subsequent analysis. Another standard feature of the LabRAM I is a height-adjustable vessel holder that accepts a wide variety of mixing vessel types within a fully enclosed, interlocking acoustic mix chamber. The controller and mixing chamber operate safely due to their inherent multi-level security and control of user access.

The ResonantAcoustics line of mixers now includes more than 14 different models ranging in load capacity, maximum acceleration force, batch scale, and ability to operate in continuous or batch mode. Several features of these different mixer models are compared in Table 15.11. Resodyn also offers various accessories that couple to the mixers to improve processing. They include systems for temperature measurement, temperature control (heated jacket), vacuum control, milling, sieving, and spraycoating.

TABLE 15.11
Comparing Features on Various ResonantAcoustics Mixers

Resodyn ResonantAcoustic Mixer Model (—)	Max. Load Mass [kg]	Max. Accel. (g's)	Batch Scale (—)	Pairs w/ CAM* (—)
LabRAM I	0.5	100	bench	N
PharmaRAM I	0.3	100	bench	N
LabRAM II, LabRAM II H, Pharma RAM II	1.0	50	pilot	N
OmniRAM, OmniRAM Pharma, OmniRAM H	5.0	10	pilot-prod.	Y
RAM 5, RAM 5 Pharma, RAM 5 H	36.0	3	pilot-prod.	Y
RAM 55, RAM 55 Pharma, RAM 55 H	420.0	1.2	production	Y

*CAM = Continuous Acoustic Mixer

15.6-II.4 ResonantAcoustics Mixing Processes and Phenomena

(a) Liquid–Liquid Mixing: Figure 15.23 shows fingers of liquid being violently injected into the adjoining gas phase to form an extensive gas–liquid interface, thus illustrating the aggressive mixing produced in a ResonantAcoustics mixer. Such phenomena are called Faraday instabilities,[66] after 18th-century scientist Michael Faraday.

FIGURE 15.23 Faraday instabilities at the liquid–gas interface produced by ResonantAcooustics.

Faraday instabilities are non-linear waves that appear on liquid surfaces due to a high-amplitude, periodic driving force. Under the proper conditions, the flat surface of the liquid becomes unstable. RAM technology is particularly effective at inducing these instabilities, and it is the first technology known to have made practical use of this phenomenon. Instabilities can be seen as spikes, or fingers, formed both above the clear liquid surface and in the cavities below it. Vast increases in the surface area of the material interfacial boundaries facilitates and greatly accelerates the materials rates of both mixing and mass transfer.

High-level agitation (typically at 80+ g's of acceleration) of low-viscosity fluids, such as gas and liquid, transition past the Faraday instabilities and form a bubbly mixing regime that can often be transitioned into the frothy state shown in Figure 15.24. In this frothy operating regime, the definition of gas–liquid and liquid–liquid interfaces are lost, virtually eliminating all boundary layers and mass concentration gradients. An extremely large interface develops between any reactive reagents in the media. This mixing regime is particularly effective in minimizing unwanted byproducts from consecutive/competitive reactions that form in zones of poorly mixed materials.

FIGURE 15.24 The frothy gas–liquid mixing regime driven by ResonantAcoustics.

(b) Liquid–Solid Mixing: When solids are added to a liquid, pastes often form (Figure 15.25) and Faraday waves no longer exist at the higher viscosities. Mixing results from sound waves traveling through the paste as a matrix, or solid particles suspended in a continuous liquid phase. As sound waves encounter materials of various densities (liquids-to-solids), they are absorbed, reflected, and transmitted through the solid media. The traveling sound waves impose body forces on the particles, thus producing vigorous and uniform mixing.

FIGURE 15.25 ResonantAcoustics-driven mixing of high-solids-laden paste.

Mixing modes for liquid–solid systems include a wetting phase, an incorporation phase, and a paste mixing phase. These same modes have been observed in both batch and continuous operations of the various Resodyn mixer configurations. Figure 15.26 depicts this mixing sequence as seen in a continuous mixer.

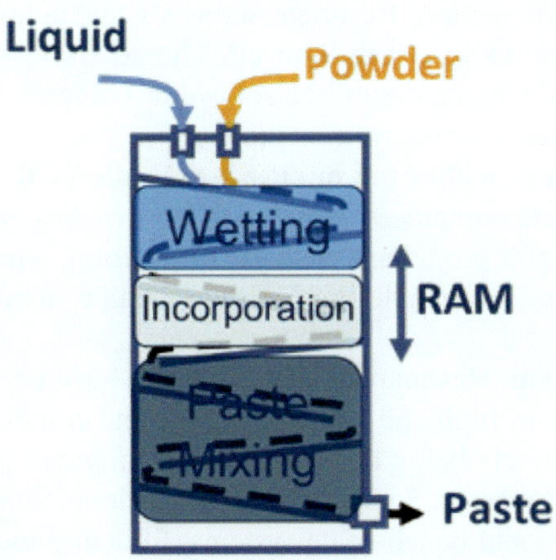

FIGURE 15.26 Mixing modes of solids and liquids in a continuous ResonantAcoustics mixer.

(c) Solid–Solid Mixing: For solid–solid materials, a ResonantAcoustics mixer creates collisions and vapor pockets that disrupt the solid–gas interface, thus enabling intense solid–solid mixing zones that are distributed and in constant motion throughout the vessel. Figure 15.27 shows a solid–solid particle mixing process containing iron oxide mixed with a white powder.

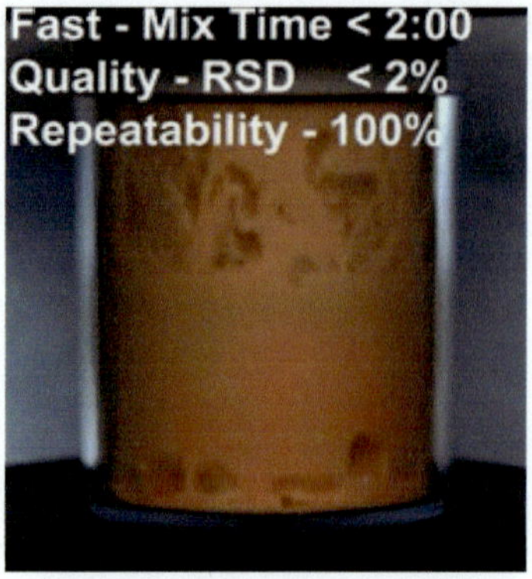

FIGURE 15.27 ResonantAcoustics mixing of solid–solid particles.

15.6-II.5 Applications

ResonantAcoustics technology has become broadly accepted and is successfully applied in more than 20 different markets/industries globally, which includes gaining global prominence as the "mixer of choice" for various applications that include mixing energetic materials (explosives, propellants and pyrotechnics). Other successful ResonantAcoustics applications include energy storage materials (batteries), pharmaceuticals, additive manufacturing materials, adhesives, technical ceramics, heavily solids-loaded materials, chemical processes, and many more.

ResonantAcoustics technology applications include its use for coating processes (liquid on solids to form solids, and solids on solids). Pharmaceuticals applications include coatings for time release and taste masking. There are energetic and materials applications that include coating of nanomaterials on micro-particles to obtain desired end-product attributes. Using the methods previously summarized in the liquid–solid section, ResonantAcoustics technology also easily mixes viscous materials ranging from <10 cP to >> 10 million cP. The solids in highly viscous materials are entrained throughout the liquid matrix, not unlike soil solids being entrained during an earthquake, which foregoes the need for conventional mixing blades.

The absence of moving parts within the mixing vessel allows ResonantAcoustics technology to mix products in their end-use containers, which can be jars or a broad range of containers like Semco tubes, artillery shells, and propellant casings. In addition, some applications can also mix materials within a specific mold to enable the highly desirable manufacturing of near-net-shape products.

Recently developed continuous ResonantAcoustics mixers have been used to manufacture products involving powders, pastes (of high and low viscosities), and combinations thereof. Applications for which the technology is actively being developed include a broad range of continuous chemical reactions and crystallization processes. Such application development currently ranges from micro-scale reactors to more conventional bench-scale systems. Pilot and industrial-scale feasibility tests have already been completed on a limited number of such processes to successfully extend the current range of scale-up feasibility.

Figure 15.28 shows a typical RAM5 continuous mixing system. Larger-scale continuous OmniRAM and RAM55 mixers are also available. Applications for which the technology is actively being developed include a broad range of continuous chemical reactions and crystallization processes. Such application development currently ranges from micro-reactors to more conventional bench-scale products. Pilot- and industrial-scale feasibility tests have already been completed on a limited number of such processes to successfully establish scale-up feasibility. The depiction of a pilot-scale continuous acoustic chemical reactor is shown in Figure 15.29.

FIGURE 15.28 Typical continuous acoustic mixer (right), with a resulting continuous-process paste (left).

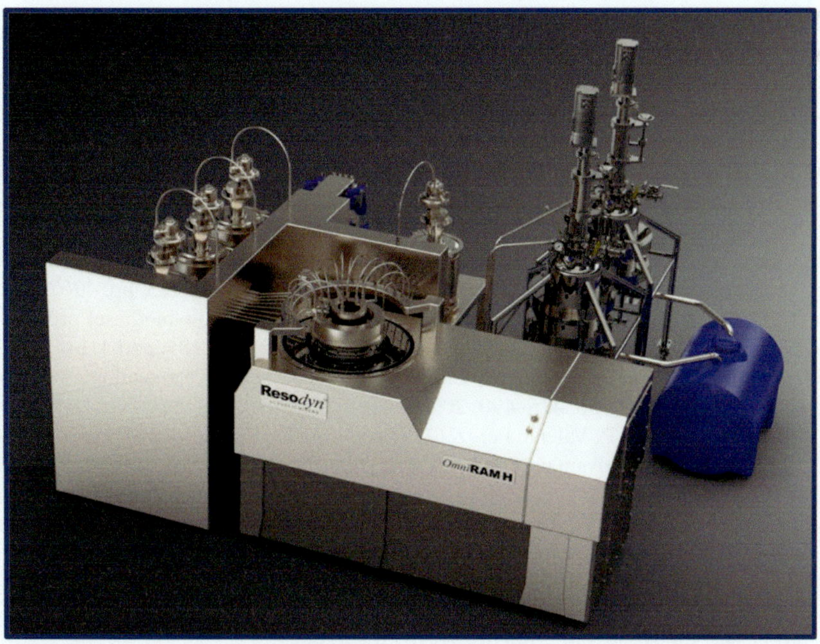

FIGURE 15.29 Depiction of a pilot-scale continuous acoustic chemical reactor system.

15.6-II.6 Design/Sizing Guidelines for R&D and Practicing Engineers

Primary operating parameters that ResonantAcoustics technology adjusts in the mixing process are acceleration (0 to 100 g), inter-vessel pressure (vacuum to several atmospheres), material temperature (controlled by jacked vessels that heat or cool the working media), and processing time. Because this mixing process is not intuitive, learning the needed combinations of process settings requires process operators to gain this specialized knowledge through experience of their own or others in operating ResonantAcoustics systems. Design-of-experiments methodologies have greatly helped to develop and optimize mix recipes.

A final word is warranted on several unique ResonantAcoustics mixing features. The first feature regards process scalability from bench, to pilot, to industrial scale, and even continuous mixing configurations. This unique capability of the ResonantAcoustics mixing process is enabled because the same acoustic field (sound energy) is imposed on, and propagated throughout, all vessels regardless of size, including continuous mixing configurations. The second feature is that required mixing power also provides information regarding the state of the in-process mixture, and is especially useful for production processes as an indicator of mix completion and mix-to-mix consistency.

15.6-II.7 Conclusions

Resodyn ResonantAcoustics mixers are a universal platform that easily scales from bench, to pilot, to plant scale, as well as from batch to continuous. This capability is unmatched by any other processing technology. Its efficient energy input makes ResonantAcoustics ideal for mixing any combination of solids, liquids, and gases to enable chemical reactions, crystallizations, coatings, milling, sieving, etc. New applications for this technology are being discovered and utilized on an ongoing basis. The technology is equally applicable to batch or continuously operating systems. While the most popular mixer model in the Resodyn suite is the RAM benchtop mixer, other available capabilities include mixing as much as 420 kg, as well as mixing on a continuous basis.

REFERENCES

1. Ni, X.-W., Fitch, A., Webster, P., *From Maximum to Most Efficient Production Using a Continuous Oscillatory Baffled Reactor*, Centre for Oscillatory Baffled Reactor Applications (COBRA), Chemical Engineering, School of Engineering and Physical Sciences, Heriot-Watt University, Edinburgh, EH14 4AS, UK. Oscillatory Baffled Reactor (OBR). https://www.sciencedirect.com/topics/engineering/oscillatory-flow-reactor
2. Phan, A. N., Harvey, A., Lavender, J., Characterization of fluid mixing in novel designs of mesoscale oscillatory baffled reactors operating at low flow rates (0.3–0.6 ml/min), *Chem. Eng. Process. Process Intensif.*, Vol. 50, issue 3, pp. 254–263, 2011.
3. Reis, N., Vicente, A. A., Teixeira, J. A., Mackley, M. R., Residence times and mixing of a novel continuous oscillatory flow screening reactor, *Chem. Eng. Sci.*, Vol. 59, issues 22–23, pp. 4967–4974, 2004.
4. Tran, D.-T., Chang, J.-S., Lee, D.-J., Recent insight into continuous-flow biodiesel production via catalytic and non-catalytic transesterification processes, *Appl. Energy*, Vol. 185, part 1, pp. 376–409, 2017.
5. Bianchi, P., Williams, J. D., Oliver Kappe, C., Oscillatory flow reactor for synthetic chemistry applications, *J. Flow Chem.*, Vol. 10, pp. 475–490, 2020.
6. Abbott, M. S. R., Harvey, A. P., Valente Perez, G., Theodorou, M. K., Biological processing in oscillatory baffled reactors: Operation, advantages and potential, *Interface Focus.*, Vol. 3, issue 1, 2013. https://doi.org/10.1098/rsfs.2012.0036
7. Munoz-Camara, J., Solano, J. P., Perez-Garcia, J., Non-dimensional analysis of experimental pressure drop and energy dissipation measurements in oscillatory baffled reactors, *Chem. Eng. Sci.*, Vol. 262, pp. 1–12, 2022.
8. Jealous, A. C., Johnson, H. F., Power requirement for pulse generation in pulse columns, *Ind. Eng. Chem.*, Vol. 47, pp. 1159–1166, 1955. http://doi.org/10.1021/ie50546a021
9. Baird, M. H. I., Stonestreet, P. J., Energy dissipation in oscillatory flow within a baffled tube, *Chem. Eng. Res. Des.*, 73, pp. 503–511, 1995.

10. Avila, M., Fletcher, D. F., Poux, M., Xuereb, C., Aubin, J., Predicting power consumption in continuous oscillatory baffled reactors, *Chem. Eng. Sci.*, Vol. 212, issue 2, 2020.
11. Mackley, M. R., Stonestreet, P., Heat transfer and associated energy dissipation for oscillatory flowing baffled tubes, *Chem. Eng.*, Vol. 50, pp. 2211–2224, 1995.
12. Baird, M. H. I., Garstang, J. H., Power consumption and gas hold-up in a pulsed column, *Chem. Eng. Sci.*, Vol. 22, pp. 1663–1673, 1967.
13. Baird, M. H. I., Rao, N. R., Power dissipation and flow patterns in reciprocating baffle-plate columns, *Can. J. Chem. Eng.*, Vol. 73, pp. 417–425, 1995.
14. McDonough, J. R., Phan, A. R., Harvey, A. P., Rapid process development using oscillatory baffled meso-reactors—A state-of-the-art review, *Chem. Eng. J.*, Vol. 265, issue 1, pp. 110–121, 2015.
15. Method and apparatus for gas induced mixing and blending, Parks, R. E., US Pat. 4,595,296, June 17, 1986.
16. Pulsair Systems Inc. https://www.pulsair.com/
17. AEA Fluidic Pulse Jet Mixer, *Innovative Technology Summary Report: DOE/EM-0447*, August 1999. https://digital.library.unt.edu/ark:/67531/metadc716945/m1/3/
18. Flexible diffuser assembly for aeration applications, Downs, E. W., US Pat. 5,330,688, July 19, 1994.
19. Aeration Products for energy efficient biological treatment. https://www.scribd.com/document/400585884/Aeration-Products-for-Energy-Efficient-Biological-Treatment
20. Sanitaire membrane diffuser. https://www.xylem.com/en-us/brands/sanitaire/
21. Wastewater technology factsheet: Fine bubble aeration, EPA 832-F-99-065, September, 1999.
22. https://www.xylem.com/en-us/brands/sanitaire/sanitaire-products/
23. Fine bubble diffused aeration system design features. https://www.webpages.uidaho.edu/ce431/Handouts/Sanitaire%20Fine%20Bubble%20Info.pdf, https://www.webpages.uidaho.edu/ce431/Handouts/Sanitaire%20Diffused%20Air%20Design%20Guide.pdf
24. https://www.xylem.com/siteassets/brand/sanitaire/resources/brochure/sanitaire-fine-bubble-aeration-grids-datasheet-a4-low-en.pdf
25. Pilo, C. W., Dahlbeck, S. W., Apparatus for intimate contacting of two fluid media having different specific weight, US Pat. 2,941,872, 1960.
26. Rao, D. P. The story of "HIGEE". *Indian Chem. Eng.*, Vol. 57, pp. 282–299, 2015.
27. Ramshaw, C., Mallinson, R. H., Mass transfer process, US Pat. 4,283,255, 1981.
28. Trent, D. L., Chemical processing in high-gravity fields, in *Re-engineering the Chemical Processing Plant: Process Intensification*, Stankiewicz, A., Moulijn, J. A., eds., Marcel Dekker, New York, pp. 46–80, 2004.
29. Trent, D. L., Tirtowidjojo, D., Intensifying the process, *Chem. Eng. (London)*, Vol. 742, pp. 30–31, 2003.
30. Rao, D. P., Bhowal, A., Goswami, P. S., Process intensification in rotating packed beds (HIGEE): An appraisal, *Ind. Eng. Chem. Res.*, Vol. 43, pp. 1150–1162, 2004.
31. Neumann, K., Konard, G., Gröβ, K., Qummar, H., Wenzel, D., Görak, A., Skiborowski, M., A guide on the industrial application of rotating packed beds, *Chem. Eng. Res. Des.*, Vol. 134, pp. 443–462, 2018.
32. Wang, Z., Yang, T., Liu, Z., Wang, S., Gao, Y., Wu, M., Mass transfer in a rotating packed bed: A critical review, *Chem. Eng. Process. Process Intensif.*, Vol. 139, pp. 78–94, 2019.
33. Xu, W., Geng, C., Hou, Z., Liu, S., Wang, L., Shao, L., Xu, W., Zhao, C., High-gravity technology-enhanced activated sludge process for municipal wastewater treatment, *Chem. Eng. Sci.*, Vol. 252, April 2022.
34. Vot, H., Mersmann, A. B., Goetz, F., Overproduction of Lipase with *Staphylococcus Carnosus* (pLipPS1) under modified gravity in a centrifugal field bioreactor, *Chem. Eng. Tech.*, Vol. 12, pp. 364–373, 1989.
35. https://www.andritz.com/products-en/group/separation/filter-centrifuges/rotating-packed-bed-centrifuge-rpb
36. Burns, J. R., Ramshaw, C., Process intensification: Visual study of liquid maldistribution in rotating packed beds, *Chem. Eng. Sci.*, Vol. 51, p. 1347, 1996.
37. Guo, F., Zheng, C., Guo, K., Feng, Y., Gardner, N. C., Hydrodynamic and Mass transfer in cross flow rotating packed bed, *Chem. Eng. Sci.*, Vol. 52, p. 3853, 1997.
38. Tung, H. -H., Mah, R. S. H., Modeling liquid mass transfer in higee separation process, *Chem. Eng. J.*, Vol. 156, pp. 588–593, 2010.
39. Munjal, S., Dudokovic, M. P., Ramachandran, P., Mass-transfer in rotating packed beds—I. Development of gas-liquid and liquid-solid mass-transfer correlations, *Chem. Eng. Sci.*, Vol. 44, pp. 2245–2256, 1989.

40. Sun, B.-C., Wang, X.-M., Chen, J.-M., Chu, G.-W., Chen, J.-F., Shao, L., Simultaneous absorption of CO_2 and NH_3 into water in a rotating packed bed, *Ind. Eng. Chem. Res.*, Vol. 48, issue 24, pp. 11175–11180, 2009.
41. Luo, Y., Chu, G.-W., Zao, H.-K., Wang, F., Xiang, Y., Shao, L., Chen, J.-F., Mass transfer studies in rotating packed bed with novel rotors: Chemisorption of CO_2, *Ind. Eng. Chem. Res.*, Vol. 51, pp. 9164–9172, 2012.
42. Sandilya, P., Rao, D. P., Sharma, A., Biswas, G., Gas-Phase mass transfer in a centrifugal Contactor, *Ind. Eng. Chem. Res.*, Vol. 40, pp. 384–392, 2001.
43. Mukherjee, R., Deb, D., Sandilya, P., Rao, D. P., Gas-phase mass transfer in a centrifugal gas-liquid contactor with a stack of flexible disks as packing, in *Process by Centrifugation*, Regel, L. L., Wilcox, W. R., eds., Kluwer Academic Publishers, Dordrecht, The Netherlands, p. 51, 2001.
44. Kelleher, T., Fair, J. R., Distillation studies on a high-gravity gas liquid contactor, *Ind. Eng. Chem. Res.*, Vol. 35, p. 4646, 1996.
45. Kumar, M. P., Rao, D. P., Studies on a high-gravity gas-liquid contactor, *Ind. Eng. Chem. Res.*, Vol. 31, p. 574, 1992.
46. Li, Y., Lu, Y., Wang, G., Nie, Y., Ying, H., Ji, J., Liu, X., Liquid entrainment and flooding in a rotating zigzag bed., *Ind. Eng. Chem. Res.*, Vol. 54, issue 9, pp. 2554–2563, 2015.
47. Lockett, M. J., Victor, R. A., Billingham, J. F., Structured packing flooding: Its measurement and prediction, IChemE, Symposium Series No. 152, pp. 400–408, 2006.
48. Lockett, M. J., Flooding of rotating structured packing and its applications to conventional packed columns, *Tran. Inst. Chem. Eng.*, Vol. 73, p. 379, 1995.
49. Rajan, S., Kumar, M., Ansari, M. J., Rao, D. P., Kaistha, N., Limiting gas liquid flows and mass transfer in a novel rotating packed bed (HiGee), *Ind. Eng. Chem. Res.*, Vol. 50, pp. 986–997, 2011.
50. Gudena, K., Rangaiah, G. P., Lakshminarayanan, S., Optimal design of a rotating packed bed for VOC stripping from contaminated groundwater, *Ind. Eng. Chem. Res.*, Vol. 51, pp. 835–847, 2011.
51. Ghadyanlou, F., Azari, A., Vatani, A., A review of modeling rotating packed beds and improving their parameters: Gas–liquid contact, *Sustainability*, Vol. 13, issue 14, p. 8046, 2021.
52. Young, F. R., *Cavitation*, Imperial College Press, London, p. 418, 1999.
53. Jyoti, K. K., Pandit, A. B., Water disinfection by acoustic and hydrodynamic cavitation, *Biochem. Eng. J.*, Vol. 7, pp. 201–212, 2001.
54. Gogate, P. R., Cavitation: An auxiliary technique in wastewater treatment schemes, *Adv. Environ. Res.*, Vol. 6, pp. 335–358, 2002.
55. Sivakumar, M., Pandit, A. B., Wastewater treatment: A novel energy efficient hydrodynamic cavitational technique, *Ultrason. Sonochem.*, Vol. 9, pp. 123–131, 2002.
56. Jyoti K. K., Pandit A. B., Ozone and cavitation for waster disinfection, *Biochem. Eng. J.*, Vol. 18, issue 1, pp. 9–19, 2004.
57. Leighton, T. G., The principles of cavitation, in *Ultrasound in Food Processing*, Povey, M. J. W., Mason, T. J., eds., Blackie Academic & Professional, London, pp. 151–182, 1998.
58. Mason, T. J., Lorimer, J. P., *Applied Sonochemistry: The Uses of Power Ultrasound in Chemistry and Processing*, Wiley-VCH Verlag GmbH, Weinheim, Germany, p. 303, 2002.
59. Majewski, W. A., Khair, M. K., *Diesel Emissions and Their Control*, United States, SAE International, ISBN 978-0-7680-0674-2584, 2006.
60. Gigas, B., Post, T., *$k_L a$ Measurement Method Presented at Mixing XV (1995) | Banff Conference Center*, Banff, AB, Canada, North American Mixing Forum (NAMF), 1998.
61. Hart, P. W., Gilboe, M., Adusei, G., Mancosky, D., Armstead, D. A., Medium consistency oxygen delignification performed with a controlled cavitation reactor, *Tappi J.*, Vol. 4, pp. 26–31, 2005.
62. Kazem, B., Armstead, D. A., Mancosky, D. G., Lien, S. J., Verrill, C. J., *Heating Black Liquor by Partial Oxidation with Controlled Cavitation as a Means to Reduce Evaporator Fouling*, TAPPI Pulping Conference, Chicago, IL, 2003.
63. Milly, P. J., Toledo, R. T., Chen, J., Kazem, B., Hydrodynamic cavitation to improve bulk fluid to surface mass transfer in a non-immersed ultraviolet system for minimal processing of opaque and transparent fluid foods, *J. Food Sci.*, Vol. 72, issue 9, pp. 407–413, 2007.
64. Howe, H. W., Warriner, J. J., Cook, A. M., Coghill, S. L, Farrar, L. C., Resodyn corporation, US Pat. 7,188,993 B1, March 2007.
65. https://resodynmixers.com/product/spray-system-bench/
66. Faraday, M., On the forms and states of fluids on vibrating elastic surfaces, *Phil. Trans. R. Soc. London*, Vol. 52, p. 31S340, 1831.

16 Numerical Methods in Fluid Mixing

Zhao Yu

16.1 INTRODUCTION

Numerical simulation approaches for solving fluid motion problems, generally known as computational fluid dynamics (CFD), have found wide applications in industrial mixing processes. As a first-principle-based model, they solve the governing flow equations using suitable computer algorithms, and can provide comprehensive and detailed flow field information that is often expensive and/or impractical to obtain experimentally. With the increased access of computational power and the improved functionality and user-friendliness of software, CFD has become a popular tool among industrial mixing practitioners. This chapter aims to briefly introduce the CFD workflow and the numerical techniques that are typically employed in fluid mixing simulations. Case studies will be provided to highlight how CFD models are applied in various scenarios that are commonly seen in industrial mixing processes.

16.2 GENERAL WORKFLOW

The process of numerically simulating a flow problem can be divided into three stages, corresponding to the three computational tools shown in Figure 16.1. These three components are usually packaged into independent software, and their communication is enabled by importing or exporting data files in compatible formats. However, many modern commercial CFD software packages today bundle all three components into a single integrated platform.

The first stage is called pre-processing. A geometric model is first constructed to represent the flow domain. The geometry can be either built from scratch or imported from other computer aided design (CAD) software. Clean-up of the geometry is sometimes required to correct the errors or remove the unnecessary features and make it more suitable for flow modeling. Then the flow domain is divided into a large number of computational cells, called the mesh. The flow field information will be computed and stored in these discrete mesh cells. There are two types of approaches to mesh a domain. The structured mesh often uses quadrilateral (in 2D) or hexahedra (in 3D) cells and can be logically mapped to a Cartesian grid. On the other hand, the unstructured mesh can utilize more varieties of element shapes, but the relative position of the elements cannot be easily identified by indexing. The structured mesh has the advantages of simple data structure, efficient algorithms, and

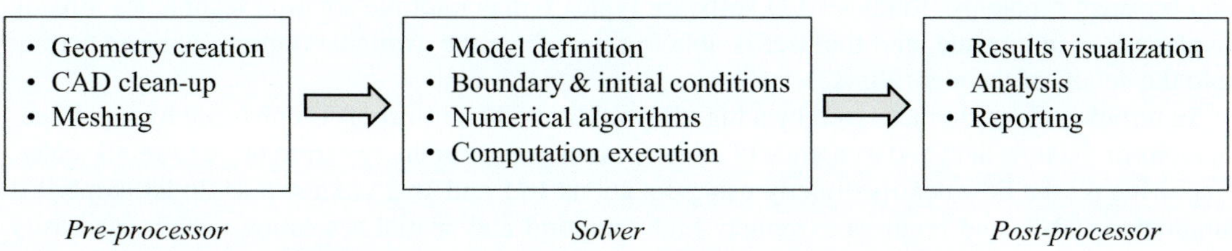

FIGURE 16.1 Typical CFD workflow.

DOI: 10.1201/9781003606390-16

straightforward code implementation. In contrast, the unstructured mesh is more flexible and can fit even complex domain boundaries. However, the algorithms using unstructured mesh are more complicated. In practice, since industrial problems usually involve complex geometries, pre-processing often turns out to be the most time-consuming and labor-intensive of the three stages.

The second stage is solving the flow problem. The physical models together with the boundary conditions are first determined, followed by specification of the solution techniques, including the numerical algorithms, discretization schemes, and model parameters. Then the algorithms execute to obtain a numerical solution in each discrete cell (volume) in the mesh. Today's typical CFD codes allow users to define and solve a flow problem without knowing the details of the theory or implementation of the numerical techniques. However, general knowledge of the numerical methods helps users understand the principles of the code, and thus, better choose appropriate models and solution techniques for the given application.

The third stage is the post-processing, where the results are analyzed and interpreted. Visualization is often used to present the simulation results in graphical formats that can be more easily understood. Several types of graphs are typically used in flow and mixing problems. For example, vector plots and streamlines are often used to show the velocity field. The degree of mixing can often be analyzed from the homogeneity of concentration or temperature field through the contour plots. Another way to assess the mixing is by tracking the trajectory and spatial distribution of discrete tracer particles. For transient simulations, a series of snapshots of the flow field can be taken at different time steps and linked in an animation to visualize the dynamic mixing process. Besides visualization, the value of primitive variables and various fluxes can also be reported in alphanumerical form. Such quantities can be either probed at specific spatial locations, or averaged, or integrated over a particular surface or volume in the computational domain.

16.3 CFD MODEL THEORIES

The governing equations for fluid motion can be written in the following general vector form:

$$\frac{\partial \rho}{\partial t} + \nabla \cdot (\rho \boldsymbol{U}) = 0 \tag{16.1}$$

$$\frac{\partial \rho \boldsymbol{U}}{\partial t} + \nabla \cdot (\rho \boldsymbol{U}\boldsymbol{U}) = \nabla \cdot (-p\boldsymbol{I} + \boldsymbol{\tau}) + \boldsymbol{g} \tag{16.2}$$

The first equation is known as the continuity equation and describes the transport of mass. The second equation is known as the Navier-Stokes (N-S) equation and describes the transport of momentum. For an incompressible flow, the density ρ is a constant, and the variables to be solved from the equations are the velocity field **U** and pressure field p. To form a closed set of equations, the constitutive relation between the stress tensor τ and the stain tensor, which is derived from the gradient of the velocity field, must be provided based on the rheology of the fluid. Body forces, such as gravity, are included in the momentum equation as the term g. The governing equations can then be solved when proper boundary conditions and initial conditions are provided. A number of numerical algorithms have been developed and are now widely used for solving these equations for both steady and transient problems. Today's CFD software typically has multiple solution techniques already implemented in its code, and the user is able to select from the available options, without getting into the details of the algorithms.

In turbulent flow, characterized by a high Reynolds number, transport is enhanced by rapid fluctuations of the flow field and a cascade of eddies, which transport energy from large to small scales. Capturing all the flow details directly using equations 16.1 and 16.2 is known as direct numerical simulation (DNS) and requires extremely high temporal and spatial resolution, which can easily become prohibitively expensive in terms of computation power. Thus, DNS is often only feasible for small-scale research problems and not a practical approach for typical industrial applications.

Instead, the Reynolds averaged Navier-Stokes (RANS) approach is often adopted for practical problems by decomposing the velocity into an equilibrium component and a fluctuating component. The resulting RANS equation is similar to the original N-S equation but with an additional Reynolds stress term that accounts for the macroscopic effect of the turbulence. Moreover, a turbulence model needs to be provided to compute the Reynolds stress.

The most widely used model is the k-ε model, which introduces two additional equations describing the transport of the turbulent kinetic energy k and the turbulent energy dissipation rate ε. Besides the standard k-ε model, modified versions such as the RNG k-ε model and realizable k-ε model have been developed to improve the predictions for swirling and separation flows, and for flows with round jets. To better capture the anisotropic turbulence, a more advanced Reynolds stress model (RSM) can be used to compute the individual components of the Reynolds stress without using the isotropic assumption. Such an approach gives rise to more accurate results for swirling flows such as those in unbaffled tanks. However, since more equations must be solved for the stress components, the RSM requires more computational resources. Large eddy simulation (LES) has also become a popular technique to account for turbulence, particularly in transient flows. It resolves the large-scale flow structures (large eddies) explicitly, and for eddies that are smaller than a typical grid size, uses sub-grid scale models to account for their effect. LES requires much finer grids compared to the RANS approach, and is therefore more computationally expensive, but provides more resolved flow structures in applications such as jets and stirred vessels.

A fluid mixing problem typically involves mass transfer or heat transfer along with the fluid motion. The mass transfer of *n* species can be expressed by transport equations for the mass fractions of *n-1* species, with the additional constraint that the sum of all *n* species fractions should add up to 1. In equation 16.3, the second term on the left-hand side describes the convective transport resulting from the flow, while the first term on the right-hand side describes the diffusive transport, which can include both the molecular and turbulent diffusion mechanisms. In addition, the last term in the species transfer equation describes the rate of reaction for a reactive flow with finite reaction rate.

$$\frac{\partial \rho Y_i}{\partial t} + \nabla \cdot \left(\rho \boldsymbol{U} Y_i\right) = \nabla \cdot \left(\rho D_i \nabla Y_i\right) - \gamma_i \tag{16.3}$$

Heat transfer can be expressed by an equation for the conservation of energy. For example, equation 16.4 expresses the conservation of internal energy *e*. It has a similar form to that of the mass transport equation (16.3), including the second term on the left-hand side and the first term on the right-hand side describing the convective and diffusive transfer mechanisms, respectively. The second, third, and fourth terms on the right-hand side of equation 16.4 describe the work done by pressure, heat dissipation due to viscosity, and heat sources such as thermal radiation, respectively. Other forms of the energy conservation equation express the energy in the form of total energy *E* or enthalpy *h*, instead of internal energy *e*, and the other terms of the equation would need to be adjusted accordingly.

$$\frac{\partial \rho e}{\partial t} + \nabla \cdot \left(\rho \boldsymbol{U} e\right) = -\nabla \cdot \mathbf{J}_q - p \nabla \cdot \boldsymbol{U} + \varphi + J_e \tag{16.4}$$

Many industrial mixing applications involve multiple phases, as characterized by their different dynamic responses to driving forces.[1] Typical examples include bubble columns where dispersed gas bubbles move through a liquid under buoyancy force, and sedimentation of solid particles in a gas or liquid under gravity. The former example involves two immiscible fluid phases, while the latter has a solid phase dispersed in a continuous fluid. To capture the different flow behavior of the individual phases, each phase needs to be described by its own mass and momentum equations. Two multiphase flow models are widely used for industrial mixing problems, namely the Eulerian-Lagrangian (E-L) method and the Eulerian-Eulerian (E-E) method. Both methods treat the continuous phase in a way similar to the single-phase flow as given in equation 16.2, but with additional terms to account for interactions with the dispersed phase. On the other hand, the E-L and E-E methods have different models for the dispersed phase. In the E-L method, the motion of the dispersed phase is tracked along the trajectory of each particle (or bubble, droplet, etc.) or particle group in the Lagrangian reference frame. This is an intuitive approach, and physical laws governing particle motion are readily applicable. Tracking a large number of trajectories is computationally expensive, and therefore this method is often limited to dilute industrial systems. In contrast, the E-E method describes the dispersed phase as a pseudo-continuum that shares the fixed space domain with the continuous phase. The E-E approach is the practical method of choice for moderately dilute to dense systems. However, due to the continuum assumption, the E-E model may not be appropriate for very dilute systems. In addition, proper constitutive relations for stresses and pressure as well as transport properties such as viscosity and diffusivity need to be provided for the pseudo-continuum dispersed phase. Compared to a single-phase model, both E-L and E-E methods need to take into account the effect of interactions between phases. For the momentum transport, drag force is typically the dominant mechanism, while other forces such as virtual mass force and lift force may become important depending on phase properties and flow characteristics. In addition, turbulence of the continuous phase may be either enhanced or suppressed by introducing the dispersed phase, and turbulent dispersion can be an important mechanism of mass transfer of the dispersed phase.

Besides the dispersed multi-phase flow, which describes the macroscopic or average behavior of a large number of particles, another type of multi-phase flow model focuses on direct resolution of the interface between two phases, which can occur in the case of a free surface in stratified flows, or surfaces in dispersed particles, bubbles, or droplets. The high computational cost to fully resolve the interface limits the use of such methods to small systems that have a limited number of particles. Examples of the numerical techniques in this category include the volume of fluid (VOF) method, front-tracking method, and level set method.

16.4 A NOVEL APPROACH: LATTICE BOLTZMANN METHOD

The lattice Boltzmann method (LBM) is a special numerical technique in flow simulation. It describes the macroscopic flow by capturing the mesoscopic motion of fluid molecules as a result of their propagation and collision. Evolution of the particle distribution function $f(x,v,t)$, which represents the probability density of the molecules (also referred to as particles) with a certain velocity v at spatial coordinates x and time t, is governed by the well-known Boltzmann equation:

$$\frac{\partial f}{\partial t} + \boldsymbol{v} \cdot \nabla f + \frac{\boldsymbol{F}}{m} \cdot \nabla_v f = Q \tag{16.5}$$

The three terms on the left-hand side of equation 16.5 represent the unsteady, propagation, and external force terms, while the right-hand side represents the particle collision. Macroscopic hydrodynamic variables including the fluid density, momentum, and energy can be directly obtained from the particle distribution function. In particular, the Navier-Stokes equation, which governs macroscopic flow behavior, can be derived systematically from the Boltzmann equation.[2] Thus, the Boltzmann equation is consistent with fluid mechanics theories on the macroscopic level.

In LBM, the probability distribution function $f(x,v,t)$ is discretized by using a finite set of velocities $\boldsymbol{c}_i$, which are defined based on a specific computation grid (also called lattice) for spatial discretization. For example, the most commonly used lattice structure in 2D problems is the D2Q9. It has Cartesian grids with uniform spacing Δx in both coordinates, and nine velocities, including a zero velocity and eight others pointing to neighboring grid points. Similarly, the D3Q19 lattice has 19 velocities, with one zero velocity and 18 others pointing to a subset of neighboring grids. Both the D2Q9 and D3Q19 lattices are illustrated in Figure 16.2.

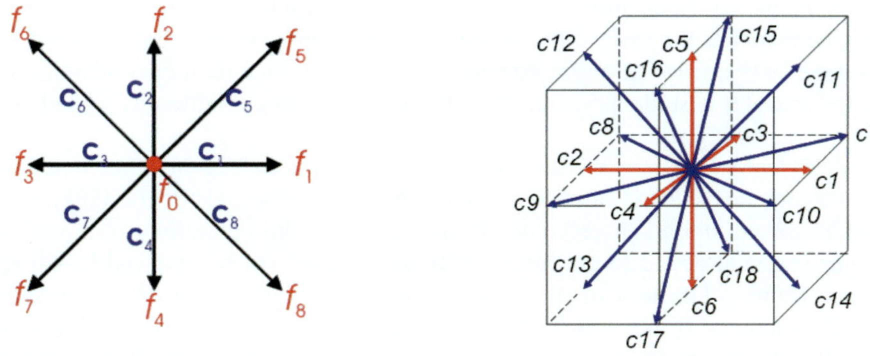

FIGURE 16.2 D2Q9 (left) and D3Q19 (right) lattice structures in lattice Boltzmann method.

The Boltzmann equation 16.6 can be discretized into the lattice Boltzmann equation (LBE), i.e., equation 16.7:

$$f_1(\mathbf{x}+\mathbf{c}_i\Delta t, t+\Delta t) - f_i(\mathbf{x},t) = -\frac{1}{\tau}\left(f_i(\mathbf{x},t) - f_i^{eq}(\rho,\mathbf{U})\right) \quad (16.6)$$

The discretized distribution $f_i(\mathbf{x},t)$ is the distribution associated with the discretized velocity $\mathbf{c}_i$. The right-hand side describes the collision as a relaxation process to an equilibrium distribution f_{eqi}.

The LBE can be solved by using a simple propagation-collision algorithm, which only involves the local lattice for time integration. Macroscopic hydrodynamic variables such as density ρ, momentum $\rho\mathbf{U}$, and pressure p are then obtained from the particle distribution functions:

$$\rho(\mathbf{x},t) = \sum_i f_i(\mathbf{x},t), \quad (16.7)$$

$$\rho(\mathbf{x},t)\mathbf{U}(\mathbf{x},t) = \sum_i \mathbf{c}_i f_i(\mathbf{x},t), \quad (16.8)$$

$$p = \rho c_s^2 \quad (16.9)$$

Mathematically, the LBM is a pseudo-compressible algorithm that is limited to low Mach numbers. Small time steps are usually needed to stabilize the simulation. It has an efficient algorithm that is easy to implement and parallelize, compared to the N-S equation-based CFD methods. By using techniques such as immersed boundary method, complicated boundaries can be handled in LBM while maintaining the simple lattice structure. This greatly simplifies the process of mesh generation and provides another key advantage of the LBM.

16.5 APPLICATIONS

16.5.1 Mixing in Stirred Tanks

Agitation in stirred tank mixers is usually driven by the rotation of an impeller, which is mounted on a shaft. In addition, there can be stationary internals, including baffles, dip tubes, spargers, coils, etc. In CFD simulations, since the computational domain needs to include both the moving impeller and other stationary parts of the tank, special numerical techniques are needed to treat the interaction between the two. Some typical techniques include:

- *Rotational Reference Frame Method.* This method is often applied to unbaffled tanks whose impeller shaft is aligned with the symmetric axis of the tank. When viewed from a moving reference frame attached to the impeller, the fluid domain is fixed and the only moving boundary is the tank wall rotating in the direction opposite to the desired impeller rotation. In this way, the entire flow field is solved within a rotational reference frame while the impeller is stationary, and the applicable boundary condition is a fixed angular velocity at the tank wall.

- *Multiple Reference Frame Method.* When the tank has baffles or other internals, or when the impeller shaft is not aligned with the symmetric axis of the tank, the flow cannot be solved in a single reference frame, since the fluid domain also moves in the reference frame attached to the impeller. However, this can be resolved by using two reference frames. The part of the fluid domain enclosing the impeller is solved in a rotational reference frame attached to the impeller, while the rest of the fluid domain is solved in the stationary reference frame. The computation is carried out using a stationary mesh, which means the impeller domain is essentially "frozen" in its reference frame. A proper transformation of momentum needs to be imposed at the interface between the two domains.
- *Sliding Mesh Method.* This method also uses multiple reference frames to represent the interaction between the impeller and the stationary parts. However, the mesh of the rotational reference frame is dynamically updated during the simulation to reflect the actual position of the impeller. The sliding mesh model yields better temporal accuracy than the multiple reference frame method and has its advantage when the interaction between the impeller and stationary part is expected to be strong, such as in the case of small clearance between the impeller tips and the baffles. However, on-the-fly mesh updates may significantly add to required computational resource and time.
- *Immersed Boundary Method.* In this method, the entire tank volume is represented in a single stationary reference frame. A separate surface mesh is embedded within the tank volume. It represents the impeller surface, and its location is updated during the simulation to reflect the impeller rotation. A special condition is imposed on the fluid cells near the impeller surface so that their velocity can be adjusted to correctly reflect the effect of the impeller.

Figure 16.3 shows an example of a lab-scale mixing vessel that was modeled with the single rotational reference frame approach.[3] The unbaffled vessel is axisymmetric about the vertical axis, and therefore the entire flow field is solved using a reference frame that rotates at the same speed as the impeller. A pressure outlet condition is imposed at the top of the vessel, and a no-slip condition is employed on the impeller and vessel walls. Because the liquid surface is close to the top of the rotating impeller, it is expected to have strong interactions with the impeller that may result in deformation of the liquid surface. Thus, the volume of fluid (VOF) method is used to model the interfacial motion between the liquid phase and the head space air. The computational domain is meshed with unstructured tetrahedral cells. The typical mesh size is about 1/40th of the impeller diameter ($D = 40\Delta x$).

In the experiment, a blue dye is initially applied at the liquid surface and is drawn down into the vessel by the impeller, which eventually reaches homogeneity within the liquid volume. In the CFD simulation, the mixing performance is visualized using the discrete particle model (DPM) to track the spatial distribution of about 20,000 tracer particles. These tracer particles are released at the beginning of the simulation from a horizontal cross-section at the top of the impeller. The transient distribution of the tracer particles is consistent with the movement of the dye in the experiment, indicating the validity of the modeling approach. Once the model is validated, it can be used to optimize the mixer design relative to impeller-to-vessel diameter ratio, impeller submergence, and off-bottom clearance.

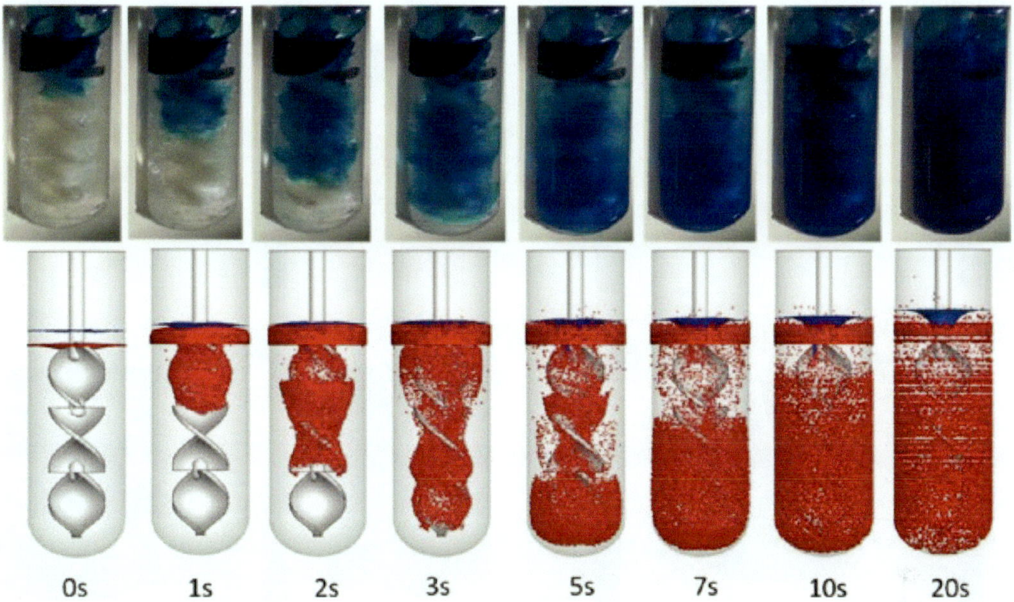

FIGURE 16.3 Dye visualization experiments and CFD simulations of a lab-scale mixing vessel for high-viscosity blending (30,000cP).

Figure 16.4 shows an example of the CFD simulation performed with the multiple reference frame approach.[4] In this case, the A310 impeller is installed at an angle to the vertical axis of an unbaffled tank. In the multi-reference frame model, a small cylindrical zone enclosing the impeller was defined as the rotational reference frame, while the rest of the tank was defined as the stationary zone. The transient CFD model was used to investigate the potential stratification between two miscible fluids with different properties in the stirred tank. Initially, the bottom 80% of the tank was filled with a high-density, high-viscosity liquid, while the top layer liquid was of a lower density and viscosity. Due to the high viscosity of the main fluid, a laminar model was selected. The mass transfer model was used to capture the spatial and temporal variation of concentrations. A mesh-independence test was performed, and the mesh with 0.75 and 1.2 million cells appeared to produce consistent results.

The stratified mixing case could be observed in CFD results in Figure 16.4 for the lower mixing speed (130 RPM), where the flow induced by the impeller was unable to effectively overcome buoyancy effects due to the density difference. A flat horizontal interface was maintained, and it divided the tank into two segregated regions, with the flow being mostly confined to the bottom portion of the tank. At a higher mixing speed (180 RPM), the impeller was able to draw down the top liquid and break the interface. The flow was able to spread to the entire tank volume in a much shorter time, and homogeneity of concentration was observed within 15 minutes. The CFD predictions were consistent with the experimentally observed mixing times shown in Figure 16.5. In the experiment, product concentration was measured in samples taken from a sampling port in the lower portion of the tank at consecutive time points. In the CFD simulation, the concentration was continuously monitored at the same location in the computational domain. The good agreement between the model and the experiment confirmed that the CFD model could be used as a reliable predictive tool to understand mixing in the stirred tank.

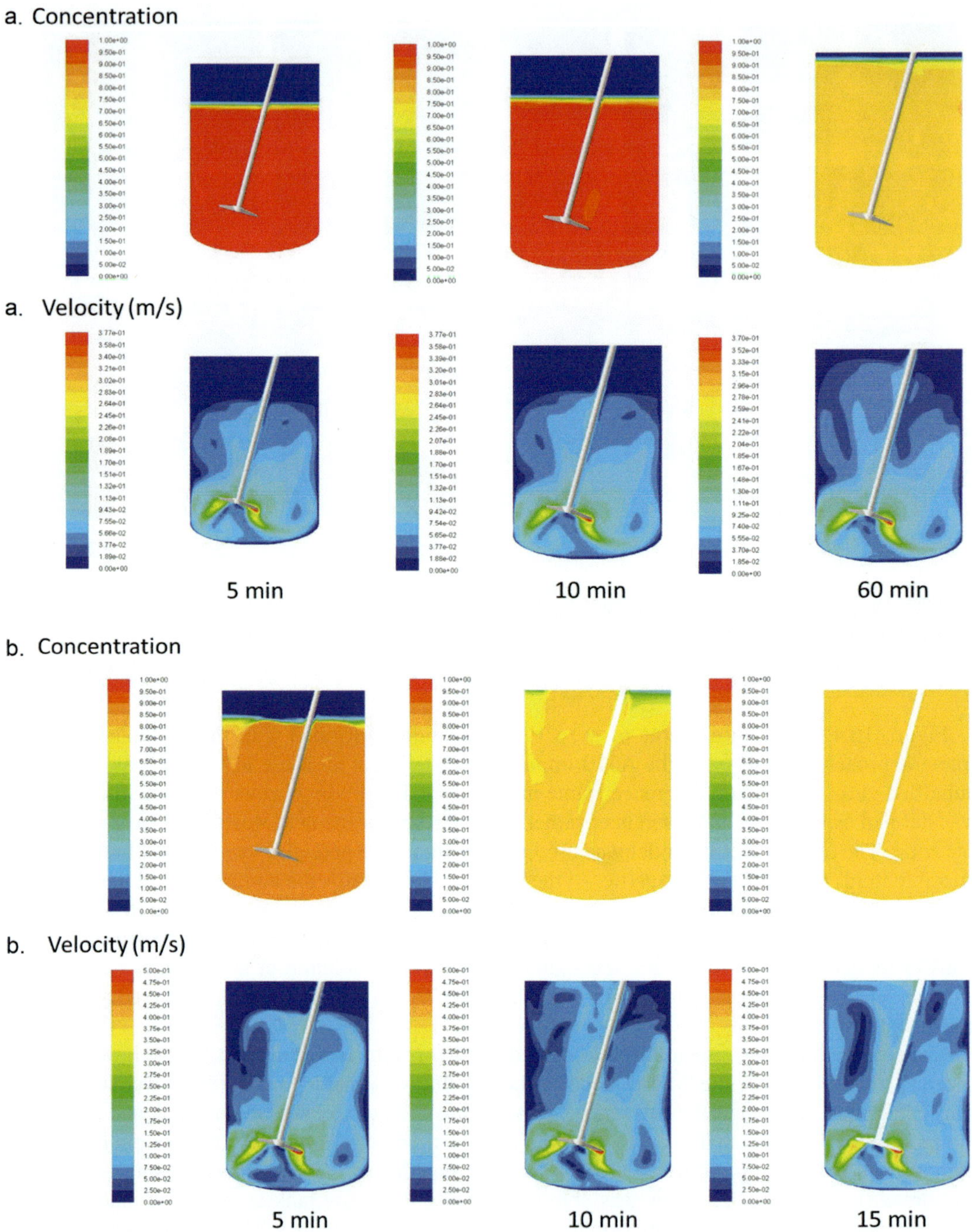

FIGURE 16.4 CFD prediction of concentration and velocity contours at different time points in drug product batches: (a) mixing at 130 rpm, (b) mixing at 180 rpm.

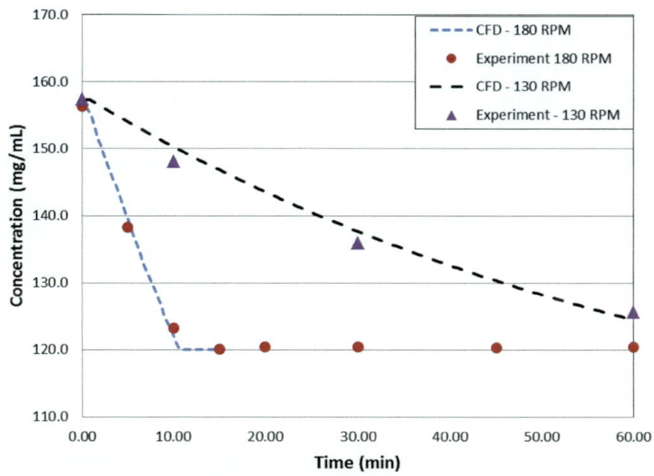

FIGURE 16.5 Change of product concentration at the sampling location with time at two mixing speeds.

16.5.2 Jet Mixing

Jet mixers are often used in process industries for storage and blending applications, especially in large tanks where mechanical agitation is less desirable due to mechanical, capital, maintenance, and/or operational issues. The location and orientation of the jet nozzle(s) usually significantly affect mixing characteristics in such systems, whose effects have been investigated by CFD simulations.[5]

One of the challenges in modeling a large jet mixing tank is the scale difference between the jet nozzles compared to the entire tank. In one example of such systems, the nozzles have an internal diameter of 2.41 cm, while the horizontal tank is 3.66 m in diameter and 10.67 m in length. The jet velocity exiting the nozzle is 8.5 m/s and decays to 0.2 m/s after traveling 5.74 m to the end of its free path.[6] Careful consideration of the mesh is needed in order to resolve the velocity gradients near the jet and to properly model the turbulence in different regions of the tank. Figure 16.6a shows a CFD model where the tank domain was discretized into about 4 million cell volumes, most of which were hexahedral. Successive mesh refinement was performed from the bulk towards the nozzle region, so that the smallest grid spacing at the nozzle is about 1/16th of the nozzle diameter. This helps resolve the detailed jet velocity profile and the secondary flow in the nozzles, as shown in Figure 16.6b.

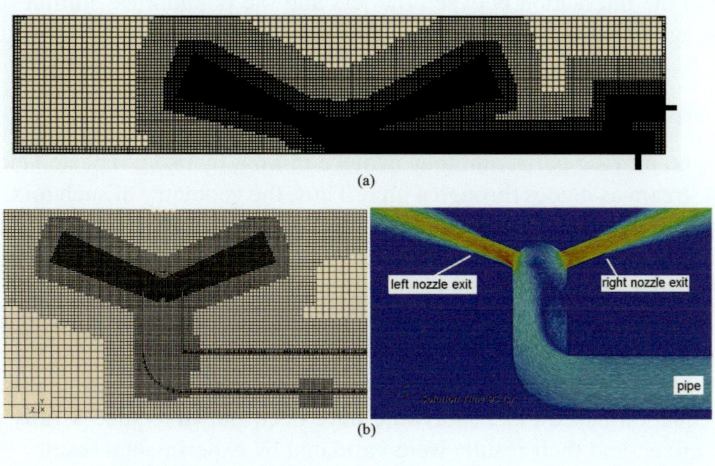

FIGURE 16.6 (a) A slice of the symmetric plane of a jet mixer showing the successive mesh refinement; (b) close-up view of the mesh and predicted velocity field near the nozzle.

A model for the entire tank volume, including both the gas volume in the head space and the liquid volume below, was developed using a volume of fluid (VOF) model. The gas–liquid interface was allowed to deform under the effect of both the gas–liquid flow and gravity; therefore, the interaction between the jets and the free surface was directly taken into account.

The simulation results were post-processed in Tecplot® software and are shown in Figure 16.7. The gas–liquid interface in the VOF model was visualized by plotting the contour where the local liquid volume fraction is 0.5. This is shown as a horizontal green plane in Figure 16.7, and indicates that there is little deformation of the liquid surface.

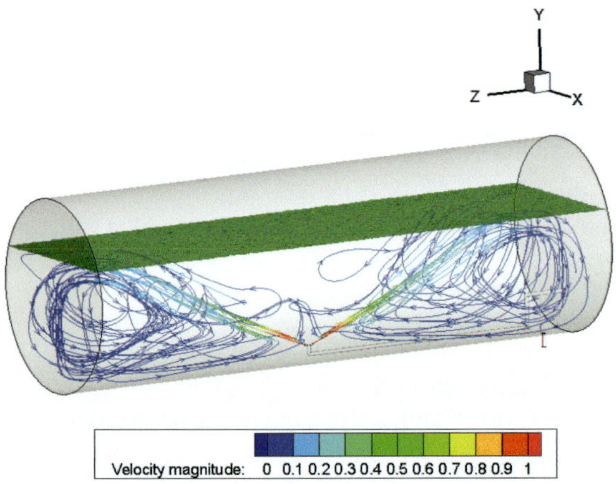

FIGURE 16.7 Liquid surface (green plane) in VOF model and streamlines colored by velocity magnitude.

Figure 16.7 also shows the recirculation flow pattern with the velocity magnitude in the liquid. When the jets reach the liquid surface, their velocity has already decreased to around 0.2m/s. Such low velocity is not expected to significantly disturb the free surface. Because jet oscillation was also observed in the simulation, a series of simulations were performed to investigate the relation between the oscillation frequency and various factors such as the jet Reynolds number, turbulence model, and boundary conditions. It was found that the oscillation was likely an intrinsic physical phenomenon due to the interaction between the jets and the boundaries, similar to the previously reported observations for jet oscillations in confined geometries.

16.5.3 Static Mixers

Static mixers have no moving parts, and they achieve mixing by using repeated elements that split and twist the fluid stream as it goes through a pipe. Thus, the geometry of such mixing elements has direct impact on the effectiveness of mixing under the intended laminar or turbulent flow condition. Common designs of the mixing elements include the helix ribbons (such as the Kenics mixer) and crossed bars (such as the SMX mixer). CFD modeling for static mixers is relatively straightforward since no moving parts are involved. However, as some of the static mixer designs have complex shapes, care should be taken to faithfully represent the geometry of the mixing elements and to use fine grid resolution to capture sufficient details of the flow structure. Both the lattice Boltzmann method and N-S equation-based Fluent software have been shown to successfully predict the flow in a Kenics static mixer, and their results were validated by experimental results. As expected, the N-S based approach was more efficient in low Reynolds number, steady flow conditions, while the LBM was faster for higher Reynolds number, transient flows.[7]

The CFD results for a static mixer are often analyzed in several aspects. The velocity field can be visualized in vector plots or streamlines, and its time characteristics can be extracted from the fluctuation of velocity components. Such information of the flow dynamics can help us to understand the mechanism of mixing under different Reynolds numbers and the onset of the transitions. The degree of homogeneity through the static mixer can be studied by tracking the trajectories of tracer particles, or by monitoring the transport of a scaler field (such as concentration) in different cross-sections along the flow direction. The homogeneity is usually measured by the CoV of the concentration field in the cross-section, and is often plotted against the mixer length to demonstrate the mixing efficiency. In addition, the pressure drop across the static mixer is another useful parameter to characterize the static mixer.

For example, a CFD study was used to demonstrate the improvement of Sulzer SMX plus mixer over the original SMX design by significantly reducing the pressure drop while maintaining the mixing efficiency.[8] Figure 16.8 shows an example of the static mixer setup in their study, and the predicted concentration distribution in the cross-section of the tube indicating increasing degree of homogeneity after each mixing element.

(a)

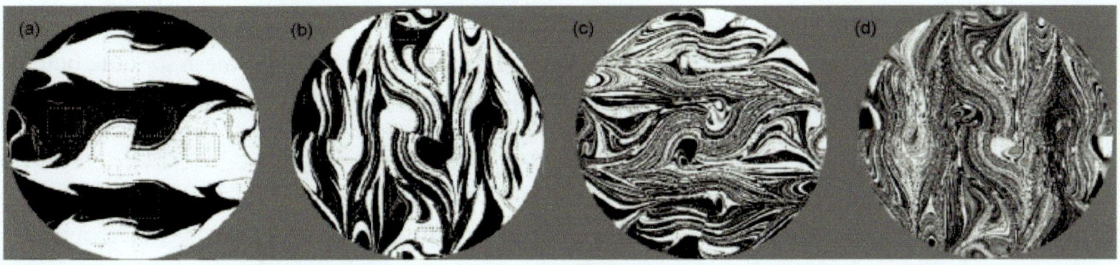

(b)

FIGURE 16.8 (a) Example of a simulation model with three mixing elements in a tube. (b) CFD predictions of concentration distributions in cross-sections, generated using the trajectory mapping scheme, for SMX plus mixing elements (ME): (a) after 1 ME, (b) after 2 ME, (c) after 3 ME, and (d) after 4 ME.[8]

16.5.4 Solid–Liquid Suspension in Stirred Tanks

For mixing in a liquid–solid system, the distribution of solid particles is an important aspect that can greatly influence the rates of mass transfer and reaction. CFD models for such two-phase systems often use the Eulerian-Eulerian model, which treats both liquid and solid phases as interpenetrating continua. The inter-phase momentum transfer is captured through various forces, with the drag force being the most significant in correctly predicting the solid distribution. Drag coefficients that are measured for a single particle settling in a quiescent fluid do not represent the cases of high solids concentrations moving in a turbulent flow field. Therefore, the well-known drag laws such as the Schiller-Nauman model have been found to be inadequate to capture key process information such as the cloud height. The drag force usually needs to be corrected by considering the local particle concentration (e.g., Gidaspow's correlation) and turbulent dispersion force.

Particle-particle interactions also play a role in determining the local solid concentration near the maximum packing limit, and can be captured through a solid pressure term in the solid phase momentum equation. With those corrections, Kubicki and Lo[9] showed that CFD models can correctly predict the distribution of solid particles in a stirred vessel by comparing the simulation results with an experimentally measured solids concentration profile. In particular, the "just suspended" condition, which is difficult to model due to large variations in solids concentration within the tank, could be modeled with their approach.

Figure 16.9 shows contours for solids concentration predicted for a down-pumping six-blade pitched-blade turbine (PBT) impeller using a multi-reference (MRF) model, followed by a sliding mesh (also known as rigid body motion, or RBM) model. Both models gave similar results, which were close to the experimental data for the condition investigated, while the MRF model required on average 18 times less computational time.

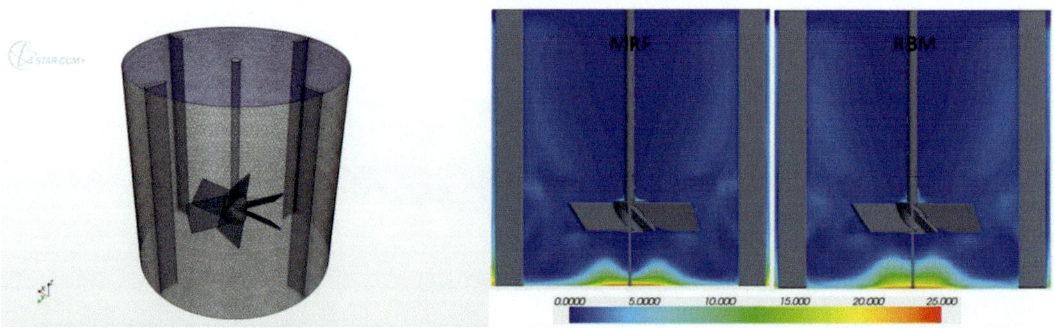

FIGURE 16.9 CFD model for solid–liquid suspension in a stirred tank: (a) surface mesh for tank boundary and impeller; (b) contour profiles of solid particle concentration predicted by MRF and RBM models.

16.5.5 Gas–Liquid Systems

Mass transfer is an important mixing consideration in many gas–liquid systems. For example, in bioreactors, gas bubbles are introduced to transport oxygen into the liquid phase where microorganisms grow. The mass flux $\dot{n}$ of a species (such as oxygen) from an individual bubble to the surrounding liquid can be described as

$$\dot{n} = k_L A\left(C_L^* - C_L\right) \tag{16.10}$$

where k_L is the mass transfer coefficient, A is the bubble surface area, C_L^* is the saturated oxygen concentration in the liquid, determined by Henry's law, and C_L is the local oxygen concentration in the liquid phase. The convective mass transfer coefficient k_L, which is a function of the energy dissipation rate ε, viscosity ν, and Schmidt number Sc, can be expressed as

$$k_L = 0.3\left(\varepsilon v\right)^{0.25} S_c^{-0.5} \tag{16.11}$$

Therefore, if the size and location of all bubbles are tracked, and the local energy dissipation rate and species concentration in the gas and liquid phases are known, in principle the total mass transfer between the gas and liquid phases can be calculated.

In practice, gas–liquid systems such as stirred bioreactors typically involve a broad spectrum of length and time scales as well as a large number of bubbles of various sizes. To reduce computational complexity, several simplifications are often employed in numerical simulations. For example, instead of tracking all the bubble trajectories in the flow field, the averaged gas phase distribution may be tracked as a scalar field in the Eulerian-Eulerian two-phase model. Similarly, the size distribution of the bubbles can be captured using the population balance model, which predicts the evolution of the bubble size distribution due to effects such as breakage and coalescence as a result of the flow dynamics and bubble-bubble interactions. The turbulence model of the liquid phase, such as the k-ε model, can be modified to account for the effect of the dispersed gas phase. Due to the low density and volume fraction of the gas phase, the turbulence of the disperse phase can be neglected and the flow of the gas phase can be considered as laminar. The simulation approach of the Eulerian-Eulerian model with a population balance model is often applied to predict gas–liquid mixing in agitated vessels. For example, it was used to study the oxygen k_La in a 50L single-use bioreactor and was validated with experimental measurements.[10] The validated model was also used to explore the effect of different design parameters, such as the sparger size and impeller diameter in order to optimize the bioreactor design.

In recent years, with the rapid progress of graphical-processing-unit (GPU) based hardware and algorithms, the computation power available for CFD simulation has also greatly increased. This makes it possible to model the physics at a more fundamental level and reduce the need to rely on simplified models or model parametrization between scales or operating conditions. For example, the lattice Boltzmann method has been used to resolve the transient flow field with a large-eddy turbulent model and to directly track the trajectories of tens to hundreds of millions of bubbles.[11] Figures 16.10 and 16.11 show the representative distributions of the velocity, energy dissipation, and bubble sizes in a 250 mL bench scale bioreactor. The model was also systematically validated against experimental data for bioreactors ranging from 5 L to 2000 L. It was reported that the GPU-based LBM model could achieve multiple orders of magnitude improvement in computation speed compared to the traditional CFD models solved on CPU clusters. This makes it possible to resolve both detailed transient flow field and bubble dynamics at timescales that are practical for industrial applications.

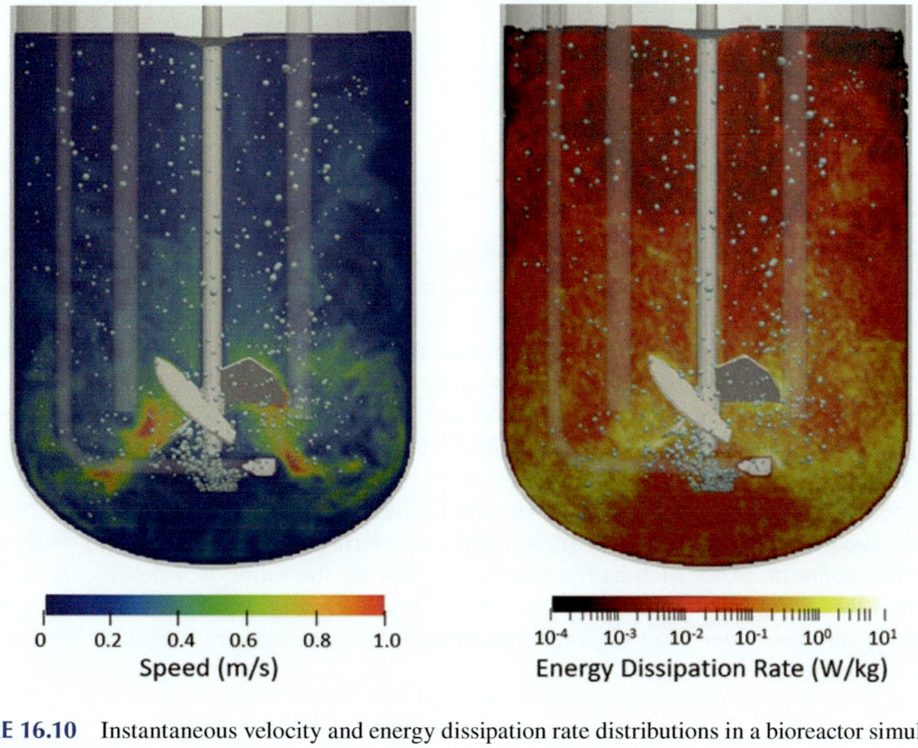

FIGURE 16.10 Instantaneous velocity and energy dissipation rate distributions in a bioreactor simulated by an LBM model.

FIGURE 16.11 Bubble size distribution in a bench scale bioreactor predicted by an LBM model.

16.6 SUMMARY

With the advance of numerical model development and computational hardware, CFD simulation has become a widely used tool in both fluid mixing research and industrial applications. Many critical mixing parameters such as mixing time and impeller power number are now routinely obtained from CFD simulations with reasonable accuracy to guide equipment and process design or investigation. Multi-phase flow modeling, which has been a more challenging topic historically, has also witnessed significant progress in recent years, with enhanced capability to predict properties such as phase distribution, particle size evolution, and interfacial mass transfer. Besides the improvement in numerical models and algorithms, the CFD software packages are also being designed to be more user-friendly in their user interface and workflow, with the aim of allowing users to focus on setting up and interpreting the mixing problem while reducing the time spent on tasks such as meshing or specifying the numerical details. Therefore, it is expected that CFD simulations will continue to find increased utility for industrial mixing applications, and see wider adoption among design and process engineers who are not necessarily modeling experts.

REFERENCES

1. Zhu, C., Fan, L., Yu, Z., *Dynamics of Multiphase Flows* (Cambridge Series in Chemical Engineering), Cambridge University Press, 2021. https://doi.org/10.1017/9781108679039
2. Wolf-Gladrow, D. A., *Lattice-Gas Cellular Automata and Lattice Boltzmann Models—An Introduction*, Springer, Berlin, 2000. https://doi.org/10.1007/b72010
3. Yu, Z., Cope, R. F., Kar, K. K., Even, R. C., Guillaudeu, S. J., Mixing performance of the novel Kar Dynamic Mixer impeller in small laboratory-scale systems, *Ind. Eng. Chem. Res.*, 51, pp. 15282–15292, 2012.
4. Yu, Z., Finch, B. A., Hale, D. A., Mixing of stratified miscible liquids in an unbaffled tank with application in high concentration protein drug product manufacturing, *Ind. Eng. Chem. Res.*, 57, pp. 3397–3409, 2018.
5. Zughbi, H. D., Numerical simulation of mixing in a jet agitated horizontal cylindrical tank, *Int. J. Comput. Fluid Dyn.*, 20, pp. 127–136, 2006.
6. Yu, Z., Kar, K., Storrs, T., Jian, C., Cope, R., Feng, J., Aglave, R., Oscillation of mixing jets in a long horizontal tank, in *Proceedings of the ASME 2014 4th Joint US-European Fluids Engineering Division Summer FEDSM2014*, Chicago, USA, 2014.
7. van Wageningen, W. F. C., Kandhai, D., Mudde, R. F., van den Akker, H. E. A., Dynamic flow in a Kenics static mixer: An assessment of various CFD methods, *AIChE J.*, 50, pp. 1684–1696, 2004.
8. Hirschberg, S., Koubek, R., Moser, F., Schöck, J., An improvement of the Sulzer SMX™ static mixer significantly reducing the pressure drop, *Chem. Eng. Res. Des.*, 87, pp. 524–532, 2009.
9. Kubicki, D., Lo, S., CFD predictions of solids distribution in stirred vessel, in *Ninth International Conference on CFD in the Minerals and Process Industries*, CISRO, Melbourne, Australia, 2012.
10. Amer, M., Feng, Y., Ramsey, J. D., Using CFD simulations and statistical analysis to correlate oxygen mass transfer coefficient to both geometrical parameters and operating conditions in a stirred-tank bioreactor, *Biotechnol Prog.*, 35, p, e2785, 2019.
11. Thomas, J. A., Liu, X., DeVincentis, B., Hua, H., Yao, G., Borys, M. C., Aron, K., Pendse, G., A mechanistic approach for predicting mass transfer in bioreactors, *Chem. Eng. Sci.*, 237, p. 116538, 2021.

17 Heat Transfer

17.1 INTRODUCTION

This chapter discusses the transfer of heat between working media inside a stirred tank and heat exchange surfaces within that tank. Although heat is sometimes extracted externally by pumping media out of the tank and through a conventional external heat exchanger, such heat exchange is not discussed here. Instead, the focus is on heat exchange surfaces inherent to stirred tanks, either as internal coils or as a jacket covering the wall and/or bottom of the vessel. Such heat exchange removes heat generated within the media by various means, including viscous energy dissipation, exothermic reaction (chemical and/or biological), mixer power input, etc. The heat removal might allow a limited rise in media temperature, or possibly maintain critical isothermal conditions. On the other hand, controlled heat addition might initiate and maintain desired reaction, as in the case of thermal polymerization of monomers while manufacturing either polyvinylidene chloride or polystyrene. Heat addition and removal surfaces typically include heat transfer jackets that have increased in sophistication and effectiveness over the years. They also include internal coils that utilize vertical and/or horizontal orientations within the vessel. There are also processes that (a) promote reaction by adding superheated steam directly to the reactants, or (b) remove excess heat from the working media by boiling a passive or sacrificial solvent within the media, such as acetic acid used in crude terephthalic acid (CTA) reactors.

17.2 HEAT TRANSFER JACKETS

Effective heat transfer often utilizes a jacket that surrounds the vessel exterior wall, as shown in Figure 17.1. Jacket design helps determine heat exchange capability, and may or may not incorporate various flow-controlling geometries, segmentations or compartmentalization, and inlet/outlet

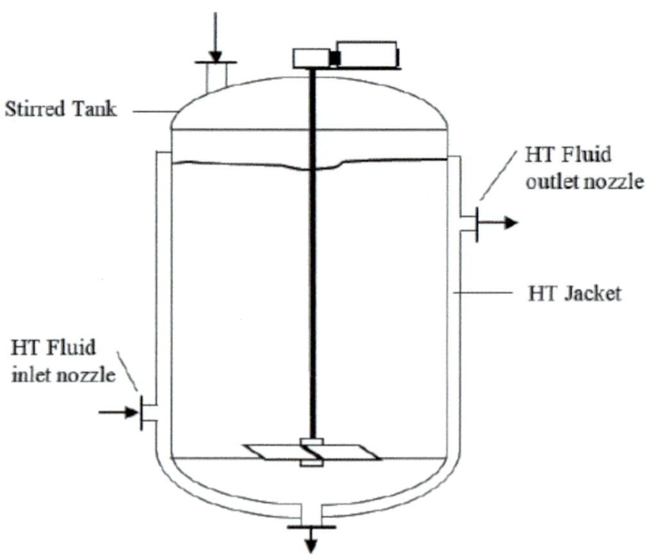

FIGURE 17.1 Schematics of a stirred tank with the heat transfer jacket.

DOI: 10.1201/9781003606390-17

nozzles to achieve and maintain a required fluid inlet-outlet temperature difference (ΔT). Insulation typically covers the jacket's outer shell to minimize heat loss to the surroundings. It should be wrap-sealed to make it impermeable to ambient moisture, and thus, better able to maintain its thermal resistance or R-value.

Figure 17.2 shows profile images of different types of jacket internals that have been created to circulate heat transfer fluids in a cost-effective manner. These different jacket types utilize (a) a helical baffle, (b) a helical coiled tube, (c) a helical coiled half-pipe, (d) surface dimples, and (e) the more recently developed pillow plate. These different jacket types are listed in ascending order of their increasing heat transfer efficiency. To improve the jacket-side heat transfer coefficient, h_o, the objective is to minimize the thermal boundary layer and stagnation domains by maintaining turbulence as the heat transfer fluid flows through the jacket. To maximize overall heat transfer through the jacket, however, the wetted surface area that contacts the jacket wall must be maximized. The jacket design that gets implemented is often a tradeoff that maximizes heat transfer while minimizing complexity and cost.

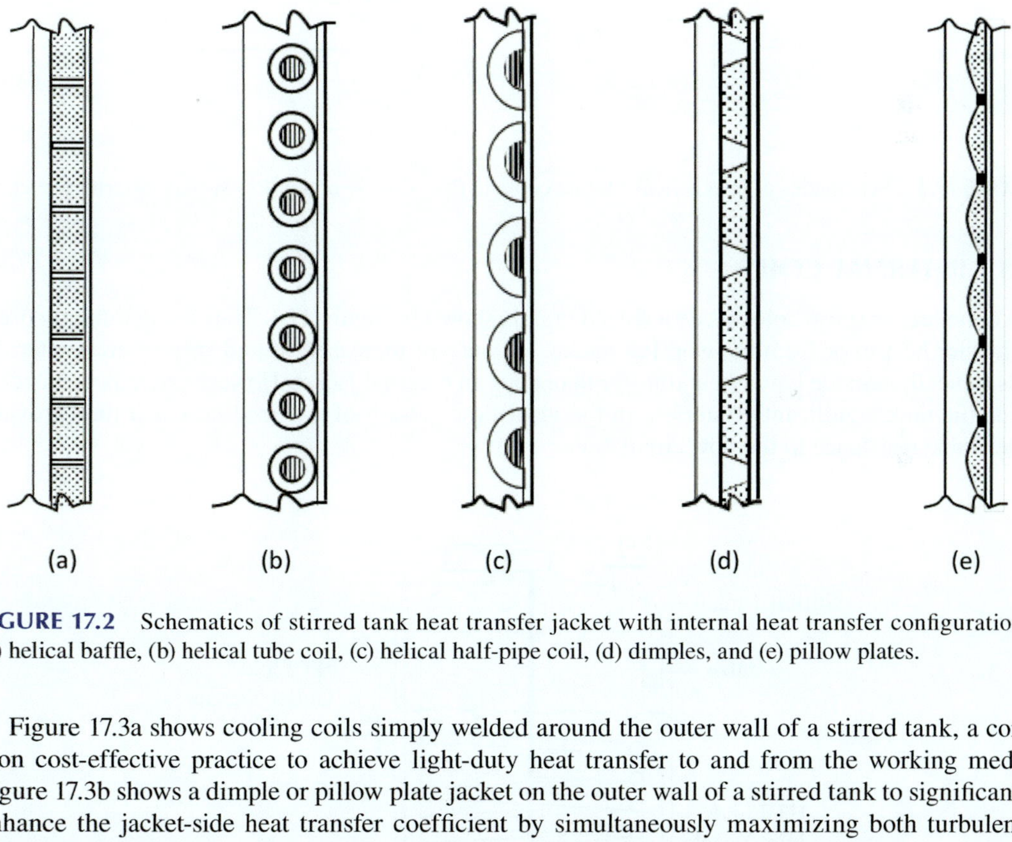

FIGURE 17.2 Schematics of stirred tank heat transfer jacket with internal heat transfer configurations: (a) helical baffle, (b) helical tube coil, (c) helical half-pipe coil, (d) dimples, and (e) pillow plates.

Figure 17.3a shows cooling coils simply welded around the outer wall of a stirred tank, a common cost-effective practice to achieve light-duty heat transfer to and from the working media. Figure 17.3b shows a dimple or pillow plate jacket on the outer wall of a stirred tank to significantly enhance the jacket-side heat transfer coefficient by simultaneously maximizing both turbulence and wetted surface area. The pillow plate design, shown previously in Figure 17.2e, is a further enhancement of the dimple concept. It provides similar enhancements as the dimples, but at lower manufacturing cost and potentially lower required heat transfer fluid flowrates (due to lower pillow height and flow cross-sectional area). Dimple plates pre-cut and pre-form dimples in one plate, which is then welded or laser fused to the other plate. Pillow plates weld or laser fuse the plates of metal together without any pre-cutting or pre-stamping of the pillow structure, although two holes are likely pre-cut into opposite corners of one plate for subsequent attachment of inlet and outlet ports. After two plates have been attached by the requisite grid of fusion points (not unlike the dimple grid), they are similarly sealed around the edges, and the inlet and outlet ports are attached.

High-pressure fluid is then injected between the plates to deform one or both to produce the characteristic inflated pillow shape. If both plates are similar, both will inflate; if one plate is sufficiently more rigid than the other, just one plate will inflate.

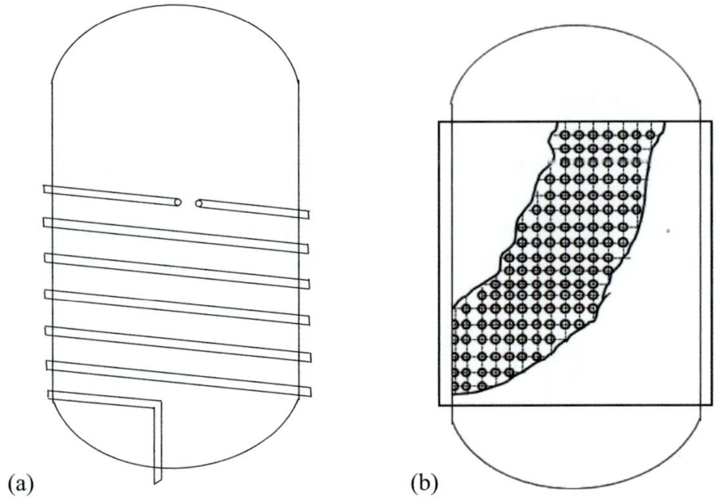

FIGURE 17.3 Schematics of (a) a helical cooling coil and (b) a dimple or a pillow surface exterior jacket.

17.3 INTERNAL COILS

Use of helical or spiral coils inside a stirred tank is shown in Figure 17.4. They are commonly used to transfer heat to or from the working media. Because of increased wetted surface area, internal coils typically provide better heat transfer than does an external jacket. However, as expected, coils inside the tank significantly interfere in the ideal flow pattern of a stirred tank and thus provide undesirable resistance to the flow circulation.

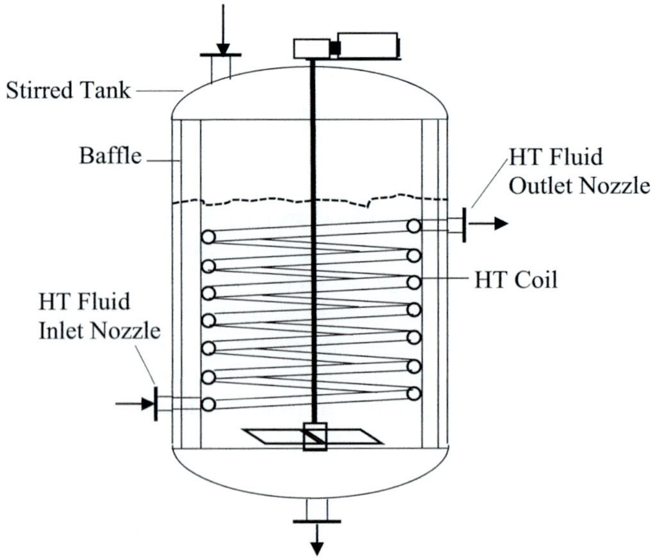

FIGURE 17.4 Schematic of a single helical heat transfer coil in a stirred tank.

The material of construction (MOC) of the coil should have high thermal conductivity, but at the same time must be compatible with the working media, e.g., corrosion and erosion resistance, and inert to any targeted chemical/biological reactions. Typical coil materials are copper, nickel, carbon steel, titanium, Inconel, stainless steel, and Hastelloy-C, in descending order of room-temperature thermal conductivity. Table 17.1 shows an approximate thermal conductivity value, λ, and an example cooling coil application for the above metals.[1] Because internal coils generally attach to the baffles, or to brackets on the tank inner wall, they experience transient hydrodynamic forces. Depending on the turbulence and blade pass frequency, fatigue failure at supporting weld points is not uncommon in plant applications. Thus, systematic design approaches for coils and supports consider their material of construction, yield stress, σ_{yield}, and weld strength. Table 17.1 also lists approximate values of room-temperature yield stress for the above metals.

In tall tanks such as fermenters, multiple helical coils are commonly used to (a) ensure uniform and adequate heat transfer with the working media at minimal ΔT, and (b) reduce the load a single long coil would produce on the wall brackets. Figure 17.5 shows three helical coils in a tall bioreactor with a separate heat transfer fluid inlet and outlet for each.

These helical coils are generally fastened to the baffles or wall brackets using U-bolts for convenience of their installation, removal, maintenance, and sanitization. The latter is particularly vital to many bioprocesses. Apart from the regular steam cleaning, hot water is periodically circulated through the coils to disinfect inaccessible areas of the coils.

TABLE 17.1
Thermal Conductivity and Yield Stress Values of Internal Cooling Coil Materials[1]

Coil MOC	Thermal Conductivity, λ $Wm^{-1}K^{-1}$	Yield Stress, σ_{yield} MPa	Application Examples
Copper	401	193	Nonaqueous media
Nickel	94	110	Corrosive media
Carbon steel	54	415	Aqueous media
Titanium	22	255	Terephthalic acid
Inconel	15	460	Corrosive media
SS-304	14	276	Bioreactors
Hastelloy-C	12	365	Corrosive media

Because specific power in an aerobic fermenter is typically high, e.g., ranging from 1 to 4 W/kg, the design of coil welds and fasteners needs extra attention. This is especially true near the gas-dispersing radial impeller that commonly draws $\gtrsim$40% of the total power under fully gassed conditions for a multi-tiered agitator like that in Figure 17.5. Although the multiple coils in Figure 17.5 are arranged in series (one above another), stirred tanks occasionally arrange multiple helical coils in parallel (nested one inside another), especially to remove excessive heat from the core of high-viscosity media and to better maintain a uniform temperature. These parallel or coaxial coils normally have their elements staggered to minimize obstruction to the circulation flow pattern. Distance between the coils should be enough (e.g., two or three tube diameters) for the elements of one coil not to be in the wake (stagnation) regions of another coil. Designs of such coil banks are discussed further in the design section of this chapter.

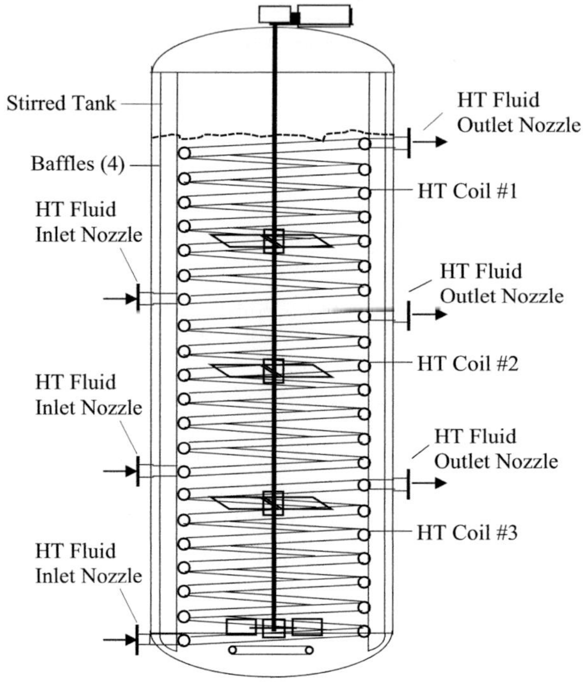

FIGURE 17.5 A schematic showing three internal helical heat transfer coils in a tall bioreactor.

The vertical coils shown in Figure 17.6 are sometimes used in stirred tanks, especially if fouling of horizontal surfaces is an issue. When located near the wall, the vertical coils partially baffle the swirling fluid. Specially designed hollow baffles, which also provide some heat transfer capacity, are common in glass-lined vessels, where lower thermal conductivity of the glass lining (e.g., 45 times less than carbon steel,[2] as seen in Table 17.2) may limit heat transfer to the jacket.

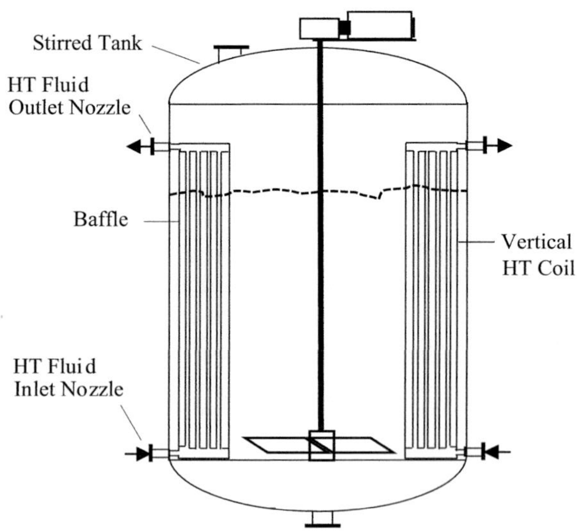

FIGURE 17.6 Schematic of a stirred tank with vertical heat transfer coils.

TABLE 17.2
Thermal Conductivities of Common Materials Used in the Construction of Vessels[2]

Vessel Material of Construction (MOC)	Thermal Conductivity, λ [W m^{-1} K^{-1}]
Carbon steel	45
Copper	377
Glass lining	1
Inconel	15
Monel (400, 404, R-405, 411)	24.7
Nickel (200, 201, 220, 225)	60.5
Stainless Steel (304,316, 321, 347)	15.7
Tantalum	55.4
Titanium	18.2

17.4 HEATED BASE PLATES

In the case of large insulated liquid storage tanks, while jet mixers are in operation, heated base plates on concrete foundations are used to transfer heat to the working media. This is a cost-effective practice to keep the aqueous contents in the vessel from freezing and degrading in cold climates. Both minor mixing due to multiple jets located near the tank bottom (see Chapter 11, Figure 11.6) and fluid buoyancy originating from the temperature gradient at the bottom (natural convection), provide adequate circulation to maintain uniform temperature in the tank media.

17.5 STEAM NOZZLES

Occasionally, nozzles located on the bottom head add superheated steam directly into the working media within a stirred tank. Such addition directly into the fluid is the fastest and most uniform heat transfer method available in some thermal polymerization processes. Nitrogen gas is often also coaxially injected to maintain an inert atmosphere and promote temperature uniformity.

17.6 HEAT TRANSFER CONCEPTS

The three common modes or mechanisms of heat transfer are (1) thermal conduction—not to be mistaken with electrical conduction, (2) convection, and (3) radiation. Their mathematical dependency on temperature is briefly outlined in this section.

17.6.1 Conduction

The energy that ultimately raises the temperature of an object first initiates random microscopic movement of electrons that leads to vibration of the associated atoms. The thermal conduction process transmits heat energy through collision of adjacent atoms or molecules. Thus, closeness of atomic or molecular particle packing dictates the extent of heat conduction. It occurs more readily in solids and liquids because the particles are closer together than in gases. Thermal conductivity, λ, the ability of a material to conduct heat, is defined by Fourier's unidirectional heat transfer law[3] as

$$\dot{q}_x = -\lambda\left(\frac{dT}{dx}\right) \tag{17.1}$$

where $\dot{q}_x$ = heat flux in x direction [W m^{-2}], $\frac{dT}{dx}$ = temperature gradient in the x-direction [K m^{-1}]

λ = proportionality constant, known as the thermal conductivity [W m^{-1}K^{-1}]

The negative sign fits the observation that temperature differences cause heat to flow from high temperature to low temperature. Total heat flow, $\dot{Q}\ or\ (\dot{q}_x \cdot A)$, through a wall with surface area A, shown in Figure 17.7, is represented by equation 17.2.

$$\dot{Q} = \lambda A \left(\frac{T_1 - T_2}{X} \right) \tag{17.2}$$

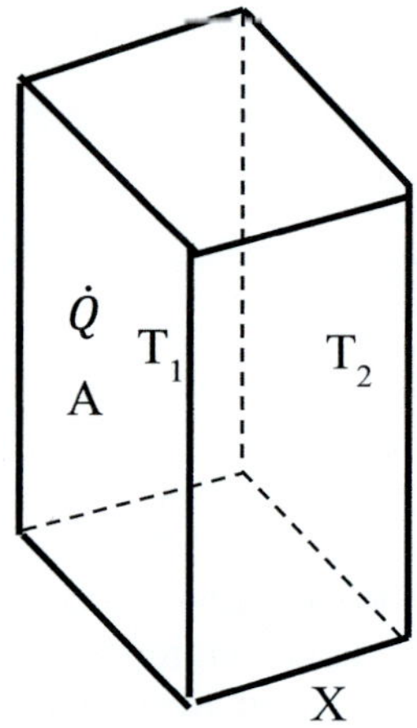

FIGURE 17.7 Schematic of a solid wall of thickness, X.

In equations 17.2–17.22, absorbing the negative sign into the temperature difference makes it reflective of the actual direction of heat flow (from high to low temperature, e.g. T_1-T_2), but also a reversal of the traditional definition of ΔT. Equations for steady-state heat conduction through either the rectangular or circular cross-sectional walls shown in Figures 17.8a–b and 17.9a–b are presented in Table 17.3.

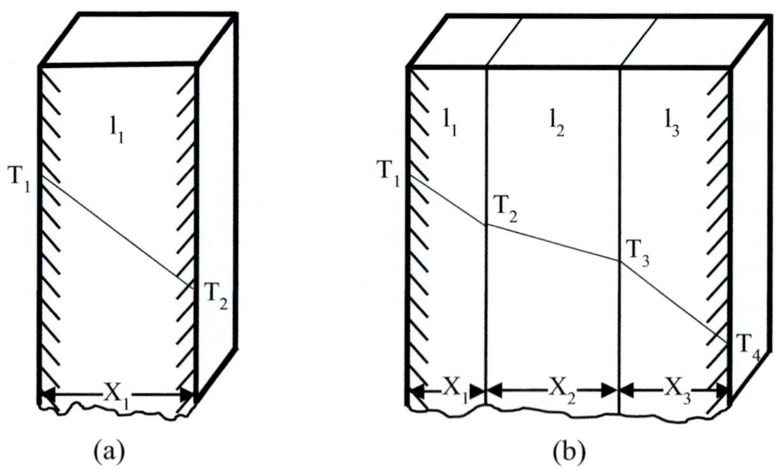

FIGURE 17.8 Schematics of the cross-section of (a) solid wall and (b) composite wall.

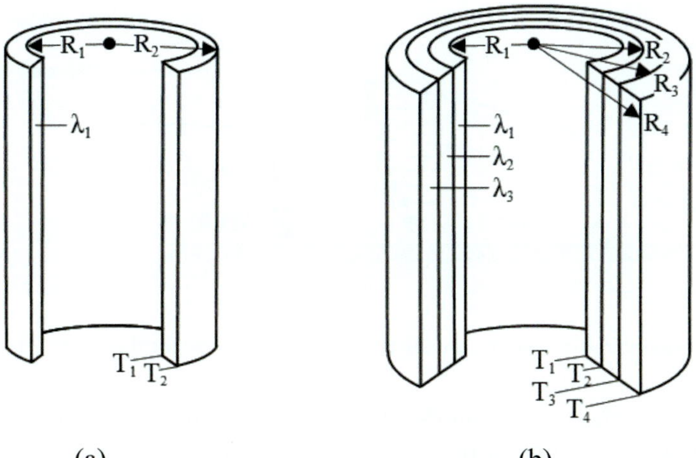

FIGURE 17.9 Schematics of the cross-section of (a) circular cylinder and (b) composite cylinder.

TABLE 17.3
Heat Transfer Equations in Rectilinear and Cylindrical Coordinate Systems

Geometry	Heat Flowrate, $\dot{Q}$ Formulae	Nomenclatures
Solid wall	$\dot{Q} = \lambda\, A\left(\frac{T_1 - T_2}{X_1}\right)$	T= temperature, [K] λ = thermal conductivity, [W m^{-1}K^{-1}] A = projected wall surface area normal to heat flow, [m^2] X = wall thickness, [m] L= length of cylinder, [m] R = Radius [m]
Composite solid wall	$\dot{Q} = A\left(\frac{T_1 - T_4}{X_1/\lambda_1 + X_2/\lambda_2 + X_3/\lambda_3}\right)$	
Solid cylinder	$\dot{Q} = 2\pi\,\lambda_1 L\left(\frac{T_1 - T_2}{\ln(R_2/R_1)}\right)$	
Composite cylinder	$\dot{Q} = \left(\frac{2\pi L\,(T_1 - T_4)}{\frac{1}{\lambda_1}\, ln(R_2/R_1) + \frac{1}{\lambda_2}\ln(R_3/R_2) + \frac{1}{\lambda_3}\, ln(R_4/R_3)}\right)$	

17.6.2 Convection

Convective heat transfer occurs due to physical transport of a fluid (gas or liquid). Depending on the transport mechanism, convection is identified as natural or forced convection. Natural convection occurs due to a density difference that leads to fluid buoyancy, typical of gaseous media. Forced convection arises when a fluid is transported by means of an external source such as a pump, fan or impeller. In this case, heat transfer is associated with the thermal mass transfer.

Fluid flow past a solid object creates both momentum (velocity) and thermal boundary layers as shown in Figure 17.10. Because of the no-slip boundary condition, fluid velocity adjacent to the wall is zero, and consequently fluid temperature remains in equilibrium with that of the solid wall, i.e., T_s. Away from the wall, fluid maintains its free stream velocity and temperature (u_0 and T_0). The thickness of the fluid layers over which the velocity and temperature change are known, respectively, as velocity and thermal boundary layers (δ_V and δ_T). Readers are advised to study

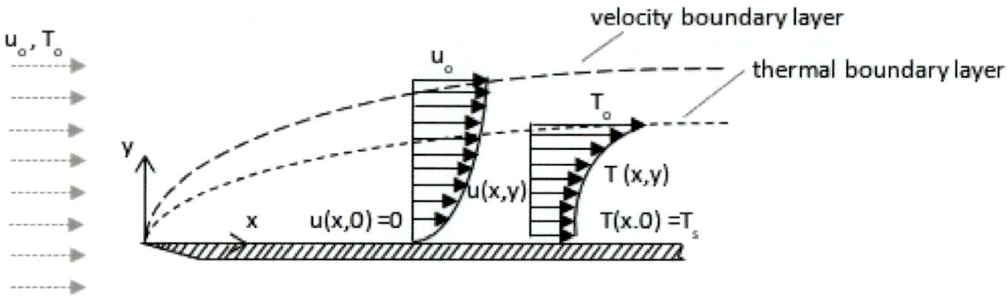

FIGURE 17.10 Development of velocity and thermal boundary layers for flow over a flat plate.

reference 4 for details. Depending on the fluid flow Reynolds number, the velocity boundary layer is characterized as laminar or turbulent and influences heat transfer to and from the solid surface. The greatest temperature drop of the fluid occurs in a thin fluid layer adjacent to the wall, δ, where mixing is insignificant and heat transfer is dominated by the thermal conductivity of the fluid, λ_f. Following the heat conduction formula presented in Table 17.3, the rate of heat flow across the fluid film is

$$\dot{Q} = \frac{\lambda_f}{\delta} A\left(T_f - T_w\right) \tag{17.3}$$

where, T_f and T_w are respectively the freestream fluid and wall temperatures.

Because film thickness, δ, is a parameter that strongly depends on fluid Reynolds number, for all practical purposes, a lumped parameter known as the heat transfer coefficient, h, is used to characterize convection.

$$\dot{Q} = hA\left(T_f - T_w\right) \tag{17.4}$$

Figure 17.11 presents a solid wall (thickness = X) separating flowing fluids 1 and 2 at respective temperatures T_1 and T_2 and velocities u_1 and u_2. Thermal conductivity of the solid wall material is λ, while thermal fluid film thicknesses for the two fluids are δ_1 and δ_2. Consider a fixed heat flow out of fluid 1, through the solid wall, and into fluid 2.

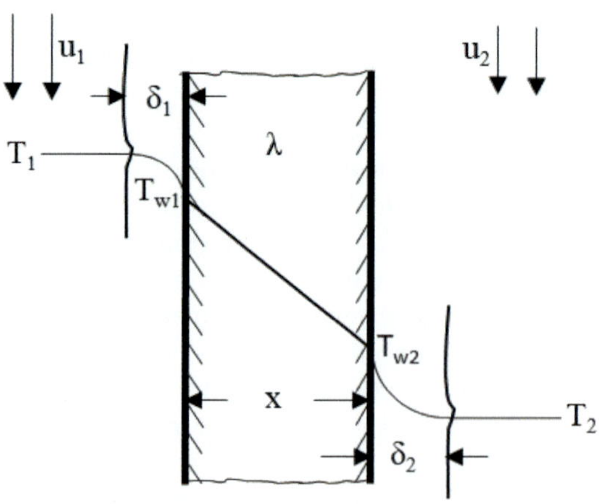

FIGURE 17.11 Heat transfer across a solid wall.

Based on Fourier's formula, the constant heat flowrate, $\dot{Q}$, is described by the following equations in the three respective phases (i.e., fluid 1, wall, fluid 2):

$$\dot{Q} = \frac{\lambda_{f1}}{\delta_1} A(T_1 - T_{w1}) = h_1 A(T_1 - T_{w1}) \tag{17.5}$$

$$\dot{Q} = \frac{\lambda}{x} A(T_{w1} - T_{w2}) \tag{17.6}$$

$$\dot{Q} = \frac{\lambda_{f2}}{\delta_2} A(T_{w2} - T_2) = h_2 A(T_{w2} - T_2) \tag{17.7}$$

where, λ_{f1} and λ_{f2} are respective thermal conductivities of fluids 1 and 2.

Thus, total heat flow across the solid wall is related to the fluid stream temperature difference as follows.

$$(T_1 - T_2) = \frac{\dot{Q}}{A}\left(\frac{1}{h_1} + \frac{x}{\lambda} + \frac{1}{h_2}\right) \tag{17.8}$$

$$\dot{Q} = U A (T_1 - T_2) \tag{17.9}$$

U is commonly known as the overall heat transfer coefficient, expressed as,

$$U = \frac{1}{\left(\frac{1}{h_1} + \frac{x}{\lambda} + \frac{1}{h_2}\right)} \tag{17.10}$$

A general form of the overall heat transfer coefficient for two flowing fluids on opposite sides of a wall composed of n sequential material layers is represented by equation 17.11.

$$U = \frac{1}{\left(\frac{1}{h_1} + \frac{x_1}{\lambda_1} + \frac{x_2}{\lambda_2} + \ldots + \frac{x_n}{\lambda_n} + \frac{1}{h_2}\right)} \tag{17.11}$$

There is no surface resistance between individual layers in this composite wall. Layer thicknesses and thermal conductivities are respectively $x_1, x_2, \ldots x_n$ and $\lambda_1, \lambda_2, \ldots \lambda_n$.

In the case of agitated vessels and heat exchangers, the overall heat transfer coefficient, U, is widely used to estimate the heat transfer rate, $\dot{Q}$, from the process fluid to the heat transfer fluid or vice versa.

1. Transfer of heat to and from the jacket or coil:
   In an idealized situation where the vessel and jacket operate under isothermal conditions, the simplest heat transfer formula is expressed in equation 17.12

$$\dot{Q} = U A \Delta T \tag{17.12}$$

where, ΔT = Temperature difference between the working media and heat transfer fluids.

2. Transfer of heat in an external heat exchanger:

$$\dot{Q} = U\,A\,\Delta T_{lm} \tag{17.13}$$

ΔT_{lm} is the log mean temperature difference (LMTD), or average temperature difference between process fluid A and heat transfer fluid B. The LMTD is calculated from equation 17.14, independent of whether fluid flows are parallel, also called co-current and depicted in Figure 17.12a, or counter-current with entrance and exit temperatures as in Figure 17.12b. The parallel or co-current arrangement produces a larger ΔT_1 and a smaller ΔT_2 than the counter-current arrangement (see Figure 17.12), but if both arrangements are in an adiabatic system and produce the same (T_{h1}-T_{h2}), then they also produce the same (T_{c2}-T_{c1}).[5]

$$\Delta T_{lm} = \frac{\Delta T_2 - \Delta T_1}{ln\left(\dfrac{\Delta T_2}{\Delta T_1}\right)} \tag{17.14}$$

where, $\Delta T_2 = T_{h2} - T_{c2}$ and $\Delta T_1 = T_{h1} - T_{c1}$.

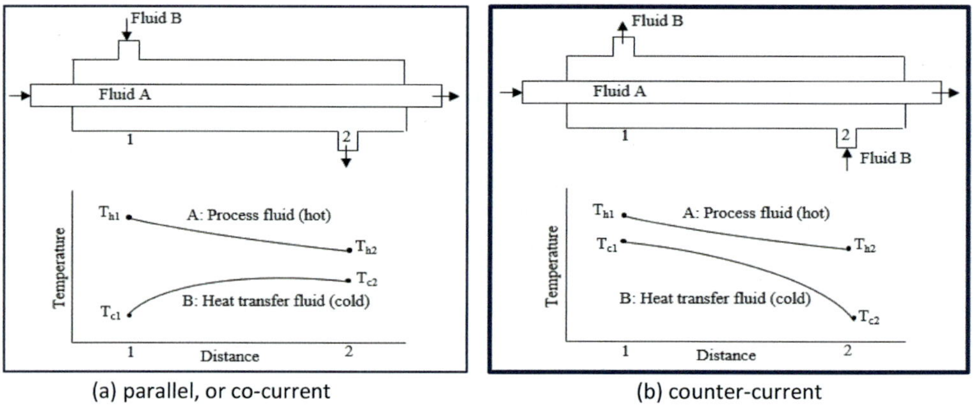

FIGURE 17.12 Double pipe heat exchanger in (a) parallel or co-current flow and (b) counter-current flow.

3. The continuous process vessels (see Figure 17.13a–b) where inlet and outlet temperatures of the working media and the heat transfer fluid are different, replicate a simple shell-and-tube heat exchanger like that in Figure 17.12. Realistically, the heat transfer equation for $\dot{Q}$ includes ΔT_{lm}.

$$-\dot{Q} = U\,A\,\Delta T_{lm}$$

And as in equation 17.14, the LMTD is represented by

$$\Delta T_{lm} = \frac{(t_2 - T_2) - (t_1 - T_1)}{\ln\left[(t_2 - T_2)/(t_1 - T_1)\right]}$$

where, t_1 and t_2 are respective process fluid entrance and exit temperatures; T_1 and T_2 are inlet and outlet temperatures of the heat transfer fluid.

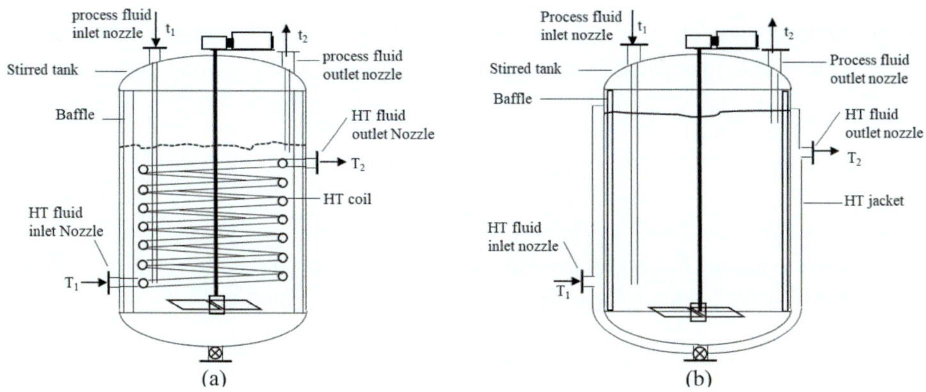

FIGURE 17.13 Continuous stirred tank reactors (CSTRs) with heat transfer by (a) internal coil and (b) external jacket.

17.6.3 Radiation

Thermal radiation transfers heat between objects by electromagnetic waves (λ = 0.1–100μm), with no intervening transfer media present. Radiating photons reaching a surface may be absorbed, reflected, or transmitted, as depicted in Figure 17.14. A black body, an idealized opaque object, absorbs all electromagnetic radiation contacting it, i.e., transmissivity (τ) = reflectivity (ρ) = 0, and absorptivity (α) = 1.0. It then emits thermal radiation in a continuous spectrum based on its temperature, i.e., emissivity (ε) = 1.0.

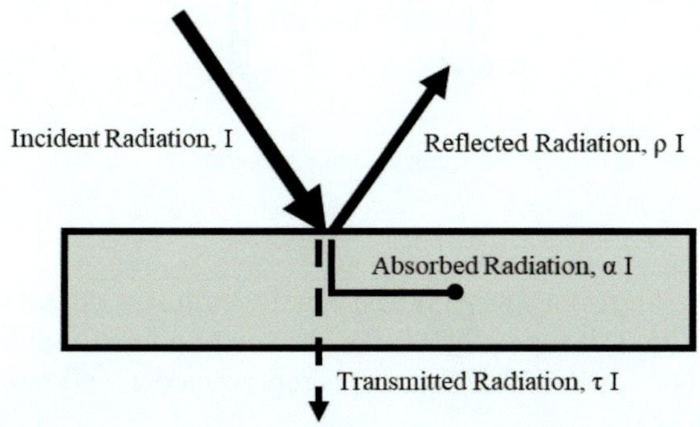

FIGURE 17.14 Radiation properties of an object.

According to Stefan-Boltzmann law, thermal radiation from the surfaces of black bodies ($\alpha = \varepsilon$ = 1.0) is described as presented in equation 17.15.

$$\dot{Q} = \sigma A T^4 \tag{17.15}$$

where, T = absolute temperature of the black body [K], A = surface area [m^2]

σ = Stefan-Boltzmann constant [5.669×10^{-8} W-m^{-2}-K^{-4}], $\dot{Q}$ = Thermal radiation [W]

When a body with surface temperature T_1 is placed in a closed environment of temperature T_2, the net amount of heat transferred depends on the temperature difference, in the form

$$\dot{Q} = \sigma A \left(T_1^4 - T_2^4\right) \tag{17.16}$$

The net radiative heat transfer, $\dot{Q}_{1-2}$, between two closely spaced black bodies, 1 and 2 as shown in Figure 17.15, is the radiation emitted from object 1, $\dot{Q}_{1\rightarrow 2}$, minus the radiation emitted from object 2, $\dot{Q}_{2\rightarrow 1}$.

$$\dot{Q}_{1\rightarrow 2} = A_1 F_{1\rightarrow 2} E_1 \quad (17.17)$$

$$\dot{Q}_{2\rightarrow 1} = A_2 F_{2\rightarrow 1} E_2 \quad (17.18)$$

Considering the view factor, F, and reciprocity between objects 1 and 2, i.e., $A_1 F_{1\rightarrow 2} = A_2 F_{2\rightarrow 1}$, the net radiation heat transfer between these objects is expressed in equation 17.19.

$$\dot{Q}_{1-2} = A_1 F_{1\rightarrow 2} \sigma \left(T_1^4 - T_2^4 \right) \quad (17.19)$$

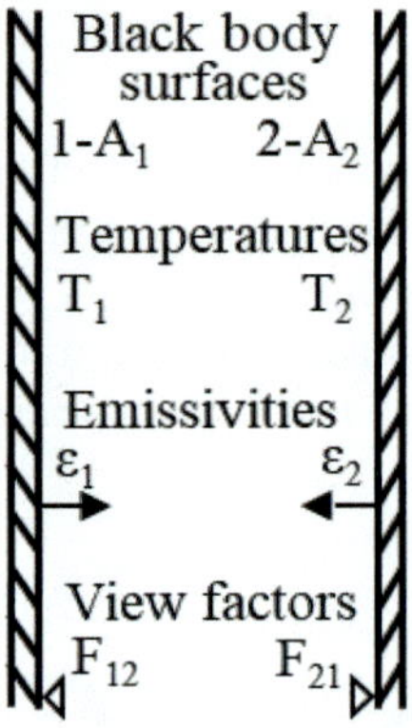

FIGURE 17.15 Radiation heat exchange between two black bodies ($\varepsilon_1 = \varepsilon_2 = 1$).

However, real surfaces (known as grey bodies) usually emit less power than the idealized black bodies, since their emissivity, ε, is typically 0.1–0.9. Accounting for emissivity, Stefan- Boltzmann's equation for grey bodies ($\varepsilon = \alpha$, Kirchhoff's Law) is represented as follows:

$$\dot{Q} = \varepsilon \sigma A T^4 \quad (17.20)$$

Thus, for two grey bodies, not being perfect radiators, equation 17.15 becomes

$$\dot{Q} = \varepsilon \sigma A \left(T_1^4 - T_2^4 \right) \quad (17.21)$$

For two grey body surfaces that form an enclosure, the heat transfer is

$$\dot{Q}_{1-2} = \frac{\sigma \left(T_1^4 - T_2^4 \right)}{\left(\dfrac{1-\varepsilon_1}{A_1 \varepsilon_1} + \dfrac{1}{A_1 F_{1\rightarrow 2}} + \dfrac{1-\varepsilon_2}{A_2 \varepsilon_2} \right)} \quad (17.22)$$

where, ε_1 and ε_2 are the emissivities of the two surfaces.[5,6]

Formulas for radiative heat transfer can be derived for more elaborate physical arrangements, such as between parallel plates, concentric spheres, and the internal surfaces of a cylinder.[5]

Forced convection is the most efficient mode of heat transfer in fluids. Along with conduction through solid walls, it is widely used to transfer heat to and from the heat exchanger fluid. Radiation is the least efficient but fastest method of heat transfer. In chemical and allied industries, radiation heat transfer is applied to dry food and fibers, as well as to melt and mold plastics in continuous processes. It's uncommon in fluid mixing applications, except for its extensive usage in furnaces and combustion chambers (temperature ≥ 1073K) where gas-solid mixing is encountered.

Sample Radiative Heat Transfer Problem: At the center of a vertically oriented, laboratory-scale furnace is a heated ceramic cylinder with 0.5 m length and 0.025 m inner diameter. The cylinder walls are held at a constant temperature of 2000 K when a porous particle of coal ash, with a temperature of 300 K and diameter of 0.025 m, is injected from top centerline with downward velocity of 0.1 m/s. If respective emissivities, ε, of the furnace walls and the particle are 0.6 and 0.9, what is the radiative heat flux (in W m^{-2}) transferred to the particle from the wall? Compare this radiative heat flux to the convective particle heat flux if the gas temperature were that of the wall (2000K), and the low airflow produced a reported[7] typical convective heat transfer coefficient, h, of 10 W m^{-2} K^{-1}.

Sample Problem Solution: Radiative heat flux is obtained by dividing both sides of equation 17.21 by the particle surface area, A:

$$\dot{Q}/A = \varepsilon\sigma\left(T_1^4 - T_2^4\right) = 0.9\ (5.669 \times 10^{-8}\ \text{W m}^{-2}\ \text{K}^{-4})\ (2000^4 - 300^4)\ \text{K}^4$$
$$= 815.9\ \text{kW m}^{-2}$$

To obtain convective heat flux, divide both sides of equation 17.4 by the particle surface area, A:

$$\dot{Q}/A = h\left(T_g - T_p\right) = 10\ \text{W m}^{-2}\ \text{K}^{-1}\ (2000 - 300)\ \text{K} = 17.0\ \text{kW m}^{-2}$$

In this high-temperature example, the dominant radiative flux is 48 times greater than convective heat flux.

17.7 HEAT TRANSFER IN AGITATED VESSELS

Figure 17.16 schematically represents heat transfer in a system where process fluid flows parallel to the left side (side 1) and heat transfer fluid flows parallel to the right (side 2) of a wall of constant thickness, x.

Table 17.4 briefly summarizes the fluid flow and the wall characteristics illustrated in Figure 17.16. Dominant mechanisms of heat transfer between the two fluids are (a) convection within the fluid domains and (b) conduction through the solid wall and its adjoining fouling layers. In practice, biological and corrosion fouling of the heat exchanger wall surfaces are quite common.

Table 17.5 presents approximate values of thermal conductivities for a few fouling layer materials commonly encountered in heat exchangers.[8] Biological fouling is caused by the growth of organisms within the fluid that deposit onto the wall surface. Corrosion products, such as deposited salt and dirt, cause fouling by building up on the wall surface. Chemical fouling is typically sticky process fluids such as latex, urethane adhesives, sealants, etc., adhering to the reactor walls. Since thicknesses (on the order of microns) of these fouling layers, δ_{f1} and δ_{f2}, are imprecisely known, they are typically combined with their respective thermal conductivities (as indicated in the general expression of equation 17.23) to form the fouling factors, f_1 and f_2, that are used in calculating the overall heat transfer coefficient, U.

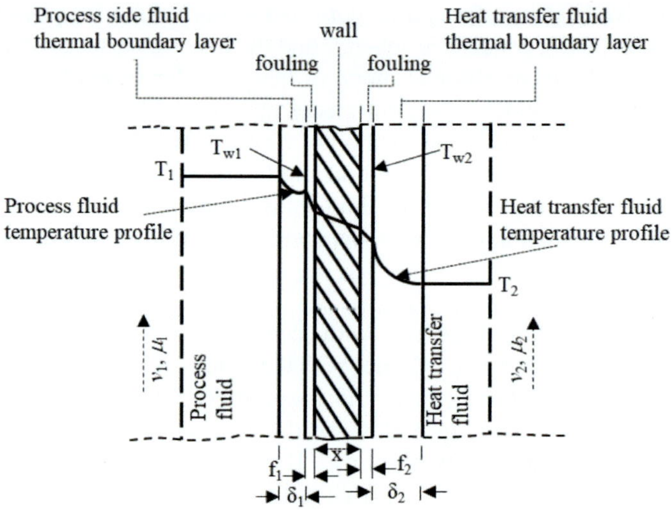

FIGURE 17.16 Parallel flow of process fluid and heat transfer fluid by the sides of a solid wall.

TABLE 17.4
Critical Parameters Related to Figure 17.16

Parameters	Process Fluid	Heat Transfer Fluid	Unit
Temperature	T_1	T_2	K
Wall temperature	T_{w1}	T_{w2}	K
Thermal boundary layer thickness	δ_1	δ_2	m
Fouling layer thickness	δ_{f1}	δ_{f2}	m
Fouling factor	f_1	f_2	m^2KW^{-1}
Heat transfer coefficient	h_1	h_2	$Wm^{-2}K^{-1}$
Physical and Thermal Properties			
Viscosity	μ_1	μ_2	Pa-s
Specific heat capacity	C_{p1}	C_{p2}	$Jkg^{-1}K^{-1}$
Thermal conductivity	λ_1	λ_2	$Wm^{-1}K^{-1}$
Wall Characteristics			
Thermal conductivity	λ		$Wm^{-1}K^{-1}$
Thickness	x		m

$$f = \delta_f / \lambda_{fo} \tag{17.23}$$

where, δ_f and λ_{fo} are the fouling layer thickness and thermal conductivity, respectively.

As noted earlier, convective heat transfer occurs in the thermal boundary layer, δ, known as fluid film resistance layer. Fluid flow Reynolds number, $Re = u_0 x / \upsilon$, dictates the thickness of the thermal boundary layer. For a specific fluid, a higher free stream velocity produces a thinner fluid film resistance layer, and consequently, a higher heat transfer coefficient, h. Convective heat transfer is a function of the Reynolds number, Re, and also the Prandtl number, Pr, which relates to physical and thermal properties of the fluid (i.e., $Pr = \mu_f c_p / \lambda_f$). In Figure 17.16, δ_1 and δ_2 represent the boundary layer film thicknesses in the process side and heat transfer (HT) fluid side, respectively. Non-dimensional analyses shows that the dimensionless Nusselt number, $Nu_1 = h_1 \delta_1 / \lambda_{f1}$, whose

TABLE 17.5
Values of Thermal Conductivities for Some Foulants[8]

Foulants	Thermal Conductivity, λ_{fo} [$Wm^{-1}K^{-1}$]
Alumina	0.42
Biofilm (effectively water)	0.6
Carbon	1.6
Calcium sulphate	0.74
Calcium carbonate	2.19
Magnesium carbonate	0.43
Titanium oxide	8.0
Wax	0.24

definition includes the process-side heat transfer coefficient, h_1, is a function of Re and Pr. The solid wall may be composed of steel and a lining of glass, PTFE (Teflon®) or PVC, in which case, the overall thermal conductivity must account for the individual conductivities of each composite material and its thickness, as mentioned in Table 17.3.

17.7.1 Jacketed Stirred Tanks

The overall heat transfer coefficient, U, for a jacketed stirred tank is determined from the thermal resistances addressed, as shown in Figure 17.16.

$$\frac{1}{U} = \frac{1}{h_1} + f_1 + \frac{x}{\lambda} + f_2 + \frac{1}{h_2} \tag{17.24}$$

For standard vessels and open impellers, the process-side heat transfer coefficient, h_1, is generally determined from the Nusselt number, as indicated in equation 17.25.

$$Nu_1 = \frac{h_1 T}{\lambda_1} = C_1\, Re_1^{\alpha_1}\, Pr_1^{\beta_1} \left(VR_1\right)^{\gamma_1} \tag{17.25}$$

where, T = tank diameter (characteristic length) [m]
λ_1 = thermal conductivity of process fluid [$Wm^{-1}K^{-1}$]

$$h_1 = \frac{Nu_1\, \lambda_1}{T} \; [Wm^{-2}\, K^{-1}] \tag{17.26}$$

The process-side impeller Reynolds number, Re_1, Prandtl number, Pr_1, and the viscosity ratio, VR_1, are defined in equations 17.27–17.29.

$$Re_1 = \frac{\rho_1\, N\, D^2}{\mu_1} \tag{17.27}$$

where, ρ_1 = density of process fluid [$kg.m^{-3}$], N = impeller rotational speed [s^{-1}]
D = impeller diameter (characteristic length) [m], μ_1 = viscosity of process fluid [Pa-s]

$$Pr_1 = \frac{\mu_1\, c_{p1}}{\lambda_1} \tag{17.28}$$

where, c_{p1} = specific heat of the fluid [J.kg^{-1}K^{-1}]

$$VR_1 = \mu_1 / \mu_{w1} \tag{17.29}$$

where, μ_{w1} = viscosity of process fluid at wall temperature, T_{w1} [Pa-s]

For non-standard agitated vessels with open impellers, the proportionality constant, C_1, depends on dimensionless geometric ratios $\frac{D}{T}$ and $\frac{Z}{T}$, as shown in equation 17.30. Furthermore, as expected, in the case of proximity impellers such as anchors and helical ribbons, C_1 is a function of the ratios of impeller clearance, e, and pitch, i, with respect to the impeller diameter (e.g., $\frac{e}{D}$ and $\frac{i}{D}$) as well. See equation 17.31.

$$Nu_1 = \frac{h_1 T}{\lambda_1} = C_1' Re_1^{\alpha_1} Pr_1^{\beta_1} (VR_1)^{\gamma_1} (D/T)^{\varphi} (Z/T)^{\psi} \tag{17.30}$$

$$Nu_1 = \frac{h_1 T}{\lambda_1} = C_1'' Re_1^{\alpha_1} Pr_1^{\beta_1} (VR_1)^{\gamma_1} (e/D)^{\zeta} (i/D)^{\xi} \tag{17.31}$$

There are numerous publications on heat transfer from and to agitated vessels. Table 17.6 presents experimental correlations for the process-side heat transfer, Nu_1, for a few open and proximity impellers, compiled in an excellent earlier review article by Bondy and Lippa.[2] Relevant references,

TABLE 17.6
Process Fluid Side Nusselt Number of Jacketed Vessels[2]

Impeller and No. of Blades	Process Fluid Flow Reynolds Number	Nusselt No., $Nu_1 = h_1 T / \lambda_1$	**
FBT (6)—baffled	$Re_1 > 400$	$0.74\, Re_1^{2/3} Pr_1^{1/3} (VR_1)^{0.14}$	1
FBT (6)—baffled and unbaffled	$Re_1 < 400$	$0.54\, Re_1^{2/3} Pr_1^{1/3} (VR_1)^{0.14}$	
FBT (6)—baffled General form	$Re_1 > 400$	$0.85\, Re_1^{2/3} Pr_1^{1/3} (VR_1)^{0.14} (Z/T)^{-0.56} (D/T)^{0.13}$	2
RBI (6)—unbaffled	*	$0.68\, Re_1^{2/3} Pr_1^{1/3} (VR_1)^{0.14}$	3
Glass-lined RBI (3)—baffled	*	$0.33\, Re_1^{2/3} Pr_1^{1/3} (VR_1)^{0.14}$	4
Alloy steel RBI (3)—baffled	*	$0.37\, Re_1^{2/3} Pr_1^{1/3} (VR_1)^{0.14}$	5
PBT (4–45° pitch)—baffled	*	$0.54\, Re_1^{2/3} Pr_1^{1/4} (VR_1)^{0.14}$	6
Paddle—baffled and unbaffled	$Re_1 > 4000$	$0.36\, Re_1^{2/3} Pr_1^{1/3} (VR_1)^{0.14}$	7
	$20 < Re_1 < 4000$	$0.415\, Re_1^{2/3} Pr_1^{1/3} (VR_1)^{0.24}$	8
Anchor—unbaffled	$30 < Re_1 < 300$	$1.0\, Re_1^{2/3} Pr_1^{1/3} (VR_1)^{0.18}$	
	$300 < Re_1 < 4000$	$0.38\, Re_1^{2/3} Pr_1^{1/3} (VR_1)^{0.18}$	9
	$4000 < Re_1 < 37000$	$0.55\, Re_1^{2/3} Pr_1^{1/4} (VR_1)^{0.14}$	10
Helical Ribbon—unbaffled	$Re_1 < 130$	$0.248\, Re_1^{1/2} Pr_1^{1/3} (VR_1)^{0.14} (e/D)^{-0.22} (i/D)^{-0.28}$	11
	$Re_1 > 130$	$0.238\, Re_1^{2/3} Pr_1^{1/3} (VR_1)^{0.14} (i/D)^{-0.25}$	

* Data in references do not provide Reynolds number ranges for Nu_1 vs. Re_1 correlations.
** See Table 17.7 for relevant information related to experiments and impeller geometries.

experimental conditions, and remarks are stated in Table 17.7. Readers are strongly advised to view the references where experimental conditions pertinent to the empirical correlations are addressed for appropriate application.

Due to thermal boundary layers, bulk temperatures of the process and heat transfer fluids, T_1 and T_2, are different from their respective wall temperatures, T_{w1} and T_{w2}, as seen in Figure 17.16. As expected, when a process fluid is heated as per Figure 17.16, the wall temperature, T_{w1}, will be higher than the bulk fluid temperature, T_1, and when cooled, vice versa, i.e., $T_{w1} < T_1$. Viscosity Ratio, *VR*, takes into account the actual viscosity of the process or heat transfer fluid adjacent to the wall at temperature, T_w, which is significantly different from the bulk fluid viscosity. In the case of a liquid, viscosity is inversely proportional to temperature, e.g., viscosity decreases with increased temperature. However, for all practical purposes, the Reynolds number, *Re*, and Prandtl number, *Pr*, are based on the bulk viscosity, μ_1, of the process fluid. In the empirical correlations, the viscosity ratio term, VR_1, indirectly introduces the heat transfer effects through the process fluid viscosity, μ_{w1}, at the wall temperature, T_{w1}. The wall temperature determination procedure uses the following trial-and-error iteration steps. Because heat flux through the solid wall remains unaltered,

$$\dot{Q}_1 = \dot{Q}_2$$

$$h_1\left(T_1 - T_{w1}\right)A_1 = h_2\left(T_{w2} - T_2\right)A_2$$

Assume there is an infinitesimally small temperature difference across the solid wall, i.e., $T_{w1} = T_{w2} = T_w$, and that areas on either side of the wall are the same, i.e., $A_1 = A_2$.

$$h_1\left(T_1 - T_w\right) = h_2\left(T_w - T_2\right)$$

or,

TABLE 17.7
References, Experimental Conditions, and Remarks Pertinent to the Correlations in Table 17.6

**	Ref.	Experimental Conditions	Remarks
1	9,10	Standard vessel geometry (See Figure 17.17) $D/T = 0.3$, $Z/T = 1.05$, $C/T = 0.3$, $B_w/D = 0.1$.	FBT—Flat-Blade Turbine
2	11	Non-standard vessel geometry—general form incl. D/T, Z/T	-do-
3	12,13	Standard vessel geometry—Jacketed unbaffled[9,14]	RBI—Retreat-Blade Impeller
4	12	Jacketed baffled vessel w/glassed-steel impeller	-do-
5	12	Jacketed baffled vessel w/alloy steel impeller	-do-
6	15	Limited data for blade pitch and vessel baffling	Use $h_1 / 1.3$ as actual process-side heat transfer coefficient[14]
7	16	Jacketed baffled and unbaffled vessels[13]	
8	17	Jacketed baffled and unbaffled vessels[13]	Vessel geometry similar to ref. 14 but w/o internal coils
9	17	Anchor to wall clearance, $C_{A-w} \leq 0.025m$	
10	15	Anchor to wall clearance, $C_{A-w} = 0.025 - 0.13m$	Vessel geometry in ref. 14. Overall coefficient[18] is inversely proportional to the anchor-wall clearance. $U \alpha \left(1/C_{A-w}\right)$
11	19	$e = (T - D)/2$ [m] and i = agitator ribbon pitch [m]	

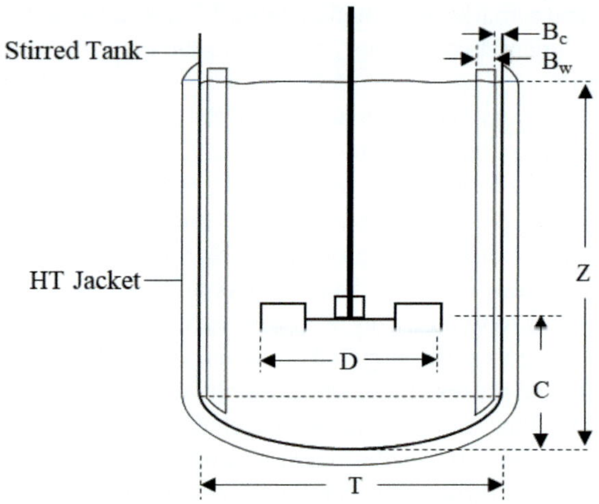

FIGURE 17.17 Jacketed stirred tank with a standard six flat-blade disk turbine impeller.[9,10]

[C = off-bottom clearance (m), B_w = baffle width (m), B_c = baffle/wall clearance (m), Z = liquid height (m)]

$$T_w = \left(\frac{h_1 T_1 - h_2 T_2}{h_1 + h_2} \right) \tag{17.32}$$

Step 1 of the experimental data correlation procedure estimates h_1 assuming $VR_1 = (\mu_1 / \mu_{w1})^{0.14} = 1$, i.e., $\mu_1 = \mu_{w1}$. Unless the term differs significantly from 1, one iteration should be sufficient to establish T_w. The viscosity value μ_{w1} at temperature T_w is used to determine the viscosity correction factor, VR_1. Else, μ_{w1} and the corresponding T_w values must be varied in stepwise fashion to achieve the appropriate process-side heat transfer correlation.[2]

Figure 17.2 shows different types of internal heat transfer configurations of a jacket. These are (a) annular jacket with and without spiral baffles, (b) full or half-pipe helical coils welded to the outer wall of the vessel, and (c) dimple surface. The jacket-side heat transfer coefficient, h_2, is calculated from the Nusselt number equations in Table 17.8. Relevant references, experimental conditions, and remarks are in Table 17.9.

$$Nu_2 = \frac{h_2 d_e}{\lambda_2} \tag{17.33}$$

where, d_e = jacket-side equivalent heat transfer diameter [m]
λ_2 = thermal conductivity of heat transfer fluid [$Wm^{-1}K^{-1}$]

The jacket-side impeller Reynolds number, Re_2, Prandtl number, Pr_2, and the viscosity ratio, VR_2, are presented in equations 17.34–17.36.

$$Re_2 = \frac{\rho_2 v_2 d_e}{\mu_2} \tag{17.34}$$

where, ρ_2 = density of heat transfer fluid [kg m^{-3}], v_2 = heat transfer fluid velocity [m s^{-1}]
d_e = characteristic length (jacket side heat transfer diameter) [m]
μ_2 = viscosity of heat transfer fluid [Pa s]

$$Pr_2 = \frac{\mu_2 c_{p2}}{\lambda_2} \tag{17.35}$$

TABLE 17.8
Jacket-Side Heat Transfer Fluid Nusselt Number of Agitated Vessels[2]

Type of Jacket	Heat Transfer Fluid Flow Reynolds Number, Re_2	Nusselt Number, Nu_2	*
Annular jacket w/spiral baffle	$Re_2 > 10^4$	$0.027\, Re_2^{4/5} Pr_2^{1/3} (VR_2)^{0.14} [1+3.5(d_e / d_c)]$	1
	$Re_2 < 2100$	$1.86 [Re_2 \, Pr_2 (d_e / L)]^{1/3} (VR_2)^{0.14}$	
	$2100 < Re_2 < 10^4$	Use above equations or determine from Figure 4 of Ref. 20	
Annular jacket w/o baffles	$Re_2 > 10^4$	$0.027\, Re_2^{4/5} Pr_2^{1/3} (VR_2)^{0.14}$	2
	$Re_2 < 2100$	$1.86 [Re_2 \, Pr_2 (d_e / L)]^{1/3} (VR_2)^{0.14}$	
	$2100 < Re_2 \le 10^4$	Use above equations or determine from Figure 4 of Ref. 20	
	Natural convection $Re_2 << 2100$	$1.02\, Re_2^{0.45} Pr_2^{1/3} (VR_2)^{0.14} (d_e / L)^{0.4} (d_{jo} / d_{ji})^{0.8} Gr^{0.05}$	3
Half-pipe coil jacket (180 or 120 deg.)	$Re_2 > 10^4$	$0.027 Re_2^{4/5} Pr_2^{1/3} (VR_2)^{0.14} [1+3.5(d_e / d_c)]$	4
	$Re_2 < 2100$	$1.86 [Re_2 \, Pr_2 (d_e / L)]^{1/3} (VR_2)^{0.14}$	
	$2100 < Re_2 < 10^4$	Use above equations or determine from Figure 4 of Ref. 20	
Dimple jacket	$Re_2 > 10^4$	$0.027\, Re_2^{4/5} Pr_2 (VR_2)^{0.14}$	5
	$Re_2 < 2100$	$1.86 [Re_2 \, Pr_2 (d_e / L)]^{1/3} (VR_2)^{0.14}$	
	$2100 < Re_2 < 10^4$	Use above equations or determine from Figure 4 of Ref. 20	

TABLE 17.9
References, Experimental Conditions, and Remarks Relevant to the Correlations in Table 17.8

*	Ref.	Experimental Conditions	Remarks
1	22	Baffle channel equivalent diameter, $d_e = 4w$ Channel cross-sectional flow area, $A_{jb} = pw$ w = width of annular space, p = pitch of the baffle Actual mass flowrate[21], $\dot{m}' = 0.6\dot{m}$	Leakage around spiral baffle = 35–50% of total mass flow rate, $\dot{m}$.[21] Fluid velocity is calculated based on actual mass flowrate.
2	22	Jacket area of cross-section, $A_j = \frac{\pi}{4}(d_{jo}^2 - d_{ji}^2)$ Equivalent diameter, $d_e = (d_{jo}^2 - d_{ji}^2) / d_{ji}$; where, d_{ji} and d_{jo} = jacket inner and outer diameters, respectively	Flow velocity is very low due to large cross-sectional area of the jacket without the baffles.
3	23	Grashof number, $Gr = d_e^3 \rho_2^2 g \beta_2 (\Delta T)_g / \mu_2^2$ Equivalent diameter, d_e = 2 × annular gap, $(d_{jo} - d_{ji})$ ρ_2 = density of heat transfer fluid β_2 = thermal coefficient of expansion of heat transfer fluid μ_2 = viscosity of heat transfer fluid $(\Delta T)_g$ = temperature difference between surface of the jacket inner wall and the bulk heat transfer fluid	Very low Reynolds number laminar flow is influenced by natural convection. *Grashof no.*, $Gr = \frac{Buoyancy\ force}{Viscous\ force}$ Correlation proposed to calculate approximate film coefficient, h_2 for water flow in the jacket annuli.

(Continued)

TABLE 17.9 *(Continued)*

References, Experimental Conditions, and Remarks Relevant to the Correlations in Table 17.8

*	Ref.	Experimental Conditions	Remarks
4	22	If coil of the half-pipe jacket has a 180-degree central angle: Equivalent diameter, $d_e = (\pi/2)\, d_i = 1.57\, d_i$ Area of half-pipe, $A_{hp} = (\pi/8)\, d_i^2 = 0.39\, d_i^2$ If coil of the half-pipe jacket has a 120-degree central angle: $d_e - 0.708 d_i$ and $A_{hp} - 0.154 d_i^2$, where d_i = id of coil tube or pipe, m A_{hp} = cross-sectional area of the half-pipe coil, m^2	Use A_{hp} to calculate velocity, v_2 in the half-pipe $v_2 = \dot{Q} / A_{hp}$ $Re_2 = \frac{\rho_2 v_2 d_e}{\mu_2}$
5	22	Since dimple geometry on outer surface of vessel wall varies, (a) use available coefficients from experimental data (b) vessel manufacturers often specify the equivalent diameter for heat transfer e.g., $d_e \approx$ 1.68cm (0.66in) and flow area, A_{dj} = 41.91 cm²/m (198 in²/ft) of vessel circumference[24]	Use A_{dj} to calculate v_2 in the dimple jacket: $v_2 = \dot{Q} / A_{dj}$ $Re_2 = \frac{\rho_2 v_2 d_e}{\mu_2}$

Note: d_e is the heat transfer equivalent diameter/hydraulic mean diameter, and different from the fluid flow hydraulic diameter described later.

where, c_{p2} = specific heat of heat transfer fluid [J kg^{-1}K^{-1}]

$$VR_2 = \mu_2 / \mu_{w2} \tag{17.36}$$

where, μ_{w2} = viscosity of heat transfer fluid at outer wall temperature T_{w2} [Pa-s]

17.7.2 Agitated Vessels with Internal Heat Transfer Surfaces

The overall heat transfer coefficient, U, for a stirred tank with internal heat transfer coils is determined from the thermal resistances addressed, as shown in Figure 17.16.

$$\frac{1}{U} = \frac{1}{h_1} + f_1 + \frac{x}{\lambda}\left(A_n / A_{avg}\right) + f_2 + \frac{1}{h_2\left(A_i / A_o\right)} \tag{17.37}$$

where, A_i and A_o = inner and outer surface areas of the tube [m²/m], respectively
A_n = surface area of outside coil or heat transfer barrier [m²/m]
A_{avg} = average surface area, $\frac{A_i + A_o}{2}$ [m²/m]

Since the circumferential area, per unit length, of a tube is related to its diameter, it follows that

$$A_i / A_o \propto \left(d_i / d_o\right) \tag{17.38}$$

where, d_i = inner diameter of the tube [m], and $d_o = d$ = outer diameter of the tube [m]

Thus, for all practical purposes, the ratio A_i / A_o can be replaced with d_i / d_o.

Several empirical correlations for the process fluid-side Nusselt number, Nu_1, are presented by Ludwig in the book[25] *Applied Process Design for Chemical and Petrochemical Plants*. As mentioned earlier, for standard vessels with open impellers, the process-side convective heat transfer coefficient, h_1, is determined from the Nusselt number equation 17.26. A general-purpose correlation for the process-side heat transfer coefficient with internal coils is presented[26] in equations 17.39–17.40.

$$Nu_1 = \frac{h_1\, d}{\lambda_1} = C_1\, Re_1^{\alpha_1}\, Pr_1^{\beta_1} \left(VR_1\right)^{\gamma_1} \left(\frac{D}{T}\right)^{\varphi} \left(\frac{d}{T}\right)^{\chi} \tag{17.39}$$

where, d = coil tube diameter [m], λ_1 = thermal conductivity of process fluid [W m^{-1} K^{-1}], D = impeller diameter [m], T = tank diameter [m], d_c = coil diameter [m]

$$Nu_1 = \frac{h_1\, d}{\lambda_1} = 0.17\, Re_1^{0.67}\, Pr_1^{0.37} \left(VR_1\right)^{\gamma_1} \left(\frac{D}{T}\right)^{0.1} \left(\frac{d}{T}\right)^{0.5} \tag{17.40}$$

See Table 17.10 for specific heat transfer conditions.

TABLE 17.10
Process Fluid-Side Nusselt Number of Agitated Vessels with Internal Coils[2,25]

Impeller and Internal Heat Transfer Surface	Process Fluid Flow Reynolds Number, Re_1	Nusselt Number, Nu_1	*
FBT (6) with multiple vertical helical coils.	$400 \le Re_1 \le 1.5\times10^6$	$0.17 Re_1^{0.67} Pr_1^{0.37} (VR_1)^{\gamma_1} (D/T)^{0.1} (d/T)^{0.5}$ γ_1 depends on the bulk viscosity of the process fluid, μ_1, and can vary between 0.1 to 1.0 for a viscosity range of 600 to 0.3cP. $\gamma_1 = 0.1\left(\mu\times8.621\times10^{-5}\right)^{-0.21}$	1
FBT (6) with single coil	$0.018 \le d/T \le 0.036$ $\mu_1 \le 10^4$	$0.17 Re_1^{0.67} Pr_1^{0.37} (VR_1)^{\gamma_1} (D/T)^{0.1} \left(d/T\right)^{0.5}$	2
FBT (4) with vertical tubes as baffles	$1300 \le Re_1 \le 2\times10^6$	$0.09 Re_1^{0.65} Pr_1^{0.3} (VR_1)^{0.14} (D/T)^{0.33} \left(2/B\right)^{0.2}$ B = number of baffles	3
FBT (4) with vertical plate coils as baffles	$Re_1 < 1.4\times10^3$	$0.1788 Re_1^{0.448} Pr_1^{0.33} (VR_1)^{0.50}$ <u>Note</u>: $Nu_1 = Nu_{1p} = \frac{h_o \cdot p_w}{\lambda_f}$ p_w = plate coil width (characteristic length) λ_f = fluid viscosity at mean fluid film temperature	4
	$Re_1 > 4\times10^3$	$0.0317 Re_1^{2/3} Pr_1^{1/3} (VR_1)^{0.5}$ <u>Note</u>: $Nu_1 = Nu_{1p} = \frac{h_o \cdot p_w}{\lambda_f}$ p_w = plate coil width (characteristic length) λ_f = fluid viscosity at mean fluid film temperature	
RBI (6) with internal helical coil		$Nu_1 = \frac{h_1 T}{\lambda_1} = 1.40\ Re_1^{0.62} Pr_1^{1/3} (VR_1)^{0.14}$	5
Propeller (6) with internal helical coil		$0.078 Re_1^{0.62} Pr_1^{1/3} (VR_1)^{0.14}$	6
Paddle with internal helical coil	$300 \le Re_1 \le 4\times10^5$	$Nu_1 = \frac{h_1 T}{\lambda_1} = 0.87 Re_1^{0.62} Pr_1^{1/3} (VR_1)^{0.14}$	7

TABLE 17.11
References, Experimental Conditions, and Remarks Pertinent to the Correlations in Table 17.10

*	Ref.	Experimental Conditions	Remarks
1	13, 26	Flat-blade turbine (six blades). Multiple vertical helical coils Tube spacing of two to four times tube diameter Four flat baffles of width, $B_w = T/10$, located either outside or inside the coil. Experiments were carried out with multiple coil banks.	Second coil with the similar heat transfer area as the first coil provides 70–90% of the heat transfer effectiveness of the first one.
2	26, 27	Flat-blade turbine (six blades) Four flat baffles located either outside or inside of the coil Single vertical helical coil. Tube spacing of two to four times d_t. $0.018 \leq d_o / T \leq 0.036$. Viscosity, $\mu_1 \leq 10^4$ cP $d_t = d_o$ = outer diameter of tube/pipe of the internal coil See Figure 17.18 for experimental setup configuration.[14,27]	Tube spacing of coil wraps of two to four tube diameters shows no appreciable effect on h_o. However, wider spacing above 50cP viscosity gives lower h_o. For example, the ratio of coefficients at 4d and 2d for water of 0.4 cp and oil of 50 cp were 0.96 and 0.88, respectively.
3	27–29	FBT (4) with vertical tube baffles Four or six banks of vertical tube baffles—three tubes per bank See Figure 17.19 for experimental setup configuration.[14,28]	This tube design gives good heat transfer without needing vertical baffles. In the turbulent regime, $h_o \alpha (shp)^{0.22}$. Agitation power required in vessel with vertical tubes is 75% that of standard wall baffles.[27]
4	27, 30	FBT (4) with vertical plate coils Coils present a solid vertical face, with the “coils” vertical but impressed in the plates for flow of the heating or cooling medium; replaces baffles.	Plate coil designs are more solid obstructions to through flow in the vessel than individual vertical tube coils. Thus, prevents the need for vertical baffles in a tank. Typically, four or six banks are used.[27]
5	12	RBI (6)—single and dual impellers See Figure 17.20 for experimental setup configuration.[13,14] Note: Nu_1 uses T as the characteristic length.	Heat transfer rate of dual impellers is ~15% greater than a single impeller. See reference 14 for vessel geometry. Correlations for three-bladed RBI seems to be unavailable.[2]
6	12	Propeller (3)	Correlation is based on limited data with regard to propellor pitch and vessel baffling.[2] Use $h_1 / 1.3$ as actual process-side heat transfer coefficient.[14]
7	16	Single Paddle See Figure 17.21 for experimental configuration.[14,16] Note: Nu_1 uses T as the characteristic length.	A narrow paddle See reference 14 for vessel details.

The viscosity ratio exponent, γ_1, depends on process fluid bulk viscosity and is experimentally determined.

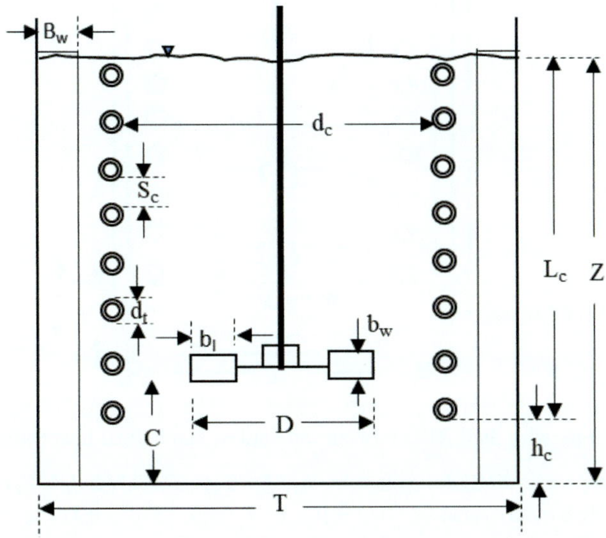

FIGURE 17.18 Stirred tank with a six-flat-bladed disc turbine and a single heat transfer coil.[14,27]

[Tank: Square vessel with flat bottom, T = 1.22m (48″); Turbine: Six flat blades mounted on a disc (Rushton); Clearances: $D/T = 0.250 - 0.583$, $C/D = 1.0$, $b_w/D = 0.2$, $b_l/D = 0.25$; Baffles: $B_w = T/10$, $B_c = T/48$, Coil: $S_c/d_t = 2 - 4$, $L_c/T = 0.650$, $d_c/T = 0.7$, $d_t/D = 0.03125 - 0.1458$, $h_c/T = 0.15$]

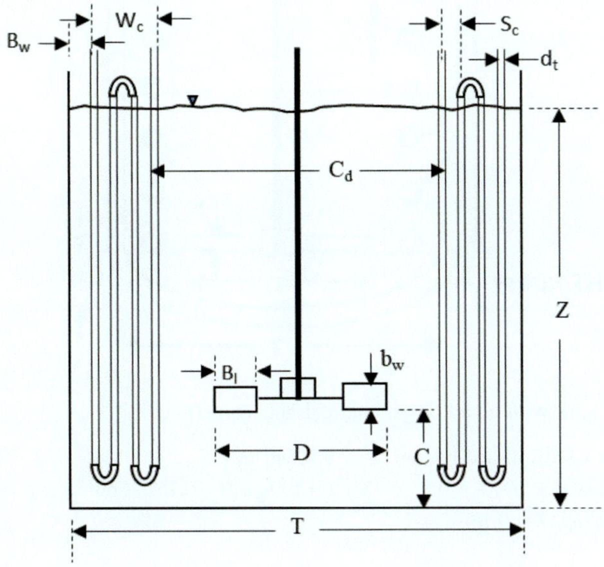

FIGURE 17.19 Stirred tank with a paddle and helical heat transfer coil.[14,30]

[Tank: flat-bottom square vessel with $T = 1.22m(48'')$, Turbine (Rushton): six flat blades and disc; Clearances: $D/T = 0.250 - 0.333$, $C/T = 0.5$, $Z/T = 1.0$, $b_w/D = 0.2$, $b_l/D = 0.25$, Baffles: Vertical coil baffles, Coil: $W_c/T = 0.23$, $S_c/T = 0.0312$, $B_c/T = 0.0208$, $d_t/T = 0.0208$]

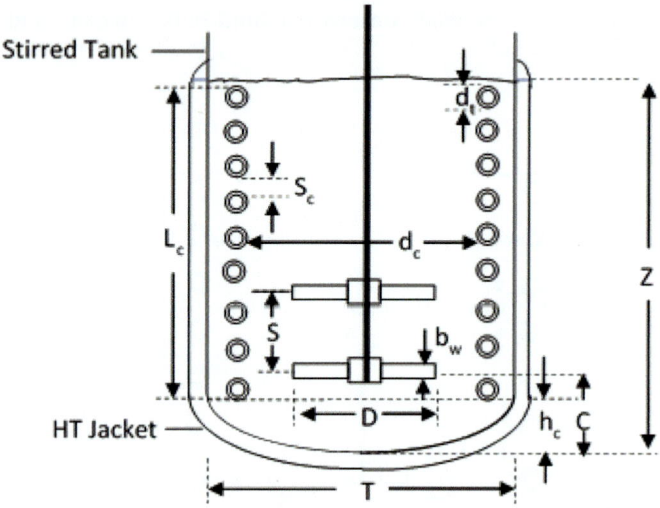

FIGURE 17.20 Stirred tank with dual RBIs (six curved blades) and helical heat transfer coil.[13,14]

[Tank: Jacketed 100-gal vessel with ASME dished bottom, $h_d / T = 0.15$, $T = 0.762m(30'')$, no baffles; Turbines: dual, retreat six-bladed impellers (RBIs) with clearances: $D / T = 0.4$, $C / T = 0.33$, $S / T = 0.33$, $b_w / D = 0.166$, Coil: 10 turns, $S_c / d_t = 1$, $L_c / T = 1.0$, $d_c / T = 0.8$, $d_t / d_c = 0.042$, $h_c / T = 0.10$]

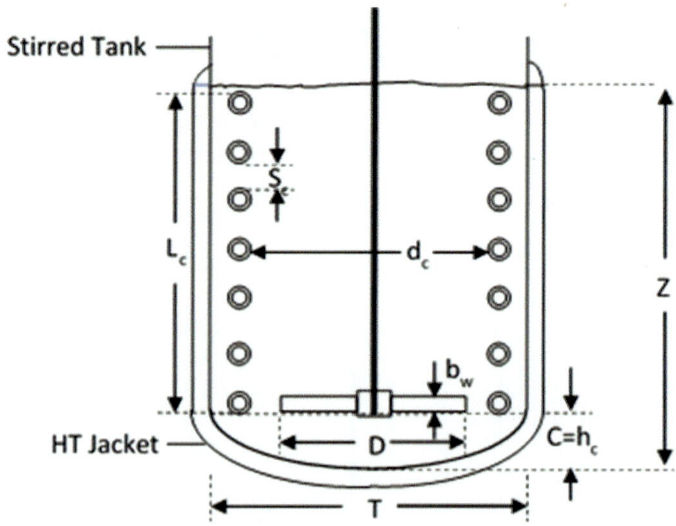

FIGURE 17.21 Stirred tank with a paddle and helical heat transfer coil.[14,16]

[Tank: Jacketed vessel with ASME dished bottom. T $= 0.305m(12'')$, $h_d / T = 0.15$, $Z / T = 0.83$, no baffles; Turbine: Single flat paddle with clearances: $D / T = 0.6$, $C / T = 0.15$, $b_w / D = 0.167$, Coil: $S_c / T = 0.0154$, $L_c / T = 0.4375$, $d_c / T = 0.8$, $d_t / d_c = 0.052$, $h_c / T = 0.15$]

Figure 17.22 schematically shows the cross-section of a heat transfer coil tube inside a vessel. Respectively, d_i, d_o, and d_m represent the inner, outer, and mean diameters of the tube. Corresponding surface areas of the coil, of length L, are A_i, A_o, and A_m. External and internal heat transfer film coefficients are represented by h_1 and h_2.

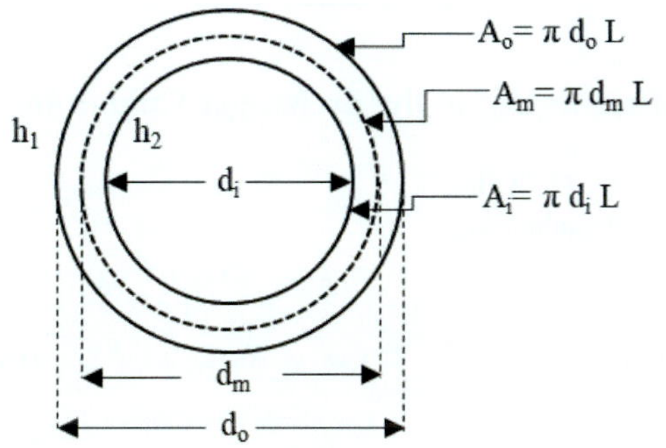

FIGURE 17.22 Cross-section of a heat transfer coil tube.

Internal coil-side heat transfer coefficient, h_2, is calculated from the equations presented in Table 17.12.

$$Nu_2 = \frac{h_2 d_e}{\lambda_2} \tag{17.41}$$

where, d_e = coil-side equivalent heat transfer diameter [m]
λ_2 = thermal conductivity of heat transfer fluid [W m^{-1}K^{-1}]

The coil-side impeller Reynolds number, Re_2, Prandtl number, Pr_2, and the viscosity ratio, VR_2, are presented in equations 17.42–17.44.

$$Re_2 = \frac{\rho_2 v_2 d_e}{\mu_2} \tag{17.42}$$

where, ρ_2 = density of heat transfer fluid [kg m^{-3}], v_2 = heat transfer fluid velocity [m s^{-1}],
d_e = characteristic coil side pipe/tube length [m], μ_2 = heat transfer fluid viscosity [Pa-s]

$$Pr_2 = \frac{\mu_2 c_{p2}}{\lambda_2} \tag{17.43}$$

where, c_{p2} = specific heat of heat transfer fluid [J.kg^{-1}K^{-1}]

$$VR_2 = \mu_2 / \mu_{w2} \tag{17.44}$$

where, μ_{w2} = viscosity of heat transfer fluid at coil wall temperature T_{w2} [Pa-s]

Turbulent flow ($Re_2 > 10^4$) inside the tubes maximizes heat transfer to and from the heat exchanger fluid. Viscous fluids may unavoidably produce laminar ($Re_2 \le 2100$) or transitional ($2100 \le Re_2 \le 10^4$) flows, however.

Assuming a linear correlation for Nu in the transition regime ($2100 < Re < 10^4$), the following simplification may be adapted to predict approximate Nu in the transition regime:

$$Nu_{transition} = Nu_{laminar} + \left(Nu_{turbulent} - Nu_{laminar}\right)\left(\frac{Re - 2100}{10^4 - 2100}\right)$$

In the case of turbulent flowing $\left(Re_2 > 10^4\right)$ water in the coil, Figure 17.23 should be used to determine the coil- or tube-side heat transfer coefficient, h_2.[2,31]

TABLE 17.12
Inside Heat Transfer Coefficients, h_2, of the Exchanger Fluid in the Internal Coils[2]

Type of Coil	Heat Transfer Fluid Flow Reynolds Number, Re_2	Nusselt Number, $Nu_2 = \frac{h_2 d_i}{\lambda_2}$	Reference
Circular cross-section	$Re_2 \geq 10^4$	$0.027 Re_2^{4/5} Pr_2^{1/3} (VR_2)^{0.14} \left[1+3.5(d_i / d_c)\right]$ *	22
	$Re_2 \leq 2100$	$1.86\left[Re_2 \, Pr_2 \, (d_i / d_c)^{0.5}\right]^{1/3} (VR_2)^{0.14}$	22
	$2100 \leq Re_2 \leq 10^4$	Use above equations or determine from Figure 4 of ref. 20.	20

* $\left[1+3.5(d_i / d_c)\right]$ is the turbulent flow coil correction factor to the straight-pipe internal heat transfer correlation[20] by Sieder and Tate. d_c = coil diameter.

As pointed out in equations 17.30 and 17.31, Nusselt number correlations involve multiple dimensionless parameters and geometric factors. In published heat transfer results, experimental ranges of the geometric variables are limited. Hence, heat transfer system designers should not expect precise results from the established correlations, and they must allow considerable design flexibility to be addressed by practicing engineers in the plant. Such agilities include but are not limited to impeller

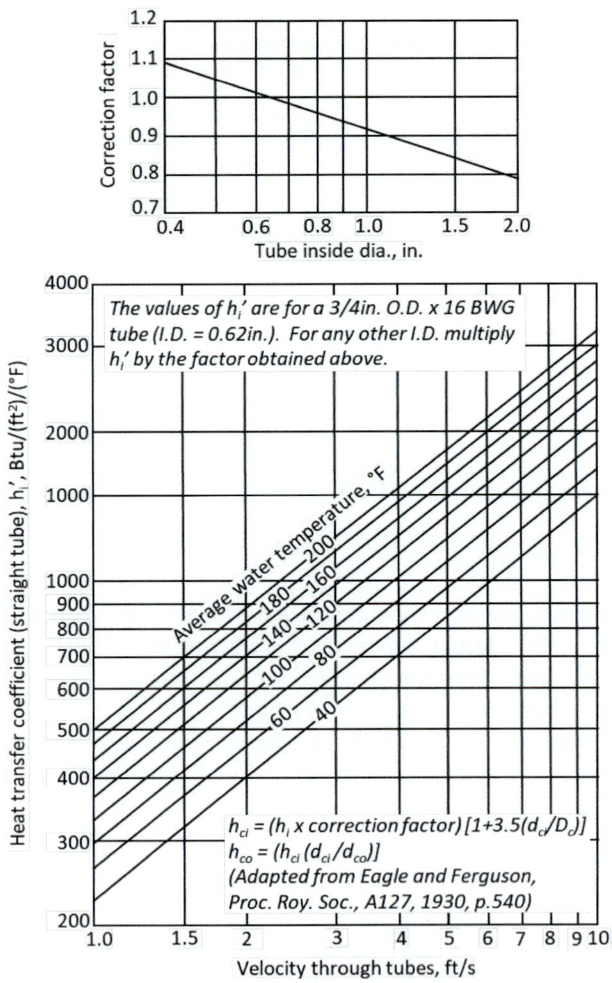

FIGURE 17.23 Film coefficients for water in tubes and coils.[31]

type, rotational speed, heat transfer fluid flowrate, and inlet temperature, etc. Furthermore, heat transfer data appear to be no better than ±20% when comparing results from several investigators and the basic fundamentals of their systems.[25]

17.8 HEAT TRANSFER AREA

17.8.1 Annular Jacket

With or without the spiral baffles in the jacket, the effective heat transfer area, A, is the shared internal and external wetted area of the process fluid and the heat transfer fluid. When the fill level of one fluid is less than that of the other, the excess wetted area of the higher-filling fluid produces losses or gains in heat.

17.8.2 Half-Pipe Coil Jacket

Standard half-pipe diameters are 2, 3 and 4 inches, while spacing between adjacent half-pipes is normally ¾ inch. Due to longitudinal heat transfer, unwetted wall surface on the jacket side (e.g., the space between half pipes) is also heated and should be included when calculating effective heat transfer area. An accepted rule of thumb adds only 60% of the nonwetted surface area between the half-pipes to the wetted surface area of the half coil. This approach yields conservative area, A', for heat transfer calculations. A' / A values are 0.9, 0.93, 0.92 for 2-inch, 3-inch and 4-inch half coils, respectively.[2]

17.8.3 Dimple Jacket

Figure 17.24 presents the geometry of a typical dimple jacket vessel, together with dimple dimensions (d_1 and d_2) and formulas for calculating the associated areas (A_{min} and A_{max}). These geometric values are used in the following correlation to determine the convective heat transfer coefficient, h_i, for a fluid being pumped through the dimple jacket.[32]

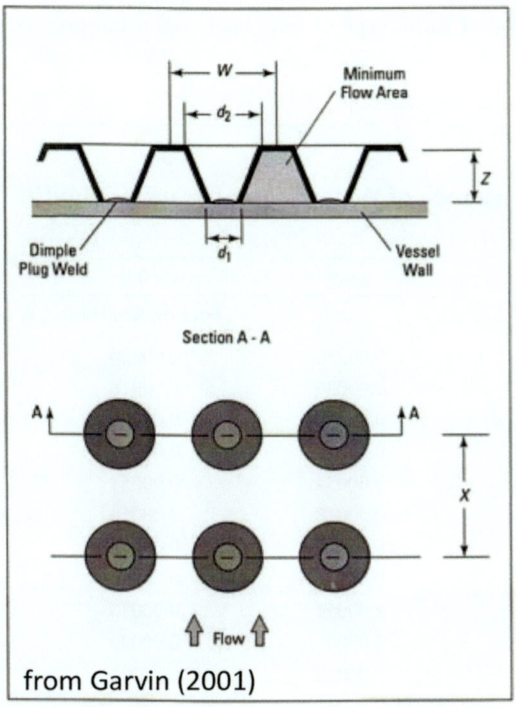

FIGURE 17.24 Geometry of a typical dimple jacket.[32]

$$h_i = \left[0.0845(W / X)^{0.368}(A_{min} / A_{max})^{-0.383} Re^{0.695} Pr^{0.33}\right] k / d_o \qquad (17.45)$$

where $A_{min} = Z(W - d_o)$, $A_{max} = ZW$; $d_o = (d_1 + d_2)/2$

$V_{max} = Flow / A_{min}$, $Re_o = (\rho d_o V_{max}) / \mu$

17.8.4 Internal Coils (Helical and Harp)

In the case of helical coils, the effective heat transfer area is the total wetted area, based on the coil outside surface. This outside transfer area, A_{co} [m^2], is calculated[2] from

$$A_{co} = \pi d_o H_c n\left[(\pi d_c)^2 + n^{-2}\right]^{1/2} \qquad (17.46)$$

where, n is the number of turns per meter of coil height, or, the inverse of the coil pitch, p
d_o = outside diameter of coiled tube, [m]
H_c = coil height, [m]
d_c = mean or centerline diameter of the coil helix, [m]

17.9 FOULING RESISTANCE

The fouling layers consisting of extraneous deposit of materials on the heat transfer surfaces usually have low thermal conductivity, as presented in Table 17.13. Fouling is a dynamic (time-dependent) phenomenon that changes with time. It simultaneously increases the overall thermal resistance and lowers the overall heat transfer coefficient in agitated vessels. Fouling also impedes fluid flow and increases pressure drop inside of heat transfer coils and plates. The magnitude and nature of fouling on a given wall is a function of the system and the equipment used. Parameters influencing the fouling rate include the type of fluid, type of heat exchanger, temperature, velocities, and material of construction.

TABLE 17.13
Approximate Fouling Resistance of Water as Heat Transfer Fluid[33]

Water Temperature, °C	≤ 52		> 52	
Water Velocity, m.s^{-1}	≤ 0.9	> 0.9	≤ 0.9	> 0.9
Heat Transfer Fluids	Fouling Resistance, f_2 (m^2 K W^{-1})			
Boiler blowdown water	0.00036	0.00036	0.00036	0.00036
Brackish water	0.00036	0.00018	0.00054	0.00036
Condensate (steam)	0.00009	0.00009	0.00009	0.00009
Cooling tower water (treated)	0.00018	0.00018	0.00036	0.00036
Cooling tower water (untreated)	0.00054	0.00054	0.00088	0.00072
City or well water	0.00018	0.00018	0.00036	0.00036
River water (minimum)	0.00036	0.00018	0.00053	0.00036
River water (average)	0.00054	0.00036	0.00072	0.00054
River water (muddy or silty)	0.00054	0.00036	0.00072	0.00054
Sea water	0.00009	0.00009	0.00018	0.00018
Treated boiler feed water	0.00018	0.00009	0.00018	0.00018

There is insufficient information to predict fouling resistance *a priori*. Overall fouling resistance that combines both the heat transfer fluid side and the process fluid side ($f = f_1 + f_2$) can be determined from production data in existing pilot plant or production-scale operating facilities. However, when production data is not available, one must rely on other sources.

The accumulated experience of heat exchanger designers and users is published in standards of Tubular Exchanger Manufacturers Association (TEMA). Over the years, approximate fouling factors have been compiled for various systems and fluids as seen in Tables 17.13–17.15. Actual values of process-side fouling resistance, f_1, depend on specific applications. Heat transfer fluids are occasionally treated with corrosion inhibitors and antibacterial agents to minimize fouling layers and subsequently to reduce their thermal resistance, f_2. Water is one of the best choices for liquid cooling applications due to its high heat capacity and thermal conductivity. It is also compatible with copper, one of the most cost-effective heat transfer materials for the fluid path.

TABLE 17.14

Approximate Fouling Resistance of Non-Aqueous Heat Transfer Fluid[33]

Heat Transfer Fluids	Fouling Resistance, f_2 ($m^2\ K\ W^{-1}$)
Refrigerants (liquid)	0.00018
Hydraulic fluids	0.00018
Industrial organic heat transfer media	0.00036
Ethylene glycol solutions	0.00036

TABLE 17.15

Approximate Fouling Resistance of Process Fluids[34]

Process Fluids	Fouling Resistance, f_2 ($m^2K\ W^{-1}$)
Transformer oil	0.00018
Engine oil	0.00018
Chlorinated hydrocarbon liquid	0.00018–0.00036
Ethylene glycol solution	0.00036
MEA and DEA solution	0.00036
DEG and TEG solution	0.00036
Emulsion polymerization media	0.00018–0.00036

17.10 AGITATED VESSEL HEAT TRANSFER SYSTEM DESIGN

As mentioned earlier, the critical parameters determining the heat transfer effectiveness of a heat exchanging system are

(a) mutually shared wetted surface area of the heat transfer and process fluids, such as A, A', or A_{co}
(b) heat transfer fluid flow Reynolds number, Re_2
(c) thermal properties of heat transfer fluid such as specific heat, C_{p2}, and thermal conductivity, λ_2

The major goals of heat exchanger designers are to increase all of these parameters in a cost-effective manner to remove or add heat to the process fluid without hindering its required mixedness in the stirred vessel.

Table 17.16 presents relative effectiveness of a few internal heat transfer systems with respect to the external jacket, assuming that the physical and thermal characteristics of the process fluid and the agitation system remain unaltered.[25,35] Figure 17.25 shows typical mixer-side heat transfer coefficients, h_1, as a function of viscosity for organic chemicals that are being mixed and heated/cooled by either internal vertical coils or a single helical coil. The h_1 values will serve to provide an "order of magnitude" estimate of h_1 to begin preliminary design of heat transfer systems. In the case of properly baffled jacketed tanks, the mixer-side heat transfer coefficient decreases by 35%.[26]

TABLE 17.16
Relative Mixer-Side Thermal Effectiveness of Heat Transfer Systems[26]

Transfer Surfaces	Relative Heat Transfer Effectiveness
Tank jacket	1.0
Vertical tubes	1.54
Helical coils (first bank)	1.54
Helical coils (second bank)	1.31
Helical coils (third bank)	1.08
Plate coils (or panel coils)	1.12

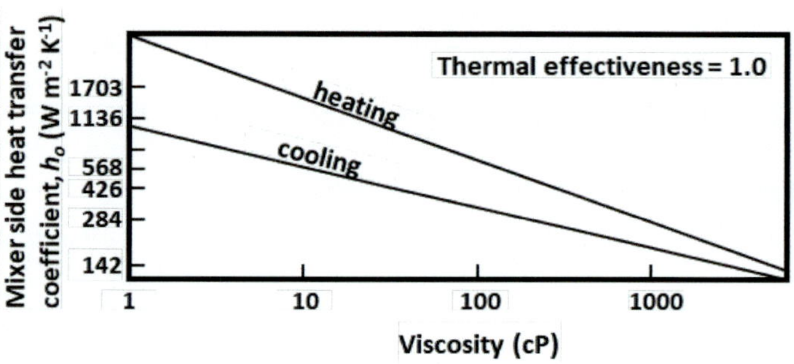

FIGURE 17.25 Typical mixer-side heat transfer coefficients of organic process fluids.[26]

17.10.1 Batch Heating and Cooling

In operating a jacketed batch vessel, it is often necessary to have a foreknowledge of the time, θ, needed to heat or cool the working media from temperature t_1 to temperature t_2. Equations 17.47–17.50 are used[36] to determine θ. The mass of the media in the vessel is M_1. The media's specific heat is C_{p1}. Jacket fluid inlet and outlet temperatures are T_1 and T_2, while T is the average jacket fluid temperature, $(T_1 + T_2)/2$. The mass flowrate and specific heat of the jacket fluid are $\dot{m}_2$ and C_{p2}, respectively.

1. For heating:

$$\theta = \frac{M_1 C_{p1}}{UA} \ln\left(\frac{T - t_1}{T - t_2}\right) \tag{17.47}$$

And for cooling:

$$\theta = \frac{M_1 C_{p1}}{UA} \ln\left(\frac{t_1 - T}{t_2 - T}\right) \tag{17.48}$$

2. If jacket-side fluid ΔT, i.e., $(T_1 - T_2)$ is > 10% of log mean temperature difference, ΔT_{LM}, the following equations are used:

For heating:

$$\theta = \frac{M_1 C_{p1}}{\dot{m}_2 C_{p2}} \ln\left(\frac{T_1 - t_1}{T_1 - t_2}\right) \frac{K}{K-1} \tag{17.49}$$

And for cooling:

$$\theta = \frac{M_1 C_{p1}}{\dot{m}_2 C_{p2}} \ln\left(\frac{t_1 - T_1}{t_2 - T_1}\right) \frac{K}{K-1} \tag{17.50}$$

K is defined by the following equation,

$$K = e^{\left(UA/\dot{m}_2 C_{p2}\right)} \tag{17.51}$$

where, A is vessel surface area in contact with the process fluid.

17.11 EXAMPLE

As an example, design of the heat transfer system of a jacketed glass-lined carbon steel vessel to produce 1600-gal of acrylic latex using emulsion polymerization process is presented. Three Schedule-40 3-inch stainless steel tubes in the form of half-pipe helical coils, as shown in Figure 17.26, are welded to the exterior of the reactor wall. One coil is wrapped around the bottom head (zone 1), whereas the other two covered the cylindrical part of the vessel (zones 2 and 3). Each coil has its own heat transfer (utility) fluid inlet and outlet nozzles. Coils are enclosed in an HT jacket whose outside surface is fully insulated. The vessel's inner-wall glass lining thickness is 80mil (0.080″).

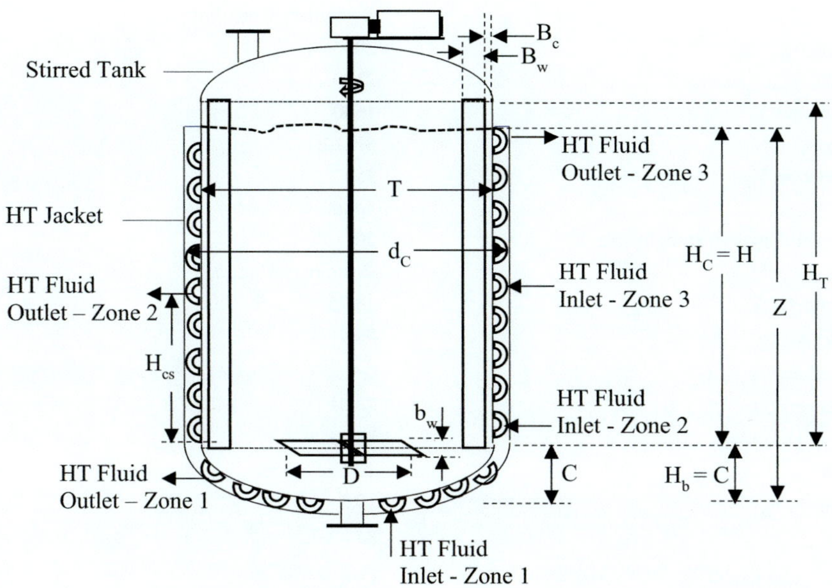

FIGURE 17.26 Schematic of a half-pipe jacketed, glass-lined reactor.

Table 17.17 summarizes the thermal and physical properties of the process fluid (acrylic latex) and heat transfer fluid (water). These properties are used to determine the process-side and utility-side Reynolds numbers (Re_1, Re_2) and Prandtl numbers (Pr_1, Pr_2) to predict appropriate Nusselt numbers (Nu_1, Nu_2) using relevant correlations expressed in Table 17.6 and Table 17.8.

In this example, without any predisposition, the basic dimensions of a De Dietrich GL-2000 gal. Hemicoil® type of glass-lined reactor is considered.[37] The bottom head of the reactor is 2:1 semi-elliptical in shape. There are four standard glass-lined carbon steel flat baffles, as shown in Figure 17.26.

TABLE 17.17

Properties of the Working Media and Heat Transfer Fluid

	Process Fluid (1)	Heat Transfer Fluid (2)
Working Media/Properties at ~298K	**Acrylic Latex**	**Water**
Viscosity, μ [Pa-s]	0.1	0.001
Density, ρ [kg m^{-3}]	1200	1000
Thermal conductivity, λ [W m^{-1}K^{-1}]	0.20	0.61
Specific heat, C_p [J kg^{-1}K^{-1}]	2095	4190

All of the design information pertinent to the tank internals and process liquid levels are presented in Table 17.18. Parameters related to the agitation system are tabulated in Table 17.19.

TABLE 17.18

GL-2000 gal. Hemicoil® Glass-Lined Reactor and Its Design Parameters[38]

Reactor type	**De Dietrich CGL-2000 Glass-Lined Carbon Steel**	
Parameters:	Customary Unit, gal	SI Unit, m^3
Working volume, V_w	2000	7.570
Total volume, V_t	2190	8.289
Process fluid volume, V	1600	6.056
Bottom head type	SEL 2:1	
Bottom head volume, V_b	254	0.961
Liquid levels	$gal.in^{-1}$	m^3m^{-1}
Liquid vol. per unit length of straight side, V_c	20	2.98
Parameters:	Customary Unit, in	SI Unit, m
Vessel outer diameter, T_o	78.75	2.000
Vessel wall thickness, t_w (x_1)	0.84	0.021
Glass-lining thickness, t_g (x_2)	0.08	0.002
Vessel inner diameter, $T = T_o - 2(t_w + t_g)$	76.91	1.954
Inner diameter of jacket, d_{ji}	82.25	2.089
Jacket width, $W_j = (d_{ji} - T_o)/2$	1.75	0.044
Bottom head depth, h_b	19.5	0.495
T-T height, H_T	84	2.134
Cylindrical part liquid height, $H = \frac{(V - V_b)}{V_c}$	67.3	1.709

(Continued)

TABLE 17.18 *(Continued)*

GL-2000 gal. Hemicoil® Glass-Lined Reactor and Its Design Parameters[38]

Reactor type	**De Dietrich CGL-2000 Glass-Lined Carbon Steel**	
Liquid height from bottom, $Z = h_b + H$	86.3	2.192
Type of baffle	Standard flat baffles	
Number of flat baffles, n	4	
Material of construction	Glass-lined carbon steel	
Baffle width, $B_w = T / 12$	6.39	0.162
Baffle clearance, $B_c = T / 48$	1.60	0.041

TABLE 17.19

Agitation System Parameters

Shaft	**Glass-Lined Carbon Steel**	
Parameters:	Customary Unit	SI Unit
Motor max power, P_{max}	15 *hp*	11.19 *kW*
Motor max rotational speed, N_{max}	125 *rpm*	2.08 s^{-1}
Agitator type	Standard glass-lined pitch-blade turbine (PBT)	
Number of agitators, n_a	1	
Number of blades, n_b	4	
Material of construction	Glass-lined carbon steel	
Parameters:	Customary Unit, *inch*	SI Unit, *m*
Impeller diameter (span), D	33.50	0.851
Impeller blade width, $b_w = D / 5$	6.70	0.170
Impeller off-bottom clearance, C	19.50	0.495
Impeller speed parameter	Customary unit, *rpm*	*rps*
Rotational speed, N (design value)	120	2

Details of the two helical half-coils designated as zones 2 and 3 inside the jacket are presented in Table 17.20. See Figures 17.26 and 17.27. For all practical purposes, the bottom coil around the bottom head, labeled as zone 1, is ignored for heat transfer calculations. This is in most instances a conservative approach; the bottom coil provides added benefit to the heat transfer to and from the working media.

TABLE 17.20

Design Parameters of Half-Coil Heat Transfer Jacket

Jacket Material of Construction	**Carbon Steel**	
Jacket internals	Helical half-pipe coil	
Coil material of construction	Stainless steel pipes	
Tube/pipe nominal diameter	Nominal 3″ SS Schedule-40	
Parameters:	Customary unit, *inch*	SI unit, *m*
Tube outer diameter, d_o	3.500	0.089

(Continued)

TABLE 17.20 *(Continued)*
Design Parameters of Half-Coil Heat Transfer Jacket

Jacket Material of Construction	Carbon Steel	
Tube inner diameter, d_i	3.068	0.078
Tube wall thickness, $t_c = (d_o - d_i)/2$	0.216	0.005
Sets of coils, s_c and s_{cc}	3 (incl. bottom head coil), 2 (on cylindrical wall)	
Each coil diameter, $d_c \approx T_{ji}$	82.25	2.089
Clearance or gap between tubes, $t_G = 0.5d_o$	1.75	0.044
Pitch of coil, $p_c = 1.5d_o$	5.25	0.133
Length of each coil wrap, $l_c \approx \pi\, d_c$	258.26	6.56
Number of coils, $n_c \approx (H / p_c) + 1$	14	
Sets of coils (cooling zones), s_{cc}	2	
Number of coils in each set, $n_s \approx n_c / s_{cc}$	7	
Each set coil length, $l_{cs} \approx n_s \times l_c$	1807	45.90
Total coil height, $H_c \approx H$	67.3	1.709
Each set coil height, $H_{cs} \approx H/2$	33.65	0.855
No. of inlet/outlet nozzles, n_i and n_o	3/3 (w/bottom head coil), 2/2 (w/cylindrical coil)	
Parameters: Mutually sharing wetted areas	$inch^2$	m^2
Wetted area of a set of coils, $A_{cws} = d_i \times l_{cs}$	5544	3.57
Total cylindrical wetted area of coils, $A_{cw} = s_{cc} \times A_{cws}$	11088	7.15
Jacket-side cylindrical area, $A_j = \pi \times T_o \times H$	16642	10.74
Actual effective heat transfer area per coil, $A = \left(\frac{0.85 \times A_j}{2}\right)$	7072	4.56

$A = E \times A_j$ where, wall coverage factor[39] $E = 0.8–1.0$

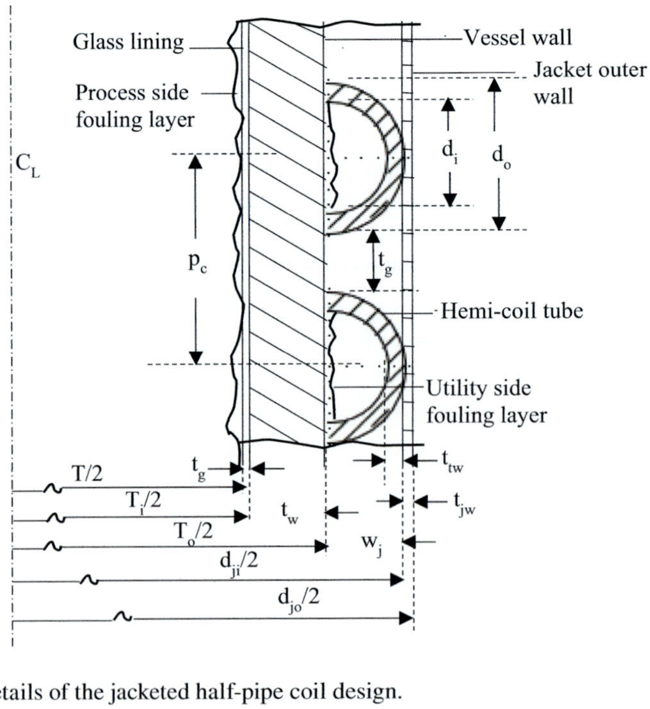

FIGURE 17.27 Details of the jacketed half-pipe coil design.

17.11.1 Design Calculations

1. The overall heat transfer coefficient, U, of a jacketed STR is determined from equation 17.24.

$$\frac{1}{U}=\frac{1}{h_1}+f_1+\frac{x}{\lambda}+f_2+\frac{1}{h_2}$$

2. The thermal conductivity, λ, of the glass-lined carbon steel composite wall is determined from the following equation. See Table 17.3.

$$\lambda=x\left(x_1/\lambda_1+x_2/\lambda_2\right)^{-1}$$

$$x/\lambda=\left(x_1/\lambda_1+x_2/\lambda_2\right)=\left(0.021/45\right)+\left(0.002/1.0\right)=0.0025W^{-1}m^2C$$

where, $x=x_1+x_2$ is the composite wall thickness, and thermal conductivities of carbon steel and glass are 45 and 1.0 $Wm^{-1}K^{-1}$, respectively. See Table 17.2.

3. Fouling factors with cooling tower water (temp $\leq 52\ C$ and velocity $> 0.9\ ms^{-1}$),

$$f=f_1+f_2=0.0003+0.00018=0.00048\ W^{-1}m^2K$$

where, f_1 and f_2 are the fouling factors of the fluid films on the process and the utility sides, respectively. See Table 17.13 and Table 17.15.

4. Process-side heat transfer coefficient (Standard PBT), h_1, is determined as follows (see Table 17.6):

$$Nu_1=0.54Re_1^{2/3}Pr_1^{1/4}\left(VR_1\right)^{0.14}$$

Process-side Reynolds number, Re_1, and Prandtl number, Pr_1, are calculated using equations 17.27 and 17.28, respectively.

$$Re_1=\frac{\rho_1 N D^2}{\mu_1}=\frac{1200kg.m^{-3}\times 2.0s^{-1}\times\left(0.851m\right)^2}{0.1Pa.s}=17380$$

$$Pr_1=\frac{\mu_1 c_{p1}}{\lambda_1}=\frac{0.1Pa.s\times 2095Jkg^{-1}K^{-1}}{0.20Wm^{-1}K^{-1}}=1047$$

Assuming viscosity of the process fluid at wall temperature, $\mu_{w1}=\mu_1$, the viscosity ratio from equation 17.29 becomes

$$VR_1=\mu_1/\mu_{w1}=1.0$$

$$Nu_1=0.54\left(17380\right)^{2/3}\left(1047\right)^{1/4}\left(1\right)^{0.14}=2068$$

$$h_1=\frac{Nu_1\,\lambda_1}{T}=\frac{2068\times 0.20}{1.954}=212\,Wm^{-2}K^{-1}\left(37.1\text{ Btu hr}^{-1}\text{ ft}^{-2}\text{ F}^{-1}\right)$$

According to Table 17.7, the actual process-side heat transfer coefficient[14] is $h_1 / 1.3$. Hence, $h_1 = 212 / 1.3 = 163 Wm^{-2} K^{-1}$ $\left(28.69 \text{ Btu hr}^{-1}\text{ft}^{-2}\text{F}^{-1}\right)$

5. Jacket-side heat transfer coefficient with half-pipe coil, h_2, is determined using the following steps. See Table 17.8. (modified Sieder-Tate correlation[20] for limpet coil)

$$Nu_2 = 0.027\ Re_2^{4/5} Pr_2^{1/3} \left(VR_2\right)^{0.14} \left[1 + 3.5\left(D_e / D_c\right)\right]\ Re_2 > 10^4$$

According to Table 17.9, for the half-pipe (180-degree central angle, i.e., semi-circular cross-section) coil in the jacket,

Equivalent diameter, $d_e = \frac{\pi}{2} d_i = \frac{\pi}{2}(0.078) = 0.122m$

Cross-sectional area of half-pipe, $A_{hp} = \frac{\pi}{8} d_i^2 = \frac{\pi}{8}(0.078)^2 = 0.0024m^2$

A_{hp} is used to calculate velocity, v_2 in the half-pipe.
Heat transfer fluid (water) flowrate in each coil

$$\dot{V}_2 = 2.53x10^{-3} m^3 s^{-1} (40 gpm)$$

$$v_2 = \dot{V}_2 / A_{hp} = 0.00253 m^3 s^{-1} / \left(0.0024 m^2\right) = 1.05 m.s^{-1} \quad \left(> 0.9 m.s^{-1}\right)$$

Mass flowrate, $\dot{m}_2 = \rho_2 \times \dot{V}_2 = 1000 kg.m^{-3} \times 2.53 \times 10^{-3} m^3 s^{-1} = 2.53 kg.s^{-1}$

Heat transfer-side fluid Reynolds number, Re_2, and Prandtl number, Pr_2, are calculated using equations 17.34 and 17.35, respectively.

$$Re_2 = \frac{\rho_2 v_2 d_e}{\mu_2} = \frac{\left(1000 kg.m^{-3} \times 1.05 ms^{-1} \times 0.122 m\right)}{10^{-3} Pa.s} = 1.28 \times 10^5 > 10^4 \left(Turbulent\ flow\right)$$

Heat transfer fluid Prandtl number,

$$Pr_2 = \frac{\mu_2 c_{p2}}{\lambda_2} = \frac{10^{-3} Pa.s \times 4190 J.kg^{-1} K^{-1}}{0.61 Wm^{-1} K^{-1}} = 6.87$$

Similarly, assuming viscosity of the heat transfer fluid (water) at wall temperature, $\mu_{w2} = \mu_2$, the viscosity ratio from equation 17.36 becomes,

$$VR_2 = \mu_2 / \mu_{w2} = 1.0$$

From Table 17.8, the jacket-side Nusselt number,

$$Nu_2 = 0.027\ Re_2^{4/5} Pr_2^{1/3} \left(VR_2\right)^{0.14} \left[1 + 3.5\left(d_e / d_c\right)\right] = 0.027 \times \left(1.28 \times 10^5\right)^{4/5}$$
$$\times (6.87)^{1/3} \times (1.0)^{0.14} \left[1 + 3.5(0.122 / 2.089)\right] = 753$$

$$h_2 = \frac{Nu_2 \lambda_2}{d_e} = \left(\frac{753 \times 0.61 Wm^{-1} K^{-1}}{0.122 m}\right) = 3765 Wm^{-2} K^{-1} \left(662 \text{Btu hr}^{-1}\ \text{ft}^{-2}\ \text{F}^{-1}\right)$$

6. Overall heat transfer coefficient:

$$U = \left(\frac{1}{h_1} + f + \frac{x}{\lambda} + \frac{1}{h_2}\right)^{-1} = \left(\frac{1}{163} + 0.00048 + 0.0025 + \frac{1}{3765}\right)^{-1}$$

$$= 107 Wm^{-2}K^{-1}\left(18.8\, Btu\, hr^{-1} ft^{-2} F^{-1}\right)$$

7. Heat transfer surface area:
   Mutually shared cylindrical wetted area between the process fluid and heat transfer fluid,

$$A_{cw} \approx 7.15 m^2$$

   Half-pipe coils' wetted area per coil, $A = A_{cw} / n_c = \frac{7.15}{2} = 3.57 m^2$

   Jacket-side cylindrical surface area up to process fluid height,

$$A_j \approx \pi T_o H_c \approx \pi \times 2.0 \times 1.709 \approx 10.73 m^2$$

   Half-pipe coils wetted area with respect to the jacket area, $A_f = \left(A_{cw} / A_j\right) 100 = 67\%$ (as expected)
   Effective vessel wall heat transfer area per each set of coils, $A = 4.56 m^2$

Case 1: Steady-state heat transfer from the process fluid at $50°C$ to the heat transfer fluid at $20°C$, i.e.:

$$\Delta T = 50 - 20 = 30°C.$$

For the steady-state operation with a temperature difference of $30°C$ between the process fluid and the heat transfer (coolant) fluid, the rate of heat transfer to each coil,

$$\dot{q} = U A \Delta T = 107 Wm^{-2}K^{-1} \times 4.56 m^2 \times 30 K = 14638 W \approx 14.6 kJ.s^{-1}$$

Total heat transfer rate to the coolant, $\dot{Q} = n_c . \dot{q} = 2 \times 14638 W = 29276 W \approx 29.3 kJ.s^{-1}$

Maximum allowable polymerization reaction heat to maintain media temperature at $50°C$,

$$\dot{Q}_{rxn} = \dot{Q} \approx 29.3 kJ.s^{-1}$$

Temperature increase of the coolant is calculated as follows:

$$\dot{q} = \dot{m}_2 C_{p2} \Delta T = 14638 W$$

$$\Delta T = \frac{\dot{q}}{\dot{m}_2 C_{p2}} = \frac{14638 W}{2.53 kg.s^{-1} \times 4190 W.s.kg^{-1}K^{-1}} \approx 1.4 K\left(1.4°C\right)$$

Coolant exit temperature, $T_2 = \Delta T + T_1 = 1.4 + 20 = 21.4°C$

Case 2: For a continuous operation where the entrance and exit temperatures, (t_1, t_2) of the process fluid are respectively, $30°C$ and $50°C$. The jacket-side coolant fluid inlet and outlet temperatures (T_1, T_2) are $20°C$ and $21.4°C$. See Figure 17.13b. The heat transfer rate is calculated based on the log-mean temperature difference. See equation 17.14b.

$$\Delta T_{lm} = \frac{(t_2 - T_2) - (t_1 - T_1)}{\ln[(t_2 - T_2)/(t_1 - T_1)]} = \frac{(323 - 294.4) - (303 - 293)}{\ln[(323 - 294.4)/(303 - 293)]} = 17.7K = 17.7°C$$

$$\dot{q} = U\,A\,\Delta T_{lm} = 107Wm^{-2}K^{-1} \times 4.56m^2 \times 17.8K = 8684W \approx 8.7kJ.s^{-1}$$

Total heat transfer, $\dot{Q} = n_c \cdot \dot{q} = 2 \times 8684W = 17368W \approx 17.4kJ.s^{-1}$

Considering only the two coil zones, i.e., the two half-coils in the jacket wrapped around the straight side of the reactor, expected heat load $\dot{q}$ on each coil is approximately $8.7kJ.s^{-1}$. The coolant mass flowrate, $\dot{m}_2$, in each coil is estimated from the equation,

$$\dot{q} = \dot{m}_2 C_{p2}\,\Delta T$$

Mass flowrate,

$$\dot{m}_2 = \frac{\dot{q}}{C_{p2}\,\Delta T} = \frac{8684W}{4190W.S.kg^{-1}K^{-1} \times 1.4C} \approx 1.48kg.s^{-1}$$

Volumetric flowrate of coolant,

$$\dot{V} = \frac{\dot{m}_2}{\rho_2} = \frac{1.48}{1000} = 1.48 \times 10^{-3}m^3s^{-1}(23gpm)$$

Iteration needed to calculate overall heat transfer coefficient based on 23gpm coolant flowrate. In the case of continuous process, required coolant flowrate significantly decreases.

<u>Case 3:</u> Batch Cooling Time

In this example, cooling time θ needed to cool down the working media temperature from $50°C$ to $25°C$ is calculated. The coolant flowrate per coil is $2.53 \times 10^{-3}m^3s^{-1}(40gpm)$.

Process fluid inlet and outlet temperatures: $t_1 = 50°C\,and\,t_2 = 25°C$

Heat transfer (coolant) fluid inlet and outlet temperatures: $T_1 = 15°C\,and\,T_2 = 20°C$

Average temperature of coolant, $T = \frac{T_1 + T_2}{2} = \frac{15 + 20}{2} = 17.5°C$

Log mean temperature difference,

$$\Delta T_{lm} = \frac{(t_2 - T_2) - (t_1 - T_1)}{\ln[(t_2 - T_2)/(t_1 - T_1)]} = \frac{(298 - 293) - (323 - 288)}{\ln[(298 - 293)/(323 - 288)]} = 15.4K = 15.4°C$$

$$\Delta T = T_2 - T_1 = 20 - 15 = 5°C$$

Since, $\Delta T > 0.10 \times \Delta T_{lm}$, equation 17.49 is used to determine cooling time, θ.

$$K = e^{UA/\dot{m}_2 C_{p2}} = e^{[(107 \times 3.57)/(2.53 \times 4190)]} = 1.04$$

Mass of the working media in the reactor,

$$M_1 = \rho \times V = 1200kg.m^{-3} \times 6.056m^3 = 7267kg$$

$$\theta = \frac{M_1 C_{p1}}{\dot{m}_2 C_{p2}} \ln\left(\frac{t_1 - T}{t_2 - T}\right)\frac{K}{K-1} = \left(\frac{7267 \times 2095}{2 \times 2.53 \times 4190}\right) ln\left(\frac{323 - 290.5}{298 - 290.5}\right)\left(\frac{1.04}{1.04 - 1}\right)$$
$$= 27377s \approx 7.6hr.$$

17.11.2 Fluid Pressure Drop in Heat Transfer Jacket

The classic Darcy-Weisbach equation 12.1 (see Chapter 12) is used to characterize the fluid pressure drop ΔP in the jacket.[2] The hydraulic diameter, d_e', depends on the heat exchanger configuration. Heat transfer fluid velocity, v_2, is determined from its flowrate, $\dot{q}_2$, and the cross-sectional area of the rectangular duct or coil. These parameters are presented in Table 17.21.

$$\text{Hydraulic radius, } d_e' = \frac{4 \times Area\, of\, cross - section}{wetted\, perimeter}$$

$$\text{Velocity, } v_2 = \frac{heat\, exchanger\, media\, flow\, rate}{Exchanger\, area\, of\, cross - section}$$

$$\text{Pressure drop, } \Delta P = \frac{f \rho_2 L v_2^2}{2\, d_e'} \text{ for } 4000 < Re < 10^8 \tag{17.52}$$

where, f and L are, respectively, the Darcy friction factor and fluid path length.

Friction factor, f, is a function of Reynolds number, Re, and is determined from the diagram (Figure 12.15) or Colebrook equations (12.3–12.4) in Chapter 12. In the case of laminar flow, $Re \le 2100$, friction factor, f, is linearly proportional to Reynolds number, e.g., $f = 64 / Re$.

For the above example case 1,

Hydraulic diameter, $d_e' = 0.61 \times d_i = 0.61 \times 0.078m = 0.048m$

Velocity of heat transfer fluid, $v_2 = 1.05ms^{-1}$

TABLE 17.21
Hydraulic Diameter and Velocity Formulae for Different Type of Jacket Configurations[2]

Configuration	Hydraulic/Fluid Flow Equivalent Diameter, d_e'	Velocity, v_2
Annular jacket w/spiral baffling	$\frac{2(w_j \times t_j)}{w_j + t_j}$	$\frac{\dot{q}_2}{w_j \times t_j}$
Annular jacket w/coil	d_i	$\frac{\dot{q}_2}{\pi d_i^2 / 4}$
Annular jacket w/half-coil	$\frac{4\left(\frac{\pi}{8} d_i^2\right)}{(d_i + \pi d_i / 2)} = 0.61 d_i$	$\frac{\dot{q}_2}{\pi d_i^2 / 8}$

Note: w_j and t_j are the width and thickness of the spiral channels in the jacket.

Reynolds number, $Re = \frac{\rho_2 v_2 d'_e}{\mu_2} = \frac{1000kg.m^{-3} \times 1.05m.s^{-1} \times 0.048m}{1 \times 10^{-3} Pa.s} = 5.04 \times 10^4$ (turbulent flow)

Typical roughness value for drawn tubes is 0.0025mm.

Roughness factor of stainless steel pipe,

$$\frac{surface\,roughness}{pipe\,hydraulic\,diameter} = \frac{\in}{d'_e} = \frac{2.5 \times 10^{-6} m}{5.1 \times 10^{-2} m} = 4.9 \times 10^{-5}$$

Darcy friction factor, $f = 0.02$ (from Moody diagram, Chapter 12, Figure 12.15)

Length of the coil, $L = l_c = 6.56m$ (see Table 17.20)

Pressure drop in each coil,

$$\Delta p_1 = \frac{f\,\rho_2\,L v_2^{\,2}}{2\,d'_e} = \frac{0.02 \times 1000kg.m^{-3} \times 6.56m \times \left(1.05ms^{-1}\right)^2}{2 \times 0.051m}$$

$$= 1.42 \times 10^3\,Pa\left(\sim 0.21psi\right)$$

Pressure drop in each set of coils,

$$\Delta p_{cs} = n_s \times \Delta p_1 = 7 \times 1.42 \times 10^3 = 9.94 \times 10^3\,Pa\left(\sim 1.44psi\right)$$

Mixit-io SS17.1 may be used to design jacketed and internal heat transfer systems.

17.12 BASIC DESIGN GUIDELINES

1. Conventional heat transfer jacket should have baffles to create tangential helical flow over the exterior wall of the vessel. Baffles help create turbulent flow and significantly improves utility-side heat transfer coefficient compared to a plain jacket.
2. In the absence of baffles, heat transfer fluid injection nozzles should be used to appropriately direct the flow towards the vessel wall.
3. Upward flow of the service fluid is preferred over the downward flow to fully flood the jacket.
4. A vent should be provided to eliminate gas pockets and/or gas blanketing, if any, in the jacket.
5. Half-pipe coil (limpet) and dimple or pillow surfaces inside the jacket enhances flow turbulence and are superior in their heat transfer performance to the conventional baffled jacket but may not be cost-effective for the desired heat duty.
6. Internal coils in the tank, if not appropriately designed, often hinder the circulation of the process fluid and consequently diminish mixing efficiency of the agitation system. If multiple banks of helical coils are used to enhance heat transfer, these coils must be staggered to avoid their tube locations in the stagnation flow zones of the adjacent coil.
7. Tubes of the internal coil must not be tightly packed to resemble a cylindrical wall. This would significantly reduce heat transfer and mixedness in the vessel. In a two-phase G-L viscous media, such as in aerobic fermenters, gas bubbles are likely to be trapped and coalesce to form bigger bubbles prior to their escape to the headspace. Each wrap of coil should have a minimum pitch, p_c, of 1.5 times the tube outer diameter, d_o.
8. As mentioned earlier, internal heat transfer coils must be sturdily fastened or welded to the wall or baffles to withstand long-term transient fluid force.
9. The service-side recommended heat transfer fluid velocities are presented in Table 17.22. Turbulent flow is always preferred unless severely limited by viscosity of the heat transfer fluid.

TABLE 17.22
Recommended Utility-Side Heat Transfer Fluid Velocity, v_2 [40,41]

Type of Jacket	$v_2\left[m.s^{-1}\right]$
Conventional unbaffled	1–1.5
Half-pipe (limpet coil)	2.3
Dimpled jacket (limited by pressure drop)	≈ 0.6
Internal coil	1.5

10. As a rule of thumb, typical overall heat transfer coefficient values are depicted in Table 17.23.

TABLE 17.23
Typical Overall Heat Transfer Coefficient, U [40–42]

Heat Transfer System	Duty	$U\left[Wm^{-2}K^{-1}\right]$
Jacketed glass-lined steel vessels	Heating	310
	Cooling	200
	Cooling (chilled service)	100
	Distillation/Evaporation	350
Jacketed carbon steel and stainless steel vessels	Heating	400
	Cooling	350
	Cooling (chilled services)	150
Internal coil (well designed)	Heating/Cooling	400–600

11. In the case of heat transfer using coils in the tank, it is generally necessary to increase the horsepower per unit volume from small-scale to full-size equipment.[24]
12. If heat transfer is important, the vessel's heat transfer surface area to the volume of the working media is considered as one of the scale-up criteria. Thus, taller tanks with higher aspect ratio compared to the pilot vessel are used in the plant. See Chapter 7, Figure 7.17.
13. For specific applications, if heat transfer correlations are unavailable, experimental methods to determine process side, jacket/internal coil side, and the overall heat transfer coefficients (i.e., h_1, h_2 and U) are presented in Chapter 18.
14. Heat transfer system designers may use design calculation spreadsheets such as TANKJKT,[43] ChemCalc,[44] and CheGuide.[45] These are based on the correlations listed in Tables 17.6–17.11. The heat transfer chapter in *Handbook of Industrial Mixing: Science and Practice*[46] or in its updated companion version *Advances in Industrial Mixing*[47] is recommended as an excellent resource for heat transfer design engineers.

REFERENCES

1. Thermal Conductivity of Common Metals, Metallic Elements and Alloys. https://www.engineeringtoolbox.com/thermal-conductivity-metals-d_858.html
2. Bondy, F., Lippa, S., Heat Transfer In Agitated Vessels, *Chem. Eng.*, p. 62, April 4, 1983.
3. Fourier, J., *The Analytical Theory of Heat*, Dover Publications, New York, 1955.
4. Schlichting, H., *Boundary-Layer Theory*, 7th. ed., Mc Graw Hill, New York, 1999.

5. Vlachopoulos, J., Strutt, D., Basic Heat Transfer and Some Applications in Polymer Processing, *Plastics Technician's Toolbox*, Vol. 2, p. 21, September 2002.
6. Yunus, C. A., Afshin, G. J., *Heat and Mass Transfer: Fundamentals and Applications*, 4th. ed., McGraw Hill, New York, February 2010.
7. https://www.sciencedirect.com/topics/engineering/convection-heat-transfer-coefficient.
8. Kuppan, T., *Heat Exchanger Design Handbook*, Marcel-Dekker, New York, 2020.
9. Holland, F. A., Chapman, F. S., Heat-Transfer Correlations in Jacketed Vessels, *Chem. Eng.*, p. 175, February 15, 1965.
10. Brooks, G., Su, G. D., Film Coefficients of Heat Transfer for Agitated Process Vessels, *Chem. Eng. Prog.*, p. 54, October 1959.
11. Dickey, D., Hicks, R. W., Fundamentals of Agitation, *Chem. Eng.*, p. 93, February 2, 1976.
12. Ackley, E. J., Film Coefficients of Heat Transfer for Agitated Process Vessels, *Chem. Eng.*, p. 133, August 22, 1960.
13. Cummings, G. H., West, A. S., Heat Transfer Data for Kettles with Jackets and Coils, *Ind. Eng. Chem.*, Vol. 42, p. 2303, 1950.
14. Holland, F. A., Chapman, F. S., Heat-Transfer Correlations for Agitated Liquids in Process Vessels, *Chem. Eng.*, p. 153, January 18, 1965.
15. Brown, R. W., Scott, R., Toyne, C., An Investigation of Heat Transfer in Agitated Jacketed Cast Iron Vessels, *Trans. IChE*, Vol. 25, p. 181, 1947.
16. Chilton, T. H., Drew, T. B., Jebens, R. H., Heat Transfer Coefficients in Agitated Vessels, *Ind. Eng. Chem.*, Vol. 36, p. 510, 1944.
17. Uhl, V. W., Heat Transfer to Viscous Materials in Jacketed Agitated Kettles, *Chem. Eng. Prog. Symposium Series*, Vol. 51, p. 99, 1954.
18. Uhl, V. W., Voznick, H. P., The Anchor Agitator, *Chem. Eng. Prog.*, Vol. 56(3), p. 172, March 1960.
19. Blazinski, H., Kuncewicz, C., Heat Transfer during the Mixing of Pseudoplastic Fluids with Ribbon Agitators, *Intl. Chem. Eng.*, 21, p. 679, 1981.
20. Sieder, E. N., Tate, G. E., Heat Transfer and Pressure Drop of Liquids in Tubes, *Ind. Eng. Chem.*, Vol. 28, p. 1429, 1936.
21. Bolliger, D. H., Assessing Heat Transfer in Process-Vessel Jackets, *Chem. Eng.*, p. 95, September 20, 1982,
22. Perry, R. H., Chilton, C. H. (ed.), *Chemical Engineer's Handbook*, 5th. ed., pp. 10–17, McGraw-Hill, New York, 1973.
23. Chen, C. Y., Hawkins, G. A., Solberg, H. L., Heat Transfer in Annuli, *Trans. ASME*, Vol. 68, p. 99, 1946.
24. Markovitz, R. E., Picking the Best Vessel Jacket, *Chem. Eng.*, p. 156, November 15, 1971.
25. Ludwig, E. E., *Applied Process Design for Chemical and Petrochemical Plants*, Vol. 1, 3rd. ed. Butterworth-Heinemann, 1999.
26. Oldshue, J. Y., *Fluid Mixing Technology*, Chemical Engineering, McGraw-Hill Publishing Co., Inc., New York, 1983.
27. Oldshue, J. Y., Gretton, A. T., Helical Coil Heat Transfer in Mixing Vessels, *Chem. Eng. Prog.*, Vol. 50, p. 615, 1954.
28. Dunlap Jr, J. R., Rushton, J. H., Heat-Transfer Coefficients in Liquid Mixing Using Vertical-Tube Baffles, *A.I.Ch.E. Symposium Series No. 5*, Vol. 49, p. 137, 1953.
29. Rushton, J. H., Lichtman, R. S., Mahoney, L. H., Heat Transfer to Vertical Tubes in a Mixing Vessel, *Ind. Eng. Chem.*, Vol. 40(6), pp. 1082–1087, June 1948.
30. Petree, D. K., Small. W. M., Heat Transfer and Power Consumption for Agitated Vessels with Vertical Plate Coils, *AIChE. Sym. Ser.*, Vol. 74(174), 1978.
31. Eagles, A., Ferguson, R. M., On the Coefficient of Heat Transfer from the Internal Surface of Tube Walls, *Proc. Roy. Soc.*, Vol. A127, p. 540, June 1930.
32. Garvin, J., Heat Transfer and Friction in Dimpled Jackets, *Chem. Eng. Prog.*, pp. 73–75, April 2001.
33. H&C Heat Transfer Solutions, *Fouling Factors*, 2021. https://www.hcheattransfer.com/fouling_factors2.html.
34. Byrne, R. C., *Standards of the Tubular Exchanger Manufacturers Association*, 10th. Ed., Standards of the Tubular Exchanger Manufacturers Association, Inc., 2019.
35. Lightnin Technology Seminar, *3rd. Ed., Mixing Equipment Co., Inc., Lightnin, a unit of General Signal*, p. 1, 1982, Section 2A.
36. Dream, R. F., Heat Transfer in Agitated Jacketed Vessels, Chemical Engineering, p. 91, January 1999.

37. De Dietrich Process Systems, Inc., *Hemicoil® Jacketed Glass-Lined Steel Reactor*, News and Notes No. 2.
38. De Dietrich Process Systems, Inc., *Reactor_ Cut_Sheet_GL2000*, D003-2000, 7/2011.
39. Kneale, M., Design of Vessels with Half Coils, *Trans. I. Chem. E.*, 47, 1969.
40. Carpenter, K. J., Agitated Vessel Heat Transfer, Thermopedia, 2022. https://thermopedia.com/content/547/
41. Fletcher, P., Heat Transfer Coefficients for Stirred Batch Reactor Design, The Chemical Engineer, April 1987.
42. Kern, D. Q., *Process Heat Transfer*, McGraw-Hill, New York, 1950.
43. Hall, S., *TANKJKT Spreadsheet Template Heat Transfer Calculations for Jacketed Tanks*, UINS-TJ-02 February 2015. chemengsoftware.com
44. CheCalc, *Jacketed Vessel Heat Transfer*, 2015, https://checalc.com/solved/jacketHeat.html
45. *Jacketed Vessel Heat Transfer (Half Pipe Coil)*, 2016, https://cheguide.com/jacket_vessel.html
46. Penney, W. R., Atiemo-Obeng, V. A., Heat Transfer. In *Chapter 14 in the Handbook of Industrial Mixing: Science and Practice*, Eds. Paul, E. L., Atiemo-Obeng, V. A., Kresta, S. M., p. 869–886, John Wiley & Son, Inc., 2004.
47. Nunez, J. R., Heat Transfer in Stirred Tanks—Update. In *Chapter 14 in Advances in Industrial Mixing: Companion to the Handbook of Industrial Mixing*, Eds. Kresta, S. M., Etchells, A. W., III, Dickey, D. S., Atiemo-Obeng, V. A., pp. 493–532, John Wiley & Son, Inc., 2016.

18 Experimental Techniques in Mixing

18.1 INTRODUCTION

This chapter is dedicated to various experimental methods commonly practiced in mixing R&D and pilot plants to address product and process developments. Experiments are carried out at different scales, ranging from 1-gal. glass vessels to 2000-gal. stainless steel tanks, depending on the sought-after design and/or performance parameters. Typically, smaller vessels (18–30gal.) suffice for simple blending and solid–liquid suspension studies, whereas larger vessels are desired for gas–liquid mass transfer studies to overcome excessive gas holdup at low gas–liquid interfacial tension and high specific power levels. Occasionally, researchers develop new mixing hardware to solve unmet manufacturing needs. Before implementing novel hardware in a production plant, tests are conducted in both the R&D and pilot scales to assess required mixing performance and structural stability. Important mixing performance parameters are: impeller power number (N_P), flow number (N_Q); flow pattern, mixing time (θ_{95}), gas holdup (ε_g), gas–liquid mass transfer coefficient (k_La), liquid–liquid dispersion drop-size distribution (d_{32}), solid suspension uniformity, fouling and internal heat transfer coefficient (h_i). Briefly, the experimental techniques to determine these characteristics are described in the following sections.

18.2 IMPELLER CHARACTERISTICS

18.2.1 Power Number

The impeller power number, N_p, is used to determine the agitation (shaft) power, P, input into working media within the mixing vessel.

$$P = N_p \rho N^3 D^5 \tag{18.1}$$

where, P = shaft power [N-m/s or W]
ρ = density of working media [kg/m^3]
D = impeller diameter [m]
N = impeller rotations per second [1/s]

Impeller power number is a function of the flow regime, which varies from laminar flow ($N_{Re} \leq 10$) to fully developed turbulent flow ($N_{Re} \geq 10^4$), as shown in Figure 5.6. In a fully baffled tank, the power number tends to have a constant value under fully turbulent flow conditions.

Impeller Reynolds number, N_{Re}, is defined as

$$N_{Re} = \rho D^2 N/\mu \tag{18.2}$$

where, ρ = density of working media [kg/m^3]
D = impeller diameter [m]
N = impeller rotations per second [1/s]

DOI: 10.1201/9781003606390-18

Shaft power, P, is expressed as a function of the torque, T, experienced by the shaft [N-m], as

$$P = 2\pi NT \tag{18.3}$$

Substituting equation 18.3 into equation 18.1 directly relates shaft torque to the impeller power number as shown in equation 18.4. Thus, an experimental program for a specific type of impeller varies its diameter and agitation speed to determine power number from the measured torque values, which can be plotted at different Reynolds numbers as shown in Figure 5.6.

$$N_p = \frac{2\pi}{\rho N^2 D^5} T \tag{18.4}$$

Figure 18.1 shows a "mixing experiment" setup that uses an 18-inch diameter (20-gal.) dish-bottom Plexiglas® tank containing water. The associated agitation system consists of a hydrofoil impeller mounted onto a ¾-inch-diameter stainless steel shaft. The long shaft is coupled with the short shaft (1-inch dia.) through the drive system, by means of a Lebow load cell (rotary torque sensor) that measures the torque experienced by the shaft. Experiments are conducted at different speeds, N, with and without the impeller on the shaft, to determine the agitation torque, T_a, and the frictional torque, T_0, (tare value) respectively. The net hydraulic torque, $T = (T_a - T_0)$, is used to determine the power number, N_p, from equation 18.4.

An alternate, quicker experimental method uses a fully baffled 5.5-inch-diameter (1-gal.) glass vessel with a Servodyne® mixer motor, as shown in Figure 18.2. The speed controller of the motor

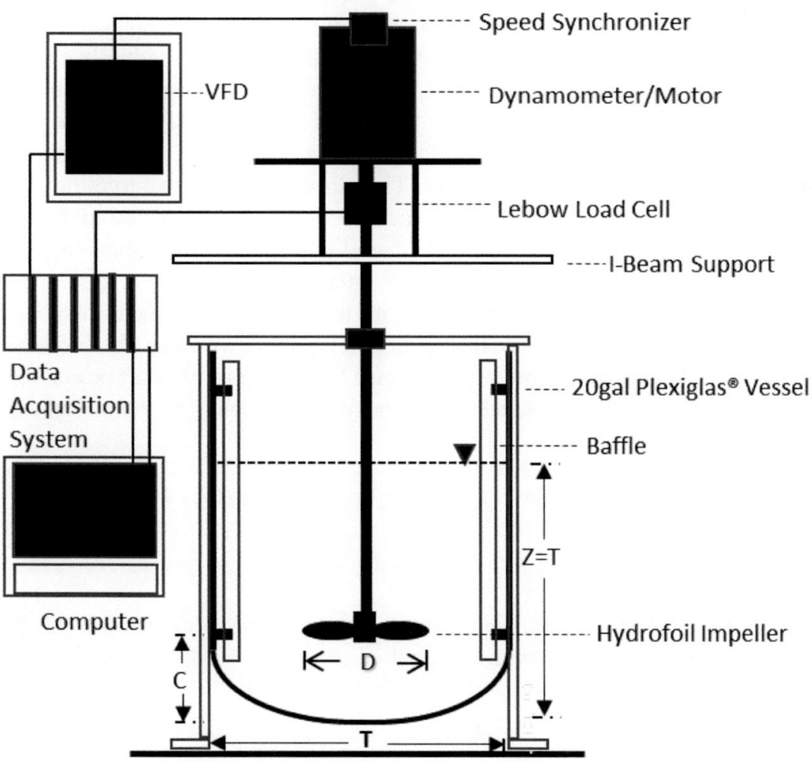

FIGURE 18.1 A 20-gallon scale experimental setup with a dynamometer for torque measurement.

[ASME Dish Bottom, T = 18″, V_{dish} = 2gal., V_{cyl} = 18gal.]

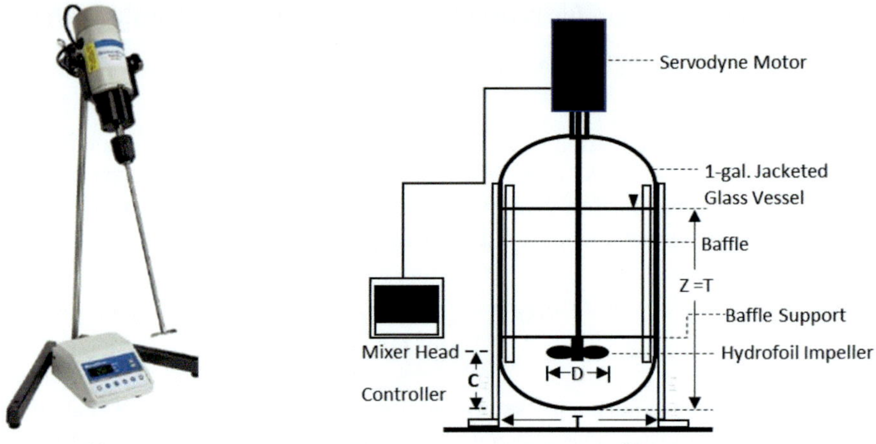

FIGURE 18.2 (a) Servodyne high-speed (60–3000rpm), low-torque (0–45in-oz) mixer head and controller and (b) 1-gallon (T = 5.5") scale experimental setup with Servodyne system for torque measurement.

has digital readouts for speed (rpm) and torque (in-oz). While this is a simple method, measured torque values in the presence and absence of the impeller (T_a and T_0), respectively, are commonly low enough that the calculated power number may contain a large ± error value. Additionally, fully turbulent flow conditions often create a high Froude number ($N_{Fr} = N^2 D / g$), which often creates a large vortex and a noticeable amount of headspace gas drawn down into the liquid. Consequently, the mixer controller sees misleadingly low torque, T_a, values.

In most cases, the fully turbulent-flow power number value of a commercially available impeller is determined by its manufacturer (see Table 5.1). In such cases, shaft power for turbulent conditions ($N_{Re} \geq 10^4$) are effortlessly calculated using equation 18.1. Graphs of the impeller power number versus Reynolds number (N_p vs. N_{Re}) relationship for proprietary impellers are not typically published, and thus, not easily accessible from the vendors. Therefore, experiments targeting a given mixing impeller operating within a given mixer flow regime are commonly conducted in the laboratory.

Handheld, clamp-on AC power meters are widely used in manufacturing plants to measure voltage, current, power factor, power, etc., at any desired point on a live single-phase or three-phase circuit. In many instances, the measurement readings are transmitted to a remote device such as a recorder for graphing, data analyses and troubleshooting. Note that an important unknown value, the frictional tare power, must be subtracted from the measured power to determine hydraulic power (shaft power) delivered to the working media. This power can be calculated using either of the following mathematical expressions:

Three-phase system: Power = 1.732 × Voltage (phase to phase) × Current × Power factor (Cos Φ)
Single-phase system: Power = Voltage × Current × Power factor (Cos Φ)

18.2.2 Flow Number

The impeller flow number, N_{Fl}, is an important parameter used to determine the fluid flow, $\dot{Q}$, induced into the working media by a rotating impeller.

$$\dot{Q} = N_{Fl} N D^3 \tag{18.5}$$

where, $\dot{Q}$ = flowrate [m³/s]
N = impeller rotational speed [rps]
D = impeller diameter [m]

Fluid flow must create a desirable flow pattern to successfully achieve required process behaviors such as mixing, mass transfer and heat transfer. A rotating impeller is basically a pump that creates a suction zone and a discharge zone, as shown in Figure 7.1a–c, for three commonly used open impellers (i.e., a hydrofoil that creates axial flow, a standard pitch-blade turbine that creates mixed flow, and a flat-blade turbine that creates radial flow). The typical turbulent flow pattern from a rotating PBT in a fluid is schematically represented in Figure 18.3. The suction and discharge zones are identified. Volumetric flowrate, $\dot{Q}$, is determined from the integration of the discharge flow velocity vectors with respect to discharge area of the swept volume, as shown in equation 18.6. Typically, the fluid velocity profiles are determined using Laser Doppler Velocimetry (LDV) technique. A detailed description of the Laser Doppler Anemometry is presented in the "Flow Pattern" section of this chapter.

$$\dot{Q} = 2\int_0^{D/2} v(x)(4b_w)dx + \int_0^{b_w} v(y)(4b_t)dy \text{ [m}^3\text{/s]} \quad (18.6)$$

where, b_w and b_t are blade width [m] and blade thickness [m], respectively, and $v(x)$ and $v(y)$ are LDA-measured x and y velocity profiles near the impeller blade.

Flow number is determined from the following equation:

$$N_{Fl} = \dot{Q} / N D^3 \quad (18.7)$$

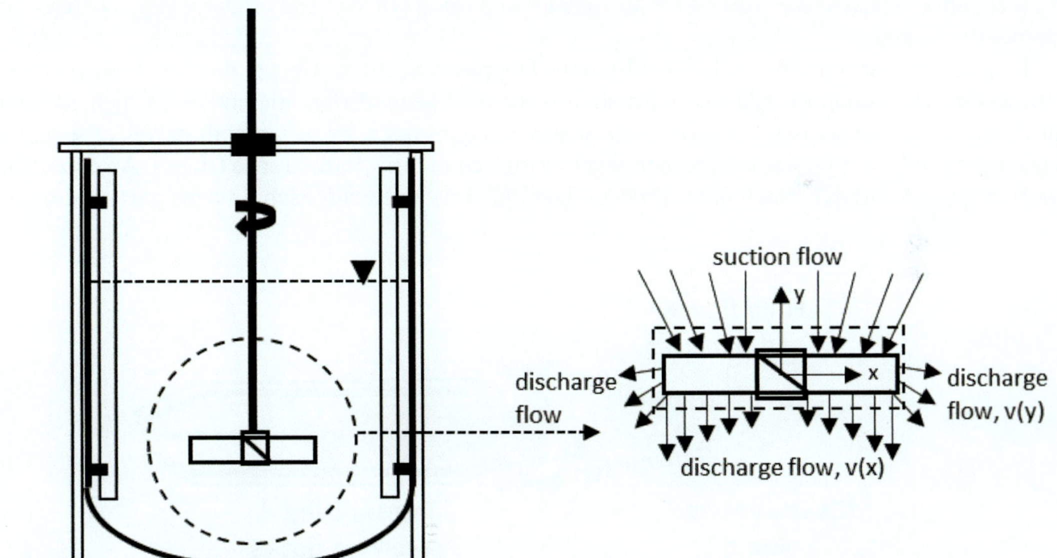

FIGURE 18.3 Schematic of a standard PBT (mixed flow) impeller operating within a mixing vessel, and its surrounding discharge and suction flow zones.

18.2.3 Flow Pattern

The turbulent fluid flow pattern created in a fully baffled agitated vessel is characteristic of a given impeller type (see Figure 7.6.) In the laminar regime ($N_{Re} \leq 10$), however, most open impellers create radial flow. Flow pattern dictates such characteristics as blend time, quality of solid–liquid and liquid–liquid dispersions, internal heat transfer rate, etc. The flow pattern from an impeller can be observed visually in a laboratory-scale glass or Plexiglas® vessel while dispersing flow tracers such as

(a) Larger solids like colored plastic (poly-acetate) beads of density nearly that of the working media
(b) Fluorescence particles that reflect laser light

In transparent laboratory vessels, non-invasive experimental techniques such as laser Doppler anemometry (LDA) and particle imaging velocimetry (PIV) are widely used for quantitative determination of fluid velocity, and subsequently, flow pattern and isoenergetic contours, as shown in Figure 5.3.

18.2.4 LDA, PIV and PLIF

Laser Doppler anemometry (LDA), also called laser Doppler velocimetry (LDV), as well as particle image velocimetry (PIV) and planar laser-induced fluorescence (PLIF), are widely used, non-intrusive, optical techniques for experimentally measuring fluid velocity and flow fields in applications such as industrial mixing and fluid dynamics. Each injects tracer species into a flowing fluid and illuminates or excites them using projected laser light. Since micron-sized particles that are injected into LDA and PIV systems are small enough to be entrained at the velocity of the flowing fluid, measurement of their velocities is equivalent to measuring the fluid velocities.

Each of these optical measurement techniques is briefly described in the following discussion. Readers are referred to the works of Watrasiewics and Rudd (1976),[1] and of Durst et al. (1976),[2] for deeper discussions.

Besides small seed particles, LDA utilizes the Doppler frequency effect to measure fluid velocity. Perhaps the most common LDA configuration is the dual-beam differential system, which, as seen in Figure 18.4, crosses two coherent laser beams of equal intensity (often split halves of a single laser beam) to form a very small measurement volume containing interference fringes. An entrained particle passing through that volume reflects laser light into a detector each time the particle crosses

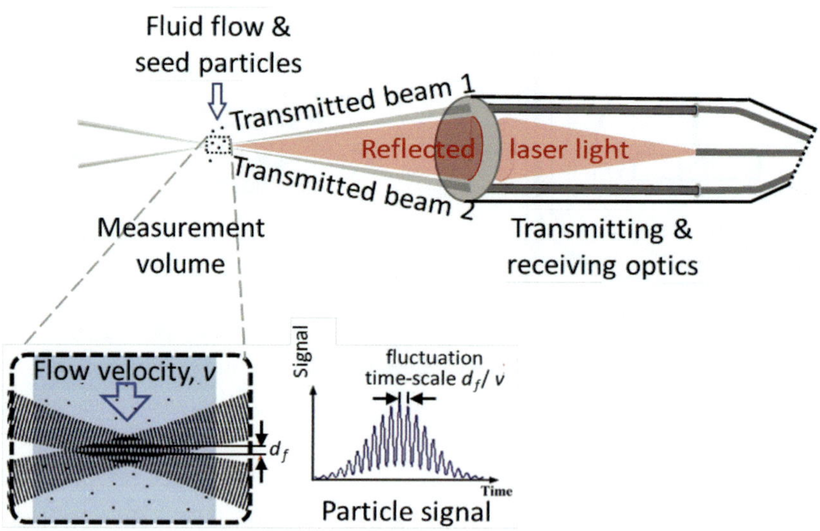

FIGURE 18.4 Schematic of laser Doppler anemometry (LDA).

(modified from www.DantecDynamics.com)

a constructive interference fringe. The distance between such fringes is known through calibration. Multiplying such distance by particle frequency, obtained by fast Fourier transform (FFT) analysis of the reflected light, yields local particle, and thus, fluid velocity magnitudes within the measurement volume. LDA commonly measures mean velocity and turbulence intensity in one direction, regardless of whether that direction is positive or negative. The probe in Figure 18.4 houses optics to transmit the two original laser beams and to capture reflected light.

Using multiple detectors can yield the three components (x, y, and z directions) of particle velocity. Because multi-component LDA can simultaneously measure velocity at multiple points in the fluid, it is a useful tool for high spatial resolution of velocity profiles within turbulent flows.

Figure 18.5 is a simple schematic of a single-camera PIV system, excluding the external trigger that synchronizes the laser and camera. The PIV technique seeds a flowing fluid with a relatively low concentration of small particles, and then illuminates downstream portions of the particle-seeded flow field with a thin, planar laser light sheet. Some light scattered by the particles is captured in sequential images of the particle field, by a high-speed camera. Captured images are processed using cross-correlation algorithms to assess particle displacements during the time interval separating two consecutive images. This quantifies velocity fields within the camera's imaging plane. The resulting two-dimensional vector map of the flow field characterizes temporal and spatial variations in vortices, turbulence, and other flow phenomena. PIV resolution can be adjusted by changing the size of the imaging plane or the particle images.

Stereoscopic PIV, which is not represented in Figure 18.5, differs in that it adds a second high-speed camera. The two cameras must have separate viewing angles to extract the third component (z-axis) of displacement. Both of these cameras must also be carefully focused on the same spot within the flow field and must be properly calibrated to truly have the same focus.

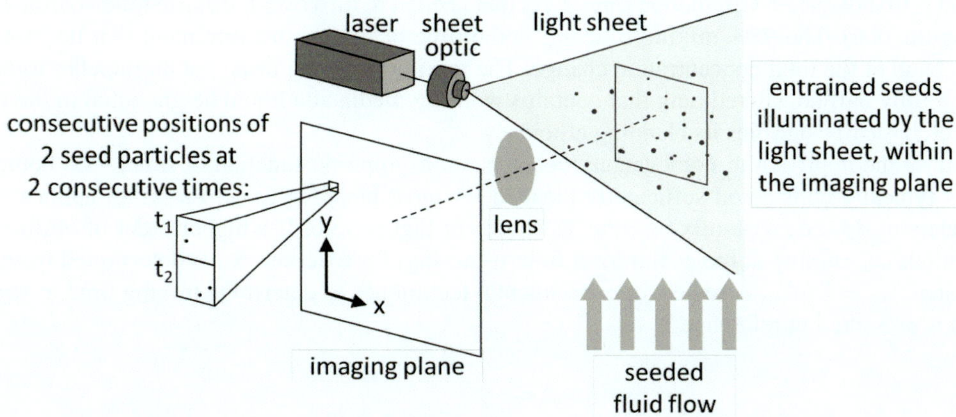

FIGURE 18.5 Particle image velocimetry (PIV) system.

PIV results differ from LDA results by being two-dimensional or three-dimensional vector fields, rather than measured velocity magnitudes at a single point. PIV particle concentration allows individual particles to be identifiable, but not with enough certainty to track specific movement of each individual particle between two consecutive images. Particle displacements are obtained using cross-correlation algorithms.

PLIF is used to visualize and quantify concentration fields within flowing fluids. It can be utilized in a PIV system like the one shown previously in Figure 18.5. PLIF relies on emissions from a fluorescent tracer molecule (can be liquid or gas) that is introduced into the fluid and excited when passing through projected laser light. The fluorescent tracer molecule must be selected for a high quantum yield, good photostability, and a high molar absorption coefficient at the excitation wavelength of the laser. It must not interfere with behaviors of interest in the background fluid (e.g., PIV). The tracer emission is at a longer wavelength than the scattered laser light. Tracer intensity, as detected by a camera or photodetector, is proportional to tracer concentration within the fluid. Thus, PLIF effectively monitors transient mixing behaviors.

These non-intrusive techniques provide powerful assessments of fluid dynamics. Each also has its own characteristic strengths and limitations, and must be selected for the specific application and flow characteristics of interest. PLIF is uniquely able to be applied in combination with the other techniques.

18.3 MIXING CHARACTERISTICS

Because the mixing time, θ, addresses the mixing efficiency of an agitation system in a vessel, it is considered the most critical design parameter for practicing engineers. An elaborate description of batch mixing time is provided in Chapter 6, "Blending of Miscible Liquids." This chapter briefly presents experimental techniques to measure the mixing time for 95% homogeneity, θ_{95}, in batch and continuous processes.

18.3.1 Batch Process

In general, batch mixing time, θ, is defined as the time interval between the instant that component two is added to component one, and the instant that the two components have reached a specific degree of uniformity. Mixing experiments often add tracers like pH carriers, dyes, and salt solutions into a geometrically scaled-down stirred tank that can contain real or simulated fluids. Colorimeters and probes for pH and/or electrical conductivity are used, with their respective analyzers, as portable detectors. If the moment of tracer addition is taken as zero, t_d, is the dead time before that addition is first detected. Circulation time, t_c, is that needed to turn over the entire fluid volume once (see Figure 18.6). The 95% mixing time, t_m, is the time of the last measurement that lies outside the 5% band of the total concentration change. The mixing times, t_m or θ_{95}, of an impeller rotating within a fully baffled, stirred tank that contains working media at a liquid height equal to the tank diameter, basically signifies its blending efficiency.

Because the approach to homogeneity becomes an asymptotic function of time, 95% homogeneity is typically considered sufficiently blended for most blending (macro-mixing) applications. This refers to θ_{95}, i.e., 95% mixing time, as shown in Figure 18.6. If a higher order of uniformity is essential, i.e., mixing at molecular level (micro-mixing) for reaction, θ_{99} is determined from the expression, $\theta_{99} = 1.6\,\theta_{95}$. A review of experimental techniques to determine mixing time in stirred vessels is presented in reference 3.

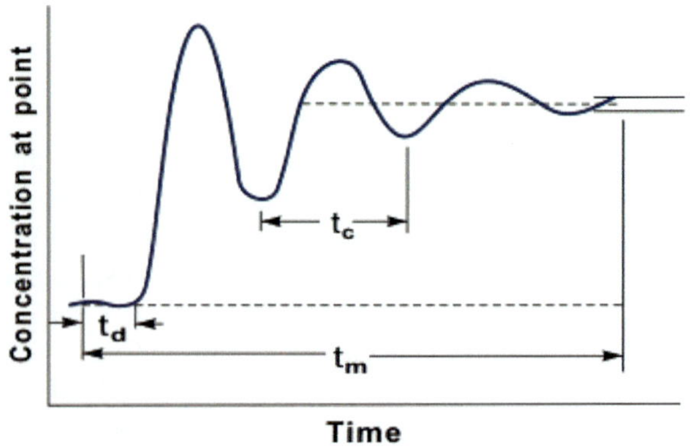

FIGURE 18.6 Results (concentration vs. time) of an idealized mixing time experiment.[4]

Figure 18.1 is the schematic of a laboratory system consisting of an 18-inch diameter (20-gal) transparent Plexiglas® tank with an axial hydrofoil impeller. The agitation system includes a Lightnin dynamometer with a Lebow load cell mounted on the shaft. The impeller speed is controlled by a variable frequency drive (VFD). For a specific impeller, mixing time, θ, is a strong function of the flow regime defined by the impeller Reynolds number. As explained earlier, this laboratory system is well equipped with the instrumentation and accessories required to measure agitation speed, N, and corresponding shaft power, P.

Colorimetry: This simple, economical experimental technique injects a dye or colorant, e.g., simply a food color, into an agitated liquid as depicted in Figure 18.7a. According to the Beer-Lambert law, colored compounds absorb light of a certain characteristic wavelength when monochromatic light passes through them. A colorimeter or a spectrophotometric probe detects temporal color changes, at its location within the liquid. Macromixing time is qualitatively determined from visual observations of color change, or more preferably, quantitatively determined by monitoring the temporally changing amount of light passing through the solution (i.e., incident flux from the source—transmitted light between detector and source [$(f_i - f_t) = f(t)$].

Initially, the light transmissivity of the instrument, f_i, is calibrated with the liquid using a colorimeter photocell that determines a given color's quantifying value in the RGB color spectrum, as visible to the human eye. A spectrophotometer, however, offers more accurate color measurement by capturing color across the entire visible spectrum, then filtering that light into very narrow color bands. Using multiple probes at critical locations in a tank alleviates many issues with local color change variations. Typically, colorimeters are placed in low-flow or stagnation regions, based on engineering intuition and/or prior knowledge of fluid flow patterns, as depicted in Figure 18.7b. Figure 18.8 illustrates light transmissivity readings from three probes (top, middle and bottom) as a function of time.

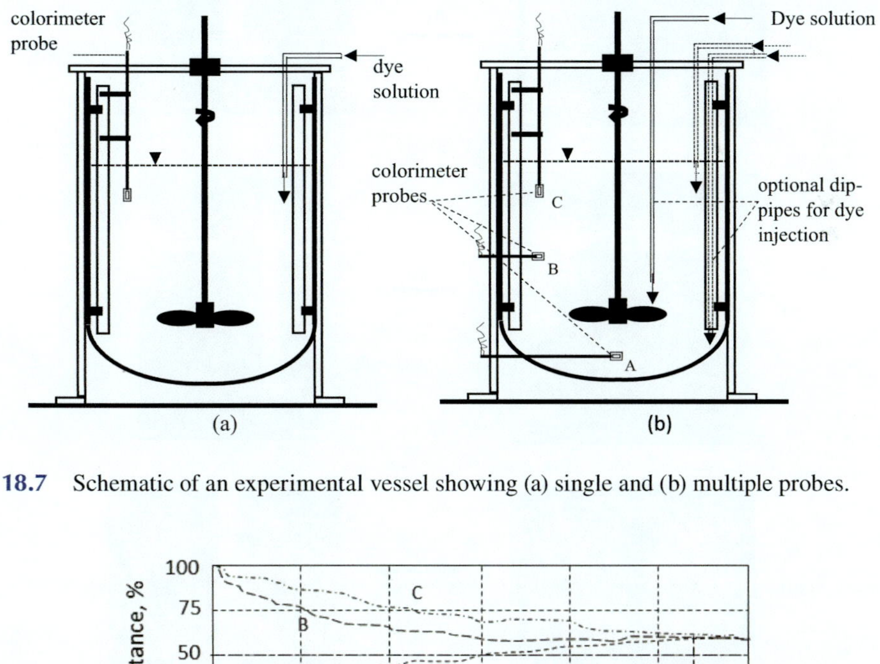

FIGURE 18.7 Schematic of an experimental vessel showing (a) single and (b) multiple probes.

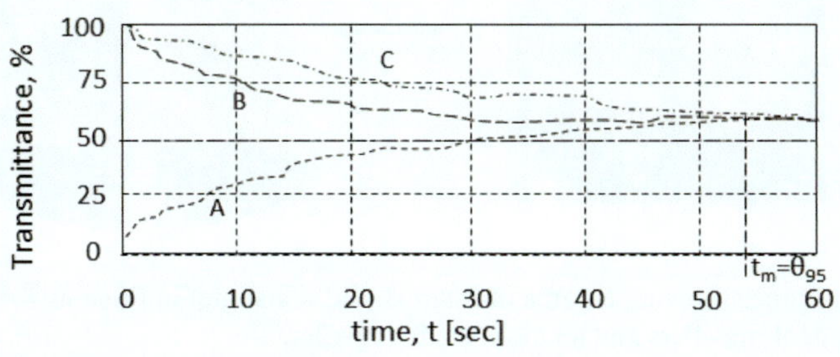

FIGURE 18.8 Light transmittance in an idealized mixing experiment with injection above a hydrofoil.

pH Metric: Acid-base neutralization reactions are a common alternative experimental approach to using a colorant. Typically, an aqueous acid solution (e.g., sulfuric acid) is quickly added to a stirred vessel containing aqueous sodium hydroxide. As the neutralization reaction occurs, the relative amounts of free hydrogen and hydroxyl ions in the solution are changing. Time-dependent evolution of this phenomenon is measured by the electrode(s) submerged in the media and transmitted to the pH meter for further analyses. Similar to Figure 18.6, the pH readings plotted as a function of time provides desired micro-mixing times, θ_{95} or θ_{99}. However, if a simple qualitative approach is sufficient, visually observed decolorization may be conducted while dispersing phenolphthalein indicator in an NaOH solution. As pH of the agitated content changes from basic to neutral and acidic, color of those contents changes from purple to colorless.

Time Lapse Image Acquisition: In lieu of the above transmittance and pH measurement techniques, an alternate approach uses a high-speed digital camera to sequentially photograph the working media's evolving color change, followed by the application of image post-processing software to extract mixing time. Such imaging methods also assist in understanding the development of laminar flow mixing patterns, with cavern formation around an impeller and potential fluid stagnation in some vessel regions (see Figure 18.9). This technique removes the subjectivity of estimating macro-mixing time by the naked eye, and yields a higher degree of reliability and repeatability.[5]

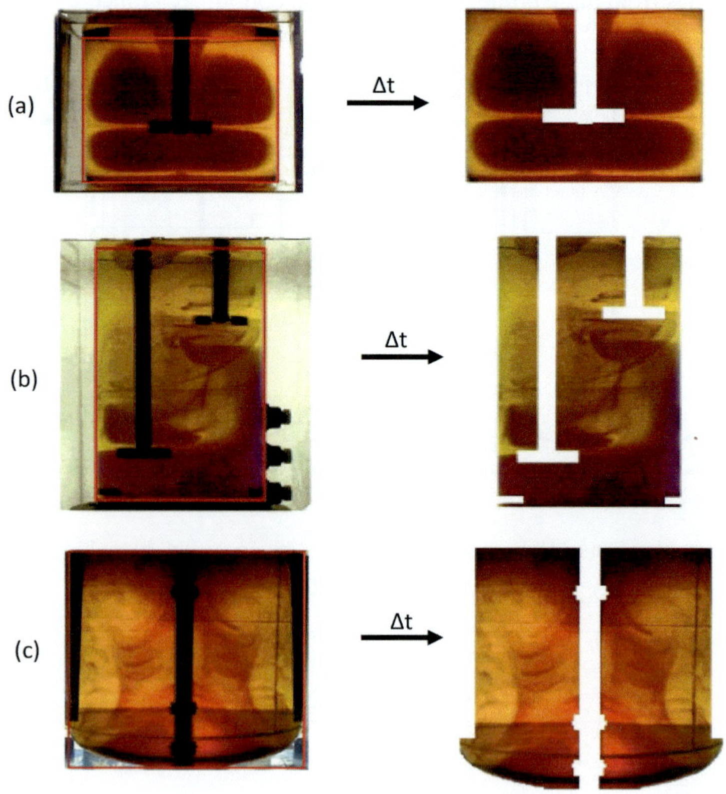

FIGURE 18.9 Video frames showing an area of interest (red rectangle) in three mixers: (a) single impeller (cavern formation), (b) dual impellers and (c) Maxblend® impeller.[5]

In line with the acid-base neutralization, an alternatively proposed technique (double indicator system for mixing time, or DISMT)[6] allows experimentalists to directly determine the 95% mixing time. The DISMT method uses two standard acid-base indicators, methyl red (red to yellow, pK ~ 5) and thymol blue (yellow to blue, pK ~ 9). These indicators are separately added to deionized water to form their respective solutions. Then the red (acidic) solution is mixed with the blue (basic) solution, and in those regions where the mixing fraction is within 5% of the mixing fraction at infinite time, the solution is yellow. In a clear mixing vessel, an observer may see distinct red and blue regions as well as the later emergence of yellow regions. Since DISMT is a reaction-based diagnostic, color changes occur only when reactants have been mixed at the molecular level. Such reaction-based diagnostics avoid the ambiguities associated with dilution diagnostics and the phenomena generally referred to as "micro-mixing." Thus, in the case of DISMT, the 99% mixing time for the entire vessel may be defined as the time for the entire liquid volume to become yellow.[6] Although the intermittent color sequence of this dual indicator system can provide substantial semi-quantitative information, the method itself is harder to follow than standard single indicator methods.

Conductivity: A simple and cost-effective, intrusive method of determining apparent mixing time places an electrical conductivity probe(s) in stirred vessels. A concentrated aqueous salt (NaCl) solution is injected into the agitated, deionized water as instantaneously as possible. Conductivity meters continuously measure media conductivity at various probe locations, while a computer logs the continuous data for subsequent analysis. Comparing this conductivity data to an initial concentration vs. conductivity calibration curve converts conductivity values to salt concentration, which fluctuates as follows:

$$C_i = \frac{C_t - C_o}{C_\infty - C_o} \tag{18.8}$$

C_o is initial tracer concentration, C_∞ is final tracer concentration, and C_t is tracer concentration at any given time "t".

Figure 18.10 shows the resulting concentration fluctuation as a function of time.[7] The 95% mixing time, θ_{95}, is determined when the amplitude of the concentration fluctuations is less than 5%.

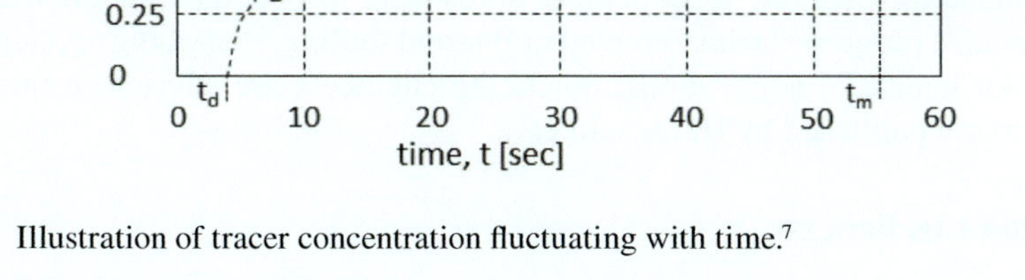

FIGURE 18.10 Illustration of tracer concentration fluctuating with time.[7]

Tomography: Electrical resistance tomographic is a three-dimensional (3D) imaging technology where multiple two-dimensional (2D) slices of the images are computationally superimposed or merged to create an object of interest. One simple example will be to sequentially capture photographs (images) of the evolution of color change during acid-base reaction in a transparent stirred vessel by means of multiple high-speed and high-resolution digital cameras. Then the color pixels (data) of the 2D images are processed using a commercial data analyzer (tomography software) to create evolution of colored objects as a function of time. Apart from the micro-mixing time prediction, visualization of this 3D tomography reveals developing flow patterns and segregated regions in the tank prior to homogeneity. The method is an analog of the medical CT, a diagnostic procedure where the x-ray image slices at different depth are computationally processed and overlaid to reconstruct a biological object of interest.

Most industrial-scale reactors are not transparent, in lieu of the above optical coherence technology (photography), but the non-intrusive and non-invasive electrical resistance tomography (ERT) can be applied to discern flow phenomena of single-phase and multi-phase (gas–liquid or solid–liquid) working media in a stirred vessel or in an in-line mixer. When an electric field is applied to a multi-component media, the local voltage output varies depending on the resistivity of the constituent. ERT is a technique of planar (cross-sectional) mapping of the potential field in a fluid using DC current, preferably by means of at least two current electrodes and two potential electrodes. Then through appropriate post-processors, the 2D voltage data (image) are superimposed to create a 3D (volumetric) domain (image) and recognized with designated RGB colors. Successive time-dependent application of the current develops transient-colored topographies that identify the evolution of mixing and multiphase interface. The real-time data qualitatively provides a 3D movie of fluid mixing and quantitatively the mixing time parameter, θ_{95}. ERT, the first technology commercialized by ITS, has many installations around the world.[8] The ITS system is generally customized to the application, such as batch or in-line mixing, with a typical setup consisting of an array of electrodes, a data acquisition system and a user-friendly interface.

In industrial chemical reactors, fouling (solid deposition and crust formation) of tank internals due to deposition of products or byproducts is not uncommon. Crystallizers and reactors processing slurries and viscous media are particularly prone to fouling. Because these reactors are typically closed systems, fiber-optic probes are used in the headspace to observe the extent of fouling. When it exceeds tolerable plant levels, appropriate in-situ cleaning procedures utilizing steam, solvents and possibly spray nozzles are employed to minimize production shutdowns. Since fiber-optic probes are also susceptible to fouling, an industrial non-disruptive and non-invasive tomographic gamma-ray scan can be implemented. This technique moves a sealed radiation source and detector along the outside surface of the equipment, with the intensity readouts generating a density profile of the equipment.

Like tomography from x-ray CT images, the 2D density profile images are reconstructed to display the interior of the vessel on a computer. Density magnitudes are tabulated using the density vs. irradiation calibration curve, with a pre-determined density criteria triggering the alarm to draw operator attention for follow-up action. Drawbacks of gamma-ray tomography include safety and general complexity (which require extensive operator training) in data gathering, tomographic reconstruction, reporting and interpretation of results.[9] To lessen the burden on users, customized gamma-ray commercial devices come with the CT scanner, data acquisition cards (DAQ), image processors and user-friendly interfacing software. Since the technology can be safety prohibitive, it may be acceptable to substitute a retractable boroscopic inspection camera, with light, in the headspace to visualize progress during removal of internal fouling. Regarding application of tomography scanning for industrial applications, practicing engineers are advised to pursue the recent comprehensive article published by Haraguchi et al.[10]

18.3.2 Continuous Process

Many chemical processes use one or more continuous stirred tank reactors (CSTRs) in series, as depicted in Figures 6.10–6.11. Components such as reactants, catalysts and solvents are continuously added and products continuously removed. Each vessel behaves like a steady-state batch process without accumulating its contents. Agitator design must still provide effective mixing inside the reactor. Preferred feed and product removal locations, with respect to the impeller position(s), are discussed in Chapter 6. Ideally, composition of the output stream should be identical to that inside the reactor, which is a function of residence time and reaction rate. The recommended mean residence time ($\tau = V / \dot{Q}$) in a well-mixed media is at least 10 times greater than the mixing time, θ_{95}, of media in the vessel.[11] V and $\dot{Q}$ are steady-state media volume and volumetric flowrate (in/out), respectively. For example, if media volume (V) is 100gal. (~380liter), and the sum of component flowrates ($\dot{Q}$) is 25gal/min (~95liter/min), then residence time (τ) is 4 minutes. In this case, ideal mixing time (θ_{95}) should be less than 24sec. If a desired reaction extent exists, it should be a complete reaction within 4min.

The required minimum media residence time dictates input and output flowrates of the reactants and products. The residence time distribution (RTD) of a chemical reactor is a probability distribution function describing the time that fluid elements could spend within the reactor. Chemical engineers use the RTD to characterize mixing and flow within reactors and to compare behavior of real reactors to that of ideal reactor models, such as the plug-flow reactor (PFR) with no back mixing. This is useful for troubleshooting existing reactors, but also for estimating reaction yield and future reactor design.[12] CSTR residence time distribution is determined in the lab using scaled-down parameters such as vessel geometry, specific power, flowrates, etc. Details of the method are discussed in Chapter 6. An experimental approach is used here to study the RTD of an ideal continuous stirred tank reactor (see Figure 18.11).

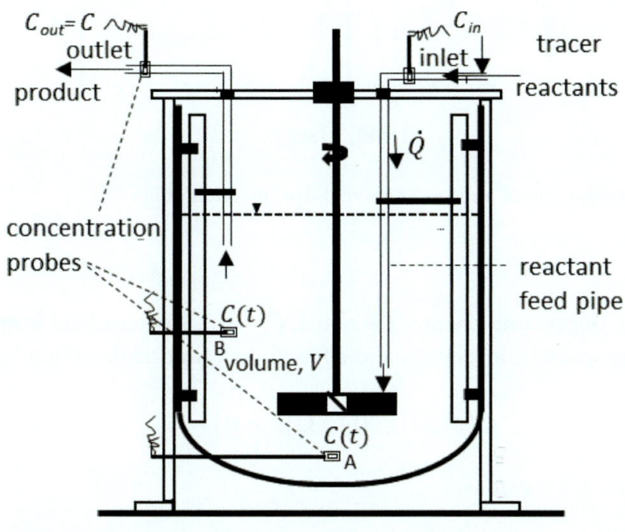

FIGURE 18.11 Schematic of an experimental system to study RTD and RTD density.

For this illustration, a 100gal. (~380liter) Plexiglas® laboratory vessel is used to simulate a production-scale reactor. DI water and a dye solution (food color) at the inlet were used, respectively, as the working media and tracer. An injection pump rapidly injects food color (in infinitesimal time) into the inflow DI water stream as a Dirac or δ-pulse (see Figure 18.12). Appropriate reactor scale-down rules (conserving both geometric and dynamic similarity) were followed. Inlet and outlet flowrate scale-down was based on the media volume in the laboratory vessel, as compared to that in the production scale. Initially, all of the concentration probes (colorimeters) are calibrated with respect to the color of the dyed-DI water solution.

In the case of a fully mixed ideal CSTR, the tracer component mass balance is expressed as,

$$C_{in}.\dot{Q} - C.\dot{Q} = V\frac{dC}{dt} \tag{18.9}$$

where, C_{in} = concentration of tracer (dye/salt) at the inlet [gm/l]
Q = inlet flowrate [l/sec]
C = concentration of tracer (dye/salt) in the reactor [gm/l]
V = volume of working media in the reactor [l]
t = time [sec]

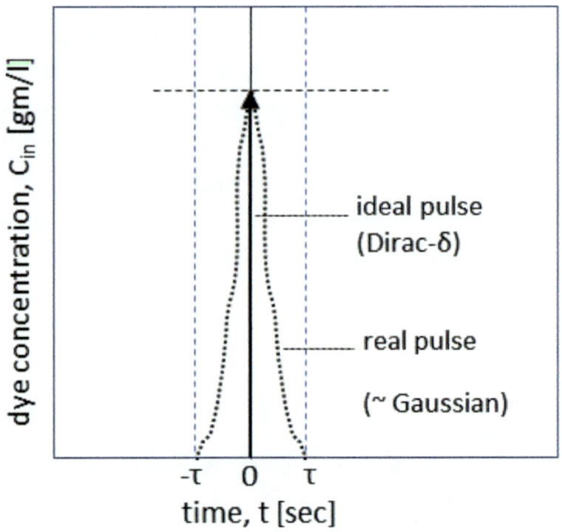

FIGURE 18.12 Representations of an idea and a real dye injection pulse.

Concentration of the tracer component at the outlet, $C_o = C$, is determined from equation 18.9 using the inlet tracer injection (pulse) concentration as a function of time as the initial boundary condition.

$$C(t<0)=0 \text{ to } C(t\geq 0)=C_{in}$$

Equation 18.9 can be rewritten as,

$$\frac{\dot{Q}}{V}dt=-\frac{dC}{C-C_{in}} \text{ or } \frac{dt}{\tau}=-\frac{dC}{C-C_{in}} \tag{18.10}$$

where, $\tau = V/\dot{Q}$ is the residence time for an ideal CSTR [sec]

$$-\int_0^t \frac{dt}{\tau}=\int_0^C \frac{dC}{C-C_{in}} \tag{18.11}$$

$$-\frac{t}{\tau}=ln\left(\frac{C_{in}-C}{C_{in}}\right) \tag{18.12}$$

and concentration of tracer at the outlet of the reactor, as a function of time, is presented in equation 18.13.

$$C(t)=C_{in}exp(-t/\tau) \tag{18.13}$$

The residence time distribution density function, $E(t)$, can be calculated by normalizing the pulse response at the outlet $C(t)$, with respect to the whole amount of tracer used.

$$E(t) = \frac{C(t)}{\int_0^\infty C(t)dt} = \frac{1}{\tau} e^{-t/\tau} \tag{18.14}$$

The dimensionless RTD is represented by equation 18.15 and graphically shown in Figure 18.13.

$$E(\theta) = e^{-\theta} \tag{18.15}$$

where $\theta = t / \tau$, t is non-dimensional process time, and τ is mean residence time.

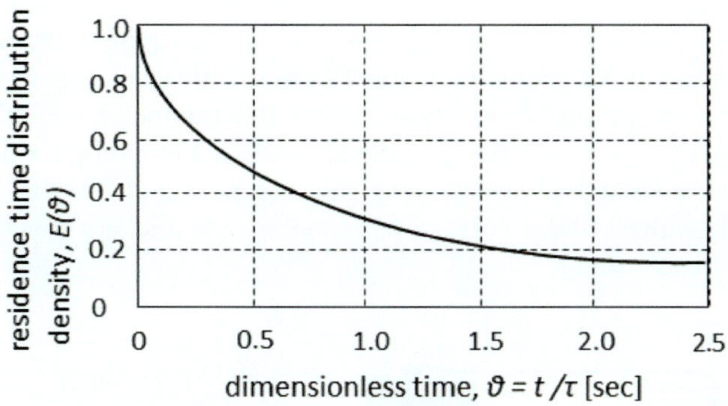

FIGURE 18.13 Graphical representation of RTD distribution for an ideal STR.[13]

Readers might review Chapter 16, "Resident Time Distributions of Chemical Reactors," in reference 13.

18.4 GAS–LIQUID CONTACTORS

Stirred tank reactors (STRs) create the intimate gas–liquid contact needed for mass transfer in aerobic fermentation, hydrogenation, phosgenation, neutralization, chlorination, organic oxidation, and numerous other chemical processes. Chapters 9 and 10 of this book, respectively, describe the theory and design principles of STRs as gas–liquid reactors and aerobic fermenters. For design purposes, characteristics of interest in gas–liquid contactors include: (a) gas holdup, $\in_g$, and (b) volumetric mass transfer coefficient, $k_L a$. These two design variables are characterized over ranges of specific power, ε_g, and superficial gas velocity, V_{sg}. Additional gas–liquid-contactor design variables include the (c) ungassed power draw, P_{ug}, measured at no gas flow (aeration number, Na, = 0.0), and (d) gassed power draw, P_g, and k-factor, (P_g / P_{ug}), which are evaluated over a range of aeration numbers. Common experimental methods used to evaluate these characteristic design variables are described as follows.

GAS HOLDUP

In vessel sizing, prior knowledge of expected gas holdup, ε_g, is critical. Gas holdup is defined as the ratio of gas volume within the media to the mixture volume, as shown in equation 18.16.

$$\varepsilon_g = \frac{V_g}{\left(V_g + V_L\right)} \tag{18.16}$$

where V_g is the gas-phase volume (m^3) and V_L is the liquid-phase volume (m^3).

As mentioned earlier, gas holdup is a function of the specific power under gassed conditions (ε_{p-gas}), volumetric flow of gas per unit liquid volume (vvm) or superficial gas velocity (v_{sg}), and liquid physical properties such as density (ρ_l), viscosity (μ_l), and interfacial tension (σ_i). Gas holdup values for aqueous coalescing and non-coalescing low-viscosity systems are typically between 10% to 50%, respectively, depending on system interfacial tension. A viscous liquid media may accumulate many tiny (<2mm) non-coalescing bubbles, however, and temporarily make gas holdup much higher than 50%. If the accumulated bubbles can gradually coalesce into large bubbles, on the underside of broad impeller blades and/or in low-flow regions, they will likely rise and burp out of the media surface at intermittent times. The following methods are used to determine the gas holdup in the laboratory and production scales.

Visual Metric Method: For clarity, this laboratory gas holdup measurement is described using a simple illustration. Figure 18.14 shows a 20-gal. (18-inch dia.) transparent Plexiglas® tank equipped with a gas sparging system. It is a geometrically and dynamically scaled-down version of a planned production reactor, with an appropriate agitation system, dynamometer, and other instrumentation. Heights of the simulated (ρ_l, μ_l, σ_i) working media are assessed at multiple values of specific shaft power and gas flowrate, with and without flowing gas, using a ruler affixed to the side of the tank. This yields gas holdup values under varying agitation speeds and gas flowrates until a relative steady-state condition is achieved.

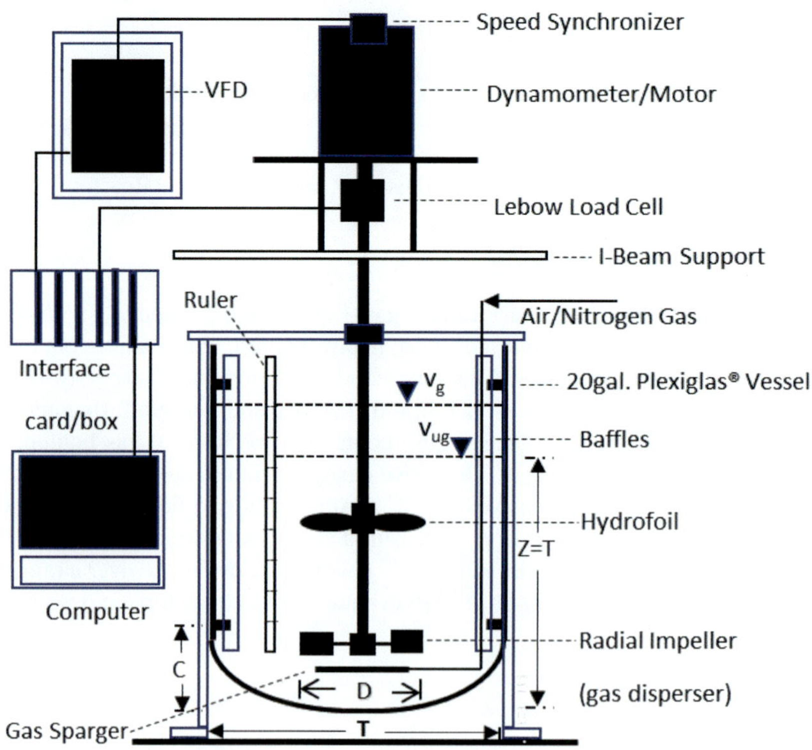

FIGURE 18.14 Laboratory gas-holdup measurement method using a ruler.

Level Detection Method: In production plants, transmitters utilizing either radar or ultrasonics are popular for measuring media fill levels. When deploying the latter non-intrusive device, tank headspaces that are reasonably free of dusts and vapors are preferred. More of these transmitters, their working principles, usage pros-and-cons, and application suitability can be learned from reference 14. When mixing non-coalescing media in stirred gas–liquid contactors, the need for high specific power (1–3 W/kg) and high gas holdup make standard 20–100gal laboratory vessels unsuitable for scaled laboratory testing. Therefore, the authors configured the available production scale (1800 gal) tank pictured in Figure 18.15 for laboratory determination of mass transfer coefficients and gas holdup values.

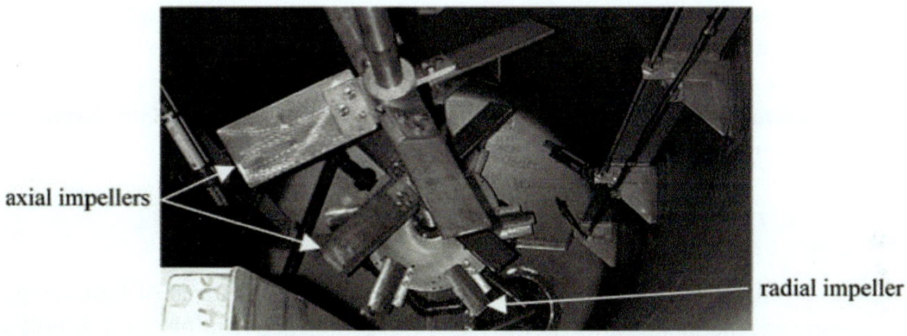

FIGURE 18.15 Gas–liquid mixing system for gas holdup and mass transfer studies.

In the above vessel, media height and gas holdup were obtained by mounting the transducer of a Rosemount level detector (see Figure 18.16) within a nozzle in the vessel headspace. Media disengagement height was detected using fast-sweep frequency modulated continuous wave (FMCW) technology. The FMCW transmitter continuously emits signal sweeps, with constantly varying frequency waves, onto the working media surface. Resulting frequency differences between the transmitted and reflected signals are proportional to the distance to the liquid surface. The signal is further processed to provide highly accurate level measurements.[15] In gassed and un-gassed conditions at different specific powers and varying superficial gas velocities, measured media heights lead to gas holdup values, per equation 18.16.

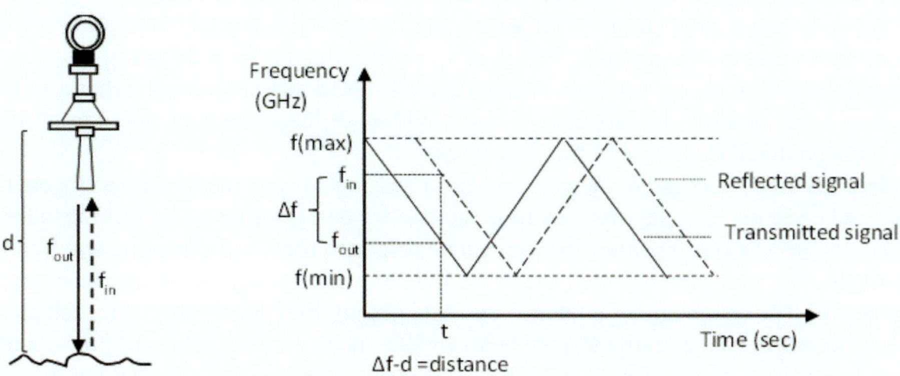

FIGURE 18.16 Rosemount transmitter and illustration of FMCW technology.[15]

18.4.1 Mass Transfer Coefficient

Knowing the needed gas–liquid mass transfer rate (MTR) is possibly the most critical parameter in designing gas–liquid contactors such as stirred tank reactors, fermenters and bubble columns. Discussions in Chapters 9 and 10 are dedicated to the design of these units with examples based on MTR correlations. Two popular laboratory experimentation methods (transient and steady-state) for determining the volumetric mass transfer coefficient, $k_L a$, are now comprehensively addressed. The mass transfer rate is defined as shown in equation 18.17. Readers should study other experimental methods associated with bioreactors, as described in the literature,[16] to better understand the relevant merits and demerits of each, and to pursue the one most pertinent to each specific application. The article by Moutafchieva et al.[17] is recommended for its succinct, tabular outline of established empirical correlations for $k_L a$ in various gas–liquid contactors, including pros and cons for each. The dynamic gassing-out techniques used in three types of common gas–liquid contactors (stirred tank, bubble columns and airlift) are addressed.

Because diffusion is much slower in the liquid phase than in the gas phase, the liquid film presents most of the resistance to mass transfer (see Figure 9.1 for details). Gas–liquid film theory defines the mass transfer rate in terms of the liquid film, as follows:

$$MTR = k_L a\left(C^* - C_L\right) = k_L a\left(C_g H_e - C_L\right) \text{ (mol/m}^3\text{ s)} \tag{18.17}$$

where k_L is liquid film mass transfer coefficient (m/s), a is interfacial area of bubbles per unit volume (m^{-1}), C^* is saturation concentration at the gas–liquid interface (mol/m^3), C_L is bulk concentration in the liquid phase (mol/m^3), C_g is transfer gas concentration in the bubble (mol/m^3), and H_e is Henry's law constant.

$$H_e = \frac{c^* RT}{p_g} \tag{18.18}$$

where, R = universal gas constant (~8.314 J/mol-K), T = temperature at the gas–liquid interface (K), and p_g = transfer gas pressure in the bubble (Pa).

The average value of k_L is typically around 0.2 ± 0.02 cm/s depending on diffusivity of the dissolved gas and relative velocities of the gas and liquid phases.[18] The value of "a" can vary by several orders of magnitude, especially in bubble systems, where it is inversely proportional to bubble size. Because of difficulties in measuring k_L and "a" individually, the lumped parameter or volumetric mass transfer coefficient $k_L a$ (i.e., k_L times "a") is typically used to determine MTR. The value of $k_L a$ can be raised by increasing superficial gas velocity and power input. A higher $k_L a$ can cut capital costs by producing a comparable MTR with a smaller vessel, a shorter batch process time, or a higher continuous process throughput. Values of $k_L a$ are evaluated for a diffusing gas (e.g., O_2), at given operating conditions, by using an appropriate dissolved gas probe, and either a transient or a steady-state experimental measurement method. Although these methods apply to all diffusing gases, they are confined to oxygen in this discussion.

Transient Method: Successive dynamic cycles of gassing a liquid media with nitrogen, then air, is a simple and common "gassing-out" aeration method for determining $k_L a$ in mixing laboratories. A dissolved oxygen (DO) probe must have adequate response time, τ_r, to capture transient oxygen dissolution rates in a well-mixed vessel, i.e., ($\tau_r < 1/k_L a$), where $1/k_L a$ is the characteristic time of mass transfer.[19,20] The schematic of a laboratory-scale (20gal. Plexiglas®) stirred vessel, along with the required accessories and equipment to measure DO under varying values of power and superficial gas velocity, is shown in Figure 18.17. It uses two fast-response ($\tau_r \le 2s$) DO probes and also an RTD to measure working media temperature, T_w.

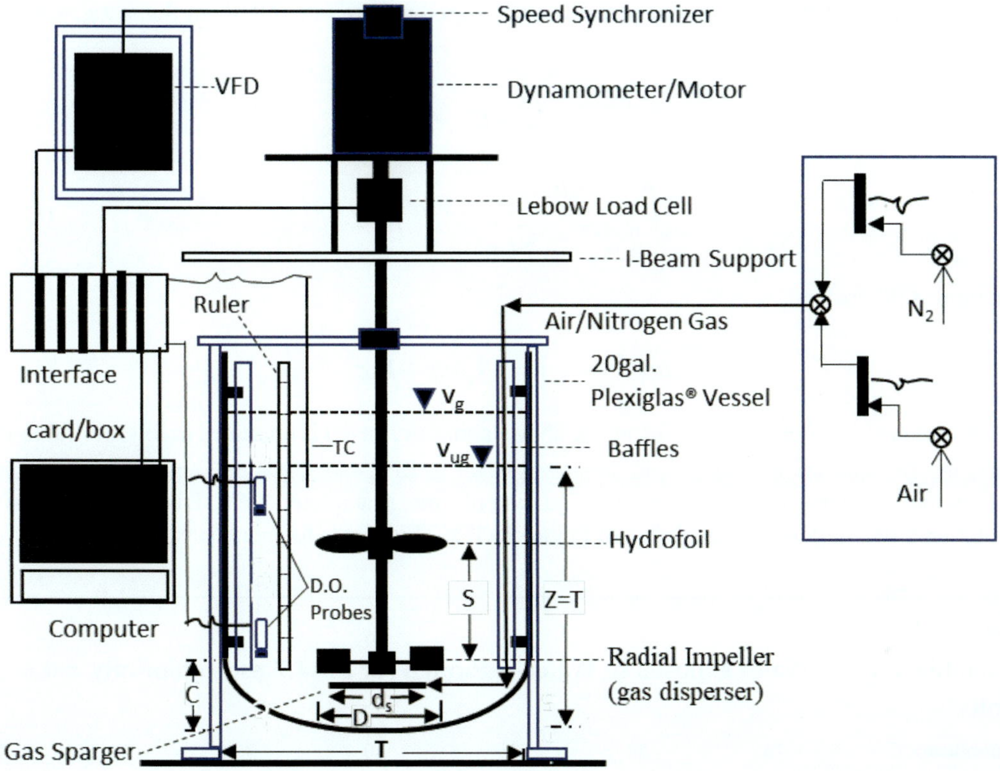

FIGURE 18.17 Laboratory setup for the transient method measurement of gas–liquid mass transfer. [Z = T = ~18″, C = 3″, D = 8″, S = 8″, d_s = 6″]

The transient method of determining k_La for oxygen, at a given set of operating conditions, begins by sparging an inert gas (e.g., N_2) into the liquid to purge it of dissolved oxygen (DO). After the DO probe reading indicates essentially zero concentration, the flowing N_2 is turned off and replaced with dry air. The increasing concentration of DO in the liquid phase, C_L, is recorded with increasing time, t (see Figure 18.18a). As the gas dissolution process can be described by equation 9.2, the corresponding value of k_La is obtained by regressing the experimental data as suggested in equations 18.19–18.22.

Since the specific power required to achieve a desired level of k_La is in the range of 1–3W/kg, the rotational speeds of the STR agitation systems are typically higher than normal single-phase low-viscosity media operations. Drawing air into the media, from the vessel headspace, is highly possible and highly undesirable because it alters measured superficial air velocity, and consequently, the resulting k_La value. To avoid this, agitation must be turned off after the DO reading is zero, and then turned back on again at the preset speed, simultaneous with beginning air sparge into the vessel. This operation needs appropriate synchronization between the mass flow meter and the motor speed controller (VFD). As noted earlier, particularly in experimental vessels smaller than 100gal, headspace gas entrainment can be almost unavoidable at high specific power levels, and thus, corresponding k_La predictions could contain significant systematic error.

A mass balance for dissolved oxygen in a well-mixed liquid phase can be written as,[17]

$$\frac{dC_L}{dt} = k_La\left(C^* - C_L\right) = OTR - OUR \tag{18.19}$$

Without reaction (or bio-respiration) in the water, oxygen uptake rate (OUR) in the static state = 0. Hence,

$$\frac{dC_L}{dt} = k_L a\left(C^* - C_L\right) \tag{18.20}$$

$$\int_0^{C_L} \frac{dC_L}{\left(C^* - C_L\right)} = k_L a \int_0^t dt \tag{18.21}$$

Equation 18.20 leads to

$$ln\left(c^* - c_L\right) = -k_L a.t + \ln c^* \tag{18.22}$$

An illustration of $k_L a$ determination using the above laboratory vessel is presented as follows. Table 18.1 depicts the DO saturation limits, c^*, in deionized water at 15°C–45°C, while Table 18.2 lists experimentally measured oxygen content, c_L, as a function of time, t, at a fixed 25°C. Dissolved oxygen (%) in water, as a function of time, is plotted in Figure 18.18a. This plot is founded on equation 18.21.

TABLE 18.1
O_2 Solubility, c^*, in Water Exposed to Water-Saturated Air at 14.7 psia, chlorinity = 0 = salinity[21]

Temperature, °C	15	20	25	30	35	40	45
Solubility, mg/L	10.08	9.09	8.26	7.56	6.95	6.4	5.93

TABLE 18.2
Measured Dissolved Oxygen Percentage (% DO), as a Function of Time, in a Vessel at 25°C

Time [sec]	0	5	10	15	20	25	30	35	40	45	50	55	60	65	70	75
C_L [mg/L]	0	1.82	3.25	4.35	5.21	5.89	6.41	6.82	7.14	7.39	7.58	7.73	7.84	7.94	8.01	8.06
time [sec]	80	85	90	95	100	105	110	115	120	125	130	135	140	145	150	
C_L [mg/L]	8.11	8.14	8.17	8.19	8.20	8.22	8.23	8.23	8.24	8.25	8.25	8.25	8.25	8.25	8.26	

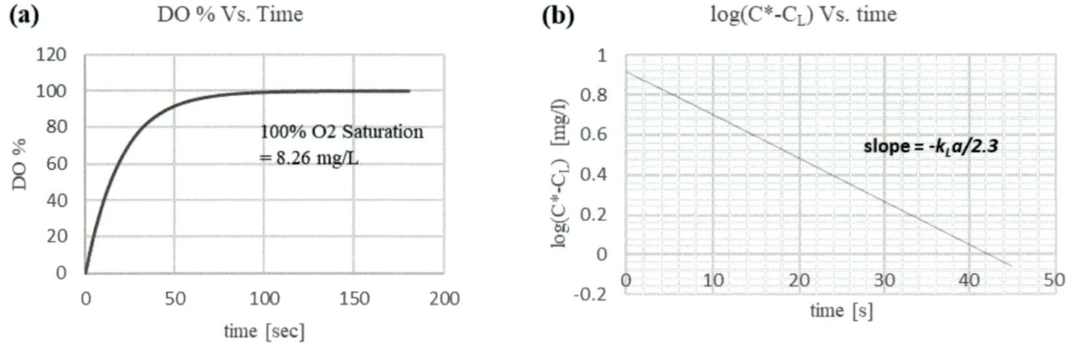

FIGURE 18.18 (a) Dissolved oxygen, DO percentage, as a function of time, and (b) Log_{10} plot of $\left(C^* - C_L\right)$ vs. time, t, to determine $k_L a$ from the negative slope.

Equation 18.22 results are plotted in Figure 18.18b, with a slope of $-k_L a$ and intercept of $\ln c^*$. From this latter plot, the volumetric mass transfer coefficient, $k_L a$, is determined to be $0.05 s^{-1}$.

ASCE standard ASCE/EWRI *2–06*[22] says experimentally determined values of $k_L a$ at any given media temperature should always be standardized to 20°C (i.e., $k_L a_{20}$) prior to establishing any empirical correlations with power per volume, ε_g, superficial gas velocity, v_{sg}, and viscosity ratio, μ_r. This is shown in the following equations, the details of which are described in Chapters 9 and 10.

$$k_L a_{20} = k_L a \cdot 1.024^{(20-t)} \tag{18.23}$$

Standardization, if $k_L a = 0.05\ s^{-1}$ and $t = 25°C$: $k_L a_{20} = 0.05 \times 1.024^{(20-25)} = 0.04 s^{-1}$

$$k_L a_{20} = \alpha \varepsilon_g^{\beta} u_g^{\gamma} \left(\frac{\mu_{susp}}{\mu_w} \right)^{\kappa} \tag{18.24}$$

where μ_{susp} and μ_w are, respectively, the dynamic viscosities of the working media and water. The coefficient α and the exponents β, γ and κ are determined from experimental data regression.

Instead of the dynamic gassing-out technique mentioned earlier, i.e., nitrogen sparging followed by aeration of the working media, a reverse procedure has also been widely used for $k_L a$ measurements due to its simplicity. In the latter technique, dissolved oxygen is removed from the liquid phase, for example, by bubbling nitrogen through the liquid phase until its oxygen concentration reaches zero.[23]

Steady-State Chemical Method: The steady-state chemical method of determining $k_L a$ involves oxygen transfer from the bulk liquid into gas bubbles, as opposed to the above transient method where the transfer process is vice versa. In the steady-state process, oxygen is generated in the liquid by altering the oxidation state of water-soluble chemicals such as hydrogen peroxide (H_2O_2) or sodium sulfite (Na_2SO_3) using appropriate catalysts such as a catalase (e.g., bovine liver enzyme) or MnO_2 etc.[24,25] For example, in the presence of ferric chloride ($FeCl_3$), the following regenerative decomposition reactions occur to produce oxygen in water:

$$2FeCl_3 + 3H_2O_2 + 2H_2O \iff 2H_2FeO_4 + 6HCl$$
$$2H_2FeO_4 + 3H_2O_2 \iff 2Fe(OH)_3 + 2H_2O + 3O_2$$
$$2Fe(OH)_3 + 6HCl \iff 2FeCl_3 + 6\,H_2O$$

$$2H_2O_2 = 2H_2O + O_2$$

The oxygen concentration in the liquid is monitored by dissolved oxygen (DO) probes. The steady-state condition is inferred when the oxygen concentration in the liquid, c_L, remains constant irrespective of time, i.e., oxygen transfer rate is equal to the oxygen production rate. With the assumptions of ideal gas behavior, a well-mixed liquid phase, and equilibrium oxygen concentration, C^*, at the gas-bubble interface, the volumetric mass transfer coefficient, $k_L a$, is determined from the oxygen transfer rate, OTR, and the oxygen concentration, C_L, in the bulk liquid phase.

$$k_L a = \frac{OTR}{\left(C_L - C^* \right)} \tag{18.25}$$

In order to circumvent the drawbacks of transient methods, such as headspace gas entrainment and restrictions on the probe response time, the dynamic method is more suitable for gas–liquid mass transfer studies using larger-scale (≥ 100gal) stirred tanks. However, cost of the required chemicals, especially that of the catalase, can be prohibitive. Because ferric chloride serves as an inexpensive catalyst, it was used in an 1800-gal industrial-scale 316L stainless steel tank containing DI water. Figure 18.19 schematically represents a large-scale experimental setup that was used. It has an ASME dish-bottomed tank with agitation system, the indicated peripherals, and associated instrumentation such as DO probes, an RTD, etc.

The Figure 18.19 mixing system includes a typical radial gas-dispersing impeller (Rushton turbine) with an upper axial flow impeller (PBT) in the downflow mode. To avoid a concentration boundary layer on the membrane surfaces, the two YSI Model 59 DO probes were directed against the direction of the tangential flow in the tank. According to the user manual, these concentration probes are calibrated based on the media conductivity and temperature. A micropump feeds 35wt.% aqueous hydrogen peroxide solution from 55-gal. drums into the tank, where it is mixed with deionized water and dissolved $FeCl_3$ at ~0.01 mole/l. The peroxide mass flowrate from the drums is obtained from a platform scale. To vary gassed-power levels, P_g, and corresponding superficial gas velocities, V_{sg}, experiments were conducted at several different agitation speeds and air flowrates (volumetric air flow, or vvm). Based on the partial pressure of oxygen in dry air, with a molar oxygen fraction of 0.2095, the initial equilibrium concentration of oxygen at the saturated surface of bubbles exiting the gas sparger, c^*, is calculated with Henry's law (see equation 18.18). At the sparger, total pressure, p_t, is the sum of atmospheric pressure, p_o, and hydrostatic pressure, p_{hs}. Thus, transfer gas pressure is

$$p_s = 0.2095 \cdot p_t \quad (18.26)$$

$$c^* = H_e \cdot p_s \text{ [mol/l]} \quad (18.27)$$

An alternate simplified approach obtains a reasonably accurate value of c^*, at the operating temperature and pressure, from the ASCE tables.[21,26] $FeCl_3$ solution, in the long run, is prone to initiate

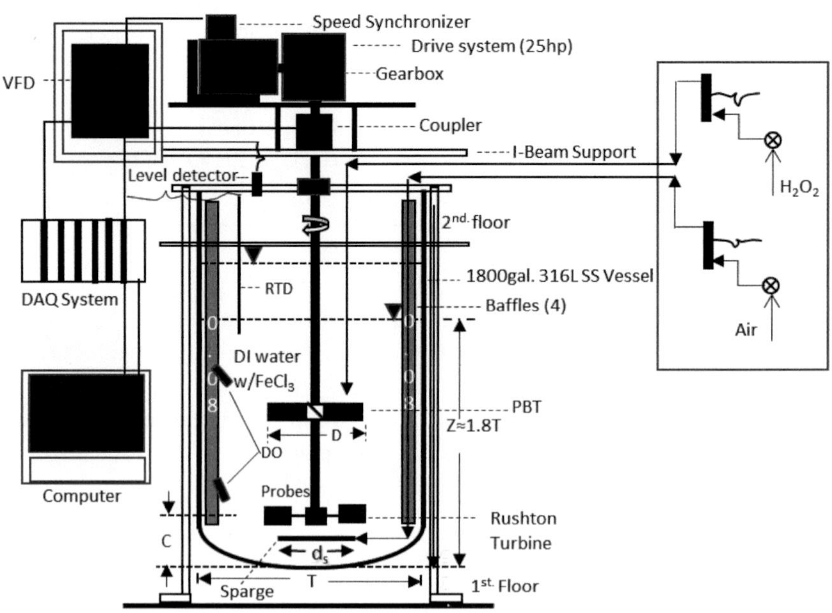

FIGURE 18.19 Schematic of the large-scale experimental setup for determination of $k_L a$ using a steady-state method. [H = 122", T = 66", Z = 71", C=21.5", D = 33", d_s = 24"]

corrosion stress cracking of the SS vessel. Thus, as a caution, experiments should use rust removers such as Zero Rust 70SP Green®, immediately followed by steam cleaning, to prevent long-term wall failure. In smaller stirred tanks, ≤100 gal, one could economically use bovine liver catalase for mass transfer studies in aerated, agitated vessels.

At steady state,

$$OTR = \frac{\dot{Q}_{H_2O_2}}{Mw_{H_2O_2}} \cdot \frac{1}{2 \cdot V_L} \tag{18.28}$$

where, OTR = oxygen transfer rate [mole/l/s]
$\dot{Q}_{H_2O_2}$ = mass flowrate of pure H_2O_2 fed [g/s]
$Mw_{H_2O_2}$ = molecular weight of H_2O_2 [34g/mole]
V_L = liquid phase reactor volume [l]

In equation 18.28, the molar feed rate of H_2O_2 is divided by 2 due to decomposition reaction stoichiometry. The liquid phase reactor volume is corrected for the increase of liquid volume accompanying the addition of H_2O_2 solution during the tests. The bulk liquid oxygen content in mg/l is measured by the YSI DO probes, and percentage O_2 saturation is obtained after correcting for the temperature and salinity of the solution.[27,28] A sample calculation of k_La is presented through the data reported in Table 18.3.

Steady-State Oxygen Balance Method: This simple method is used in aerobic fermenters, where the oxygen transfer rates are determined from measured oxygen concentrations in the inlet

TABLE 18.3
Example Showing Calculated Values Leading to a Volumetric Mass Transfer Coefficient, k_La_{20}

Parameters	Nomenclature	Unit	Values
Working media volume	V_L	liters	3790
Aqueous H_2O_2 mass flowrate	$\dot{Q}_{H_2O+H_2O_2}$	gm/s	28.35
H_2O_2 mass flowrate	$\dot{Q}_{H_2O_2}$	gm/s	9.92
Oxygen transfer rate (equation 18.28)	OTR	mole/l/s	3.85×10^{-5}
Oxygen transfer rate (equation 18.28)	OTR	mg/l/s	0.616
Volumetric flowrate of air to volume of liquid	vvm	m^{-1}	1
Dry air flowrate at STP	$\dot{Q}_{air}$	s^{-1}	63.17
Tank area of cross-section	A_{cs}	m^2	2.21
Superficial gas velocity	V_{sg}	m/s	0.029
Working media temperature	t_m	°C	15.5
Working media conductivity	K_m	μS/cm	4.0
Working media salinity	S	ppt. [g/kg]	2.0
YSI reading	C_L	mg/l	15.53
YSI calibration oxygen saturation value	C^*	mg/l	9.85
Concentration driving force	$C_L - C^*$	mg/l	5.68
Volumetric mass transfer coefficient (equation 18.25)	k_La	s^{-1}	0.11
Agitation speed	N	rps [s^{-1}]	2
Motor power	P_g	kW	8.87
Gassed power per mass	ε_g	W/kg	2.34
k_La at 20°C (equation 18.23)	k_La_{20}	s^{-1}	0.12
k_La at 20°C per Van't Riet[29]/Dow (Table 9.1*)	$k_La_{20-VR/D}$	s^{-1}	0.10

The value of a can change significantly for different working media, i.e., $k_La_{salt\ water} = 1.5\text{–}2 \times k_La_{pure\ water}$. * See Chapter 9.

and outlet air, using on-line oxygen gas phase analyzers. Simultaneously, oxygen concentration in the liquid, C_L, is measured with a DO probe such as the Clark cell O_2 sensor.[19] At steady-state, the oxygen mass balance yields equation 18.29.

$$\dot{Q}_{O_2-in} - \dot{Q}_{O_2-out} - V \cdot OUR = 0 \tag{18.29}$$

where, OUR is the oxygen uptake rate, $\dot{Q}_{O_2-in}$ and $\dot{Q}_{O_2-out}$ are, respectively, the molar flowrates measured at bioreactor inlet and outlet, and V is the working media volume.

The steady-state rate of oxygen transfer from the bubbles equals that of oxygen consumption by the cells

$$k_L a\left(C^* - C_L\right) = OUR \tag{18.30}$$

Thus, it is possible to establish $k_L a$ directly from the following equation:

$$k_L a = \frac{\dot{Q}_{O_2-in} - \dot{Q}_{O_2-out}}{V\left(C^* - C_L\right)} \tag{18.31}$$

where C^* is oxygen concentration in the liquid, at the bioreactor temperature and pressure.[30]

If the bioreactor is too large, the variation in the driving force $C^* - C_L$ could be significant. Then the logarithmic average value between the inlet and outlet of the gas stream can be a good approximation for the driving force. Because this technique relies on the measurement of oxygen content of inflow and outflow air, accurate modeling of gas phase mixing is required for the correct interpretation of the oxygen transfer rate measurement.[19,30]

18.4.2 Gas–Liquid Interfacial Area

In a stirred gas–liquid media, separation of specific interfacial bubble surface area, "a" from the volumetric mass transfer coefficient, $k_L a$, reveals the intrinsic characteristic of the liquid-film mass transfer coefficient, k_L, irrespective of the bubble size distribution. Using equation 18.32, interfacial area per unit volume of spherical bubbles can be determined from gas holdup, $\in_g$, and Sauter-mean bubble diameter, d_{32}.

$$a = \frac{6 \in_g}{d_{32}} \tag{18.32}$$

where d_{32} is defined as,

$$d_{32} = \frac{\int_{i=1}^{n} d_i^3}{\int_{i=1}^{n} d_i^2} \tag{18.33}$$

Bubble diameter in an agitated vessel depends primarily on physical properties of liquid phase, such as interfacial tension and viscosity, and the specific power imparted to the media under gassed condition. For example, the bubbles in a viscosity-dominated laminar flow are expected to be stable but much smaller than those in an inertia-dominated fully turbulent flow regime. Most gas–liquid-contacting applications employ stirred tanks under fully turbulent conditions. Typically, bubbles trapped in the continuous phase follow the fluid path—with their inherent surface transience due to turbulent shear. Latter phenomena lead to continuous bubble collapsing and reformation phenomena until these are gassed out of the working media. Thus, in the case of a dense-phase gas–liquid turbulent flow system, it's an insurmountable task to measure the bubble size distribution. However, in spite of their lack of reliability in dense phase applications, a few experimental methods based on photography and light transmission are briefly described as used to determine apparent bubble diameter, d_{32}.

The simplest approach determines bubble size distribution by analyzing photograph images of a gas–liquid dispersion, taken at various locations within a transparent stirred tank. The inherent limitations are that the media must be a light-penetrating, slow-motion dispersion to circumvent bubble overlapping and blurriness, which is a significant challenge to differentiate dispersed phase from continuous phase. It has been pointed out that bubble size distribution curves from fine bubbles (d < 10mm) dispersions exhibit classical log-normal function traits within ± 3%.[31]

In industrial reactors (mostly non-transparent vessels), several direct but intrusive methods withdraw samples through suction tubes (aspirators) that are inserted into the dispersed phase, within the vessel[32–35] or a recirculation line. Samples are subsequently analyzed to determine gas holdup and bubble size distribution. Capillary suction tubes are able to capture bubble sizes from as small as 300μm to as large as 8mm.[36] This approach is seriously limited by the stability of bubbles dispersed in a quiescent media. Hence, it is hardly practiced in an industrial scale. An alternative invasive approach coupling endoscopy (using a borescope), a high-speed camera,[32,37–39] and subsequent image analysis, is also restricted from the harshest reactor environments (extreme temperatures, pressures, and corrosive media), but ideally applicable to low-concentration (dilute) dispersed phases. Optical densitometry probes are based on refractive index differences (which are proportional to density differences) between two fluids (gas and liquid), and can capture similar bubble dynamics to those of the endoscopy approach. Because the optical probes are less susceptible to the aforementioned punitive destruction, this method is more appealing for industrial reactor applications. Perhaps such application could be advanced by applying fiber-optic probes. To this end, the dissertation of Mueller describes operating principles and designs of fiber-optic probes.[32]

Non-invasive laser techniques to determine fluid dynamics and bubble size in gas–liquid dispersions, such as phase Doppler anemometry (PDA), have been reported in numerous works. Unless a laboratory already has an LDA system, however, purchasing a PDA only for bubble size distribution studies is a costly venture. Furthermore, limitations of the technique are similar to those of the photography approach described earlier. The schematic of a typical PDA lab setup is presented in Figure 18.20.

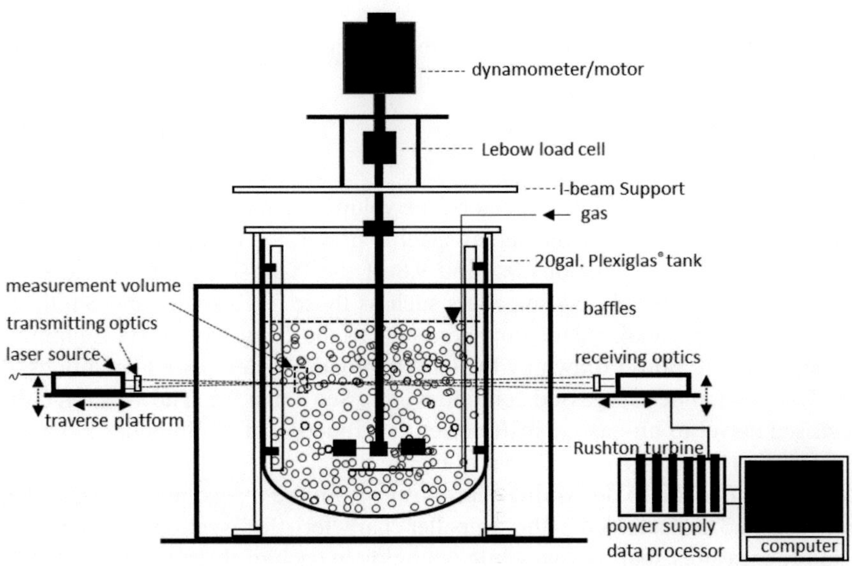

FIGURE 18.20 Laboratory system using phase doppler anemometry (PDA) for bubble size distribution.

Direct measurement of specific interfacial surface area by light transmission is more attractive than the prediction based on particle size distribution and Sauter mean diameter, d_{32}. An approach used for bubble columns could be adopted to stirred tanks provided the dispersion media contains a low concentration of uniformly distributed stable bubbles that can be unaltered by withdrawal from the tank and passage through a transparent rectangular cell similar to a bubble column. This approach is exemplified by a simple experimental optical system consisting of a light source and a photocell receptor with lenses and peripherals, as shown in Figure 18.21.[40] When a parallel beam from the light source is passed through transparent suspensions in the media, only light transmitted through the bubbles is captured by the photocell. The optics are positioned so that scattered reflection and refraction are differentiated and removed from the transmitted light that is detected. According to the optical analyses by Calderman,[41] McLaughlin and Rushton,[42] the total light transmission is a unique exponential function of the non-dimensional group "*a l*" shown in equation 18.34. In this equation, "*a*" is interfacial area per unit volume [m⁻¹] and "*l*" is pathlength [m] of light through the dispersion. I_o and I are incident and transmitted light intensities in volts, respectively.[40]

$$log_{10}\frac{I_o}{I}=\frac{a\cdot l}{9.21} \tag{18.34}$$

$$a=\frac{9.21}{l}log_{10}\frac{I_o}{I}\ [\text{m}^{-1}] \tag{18.35}$$

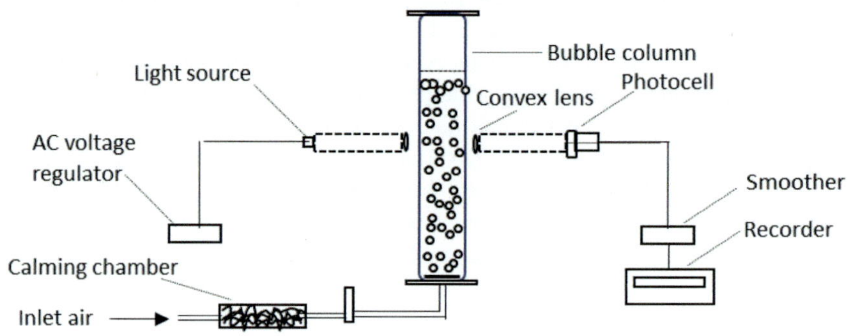

FIGURE 18.21 Experimental bubble column set-up to determine "*a*" using a light transmission method.[40]

18.4.3 *κ*-Factor

In stirred tanks serving as gas–liquid contactors, the volumetric mass transfer coefficient, $k_L a$, and the gas holdup, ε_g, are functions of the actual specific power, ε_g, imparted to the working media under gassed conditions. See equations 9.5 and 9.8 of the "Gas–Liquid Contactors" chapter. In the case of gas-dispersing radial flow impellers such as those of Rushton, and Smith turbines, or even the more advanced parabolic blade impellers such as the Chemineer BT6, Lightnin R135, and Philadelphia Mixer GDX, power drawn by the working media under gassed conditions, P_g, significantly differs from that under un-gassed conditions, P_{ug}. See Table 9.5. The *k*-factor, defined as the ratio of the gassed power to un-gassed shaft power, is a function of the impeller Reynolds number, N_{Re}, and the Aeration number, *Na,* as shown in Figure 9.6.

The experimental method to determine shaft power from the torque measurement, followed by the use of equation 18.3, is explained in the "impeller characteristics" section of this chapter. The test procedure simply measures shaft torque while operating at each of three or more agitation speeds. Gas flowrate into the working media is varied over a range of values (starting at a zero-gas-flow, un-gassed condition), and then corresponding un-gassed and gassed power draw values are determined using equation 18.3.

Aeration number (N_a), the ratio of gas flow to the gas-dispersing impeller discharge flow, is

$$N_a = \dot{Q}_g / ND^3 \tag{18.36}$$

where $\dot{Q}_g$ is gas flowrate (m³/s), N is impeller speed (s⁻¹), and D is impeller diameter (m).

The impeller Reynolds number, N_{Re}, and the Aeration number, N_a, are respectively calculated using equations 18.2 and 18.36. Typically, at constant N_{Re}, the *k*-factor (P_g / P_{ug}) is plotted at different N_a values, as seen in Figure 18.22. This information is critical for practicing engineers to specify the agitation speed under gassed conditions, and thus, to attain the desired power level for required mass transfer. Typical plant procedure calls for higher operating speed under gassed conditions than under un-gassed conditions.

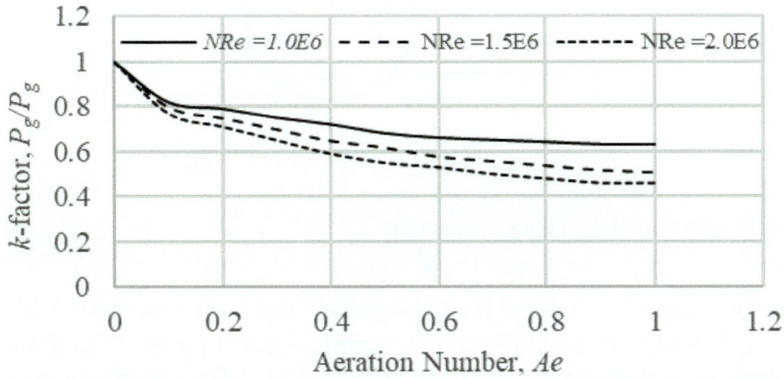

FIGURE 18.22 Dependency of the k-factor on aeration number, N_a, and Reynolds number, N_{Re}.

18.5 LIQUID–LIQUID AND SOLID–LIQUID DISPERSIONS

A comprehensive discussion of immiscible liquid–liquid dispersions using a stirred tank is presented in Chapter 8. Figure 8.9 shows dispersion classification criteria based on dispersed-phase drop size.[43] Photographs illustrating the influence of elongation and shear forces on drop formation are presented in Figures 8.12a–b.[44] There are numerous emulsion and polymerization processes that encounter liquid–liquid dispersion in stirred tank reactors, wherein interphase reaction and mass transfer are controlled by the interfacial area. The outcome of reaction is often solid particles whose size distribution determines the quality and downstream processibility of the reaction product. In most instances, well-established correlations relate liquid drop-size distributions (dsd) to the expected solid particle-size distribution (psd), based on shrinkage or expansion factors.

In the case of non-reactive micro-emulsion formulations such as those in food, or in homecare and healthcare products, immiscible dispersed-phase droplets must remain in suspension for the desirable stability, longevity and performance efficacy. Thus, characterization and control of the dsd is critical to reactive and non-reactive processes, which have been studied extensively by multiple researchers. Invasive in-situ (intrusive) and external offline (non-intrusive) methods include several broad classifications, which are briefly described here without elaborating on their working principles. Except for sieve screening,[45] most experimental psd and dsd techniques, and even many of their instrumentational configurations, are similar enough in appearance and in resulting size distributions for this discussion to address them interchangeably.

18.5.1 In-Situ Methods

Fiber-optic probes in the working media, coupled with a high-speed camera (1000fps max) or a charge-coupled device (CCD w/50fps max), and associated image processor are used in dilute dispersions. This optical microscopy technique is similar to the determination of bubble size characteristics and lucidly described with its inherent limitations in the "Gas–Liquid Interfacial Area" section. Such in-situ methods are conducted at the operating conditions (temperature and pressure) of the process media. They are able to produce online, real-time measurements of psd or dsd that directly captures droplet or particle evolution during transient phenomena such as drop coalescence or breakup; particle nucleation, particle growth, and even variations in some morphological features.

Figure 18.23 depicts the optical effects of a beam of light striking a particle or droplet.[46]

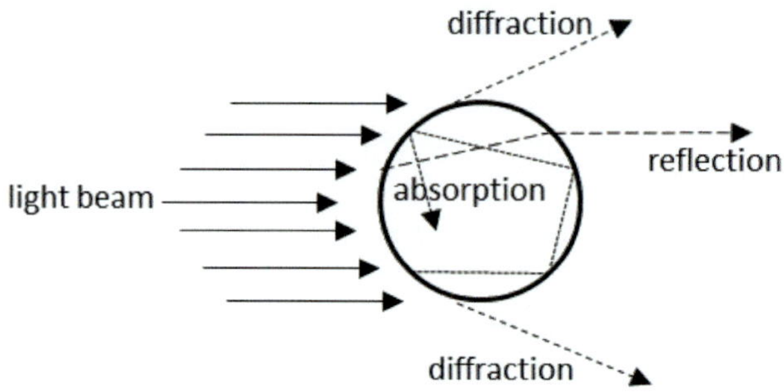

FIGURE 18.23 Light scattering by a spherical particle.[46]

An endoscope with guided fiber-optical light and particle video microscope (PVM), with illuminating lasers, are special in-situ probes used widely in chemical plants for their durability and cost effectiveness. These probes are schematically shown in Figure 18.24.[47] The camera located at the upper end of the endoscope captures images of drops in the vicinity of the glass window. An endoscope provides reliable and accurate values of drop size in the range 5–5000μm, even in a reasonably high-dispersed phase (45%), and is considered the "standard" when comparing other probes.[47] In the case of PVM, particles are illuminated by lights using near-IR lasers, which are focused using a hexagonal array of lenses. Drops passing through the illuminated region scatter light. The backscattered light is collected using a lens system inside the probe and conveyed to the CCD for spectroscopic image analyses.

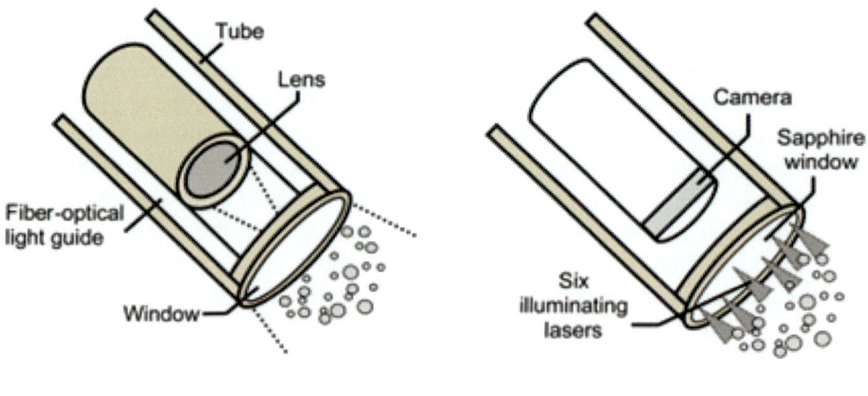

Endoscope PVM

FIGURE 18.24 Illustrations of endoscope and PVM probes for PSA.[47]

Laser-based optical techniques such as the focused beam reflectance measurement (FBRM) and the two-dimensional optical reflectance measurement (ORM) are used as indirect real-time in-situ dsd and psd measurements, based on the principle of light backscattering (see Figure 18.25).[47] In the case of FBRM, near-IR light is transmitted through fiber optics onto an optical lens that rotates at high speed and focuses the laser beam near the outer surface of the sapphire window at the probe tip. The focused beam scans a circular path (creating a chord) at the interface between the window and the particle system, where it will be reflected when it scans across a given particle's surface. The probe measures the length of chord required to traverse edge to edge across the particle (its characteristic length scale), as determined from the measured reflectance time and the velocity of the laser scan (2–16ms^{-1}).[47] Mettler-Toledo Particle Trac™ FBRM®[48] measures particle chord lengths computationally and uses an algorithm to convert those measured chord lengths into the diameters of spherical particles with equivalent surface areas.[49]

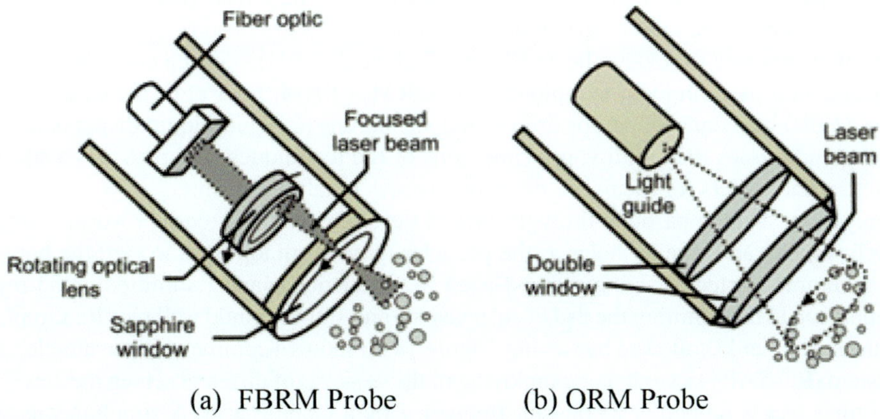

FIGURE 18.25 Illustrations of (a) FBRM and (b) ORM probes.[47]

The authors used the FBRM, in a couple of process applications, to study temporal evolution of particle size distribution in a crystallizer. The ORM probe uses a rotating laser beam focused in the dispersion (focal point diameter of 0.6μm) at very close proximity (≤ 1mm) to its unprotected window. Each time the beam, which is rotating at a known speed, intercepts a drop, light is scattered back into the beam path and transmitted to a detector, where it is transformed into a proportional electrical signal. Chord length is determined from the electric signal duration and the speed of the rotating beam.

18.5.2 External Offline Methods

As pointed out earlier, inherent issues with optics make the use of in-situ dsd or psd measurement probes untrustworthy within dense phase dispersions. If droplets or solids can be made stable from further breakage, growth, or aggregation, samples might be extracted at various times and locations in an STR, and tested externally in commercially available particle size analyzers (e.g., Anton-Parr PSA[50] (d_p= 0.04μm –2.5mm), Horiba Scientific Partica®[51] (d_p = 1.0nm-30mm), Malvern Mastersizer®[52] (d_p = 0.1μm-3mm) etc.). These instruments are commonly used in a production plant's analytical/R&D lab. It is important to select the most suitable method for a given application, since different methods can produce different results from the same material.[53] Hence, practicing engineers should consult with the manufacturers of droplet or particle size analyzers to assess an instrument's suitability in terms of such expected properties as particle morphology, psd, and corrosion compatibility. Prior to either purchase or implementation, onsite size analyzer demonstrations with the equipment is generally warranted.

A detail description of the principle of laser-diffraction particle size analysis is illustrated with figures in reference 54. Typically, laser diffraction analyzers cover a particle size range of 10nm to 4mm. This corresponds to a factor of 400,000 between the smallest and the largest measurable particles. In most applications, laser diffraction is used for particle size distributions between 30nm to 1mm.[54]

Size analysis based on laser diffraction measures particle size distributions by measuring angular variations in the intensity of scattered light, as a laser beam passes through a dispersed particulate sample. The angle of light scattering is inversely proportional to particle size (i.e., the smaller the particle size, the larger the angle of light scattering). Angular-scattering intensity data is then analyzed to calculate the size of the particles responsible for the observed scattering pattern, using the Mie theory[46,54] of light scattering. Particle size is generally reported as a volume-equivalent spherical diameter. In most laser diffraction methods, the volume of the particles is measured, and then the diameter of an equivalent volume of sphere is determined to address the morphology of non-spherical particles. Laser diffraction is claimed to be a very fast, reliable, and reproducible technique that can measure over a very wide size range.[46]

Readers are suggested to review the article by Maab et al. (2011),[55] which compares the four different experimental measurement techniques (2D-ORM, FBRM, fiber optical forward-backward-ratio sensor (FBR), and endoscope) for drop size distributions in liquid–liquid dispersions in STRs. This reference addresses the validity, pros and cons of the techniques, and provides a summary of the parameters and results for the users' decision process in tabular form.

Size analysis by shaking particles through stacked sieve screens (wire mesh or woven wire), having progressively smaller and smaller holes as the particles move from top-most toward the bottom-most screen, is a common external sizing method used in production plants. Admittedly, the most cost-effective approach to determining the psd of suspensions and slurries could well require sample extraction from the stirred tanks followed by sieving. Figure 18.26 shows a commercially available vibratory/tapping system (RO-TAP® sieve shaker) employing multiple sieves of different screen meshes.[56] A cover on the top filter pan is tapped at a constant frequency by a tapping arm. A simultaneous vibration mechanism spreads particles over the screen surface, where they may or may not fall through, depending on their sizes relative to that of the holes. The base pan collects the finest particles that have passed through the bottom-most screen. A practical guide to performing wet sieving, including a video, is described in reference 57. According to ASTM, for accurate size determination of materials finer than the 75μm (No. 200), instead of ASTM C136-17[45] (dry sieving), the wet sieving (washing) procedure described in ASTM C-117-17[58] must be employed. Particle sizes, in microns, and corresponding US mesh numbers (number of openings per linear inch of screen) are presented in Table 18.4. Determining a psd by using a wet sieving method is not ideal for elliptical or needle-like particles.

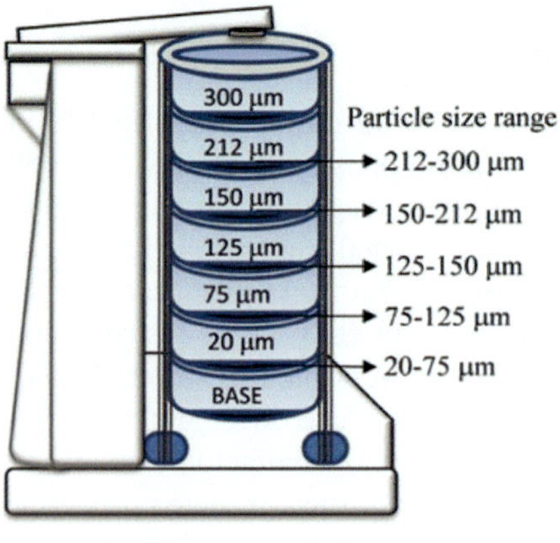

FIGURE 18.26 Schematic of a sieve setup on a Ro-Tap sieve shaker.[56]

TABLE 18.4
Particle Size in Microns vs. US Sieve Mesh Number

US Mesh No.	Particle Size [microns]	US Mesh No.	Particle Size [microns]
40	400	170	88
45	354	200	74
50	297	230	63
60	250	270	53
70	210	325	44
80	177	400	37
100	149	450	32
120	125	500	25
140	105	635	20

Fineness of grind gage, also known as Hegman gage, is widely used in plants as a simple quality control tool identifying fineness of solid–liquid dispersions such as pigments in inks, paints, coatings and carbon-black fillers in automotive and construction sealants, etc. It's a traditional, fast, simple, and cost-effective method to determine the presence of coarse particles or agglomerates in the dispersion. It is especially ideal for high-solid-content working media such as viscous slurries. It does not determine particle size or particle size distribution.[59] The standard test procedure is covered in ASTM D1210-05 "Fineness of Dispersion of Pigment-Vehicle Systems by Hegman-Type Gage.[60]

There are various configurations for the Hegman or "fineness of grind" gage.[61] Figure 18.27 shows a Hegman-type single-notch stainless steel drawdown block and a scraper gage (0–100μm).[62] The notch or groove decreases in depth from one end of the block to the other, according to a scale stamped next to it. Approximately 10cc of dispersion is extracted from the STR or slurry tank and is placed at the top edge (deepest end) of the notch (~100μm). The flat edge of the scraper is used to draw down the slurry in the notch towards the bottom edge (depth 0μm) in approximately 3 seconds. Typically, tests are replicated thrice. Particles or grinds bigger than the depth of the notch produce visible continuous scratches or grooves on the media that represent their relative sizes with respect to the depth. The graduated scales on the left and right of the block, respectively, indicate the coarseness or fineness of the particles measured in terms of dimensionless Hegman unit (0–8) and standard unit based on notch depth (0–100μm). The location where a regular, significant "pepperiness" in the appearance of the coating appears marks the coarsest-ground dispersed particles.[63] This is the point at which oversized particles start to appear, in high density, and determines the rating for the test material.[64] A detail description of the procedures and assessment of "fineness of grind" can be found in references 61 and 63.

FIGURE 18.27 Measuring block and scarper: Hegman MXBAOHENG single-channel grind fineness gage.[62]

18.6 HEAT TRANSFER COEFFICIENT

A comprehensive discussion of heat transfer to and from heat exchange surfaces exposed to working media in a STR is presented in Chapter 17. In a stirred tank, the media is typically heated or cooled by means of external heat transfer jackets and/or internal coils, as shown in Figures 17.1–17.6. The overall heat transfer coefficient, U [W m^{-2} C^{-1}], is widely used to estimate the heat transfer rate, $\dot{Q}$ [J s^{-1}], from the process fluid to the heat transfer fluid or vice versa. In an idealized situation where the vessel and jacket operate under isothermal conditions, the simplest heat transfer formula is expressed in equation 18.37.

$$\dot{Q} = U\,A\,\Delta T \tag{18.37}$$

where, ΔT = Temperature difference between working media and heat transfer fluid, °C.

The overall heat transfer coefficient, U, for a jacketed stirred tank is determined from the thermal resistances as shown in equation 18.38.

$$\frac{1}{U} = \frac{1}{h_1} + f_1 + \frac{x}{\lambda} + f_2 + \frac{1}{h_2} \tag{18.38}$$

where, h_1 = process-side heat transfer coefficient [W m^{-2} C^{-1}]
h_2 = jacket-side heat transfer coefficient [W m^{-2} C^{-1}]
λ = material-dependent thermal conductivity of the vessel wall [W m^{-1} C^{-1}]
x = vessel wall thickness [m]
f_1 and f_2 = process-side and jacket-side fouling factors [m^2 C W^{-1}].

The process-side heat transfer coefficient depends on flow characteristics, and thermophysical properties of the fluid near the internal wall, as seen in Figure 17.16 and Table 17.4 of the "Heat Transfer" chapter. This is expressed in the form of the dimensionless Nusselt number (Nu) as follows:

$$Nu_1 = \frac{h_1\,T}{\lambda_1} = C_1\,Re_1^{\alpha_1}\,Pr_1^{\beta_1}\left(VR_1\right)^{\gamma_1} \tag{18.39}$$

where, T = tank diameter (characteristic length) [m]
λ_1 = thermal conductivity of process fluid [W m^{-1} °C^{-1}]

$$h_1 = \frac{Nu_1\,\lambda_1}{T}\ \text{[W m}^{-2}\ \text{°C}^{-1}\text{]} \tag{18.40}$$

The process-side impeller Reynolds number, Re_1, Prandtl number, Pr_1, and the viscosity ratio, VR_1, are defined in equations 18.41–18.43.

$$Re_1 = \frac{\rho_1\,N\,D^2}{\mu_1} \tag{18.41}$$

where, ρ_1 = density of process fluid [kg m^{-3}]
N = impeller rotational speed [s^{-1}]
D = impeller diameter (characteristic length) [m]
μ_1 = viscosity of process fluid [Pa-s]

$$Pr_1 = \frac{\mu_1 c_{p1}}{\lambda_1} \tag{18.42}$$

where, c_{p1} = specific heat of the fluid [J kg^{-1} °C^{-1}]

$$VR_1 = \mu_1 / \mu_{w1} \tag{18.43}$$

where, μ_{w1} = viscosity of the process fluid at the wall temperature, T_{w1} [Pa-s]

Non-dimensional empirical correlations for jacketed stirred tanks with various types of agitation systems are presented in Tables 17.6–17.9. From a mixing perspective, internal heat transfer coefficient, h_1, and subsequently the constant, C_1, are mostly influenced by impeller geometry, working media thermophysical properties (i.e., viscosity, density and thermal conductivity), flow pattern (axial, tangential or radial) and velocity magnitude near the inner wall. Similarly, jacket-side heat transfer coefficient, h_2, is most dependent on jacket configuration and thermophysical properties of the heat transfer fluid. These properties dictate flow regime (laminar, transition or turbulent) adjacent to the exterior wall surface. Generally, vessel and jacket manufacturers have in-depth knowledge of expected h_2 values for such HT fluids as water, glycol and steam for internal design of various jackets. Thus, explicit understanding of the internal heat transfer coefficient, h_1, is critical to design engineers for their specific process applications.

Most of the many methods for measuring heat flux are based on temperature measurements on or near the surface of a solid material. The authors utilized a simple, reliable experimental method to determine the heat transfer coefficient, h_1, for a well-mixed STR. Heat transfer experiments were carried out in a geometrically scaled-down agitated tank constructed of a similar material to the production reactor. During the experiments, the process fluid's dynamic and thermal parallels (e.g., Reynolds number, Re_1, and Prandtl number, Pr_1) were stringently maintained.

Figure 18.28 is the cutaway schematic of an insulated 20-gal. stainless steel jacketed vessel with a heat transfer tube spirally wound around its exterior wall. Multiple RdF thin-foil (<250μm)

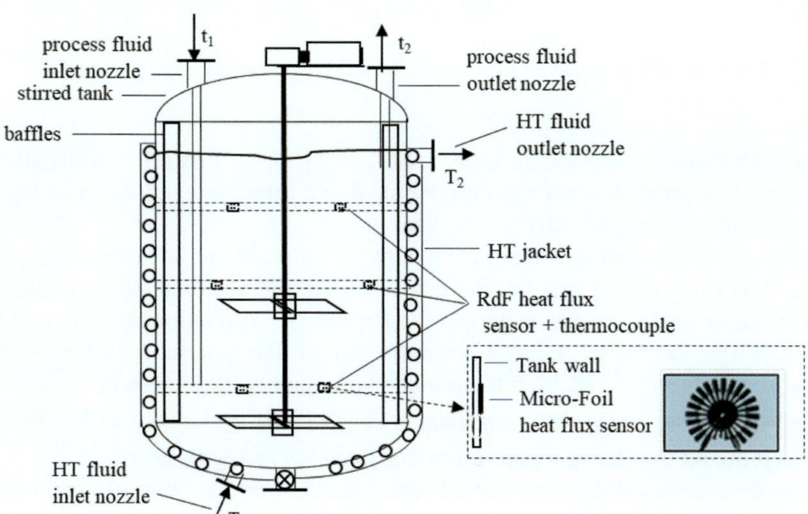

FIGURE 18.28 Schematics of the lab heat transfer experimental setup and Micro-Foil® heat flux sensor.[65]

micro heat flux sensors are flush mounted onto the vessel's interior surface by means of a light amount of high thermal conductivity epoxy cement. The insert in Figure 18.28 shows the pictorial view of a flexible RdF Micro-Foil® heat flux sensor[65] used for precise measurement of heat transfer (loss or gain) through any material surface. This surface-mounted model has ultra-fast response (0.02–0.50 s) and is integrated with a thermocouple for its surface temperature measurement up to 250°C.

The RdF heat flux sensors are thin and flexible enough to essentially become a "component" of the subject surface when properly mounted.[66] RdF heat flux sensors are addressed here, without any predisposed bias, for reader enlightenment. Several heat flux sensor manufacturers, their heat flux products, and associated instrumentation are succinctly tabulated in reference 67. Practicing engineers might consider contacting these or similar manufacturers to select the sensors best suited to their specific applications.

In a batch process that is subjected to heating or cooling with an external jacket, the process-side heat transfer coefficient, h_1, is determined from the RdF heat flux sensor using equation 18.44. Based on the calibration curve, or the constant supplied by the manufacturer, experimentally measured millivolts are converted to heat flux (W m^{-2}) and temperature (°C).

$$h_1 = \frac{\dot{q}}{T_s - T_f} \quad [\mathrm{W\ m^{-2}\ C^{-1}}] \qquad (18.44)$$

where, $\dot{q}$ is heat flux into the fluid [W m^{-2}], T_s and T_f are temperatures of the sensor surface and of fluid near the inside tank wall [°C], as schematically shown in Figure 18.29.

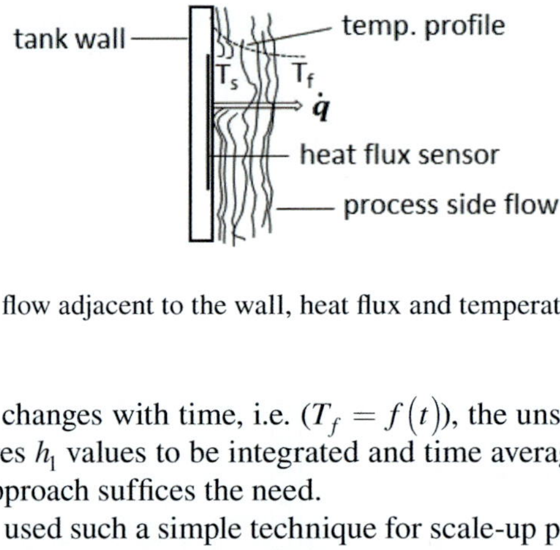

FIGURE 18.29 Diagram of flow adjacent to the wall, heat flux and temperature profile.

Since fluid temperature changes with time, i.e. ($T_f = f(t)$), the unsteady nature of the process-side heat transfer, $\dot{q}$, requires h_1 values to be integrated and time averaged. In most instances, such a direct but rudimentary approach suffices the need.

In one instance, authors used such a simple technique for scale-up purposes. Internal heat transfer coefficient, h_1, as a function of impeller Reynolds number, *Re*, and media Prandtl number, *Pr*, are experimentally determined to establish the *Nu* vs. *Re* and *Pr* correlation specific to a given agitation system. Tests were carried out using a geometrically similar externally jacketed stainless steel vessel like that shown in Figure 18.24, but equipped with dual A310® low-solidity hydrofoil impellers. Tests were carried out in the fully turbulent regime $\left(2\times10^5 \le Re_1 \le 6\times10^5\right)$. Process fluid in the tank was well mixed to achieve temperature homogeneity at any time, t.

The more appropriate non-intrusive method where heat balance is normally practiced uses continuous flow of the jacket-side fluid around a vessel containing a working media. Ideally, the jacket outer surface is fully insulated to prevent any heat exchange with the atmosphere. Except from the mutual heat transfer surface with the external jacket, the vessel is also fully insulated, and at any time, its contents are fully mixed to provide uniform temperature, $T_f(t)$. Figure 18.30, as an illustration, shows a mechanically agitated experimental vessel with a half-coil external cooling jacket.

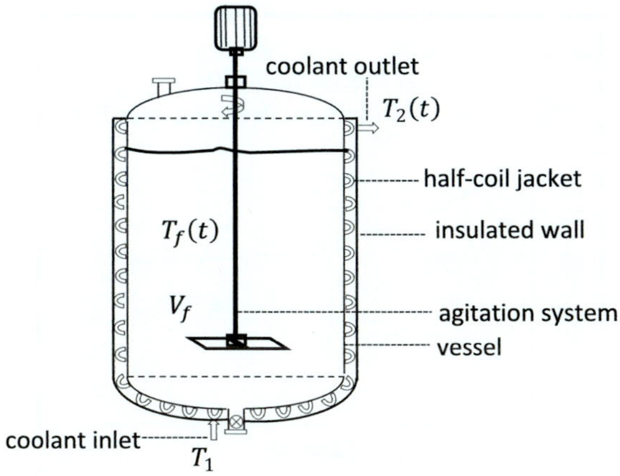

FIGURE 18.30 Schematic of an externally jacketed experimental vessel for process-side heat transfer study.

Since the heat flowrate, $\dot{Q}$, is the same in all resistances, at any time, t, heat loss of process fluid can be expressed as,

$$\dot{Q} = -\left(\rho_f V_f\right) c_{pf} \frac{dT_f}{dt} = h_1 A_1 \, \Delta T_{lm} \tag{18.45}$$

where, $\dot{Q}$ = heat transfer rate, W
A_1 = internal wetted wall surface area of the vessel [m^2]
V_f = volume of the process fluid [kg-s^{-1}]
ρ_f = process fluid density [kg-m^{-3}]
c_{pf} = specific heat of process fluid [J-kg^{-1}-C^{-1}]
t = time [s]

and the log mean temperature difference (LMTD), which accounts for the time-dependent driving force for heat transfer, $\dot{Q}$ is presented in equation 18.46.

$$\Delta T_{lm} = \frac{\left(T_f - T_1\right) - \left(T_f - T_2\right)}{ln\left(\dfrac{T_f - T_1}{T_f - T_2}\right)} \tag{18.46}$$

where, process fluid temperature, $T_f = f\left(t\right)$ [C],
Jacket outlet heat transfer fluid temperature, $T_2 = f\left(t\right)$ [C],
Jacket inlet heat transfer fluid temperature, T_1 [C]

Process media density, ρ_f, and specific heat, c_{pf}, are considered at bulk fluid temperature, $T_f\left(t\right)$, unless their temperature dependency is insignificant. Figure 18.31 illustrates the temporal temperature profiles of the process fluid and that of the jacket-side heat transfer coolant.

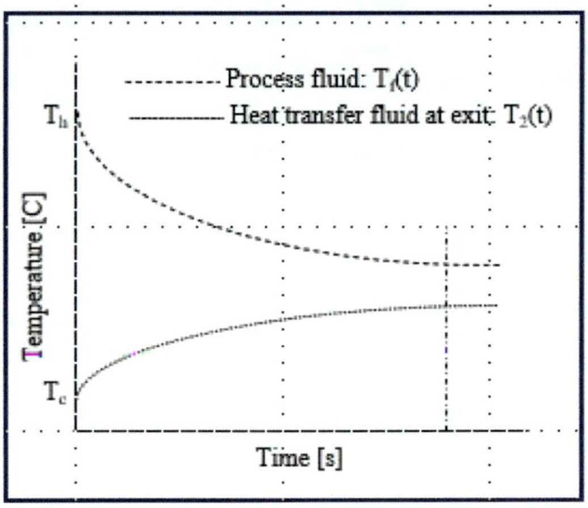

FIGURE 18.31 Illustration of idealized temperature profiles as a function of cooling time, t.

Substituting equation 18.46 into equation 18.45, one gets:

$$\frac{dT_f}{dt} = \frac{h_1 A_1}{(\rho_f V_f) c_{pf}} \left[\frac{(T_f - T_1) - (T_f - T_2)}{ln\left(\frac{T_f - T_1}{T_f - T_2}\right)} \right] \tag{18.47}$$

Equation 18.47 is an ordinary differential equation (ODE), where both process fluid and coolant exit temperatures T_f and T_2 are time-dependent functions, as shown in Figure 18.31. There are multiple commercially available numerical solvers, such as MATLAB®- ode45, Maplesoft- rkf45 for non-stiff ODEs. These codes could use Runge-Kutta method with a variable time-step function to determine temperature-time profiles of T_f' and corresponding h_1' values. The numerical curve that offers the least deviation from the experimentally measured T_f vs. t curve is considered as the ideal solution to predict h_1, where its value minimizes the sum of squares of the deviations expressed in equation 18.48. Using this condition,[68] one gets an optimal value of the process-side heat transfer coefficient, h_1, which best fits the experimental data.

$$\sum_{i=1}^{n} (T_f' - T_f)^2 = min$$

A basic alternative method to the above numerical solution would be to acquire and tabulate the data of T_f, T_1 and T_2 at small increments of n number of time steps, and calculate h_1' values based on corresponding LMTDs.[69] See equation 18.48. The mean of the h_1' values is considered as the process-side heat transfer coefficient, h_1, represented by equation 18.49. As stated earlier, the specific heat, c_{pf}, and density, ρ_f, of the working media may be determined at varying process fluid temperatures, T_f.

$$h_1' = \left(\frac{\Delta T_f}{\Delta t}\right) \left[\frac{A_1}{(\rho_f V_f) c_{pf}}\right]^{-1} \left[\frac{(T_f - T_1) - (T_f - T_2)}{ln\left(\frac{T_f - T_1}{T_f - T_2}\right)} \right]^{-1} \tag{18.48}$$

$$h_1 = \frac{1}{n}\sum_{i=1}^{n} h_{1i}' \tag{18.49}$$

Although most of the experimental techniques described in the chapter are devoted to mixing in mechanically agitated tanks, these are equally applicable to jet mixers, high shear mixers such as rotor-stators and motionless (static) mixers with simple variations to their applications.

REFERENCES

1. Watrasiewics, B. M., Rudd, M. J., *Laser Doppler Measurements*. Butterworths & Co. Ltd., 1976.
2. Durst, F., Melling, A., Whitelaw, J. H., *Principles and Practice of Laser-Doppler Anemometry*. Academic Press, 1976.
3. Ascanio, G., Mixing time in stirred vessels: A review of experimental techniques. *Chin. J. Chem. Eng.*, 23(7), pp. 1065–1076, 2015.
4. Smith, J. M., Agitation devices. http://www.thermopedia.com/content/549/
5. Cabaret, F., Bonnot, S., Fradette, L., Tanguy, P. A., Mixing time analysis using colorimetric methods and image processing. *Ind. Eng. Chem. Res.*, 46(14), pp. 5032–5042, 2007.
6. Melton, L. A., Lipp, C. W., Spradling, R. W., Paulson, K. A., Dismit—Determination of mixing time through color changes. *Chem. Eng. Commun.*, 189(3), pp. 322–338, 2002.
7. Tatterson, G. B., *Fluid Mixing and Gas Dispersion in Agitated Tanks*. McGraw-Hill, USA, 1991.
8. Industrial tomography systems LLC. https://www.itoms.com/technologies/electrical-resistance-tomography/
9. Haraguchi, M., Kim, H. Y., Calvo, W. A. P., Imaging technique for troubleshooting of industrial equipment by gamma-ray absorption scans. *Conference: Nutech International Conference on Development and Applications of Nuclear Technologies*. Kraków, Poland, September 2011.
10. Haraguchi, M. I., Calvo, W. A. P., Kim, H. Y., Tomographic 2-D gamma scanning for industrial trouble shooting. *Flow Measurement and Instrum.*, 62, pp. 235–245, 2018.
11. Roussinova, V., Kresta, S. M., Comparison of continuous blend time and residence time distribution models for a stirred tank. *Ind. Eng. Chem. Res.*, 47(10), pp. 3532, 2008.
12. https://www.chemeurope.com/en/encyclopedia/Residence_Time_Distribution.html
13. http://www.umich.edu/~elements/5e/16chap/Fogler_Web_Ch16.pdf
14. https://www.transmittershop.com/blog/radar-level-transmitter-vs-ultrasonic-level-transmitter/
15. https://www.emerson.com/documents/automation/product-data-sheet-rosemount-5408-5408-sis-level-transmitters-en-202468.pdf. Rosemount 5408 Series, April 2022.
16. https://www.biologydiscussion.com/cell-biology/assessment-of-kla-oxygen-transfer-coefficient-6-methods/7681
17. Moutafchieva, D., Popova, D., Dimitrova, M., Tchaoushev, S., Experimental determination of the volumetric mass transfer coefficient. *J. Chem. Technol. Metall.*, 48(4), pp. 351–356, 2013.
18. Krishna, R., Sie, S. T., Strategies for multiphase reactor selection, *Chem. Eng. Sci.*, 49, pp. 4029–4065, 1994.
19. Van't Riet, K., Review of measuring methods and non-viscous gas liquid mass transfer in stirred vessels. *Ind. Eng. Chem. Process Des. Dev.*, 18, pp. 357–64, 1979.
20. Mueller, J. A., Boyle, W. C., Lightfoot, E. N., Effect of the response time of a dissolved oxygen probe on the oxygen uptake rate. *Appl. Microbiol.*, 15(3), pp. 674–676, May 1967.
21. Dissolved oxygen tables—calibration and oxygen solubility tables. *Technical Instructions on YSI*. https://www.ysi.com/File%20Library/Documents/Technical%20Notes/DO-Oxygen-Solubility-Table.pdf
22. Measurement of oxygen transfer in clean water. ASCE/EWRI 2-06, ASCE Standard, 2007.
23. Cerri, M. O., Esperança, M. N., Badino, A. C., de Arruda Ribeiro, M. P., A new approach for k_La determination by gassing-out method in pneumatic bioreactors. *J. Chem. Technol. Biotechnol.*, 91(12), pp. 3061–3069, December 2016.
24. Hickman, A. D., Gas-liquid oxygen transfer: A novel experimental technique with results for mass transfer in aerated agitated vessels. *Proc. 6th Eur. Conf. Mixing, Pavia*, pp. 369–374, 1988.
25. Vasconcelos, J. M. T., Nienow, A. W., Martin, T., Alves, S. S., McFarlane, C. M., Alternative ways of applying the hydrogen peroxide steady state method of k_La measurement. *Chem. Eng. Res. & Des.*, 75(5), pp. 467–475, July 1997.

26. Stenstrom, M. K., Leu, S., Jiang, P., Theory to practice: Oxygen transfer and the new ASCE standard, water environment foundation. *WEFTEC®*.06, pp. 4838–4852, 2006.
27. YSI model 59 dissolved oxygen meter instructions. https://www.ysi.com/File%20Library/Documents/Manuals%20for%20Discontinued%20Products/59-Dissolved-Oxygen.pdf
28. https://www.ysi.com/File%20Library/Documents/Tips/YSI-Calibration-Maintenance-Troubleshooting-Tips-6-Series-Sondes-2-8-10.pdf
29. Van't Riet, K., Turbine agitator hydrodynamics and dispersion performance, Ph. D thesis. *Delft University of Technology*. The Netherlands, 1975.
30. Garcia-Ochoa, F., Gomez, E., Bioreactor scale-up and oxygen transfer rate in microbial processes: An overview. *Biotechnol. Adv.*, 27, pp. 153–176, 2009.
31. Senouci-Bereksi, M., Kies, F. K., Bentahar, F., Hydrodynamics and bubble size distribution in a stirred reactor. *Arab. J. Sci. Eng.*, 43, pp. 5905–5917, 2018.
32. Mueller, S. G., Optical measurements in gas-liquid stirred tanks, 2009. https://openscholarship.wustl.edu/etd/253
33. Barigou, M., Greaves, M., Bubble size distributions in a mechanically agitated gas–liquid contactor. *Chem. Eng. Sci.*, 47, pp. 2009–2025, 1992.
34. Wang, W., Mao, Z., Chao, Y., Experimental and numerical investigation on gas holdup and flooding in an aerated stirred tank with Rushton impeller. *Ind. Eng. Chem. Res.*, 45, pp. 1141–1151. 2006.
35. Garcia-Salas, S., Alfaro, M. E. R. P., Porter, R. M., Thalasso, F., Measurement of local specific interfacial area in bubble columns via a non-isokinetic withdrawal method coupled to electro-optical detector. *Chem. Eng. Sci.*, 63, pp. 1029–1038, 2008.
36. Alves, S. S., Maia, C. I., Vasconcelos, J. M. T., Serralheiro, A. J., Bubble size in aerated stirred tanks. *Chem. Eng. J.*, 84, pp. 109–117, 2002.
37. Peters, M. H., Fan, L. S., Sweeney, T. L., Study of particle ejections in the freeboard region of a fluidized bed with an image carrying probe. *Chem. Eng. Sci.*, 38, pp. 481–485, 1983.
38. Angst, R., Kraume, M., Particle distribution in stirred tanks: Experimental and numerical investigations in different scales. *12th European Conference on Mixing*, p. 8. Bologna, June 27–30, 2006.
39. Angst, R., Kraume, M., Experimental investigations of stirred solid/liquid systems in three different scales: Particle distribution and power consumption. *Chem. Eng. Sci.*, 61, pp. 2864–2870, 2006.
40. Ghiassi, S. H., Safekordi, A. A., Shareh, F. B., Determination of interfacial area in gas-liquid two phase by light transmission. *Iran. J. Chem. Eng.*, 31(1), 2012.
41. Calderbank, P. H., Physical rate process in industrial fermentation, part 1: The interfacial area in gas-liquid contacting with mechanical agitation. *Trans. Inst. Chem. Eng.*, 36, pp. 443, 1958.
42. McLaughlin, C. M., Rushton, J. H., Interfacial area of liquid-liquid dispersion from light transmission measurements. *AIChE J.*, 19, p. 187, 1973.
43. Hao, T., *Electrorheological Fluids: The Non-aqueous Suspensions*, p. 2. Elsevier, 2005.
44. Padron, G. A., BHRG, liquid-liquid dispersion in stirred tanks. *Short Course Lecture. Mixing XXIV, 24th Biennial North American Mixing Forum's Conference*. Lake George, NY, USA, NAMF (AIChE), June 2014.
45. Standard test method for sieve analysis of fine and coarse aggregates. ASTM C136-06, February 2015. https://www.astm.org/standards/c136
46. https://www.beckman.com/resources/technologies/laser-diffraction/mie-fraunhofer-theories
47. Abidin, M. I. I. Z., Raman, A. A. A., Nor, M. I. M., Review on measurement techniques for drop size distribution in a stirred vessel. *Ind. Eng. Chem. Res.*, 52, pp. 16085–16094, 2013.
48. Particle track with FBRM technology: Real time particle size and count analysis. https://www.mt.com/int/en/home/products/L1_AutochemProducts/particle-size-analyzers/particletrack-fbrm.html
49. Kumar, V., Taylor, M. K., Mehrotra, A., Stagnant, W. C., Real-time particle size analysis using focused beam reflectance measurement as a process analytical technology tool for a continuous granulation-drying-milling process. *AAPS PharmSciTech*, 14(2), pp. 523–530, 2013.
50. Particle size analyzer: PSA. https://www.anton-paar.com/us-en/products/details/psa-series/
51. Particle size: Particle size analyzers. https://www.horiba.com/ind/scientific/products/particle-characterization/particle-size/
52. Mastersizer® 3000. https://www.malvernpanalytical.com/en/products/product-range/mastersizer-range/mastersizer-3000
53. Basic principles of particle size analysis. https://www.atascientific.com.au/basic-principles-of-particle-size-analysis/

54. Laser Diffraction (LD) characterization down to the nano range. https://www.microtrac.com/products/particle-size-shape-analysis/laser-diffraction/
55. Maaß, S., Wollny, S., Voigt, A., Kraume, M., Experimental comparison of measurement techniques for drop size distributions in liquid/liquid dispersions. *Exp. Fluids*, 50, pp. 259–269, 2011.
56. Nagarajan, V., Mohanty, A. K., Misra, M., Biocomposites with size-fractionated biocarbon: Influence of the microstructure on macroscopic properties, *ACS Omega*, 1, pp. 636–647, 2016.
57. https://www.globalgilson.com/blog/wet-sieving-practical-guide
58. Standard test method for materials finer than 75-µm (No. 200) sieve in mineral aggregates. ASTM C117-17, November 2023. https://www.astm.org/c0117-17.html
59. BYK-Gardner grind gage for grindometer. www.labequip.com/bykgardner-grindometer-grind-gauge
60. Standard test methods for fineness of dispersion of pigment-vehicle system by Hegman-type gage, ASTM D1210-05, 2022. https://www.astm.org/d1210-05r22.html
61. Gotro, J., Dispersion of highly filled thermosets—part ten: Measuring fineness of grind using a Hegman gage, March 25, 2019. https://polymerinnovationblog.com/dispersion-of-highly-filled-thermosets-part-ten-measuring-dispersion-quality-using-a-hegman-gauge/
62. https://www.amazon.com/Fineness-Grindometer-0-100um-Single-Channel-fineness/
63. McKay, R. B., Technological applications of dispersions, p. 43. citing ASTM D1210-79 1988. Marcel Dekker, Inc., 1994.
64. Test method for fineness of grind of printing inks by the NPIRI grindometer, ASTM D1316—20, ASTM International, 2020. https://www.astm.org/d1316-20.html
65. Heat flux sensors. https://www.rdfcorp.com/products/hflux/hflux.shtml
66. HFS-A//Heat flux sensors. https://www.rdfcorp.com/products/hflux/hfs-a_02.shtml
67. Diller, T. E., Practical heat flux measurement. In *The Art of Measuring in the Thermal Science*, Eds. Meyer, B., De Paepe, M. CRC Press, 2000.
68. Mahir, M., El Maakoul, A., Khay, I., Saadeddine, S., Bakhouya, M., An investigation of heat transfer performance in an agitated vessel. *Processes*, 9(3), p. 468, 2021.
69. Pydisetty, Y., Jayakumar, N. S., Cooling of a viscous liquid using half-coil jacket. *Indian J. Chem. Tech.*, 1, pp. 305–307, September 1994.

19 Mechanical Design

Andrew Creathorn

19.1 INTRODUCTION

Designing a mixer represents a unique challenge to the design engineer. At first glance, a typical mixer looks like a relatively straightforward piece of equipment comprising an electric motor, gearbox, mechanical seal (for closed tanks) and "in tank" or "wetted" parts consisting of the rotating shaft and one or more mixing impellers attached to it. Where the shaft enters the tank defines three fundamentally different mixer classifications, namely top entering, side entering, and bottom entering, as will be discussed. This chapter will illustrate the design process for each key mixer component and will offer practical advice on design considerations that ease installation and equipment maintenance. The primary units used in this chapter are imperial, although metric units are also often used.

19.2 PRIMARY DESIGN CONSIDERATIONS

Before discussing the mechanical design process, two primary design concerns will be examined:

- The mechanical loads acting on the mixer
- The natural frequency or critical speed of the shaft and impeller system

19.2.1 Overhung vs. Supported Shaft System

Most mixers designs use an "overhung" shaft arrangement, as schematically shown in Figure 19.1 and further discussed in the section "Machine Configurations" with Figures 19.16 (a–c, e). This type of configuration is preferred because the use of an "in-tank" support (usually in the form of a bushing or journal bearing) requires periodic "in-tank" maintenance, which most operators typically want to avoid. At the same time, however, an "in-tank" support nearly always results in smaller and less expensive equipment. The reasons will become apparent while exploring more detailed design considerations. Some overhung designs infringe on a common, but non-absolute rule of thumb discouraging overhung shaft lengths (L) and diameters (D) producing an L/D ≥ 60.

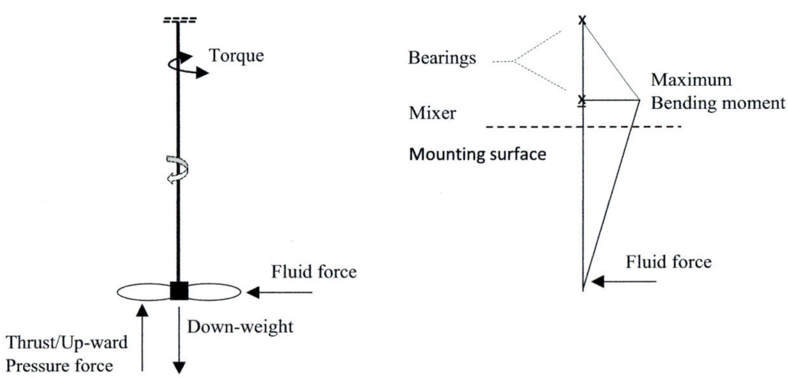

FIGURE 19.1 Diagram of the basic loading system and location of the mixer mounting surface.

DOI: 10.1201/9781003606390-19

19.2.2 Loading System

In step 1, the designer must not only calculate the system's mechanical loads but also properly account for them in the equipment design. As shown in Figure 19.1, there are three primary mechanical loads.

- *Fluid Force*: This force is considered to act perpendicular to the mixer shaft, at the centerline of the impeller. A side load is generated as a rotating impeller system causes liquid in the vessel to "flow." Whilst this flow is considered to be uniform at the macro level, its magnitude and direction vary in non-uniform ways, and this asymmetric flow produces an asymmetric force at the impeller. The radar plot in Figure 19.2 shows magnitude and direction of the net side load generated by a 16-inch pitch-blade turbine operating under full water coverage at 105 rpm. The load direction and magnitude vary significantly, with

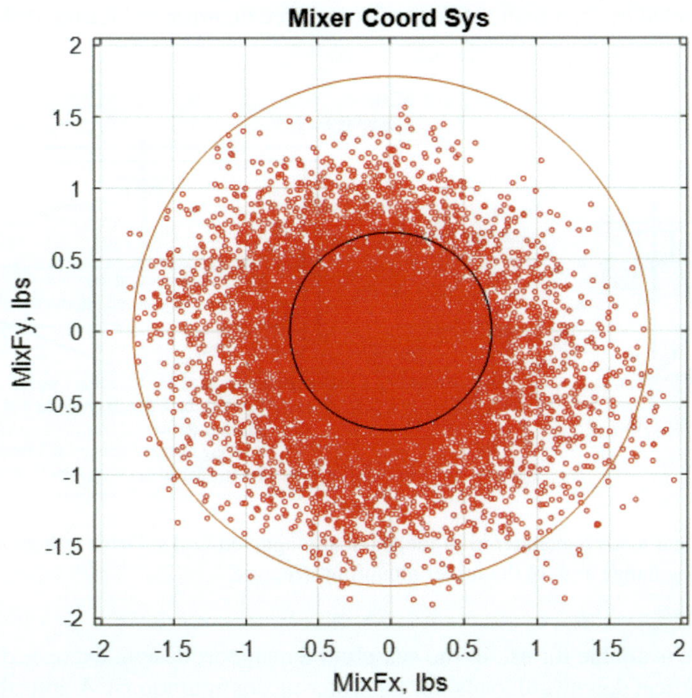

FIGURE 19.2 Transient x and y components of fluid forces experienced by an operating mixer.

  intermittent transient values sometimes exceeding baseline values by more than two to three times. The associated forces produce a maximum bending moment (BM) at the first support location of the shaft, as shown in the BM diagram (right image) of Figure 19.1. This is usually characterized as a maximum force value since it is a primary means for fatigue loading on the mixer. Mixer manufacturers typically develop numerical methods to calculate fluid force magnitudes as a function of impeller speed and size.
- *Torque*: Resistance to the impeller blades moving through the liquid generates torque, which reaches its maximum above the highest located impeller. This also has a variable nature due to the flow asymmetries described earlier. When certain conditions create torque levels above nominal operational levels, such as startup when the lower mixing

impeller is embedded in settled solids, additional consideration of various mixer components is required. In this case, the torque in the mixer shaft and loading on the impeller components will experience the torque associated with the motor start torque unless a torque-limiting system is used. Torque is often limited by using either a frequency inverter of a fluid coupling. It is particularly important to anticipate the potential range of operating conditions. For example, gassed applications will draw lower powers when that gas is sparged into the vessel. Commissioning will often take place without gas, resulting in a much higher power draw, and designers need to carefully consider the potential for loading on the gearbox, shaft and impellers beyond yield conditions.

- *Thrust*: The force due to pressure acting at the end of the shaft must be considered, especially in closed tank applications, where it attempts to push the shaft out of the vessel. This force acts opposite to the weight of the shaft and impellers. Operating impellers develop upward- or downward-directed thrust (depending on configuration), which must also be considered.

The load due to the mixer weight must also be considered by the vessel designer, along with the reacted loads generated by the mixer, as shown in the force diagram in Figures 19.1 and Figure 19.3.

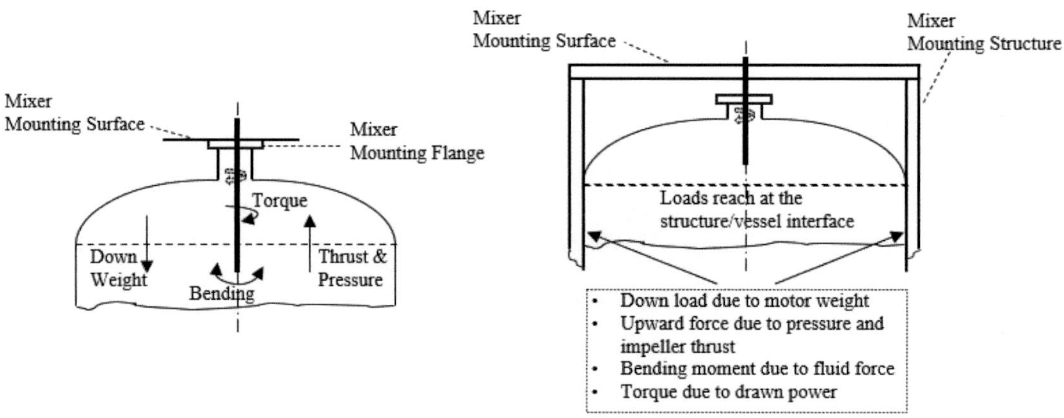

FIGURE 19.3 Typical loads encountered in mechanically agitated tanks. Drive system mounted on top of the (a) mixer mounting flange and (b) the mixer mounting structure.

Design loads are available for use in the vessel/mixer support design, expressed in torque, bending moment and vertical downward loads (for top-entering configurations). A note about mixer loads: they are fluctuating and lead to fatigue loading. Vessel design according to ASME code,[1] for example, is primarily based on allowable stresses and the item involved (2:1 semi-elliptical head, side wall, baffles, etc.). As such, additional considerations need to be incorporated to account for fatigue loading, which is usually done through reduced allowable stress or by using methods for handling the fatigue loads—i.e., ASME code[1] special rules, Finite Element Analysis (FEA), or other techniques.

Deflection is not usually part of a vessel analysis or structural support analysis, yet it can play an important role for close clearance situations such as impeller blades running close to stationary baffles, cooling coils, etc. A properly designed steady bearing is also important in many applications.

ASME code includes vessel tolerances for concentricity, angularity and extension in flanged connections, which are usually reduced in steady bearing applications. Mounting flange angularity may be reduced to 0.25-degree, not the 1-degree standard, and it needs post-weld heat treatment and machining.

Close Clearance: Some impellers (helical spirals, anchor type designs and others) depend on very close operating clearances, for example, as means of improving the surface heat transfer coefficient. But this usually requires reducing vessel wall concentricity, as well as reducing diameter tolerance beyond ASME standard values. This is best handled before finalizing vessel specs and purchase order placement.

Recommended support beams, such as those spanning the top of an open tank, are commonly available from a supplier. These are suitable for overhung shaft designs, but they must be properly supported at the tank wall to be effective. Braced columns, moment carrying connections, etc. are commonly used, especially where the tank sidewall has little additional capacity beyond resisting liquid head of the tank contents.

How the designer should apply these loads will be examined later in this chapter.

Most tank configurations include baffles. The requirement for adequate baffling has already been discussed in previous chapters of this book. The mechanical mixer designer is not usually responsible for baffle design, but it is important that baffle loads are fully defined and provided to the vessel designers. Figure 19.4 shows a typical method to assess baffle load, which is calculated using the torque associated with the impeller, as follows:

$$\textit{Baffle load} = 1.5 \times \frac{\textit{impeller torque}}{\textit{distance}: \textit{shaft centerline to baffle inner edge} \times \textit{number of baffles}}$$

where, 1.5 is considered as the factor of safety for dynamic loading.

The baffle load should be treated as a point load at the baffle center, opposite the impeller centerline.

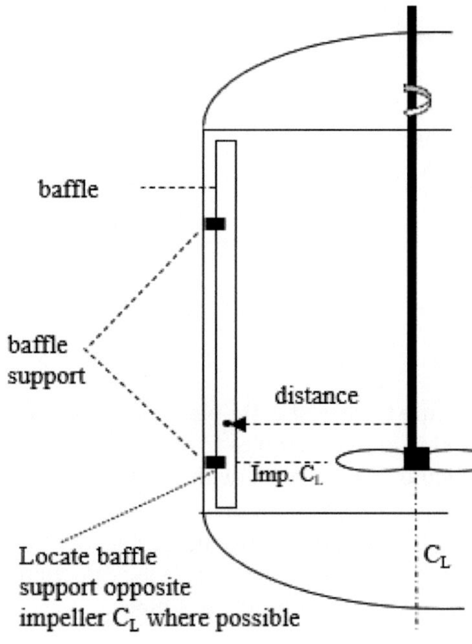

FIGURE 19.4 Diagram showing location of baffle support and baffle loading system.

The location of the baffle supports can be important. The majority of the load on a baffle occurs opposite the impeller. If the location of the baffle support coincides with the centerline of the impeller, the designer can simply consider the load on the support to act as a point load with a moment arm equivalent to the distance of the baffle support from the tank wall. If the baffle support is not located opposite the impeller centerline, the designer should consider the twisting moment, as shown in Figure 19.5.

In addition, high-pressure vortex shedding from the impeller blade tips can also induce vibration in the baffles, possibly leading to shortened life. The most dominant frequency from this effect is the blade–baffle interaction frequency. As a practical matter, the baffle design natural frequency should be stiff enough to avoid this frequency by a ± 20%. Additionally, the blade passing frequency, which is the frequency with which the blade passes the baffle, is considered as a potential "exciting" frequency, which could result in force amplification. It is common to have four baffles in a vessel and an impeller with four blades. In this configuration, there is a "pulse" each time the blade passes the baffle. If this excitation frequency is close to or coincidental with a natural frequency of the system, this should be "de-tuned" by either changing the number of baffles or using an offset different to 90 degrees.

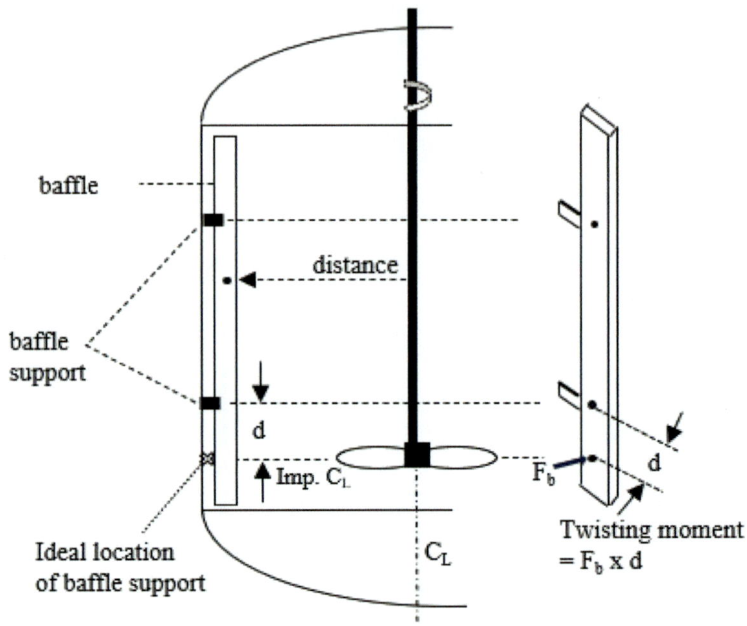

FIGURE 19.5 Schematic of expected twisting moment due to non-ideal location of the baffle support.

Similarly, some processes require the use of a dip-pipe. Most commonly, they're used to charge a reactor below the liquid level and to extract samples from the reactor as part of a sampling system. They can also be used to drain liquid from a vessel, without the need for a side or bottom outlet. The loads exerted on a dip-pipe due to the flow pattern in the vessel are often overlooked. The mechanical designer should work closely with the vessel designer to ensure that the dip-pipe and its mounting flange can withstand the bending loads associated with the direction and magnitude of flow. See Chapter 4 for design calculations related to overhung dip-pipes.

19.2.3 Critical Speed

The critical speed, or natural frequency, is the frequency at which the mixer shaft vibrates (see Figure 19.6). The first critical speed is of primary interest because most of the energy exists at this lower frequency, as compared to the harmonics. Operating at or near the natural frequency causes fluid force amplification and results in catastrophic failure of the shaft. Probably the most famous example of a mechanical system being impacted by force amplification is the Tacoma Narrows bridge (known as Galloping Gertie). The force of a relatively light wind load was amplified, and the structure started to vibrate at its natural frequency. The force amplification resulted in massive deflection and ultimate failure of the bridge. As a general rule of thumb, operating at or below 0.8 times the speed associated with the first natural frequency will dampen rather than amplify both the force and the deflection.

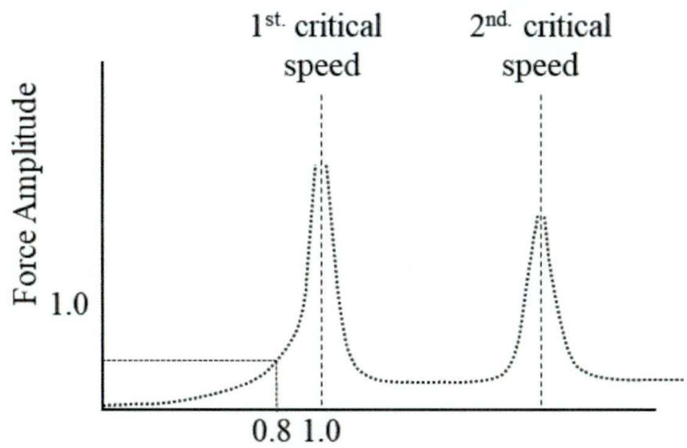

FIGURE 19.6 Diagram showing force amplification as resonance is approached.

Note that the value of the second critical speed is usually high enough not to be excited by the mixer operating frequencies, except when the mixer support is not sufficiently stiff. This is usually most prevalent on low-pressure closed tank applications where small wall thicknesses meet requirements for the pressure, but not for mixer support. In some cases, manufacturers will design equipment to operate above the first critical speed, but below the second. This tends to be for smaller equipment where the system can quickly accelerate through the first natural frequency on start-up, thus reaching a stable speed between the first and second natural frequencies.

19.3 MATERIAL SELECTIONS

A mechanical designer needs to look at stress, generated by the applied loads, in all of the mixer components (shaft, impeller, etc.) and must make sure that the stress is below acceptable levels. Mixer design has added complexity because loads generated by the mixing action varies in magnitude and direction. This creates a "fatigue" environment, and as such, needs to be considered when determining allowable stress. The vessel designer should use the loads provided by the mixer manufacture to design the vessel in accordance with the appropriate vessel design standards. The vessel designer should ask the mixer manufacture to provide the following information:

Download—due to the weight of the mixer and impeller thrust if the impeller is up-pumping
Bending Moment—due to fluid force
Torque—due to absorbed power
Thrust—due to pressure and impeller thrust
Forcing Frequencies—the operating frequencies (shaft speed, baffle passing frequency, blade passing frequency, motor speed) that could excite the system natural frequency.

After the gearbox and motor selection are known, the mechanical designer often requires no additional help from the process designer. The mechanical design can proceed as described in the following sections.

19.4 SHAFT DESIGN—DESIGN FOR STRENGTH

The shaft is subject to the combination of bending due to the fluid force and torque due to the liquid resisting impeller rotation. In cases where the shaft is "overhung" (see configuration later in this chapter), the bending load usually dominates the design. The designer must consider stress in the shaft, due to the combination of torque and bending loads, and make sure that the stress is kept below allowable limits.

An initial selection for the shaft size can be made based on torque only (as previously mentioned), bearing in mind that this will likely change when the bending loads and critical speed requirements are determined later. An initial estimate for shaft diameter, D, can be determined as follows:

$$D_{min} = \left[\frac{(32 * 63000 * MHP)}{2\pi N \tau}\right]^{1/3} \tag{19.1}$$

where D_{min} = minimum shaft diameter (typically at the mixer mounting surface location) [in]
MHP = motor nameplate horsepower [hp]
τ = allowable shear stress [psi]
N = shaft speed [rpm]

An allowable shear stress of 6,500 psi can be used for common materials with yield strengths of about 30,000 psi (typical of 300-Series stainless steels).

The equivalent bending moment, M_{eq}, takes into account the combined effect of torque/twist (T) and bending moment, M, on a structure. It is given by,

$$M_{eq} = \frac{1}{2}\left(\mathrm{M} + \sqrt{(M^2 + \mathrm{T}^2}\right) \tag{19.2}$$

where, M = bending moment due to the fluid forces
T = torque due to resistance to the impeller's rotation through the vessel contents

The equivalent twist is given by,

$$T_{eq} = \sqrt{M^2 + T^2} \tag{19.3}$$

For shaft design purposes, respectively, the maximum normal (principal) and shear stresses, σ_1 and τ_{max}, experienced by the shaft are determined from equations 19.2 and 19.3, as shown in equation 19.4.

$$\sigma_1 = \frac{32}{\pi d^3} M_{eq} \text{ and } \tau_{max} = \frac{16}{\pi d^3} T_{eq} \tag{19.4}$$

where, d = shaft diameter

Where the gearbox is used to support a low-speed shaft (Figures 19.16 a–b, g), it must be of sufficient strength to accommodate the combined torque and bending loads. It is essential that the gearbox be designed specifically for mixing loads in overhung arrangements. Most commercially available gearboxes are designed for torque loads only. When using a commercial gearbox, the mechanical designer should pursue one of the following options:

1. Work with the commercial gearbox provider to ensure that (a) the gearbox can withstand mixing loads without excessive housing deflection (which leads to gear misalignment and failure), and (b) the output shaft and bearings have sufficient strength for the bending and thrust loads.

or,

2. Design the pedestal with an arrangement that includes an independent low-speed shaft and bearing arrangement to de-couple the loads from the gearbox (as shown in Figure 19.7).

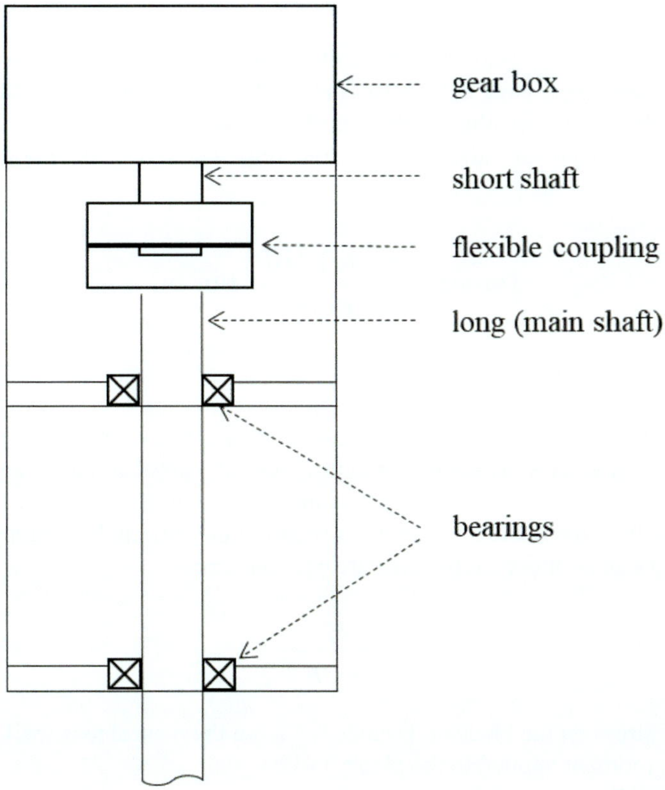

FIGURE 19.7 An independent low-speed shaft and bearing arrangement to de-couple gearbox loads.

19.5 DESIGNING THE IMPELLERS

The process designer provides the type of impellers required to achieve the desired mixing process, and when there are two or more impellers, the details of how the power splits between the impellers. The mechanical designer can use this information to size the impeller components by calculating the load acting on the various impeller components. For example, in the case of a simple four-bladed mixed flow impeller (see Figure 19.8), each impeller blade is bolted to the "ear" of the hub, and the three primary variables must be examined.

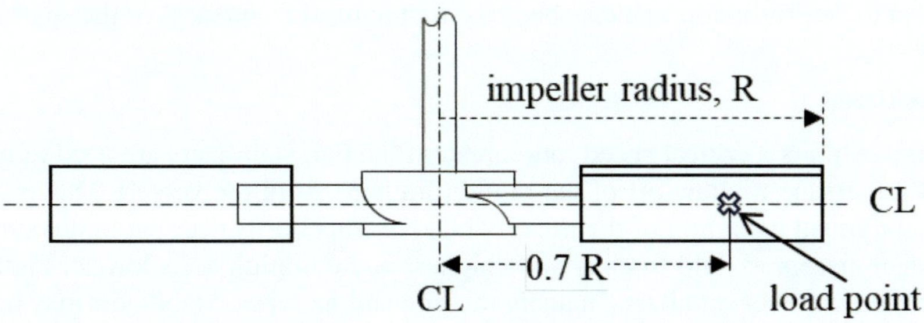

FIGURE 19.8 Diagram of a pitched-blade turbine (PBT) with the loads shown.

19.5.1 Blade Thickness

The designer wants to minimize blade thickness, thus minimizing weight and cost. Impeller weight is very important because increasing it decreases natural frequency (critical speed), which increases the likelihood of shaft speed coinciding with natural frequency.

Once shaft speed and power "absorbed" by the impeller (SHP) are known, torque can be calculated.

$$\text{Torque} = \frac{63{,}000^{*}(\text{HP})}{(\text{RPM})}\,[\text{lbf-in}] \tag{19.5}$$

where, Torque = Force × Distance

A good rule of thumb considers the torque generated by a force acting at 70% of the impeller radius, as shown in Figure 19.8. Therefore, the designer can calculate the force on each impeller blade. This calculated force can be used to determine the bending load at the bolted connection. This is the point at which the blade has its lowest section modulus, and therefore, highest bending stress. The following equation is used to calculate bending stress:

$$\sigma = \frac{M \times c}{I} \tag{19.6}$$

where, σ = bending stress on the blade at distance "c" from the neutral axis [psi]
M = bending moment applied to the blade [in-lbf]
c = perpendicular distance along the blade cross-section from its neutral axis [in]
I = area moment of inertia of the blade [in^5]

19.5.2 Ear Thickness

A similar approach is used in calculating thickness of the ear attachment to the hub. Some hubs have integrally cast ears, whereas others use a welded fabrication. The welded version is carefully designed based on the weld throat size, effects of undercuts, porosity, root gap, etc.

19.5.3 Hub Wall Thickness

Hub wall thickness is based on the diameter needed for full contact of the ear. It must also be ensured that there is sufficient wall thickness, when the impeller is driven by a key, to keep stress within the hub wall below an acceptable limit. See later comments, in this chapter, regarding allowable stress limits.

Once a design has been completed for each of the impellers, the weight and location of the impellers can be used to determine the critical speed (or first natural frequency) of the shaft system.

19.5.4 Stabilizers

If a shaft design exhibits a critical speed ratio greater than 0.4, stabilizers are used to minimize the elevated fluid forces due to draw-off of the liquid (drainage from the vessel). This is of particular concern with continued operation of the mixer while the impeller is near the liquid surface, as this is typically when measured fluid force is at its highest and damping at its lowest. Figures 19.9a–b show two common impeller stabilizers, namely the ring and fin types. Stabilizers may be attached to the blades, as in the figure, or bolted to the ears, or even supported from the disc of a radial turbine-type impeller.

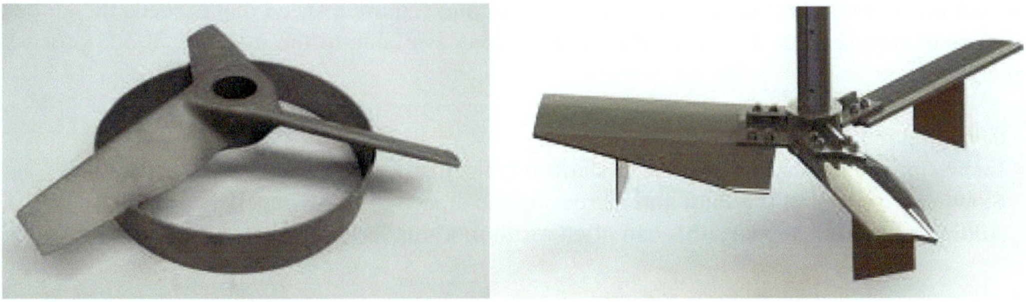

(a) (b)

FIGURE 19.9 High-efficiency hydrofoils with flow stabilizers consisting of (a) a ring and (b) fins.

Figure 19.10a plots mixer forces measured while operating an A310 hydrofoil impeller with an attached stabilizer ring. This plot is somewhat similar to that shown previously in Figure 19.2, but it displays greater asymmetry because this impeller is operating very near to the liquid surface. Figure 19.10b is a separate plot of force measurements from the same mixer, but while operating without the stabilizer ring on the impeller. Asymmetric loading has increased significantly in Figure 19.10b, with transient force magnitudes sometimes exceeding those with the impeller stabilizer ring (Figure 19.10a) by seven times or more.

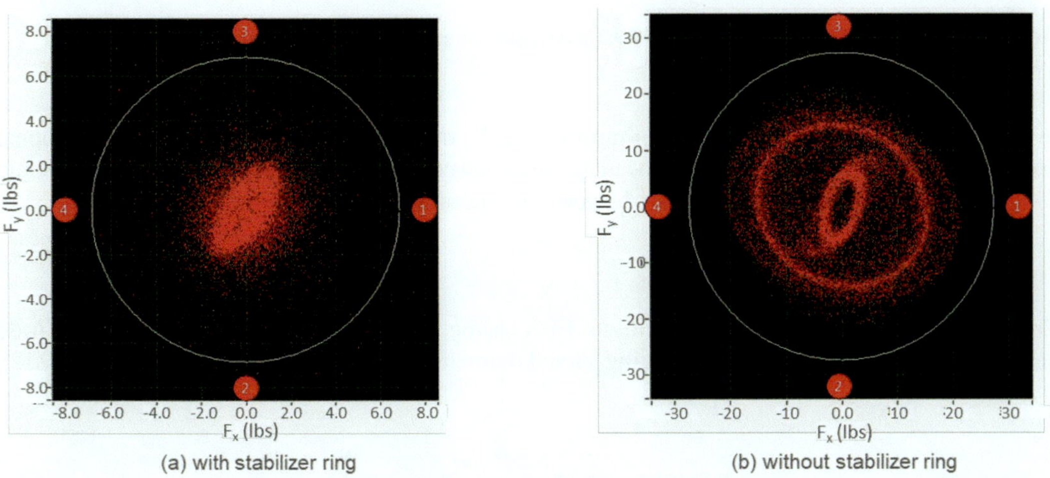

FIGURE 19.10. Transient *x* vs. *y* components of fluid force on a mixer running one A310 hydrofoil impeller near the liquid surface, after (a) and before (b) a stabilizer ring was added.

19.6 CRITICAL SPEED/RESONANT FREQUENCY

Because calculating the critical speed of a mixer shaft is complex, most mixer manufacturers employ numerical methods. For example, the structure of typical finite element analyses (FEA) models, which reliably predict critical speed, is shown in Figure 19.11.

It is not uncommon for the initially proposed design to require a speed that exceeds the guideline of less than 80% of first critical. Redesign options available to the mechanical and process designers include:

- Changing the number and location of impellers
- Increasing the diameter of the mixer shaft to make it stiffer
- Slowing the speed of the shaft and increasing the diameter of the impellers
- Adding an in-tank support (this can often result in a smaller diameter shaft)

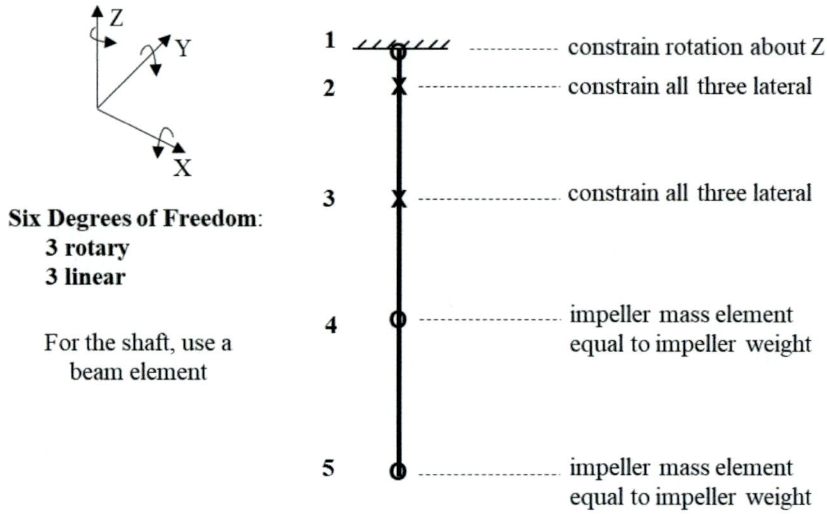

FIGURE 19.11 The structure of typical FEA models for determining shaft critical speed.

In all instances, the mechanical designer must repeat the mechanical design process. While torsional resonant conditions can also exist, they need not be reviewed in the vast majority of cases. They should be reviewed by the mechanical designer, however, in the case of large and complex equipment.

19.6.1 Simple Numerical Methods

Critical speed can be approximated without FEA, using correlations reported in Oldshue.[2] Those presented here are for a simple shaft having a fixed diameter along its entire length (see Figure 19.12).

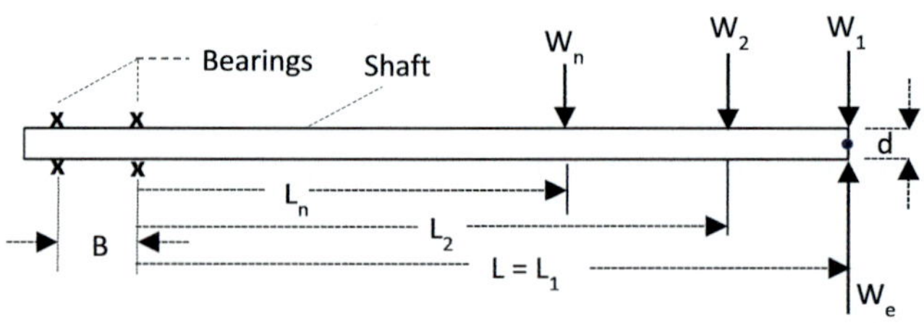

FIGURE 19.12 A fixed diameter shaft, drawn to accompany the discussion of critical speed[2] correlations.

Critical speed, N_c [s^{-1}], is calculated for a rotating shaft of fixed diameter using the following equations:[2]

$$W_s = \frac{\pi}{4} d^2 \gamma \tag{19.7}$$

where, W_s [lb_f] is shaft weight for length L [in], and γ [lb_f-in^{-3}] is shaft density.

$$W_e = W_1 + W_2 \left(\frac{L_2}{L}\right)^3 + W_3 \left(\frac{L_3}{L}\right)^3 + \ldots + W_n \left(\frac{L_n}{L}\right)^3 \tag{19.8}$$

where, W_e [lb_f] is the equivalent weight of the impellers acting at the end of the shaft, as shown.

$$N_c = 146.4 \frac{d^2}{L^2} \sqrt{\frac{E}{\left(\frac{L+B}{L}\right) + \left(W_S + \frac{4.13 W_e}{L}\right)}} \tag{19.9}$$

where, E [lb_f -in^2] is shaft MOC modulus of elasticity, and B [inch] is distance between bearings.

These correlations give the mechanical designer a reasonable critical speed estimate that approximates the value obtained by the more complex numerical methods used by mixer manufacturers. The estimated value must still be verified through comparison to the actual shaft rotational speed, thus ensuring that the critical speed ratio is below the typical maximum value of 0.8. Mechanical designers may use mixit-io SS19.1 for shaft critical speed calculation.

Table 19.1 compares correlation values to the numerical results obtained by one of the formerly mentioned mixer manufacturers. Listed impeller weights are doubled and quadrupled for illustration purposes, but they all correspond with only one impeller setting in one fixed position on the shaft.

Note that critical speed decreases as shaft diameter decreases. Also note that some applications may require the designer to consider a critical speed ratio below 0.8, for example, where high volumes of gas are sparged into the vessel, or high volumes of gas are evolved by the process chemistry.

A sample calculation now uses the same correlations (equations 19.7–19.9) to estimate the critical speed of a multitier agitation system consisting of three SPX FLOW Lightnin A320 impellers (see Figure 19.13). The results in Table 19.2 again show decreased critical speed with decreased shaft diameter.

TABLE 19.1

Critical Speeds Calculated with the Figure 19.12 Correlation, or with Numerical Software

Agitation System Parameters				Calculated Critical Speed	
Shaft Length (in)	Shaft Diameter (in)	Bearing Spacing (in)	Impeller Weight (lb)	Figure 19.12 Method (rpm)	Numerical Software (rpm)
120	2″	15″	50	158	155
150	3″	20″	100	171	168
180	4″	30″	200	169	166

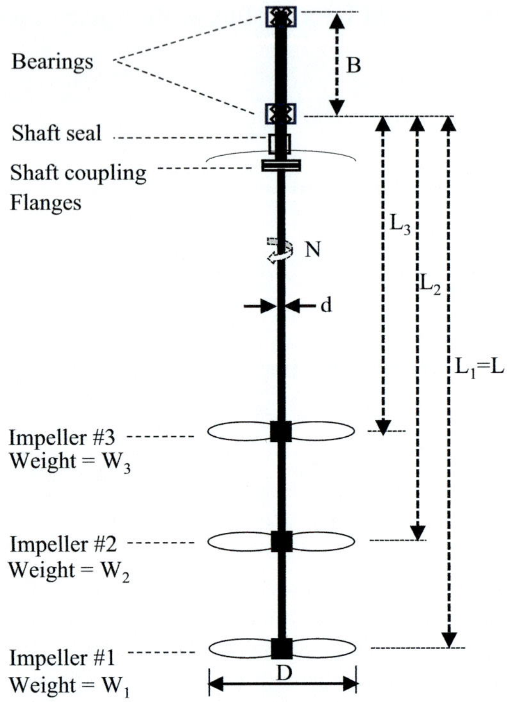

FIGURE 19.13 Schematic of a three-tier agitation system.

TABLE 19.2
Critical Speed Calculation of a Three-Tier Agitation System Based on Equations 19.7–19.9

Type of Impeller		**Hydrofoil A-320**	
Impeller MOC		316 SS	
Shaft diameter [in]	d	**2.5**	**2.0**
Impeller diameter [in]	D	24	24
Number of impellers [-]	n	3	3
Weight of impellers [lbf]	W_i		
	W_2	36	36
	W_3	36	36
	W_3	36	36
Location of impellers from lower bearing [in]	L_n		
Distance of impellers [in]	L_1	102	102
	L_2	78	78
	L_3	54	54
Length of shaft from lower bearing [in]	$L = L_1$	102	102
Elastic modulus {psi]	E	2.9×10^7	2.9×10^7
Specific weight of shaft {lbf-in^{-3}]	γ	0.29	0.29

(Continued)

TABLE 19.2 *(Continued)*
Critical Speed Calculation of a Three-Tier Agitation System Based on Equations 19.7–19.9

Type of Impeller		Hydrofoil A-320	
Weight of shaft per unit length [lbf]	W_s	1.42	1.42
Equivalent weight [lbf]	W_e	57.44	57.44
Critical speed [rpm]—Approximate value	N_c	**220**	**158**

19.7 ALLOWABLE STRESS LIMITS

A mechanical designer must limit operational stress to prevent component failure. As mentioned previously, mixers operate in a fatigue environment wherein all mechanical components experience cyclical loading that varies in both magnitude and direction during operation.

Under static loading systems, a mechanical designer can allow stress from the loading system to approach the material yield stress. This is not true, however, for materials that are operating in a fatigue environment. Materials commonly used in mixers include carbon steels and stainless steels. Although many such materials have published stress limits for operation within a fatigue environment, many do not have such published data. Mechanical designers must carefully consider allowable limits for these materials, including operating temperature, since their rise can significantly reduce normal material stress limits.

The reader is directed to mechanical engineering texts for further information concerning the Goodman diagram,[3] a commonly used tool for evaluating the strength of components subjected to alternating and static stresses. A Goodman analysis is usually performed on ductile metals. Although composite materials incorporate different techniques beyond the scope of this text, safe ratings are confirmed from test programs, which can assure that they have adequate strength for the given application.

19.8 STRESS CONCENTRATION CONSIDERATION

In addition to selecting allowable stress limits, the mechanical designer must also consider the geometry of each component. A sharp change in section modulus, like a step in the shaft, results in stress concentration and a significant change in the stress gradient. This must be considered. The behavior of materials (notch sensitivity) and the geometry (form factor) allow for a defined fatigue strength reduction factor that is used in the Goodman analysis. As a practical matter, every component will have some level of fatigue strength reduction factor, so every change in section has to be designed in detail.

19.9 COMMERCIAL MATERIAL STANDARDS

Materials are commonly available in round bars of varying diameters and in plates of varying thicknesses. ASTM standards, which contain standardized increments for bar, plate and sheet, are used to control chemistry, tolerances, minimum strength, sizes, etc. This is where imperial dimensions play a role for design purposes, due to their availability and lowest cost considerations.

19.10 BOLTED CONNECTIONS

While the shaft and impeller system would be manufactured from a solid piece of metal in a perfect world, this is obviously impractical in reality. Thus, readily available materials must be used in their manufacture. Round bars of varying diameters and plates of varying thicknesses are commonly available. Bolting systems readily connect mixer components together, with typical connections shown in Figure 19.14.

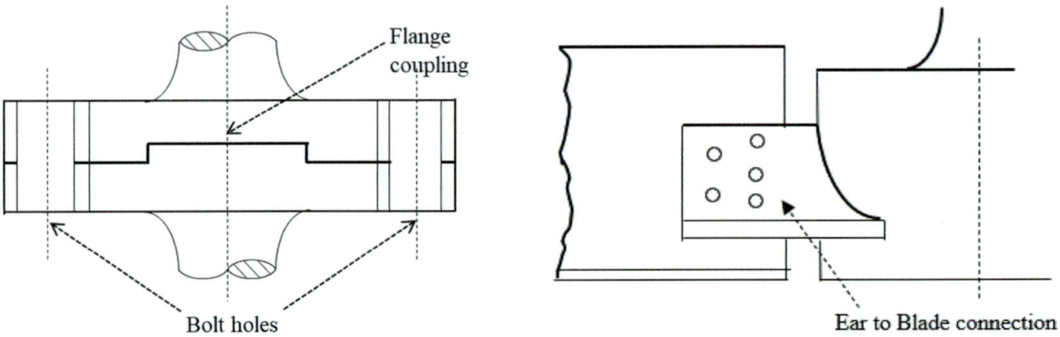

FIGURE 19.14 Flanged coupling and impeller blade-to-ear connection diagrams.

When designing bolted connections, it is critical to make sure that the "clamp load" (the load caused by bolt stretching and the associated spring action) is sufficiently large to generate needed friction between the contacting surfaces. This eliminates the possibility that loads acting on the system will be sufficient to separate the bolted components during operation. In this way, a properly sized bolt, which is properly tightened, experiences only the static preload stress and not alternating stresses. This is key because the hardware threads produce some of the highest stress concentration factors and, if used to calculate bolt stress, will vastly reduce the capacity of the hardware.

It is essential that the two bolting surfaces are flat and in full contact after being bolted together. Moreover, the hardware must be tightened using a calibrated torque wrench set to the levels recommended by the mixer manufacturer. Additionally, plain washers are recommended underneath the head of both the bolt and the nut because "spring washers" tend to relax/creep and lose preload.

19.11 HIGH-PRESSURE DESIGN

High pressure represents a challenge for the mechanical designer. A mechanical seal is used to prevent leakage between the vessel and the shaft (see Figure 19.15). Because pressure in the vessel acts on the cross-sectional area of the mixer shaft, the equipment designer must properly account for this force trying to push the shaft out of the vessel. In some cases, the gearbox bearing will be sufficient, but in many cases a bearing must be used above the mechanical seal, selected specifically to accommodate the pressure. A bearing placed close to the mechanical seal has the added benefit of limiting shaft defection (due to the fluid force) at the mechanical seal. Mechanical seals often have very limited radial movement capability.

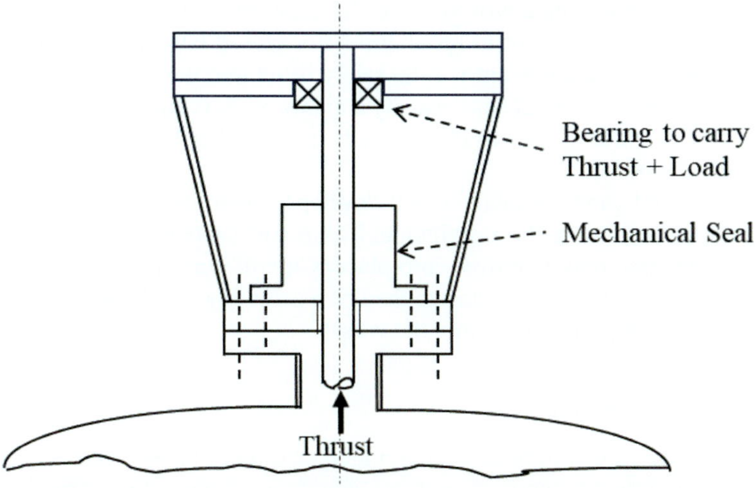

FIGURE 19.15 A schematic showing the typical locations of the mechanical seal and the load bearing.

19.12 HIGH-TEMPERATURE DESIGN

The mechanical properties of most metals degrade as temperature increases, so the mechanical designer must consider reduced allowable stress levels when determining component size.

19.13 PRACTICAL DESIGN CONSIDERATIONS

19.13.1 Bolted Connections vs. Welded Connections

As previously highlighted, connections of any type would be avoided in the perfect world. This is not practical, however, since mixers can often comprise large components that are individually difficult to handle and require on-site assembly during mixer installation. In general, welded connections are more reliable than bolted connections, due to bolted-connection variability. Variability can result from inconsistent bolt torque values or, in poorly designed equipment, lack of flatness of the surfaces being bolted together. That said, on-site welded connections include the physical challenges of safely working inside a vessel. This often makes bolted connections essential. Under these circumstances, the following general guidelines should be considered:

1. Design the bolted connection with the longest possible bolt, so it stretches during tightening/torquing. This keeps bolt unloading, during operation, from completely eliminating the clamp load.
2. Bolts must be tightened (with a calibrated torque wrench) to the mixer manufacturer's recommended values.
3. Bolts tightening must be in the sequence recommended by the mixer manufacturer.

19.13.2 Individual Component Size and Weight

The weight of individual mixer components very often requires a crane or similar lifting device. For mixer components outside of the vessel (motor/gearbox assembly and mechanical seal pedestal), this is usually straightforward. "In-tank" mixer components (shaft, impeller hub, impeller blades, impeller fins, steady bearing, etc.) are often much more difficult to install.

Most mixer manufactures include provisions for lifting each component. It is important to carefully review lifting requirements prior to installation. For closed tank applications, it is important to fully understand how each component will be safely lifted into the vessel, especially when they pass through an opening on the vessel side (usually a manway). It is equally important to know how components will be manipulated for installation (how are impeller blades matched to the hub for bolting), once inside the vessel.

Preparation is key, and should include a well-developed plan identifying all needed tools and operator skills prior to beginning the installation. The mixer shaft will usually be located in the vessel before the drive assembly is moved into place. Careful consideration, at the design stage, determines how the shaft will be connected to the drive assembly. For "open tank" installations, a crane can safely lower most components into the vessel.

19.13.3 Erosion and Corrosion

The mechanical designer should develop a clear understanding of the environment inside the mixing vessel. Normally, the process designer will provide the materials of construction for "in-tank" components based on an understanding of the potential for corrosion. The mechanical designer should pay very careful attention to where welds have been used during the manufacturing process. The performance of a welded connection, within the heat-affected zone (HAZ), can vary significantly if the proper welding process is not carefully defined and followed. The mechanical designer should carefully select and control all welding that occurs during the manufacturing process to avoid phenomena such as stress corrosion cracking.

Mixing applications often involve suspending solids. In the mixing vessel, these solids impinge on the surfaces of impeller blades. In the areas of highest velocity (leading edge of the impeller), this can lead to erosion of the impeller blade. There are several methods available to the designer to prevent/minimize erosion. The most common method is to apply rubber covering, which can also protect against corrosion. Typically, 0.25-inch (3mm) thick rubber is applied to the surfaces by specialist vendors. A second layer of rubber can be applied to the highest-wear areas for additional protection.

19.13.4 Sanitary/Hygienic Environments

The production of food, beverages or pharmaceuticals requires that the process system be thoroughly cleaned and sterilized between batches. The mechanical designer must carefully eliminate crevices where process materials can accumulate or become difficult to remove during vessel cleaning. Moreover, the designers should design surfaces to naturally "drain" so that any materials naturally "run off" the surfaces once the mixer has been stopped. Clean-in-place (CIP) provisions involving sprays of cleaning fluids are usually built into the mixing vessel during design. The design of "in-tank" parts should allow all surfaces to be exposed to the cleaning fluids. For mixers used in FDA-regulated processes, the mechanical designer should develop a working knowledge of Current Good Manufacturing Practice[4] (cGMP) and the CFR—Code[5] of Federal Regulations Title 21.

19.14 COLLABORATIVE DESIGN

The design process requires collaboration between the process designer and the mechanical engineer. Most of this book is rightly dedicated to both the mixer and the process designs required for solutions to various mixing challenges. The process designer's job is to select the most appropriate impeller or combination of impellers, to specify impeller location(s) and diameter(s), and to determine the safe mixer rotation speed that achieves the needed result. In addition, the process designer specifies the mixer mounting configuration. This is usually a top-entering mixer installed vertically on the vessel centerline. Variations to this include an angled offset, or shaft entry from the side or bottom (as opposed to the top), all of which affect mixer flow patterns and, therefore, appropriate selection of the mixer design.

Additional factors impacting the approach of the mechanical designer are:

Type of Process: This refers to basic objectives the mixer must achieve, be it blending, gas dispersion, solids suspension, heat transfer, etc., and includes specific inlet and outlet flows for either continuous or batch operations. This understanding of the environment inside and outside of the vessel, complemented by technical experience, helps the mechanical designer specify allowable stress limits for the system.

Impeller Environment: Part of the process design involves determining the fluid's character at the impeller. Standard side wall baffles are very good at distributing mixer power and flow within the vessel (normal conditions), but they can be affected by liquid level relative to the impeller(s); see the next item.

Draw-Off: As the liquid level decreases, it reaches a point where the normal flow pattern creates a vortex. As the liquid level continues to drop, it passes through the impeller. Normal mixing stops, perhaps only temporarily. The liquid surface becomes chaotic (draw off), elevating loads on the impeller blades/ears, the supporting shaft, and the machine. These conditions must be included in the mechanical design process, even though they are transient and likely to occur quickly and non-continuously.

Gassing: The impact of injecting gas into the mixing liquid must be understood, since it greatly impacts the severity of mixing impeller loads. Part of that impact depends on the method and location of that gas injection (open pipe, sparge plate, sparge ring), the gas quantity, etc. It may be necessary to reduce the mixer speed at certain gas rates to keep the elevated loads within safe limits. The use of two-speed motors or a variable frequency drive (VFD) allows the needed mixing speed change.

Impinging Flow: For process reasons, it may be necessary for a process stream to impinge directly onto a rotating impeller. Because this can elevate impeller loads, it is likely an important design consideration.

Starting Transience: As a mixer accelerates from rest, the primary inertia seen by the motor is its own rotor inertia. Intuitively this may not be apparent, but the high gear-drive ratio converts shaft and impeller inertias into small values of minimal concern—most mixers achieve full operating speed in a matter of seconds. Using appropriate-strength factors of safety while designing the mixer will account for the possible momentary pull-out torque of a National Electrical Manufacturing Association-B (NEMA-B)-designed electric motor.

These factors are characterized by the process designer, so the mechanical engineer can develop a mechanical design utilizing mixer know-how and proprietary techniques beyond the scope of this chapter. Having and using the right information is essential for the mixer to have a proper design.

19.15 DESIGN CONSOLIDATION

Successfully utilizing the basic principles outlined thus far, together with other practical design considerations not discussed here, allows the mechanical designer to confidently develop an effective mixer specification. The following guidelines summarize specifications required for the various key mixer components.

19.15.1 Electric Motor and Gearbox

Selecting the motor and gearbox is an iterative process. The process designer provides the speed of the mixer shaft and power draw at that speed. Using this data, the designer turns to gear drive selection tables for a gearbox whose output speed most closely aligns with the required shaft speed and whose torque capacity delivers the necessary power at that speed. An electric motor is selected that has a slightly higher power capacity than is required to achieve the desired mixing result.

The motor nameplate rating is usually based on a minimum 10% margin above the total shaft power, but this seldom matches common motor sizes. Therefore, the next larger motor power size is selected. Once the motor and gearbox sizes have been determined, the process designer will typically increase the initial diameter of the impellers to achieve an 80% loading of the selected motor power. This gives some "cushion" to the mixer design and better ensures that the mixing results will be achieved.

Note that the most common motor speed used for mixer applications is 1800 rpm (1780 rpm with motor slip). The speed of a motor can be calculated as follows:

Speed [RPS] = Frequency of Electricity (Hz) / Number of Pole Pairs

Using the frequency of the electricity supply in the US (60Hz) and a four-pole motor (two pole pairs),

Speed = (60 / 2) = 30 RPS = 1800 RPM

Other motor speed options are available to designers, including two-pole and six-pole motors. The former are commonly used in pump applications, but they often turn too fast for mixer applications. The latter, although available, are commercially less common.

Motors are typically attached to the gear drive by either a flange mount or a foot mount. Flange-mounted motors are typically preferred because they eliminate manual alignment of motor and gearbox shafts before connecting them with what is usually a flexible coupling. Because several different types of flexible couplings are available, the designer should carefully decide which kind to use. Couplings with rubber elements are typically more forgiving of slight misalignment than are those configured with a steel grid. Misaligned steel grid couplings can add significant unanticipated loads to bearings on the motor and the high-speed gearbox shaft. This can produce premature failure.

19.15.2 Output Shaft Speed and Gear Drives

As indicated previously, there are nominal gear ratios for drives used in mixer service. Most mixer gear drives have either helical/helical or helical/bevel designs. Part of the AGMA criteria used to design gear drives allows for the exact ratio to vary by up to 4% from nominal. Thus, known values of input power and speeds (both input and output) enable the required gear drive to be selected. Additional considerations that refine the selection of a gear drive ratio include gear service factors, bearing life requirements, and thermal ratings. These considerations combine with the actual motor speed (including slip) to identify the mixer speed required for design purposes.

19.15.3 Machine Configurations

As noted earlier in the chapter, there are a number of different mixer configurations. The most common are described in Table 19.3 and shown in Figure 19.16. One of the aforementioned mixer configurations is chosen based on the application requirements and then specified by the process designer.

TABLE 19.3
Details of the Various Mixer Configurations Shown in Figure 19.16

Sub-Figure	Gearbox Type	Tank Type	Mixer Shaft Type	Shaft Supported by
a	Right-angled	Open	Overhung shaft	Gearbox
b	Parallel shaft	Open	Overhung shaft	Gearbox
c	Right-angled	Open	Overhung shaft	Intermediate bearing system
d	Right-angled	Open	Supported shaft	In-tank steady bearing
e	Right-angled	Closed	Overhung shaft	Mechanical seal
f	Bottom entry	Either	Bottom flange	Gearbox
g	Side entry	Either	Sidewall flange	Gearbox

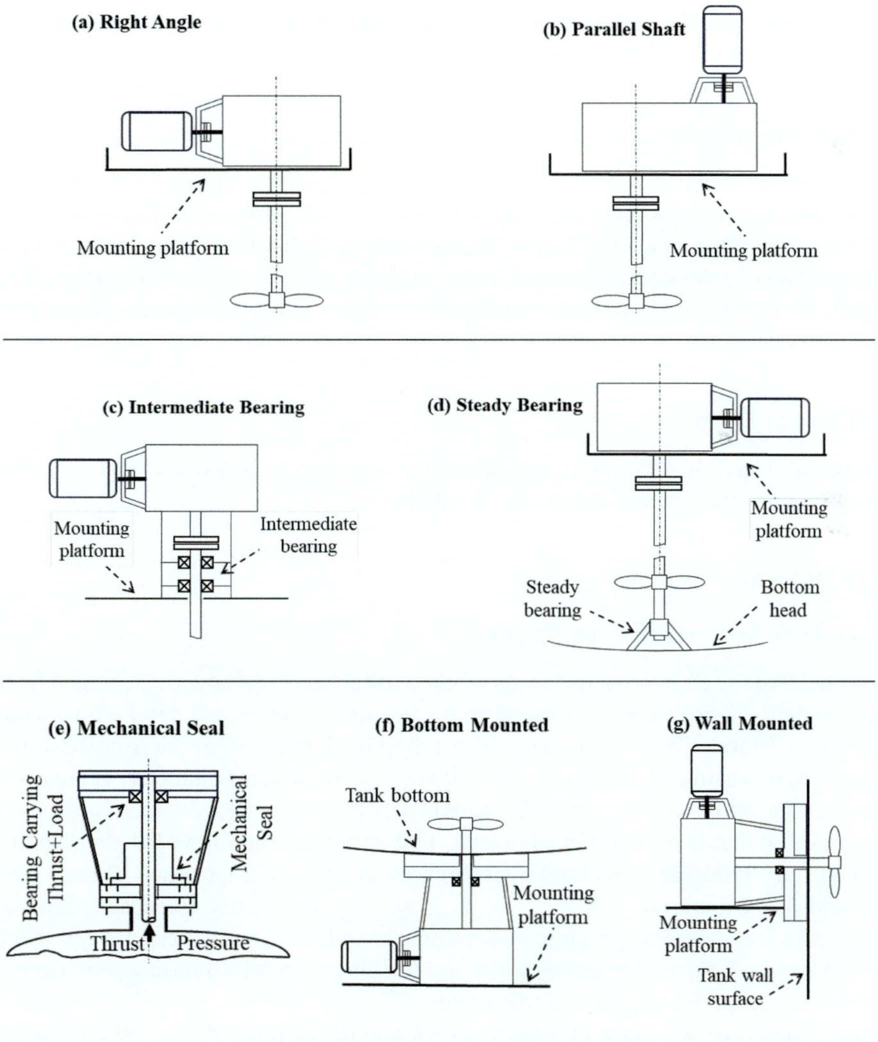

FIGURE 19.16 Most common mixer configurations, including: (a) right-angled shafts (motor and wet end), (b) parallel shafts (motor and wet end), (c) shaft with intermediate bearing(s), (d) shaft with steady bearing, (e) mechanical seal, (f) bottom mount, and (g) sidewall mount.

19.15.4 Impeller Design

The process design defined needed impeller type(s), diameter(s), and location(s) within the tank. Because specifics of impeller design use proprietary know-how developed by each mixer manufacturer, it is not possible to detail the exact steps or the other criteria (generally described earlier). The end result is that the impeller weight is now defined, along with the method of impeller construction (bolted or welded).

19.15.5 Impeller Shaft Loads

As described earlier, mixing impellers typically operate in an open flow field. In other words, they do not operate in close proximity to a housing (control volume) as a pump does. Neither do mixers have an inlet stator. Thus, mixers can be described as a low-head, high (flow)-volume, high-torque device. Because of the open flow environment, the direction of inlet flow into the impeller is variable in both direction and velocity. The net result is a variable load on the impeller blade, which in turn resolves into variable side loads (often called fluid forces) exerted on the supporting shaft. Manufacturers have developed means of quantifying these various effects.

19.15.6 Bending Moment

As described earlier, impeller fluid forces create bending moments on the lower shaft and on the machine that must be carefully analyzed. It is common to ignore centrifugal effects, as they are small in comparison to the dominant fluid forces. Section torque, bending moment, material properties, and geometry effects (stress concentration factors) are used in a Goodman type analysis to evaluate shaft strength. More information on the Goodman technique is provided in engineering textbooks or the AGMA standards.

19.15.7 Critical Speed

After having established the minimum shaft diameter, impeller weight(s), and machine size, a natural frequency (critical speed) calculation can be performed.

19.16 OPTIONAL FEATURES

19.16.1 In-Tank Support (Steady Bearing)

As outlined earlier in the chapter, in instances where a suitable critical speed value can't be obtained, supporting the shaft via a "bearing" located in the bottom of the vessel can be a valuable design option (see Figure 19.16d). This will reduce the bending load on the shaft and significantly increase the critical speed, which in turn allows the designer to reduce the shaft diameter. Typically, this support is located in the bottom of the vessel. It consists of a shaft-mounted sleeve, which runs inside of a stationary bushing that is retained in a housing. This arrangement allows the sleeve and bushing, both wear items, to be replaced without disturbing other mixer components. Careful selection of the bushing and sleeve materials is required to promote long service life. Because specific materials usually include limits on contact pressure and running speed, designer experience often determines what materials are used. Some designs may also use flushing to expel solids and provide greater life, if the process can accept the required dilution rates.

In all cases, these are not plain journal bearings due to the large diametral clearances necessary to prevent binding when the shaft deflects. As such, they operate under boundary lubrication conditions and need to be replaced when the clearance reaches the wear limit. This is usually established in an empirical way, based on monitoring wear versus time in each specific application.

19.16.2 Shaft Seals

Where the vessel needs to be sealed from the external environment, the use of a shaft seal is employed. Numerous methods are used to seal the mixer shaft: lip seals, vapor seals, stuffing boxes, single seals, double seals, and tandem seals. Commonly, end face seals of the single, double, and tandem types are supplied by third-party suppliers whose sole business is mechanical seals. These specialized assemblies require considerable engineering and special manufacturing. Thus, a complete rundown is beyond the scope of this chapter. Designers must understand likely deflection of the shaft at the location of the mechanical seal. Most seals are designed for pump applications, where shaft deflection is significantly lower than that in mixer applications, and the seal suppliers must be informed of the operational deflection.

Most seals are "wet," which means they require some sort of lubricant. Some seals are designed to operate without a lubricant, and some are designed to use a gas barrier. Once again, many variables are involved in successful seal design and selection. These include pressure, temperature, flow, and lubricant type. Thus, a detailed discussion is beyond the scope of this section and chapter.

19.16.3 Seal Lubricators

When a double mechanical seal is selected, required lubrication necessitates the use of a seal lubricator. A "standard API design[6] can be selected directly, or adapted for use, or selected from custom pre-engineered configurations provided by the seal manufacturers. It is common for wet seals to use a compatible liquid such as mineral oil or a 90/10 glycerin/water mix. Water can be used, but it is a poor lubricant that restricts the seal operation. Wet seals typically operate at 10 psi or 10% above the maximum vessel pressure, whichever is greater. Seals are designed to leak slightly with a required lubricant flow across the seal faces. Seal leakage is usually measured in drops per minute and may be so low that the seals appear not to leak at all. Additionally, the lubricant sometimes evaporates as fast as it leaks. Consequently, small leakage reservoirs (e.g., 1 to 2 gal.) usually are used with provisions to refill on the fly without stopping the mixer.

19.16.4 Mixer Design Loads

A set of mixer design loads can be established at this point. They will, of course, be most important to the vessel supplier, and they are based on the maximum conditions of torque and bending for which the mixer is rated. Sometimes, actual design loads are requested for a given application, but caution is required to avoid excessive vibrations or possible damage to the vessel during an upset. These issues require the parties to explore the added risks and ensure that everything is properly considered in the final design.

19.17 ACKNOWLEDGMENT

The author (AC) gratefully acknowledges the input and able assistance of SPX Flow colleagues Bob Blakley, Shelby Hu, and Joel Berg while creating this chapter.

REFERENCES

1. 2023 ASME Boiler and Pressure Vessel Code (BPVC). *Section VIII, Division 1: Design & Fabrication of Pressure Vessels*. www.asme.org/codes-standards/bpvc-standards/bpvc-2023
2. Oldshue, J.Y. *Fluid Mixing Technology*, McGraw-Hill, p. 412, 1983.
3. Boyce, M. P. Goodman Diagram. *Gas Turbine Engineering Handbook*, 4th. Edition, Butterworth-Heinemann, 2012. www.sciencedirect.com/topics/engineering/goodman-diagram
4. WHO Good Manufacturing Practices for Biological Products. *Annex 2*. TRS No. 999, August 2016. https://www.who.int/publications/m/item/annex-2-trs-no-999-WHO-gmp-for-biological-products
5. CFR—Code of federal Regulations Title 21. *US Food and Drug Administration*, December, 2023. https://www.accessdata.fda.gov/scripts/cdrh/cfdocs/cfcfr/cfrsearch.cfm
6. API Standard 682. *Pumps—Shaft Sealing Systems for Centrifugal and Rotary Pumps*, 4th. Edition, Product No. C68204, May 2014. www.api.org/~/media/files/publications/whats%20new/682%20e4%20pa.pdf

20 Technology Implementation and Cost Analysis

Timothy Geiger

20.1 INTRODUCTION

As a part of supply chain management, companies in the large-scale chemical processing industry (e.g., chemical, petrochemical, food, pharmaceutical, etc.) typically manufacture their products at multiple sites. Although this is not always the case, all sites manufacturing the same product would ideally use the same most effective technology (MET) available at the time. Individuals overseeing the product portfolio would use prior implementation experience, together with feedback from R&D and process engineering (PE), to develop global manufacturing guidelines and to identify METs. For a company to be competitive in the marketplace, its METs must encompass cost-effective manufacturing processes without sacrificing product quality, yield, and selectivity. Foreseeable capital expenses (CapEx) and operating expenses (OpEx) must also be considered in selecting MET. Technology assessment and adaptation must include batch versus continuous operation, minimal manufacturing footprints (real estate), operational and occupational safety, and product sanity (i.e., as-expected performance at a basic level). Besides reactors (including crystallizers, fermenters, etc.), other vital unit operations and associated peripherals such as pumps, external heat exchangers, condensers, vents, etc. must also be included.

With that brief introduction, and true to the theme of this book, this final chapter will briefly describe the typical CapEx and OpEx associated with agitated tanks. Significantly, agitated tanks typically account for more than 80% of all industrial-scale reactors. Chapter 4 previously presented an overview of internal components within a mechanically agitated tank. Such use of many components is application specific.

20.2 INSTALLATION INFRASTRUCTURE

Large-scale batch reactors traditionally require large multi-storied facilities with platform grids, while more modern continuous mixed-flow reactors (MFRs) are ideally both shorter and narrower. The process intensification (PI) initiative, in fact, seeks to replace the former with the latter, often at the cost of higher OpEx. The fact that multi-storied facilities provide more space to gravity feed raw materials, to gravity drain the working media, and to place reactor peripherals is too easily overlooked. Proper reactor support structures accommodate not only the tank but also the prime mover (motor) and drive system (gearbox) to minimize undesirable vibration and noise, and thereby avoid mechanical failure.

20.3 MECHANICAL AGITATION SYSTEM COMPONENTS AND IMPLEMENTATION

The design features of several components in stirred tanks are described in Chapters 4 and 19. The following sections briefly address cost-effective implementations of a few major components noted in Table 4.1, as well as essential instrumentation; associated peripherals such as pumps and heat exchangers are described elsewhere in the book.

 DOI: 10.1201/9781003606390-20

20.3.1 Mixing Vessel

The cost of a reactor tank varies significantly with materials of construction (MOC), size, and internal surface finish. Approximate tank costs in Table 20.1 indicate that electropolished tanks (RA ≈ 5μm) are typically 10% more expensive than standard mechanically polished (RA ≈ 25μm) tanks. Accordingly, the latter are actually more cost effective for most applications. In the United States, cylindrical tank diameters are commonly below 13 feet because the US Department of Transportation (USDOT) restricts the transport of larger diameter tanks on public highways with underpasses/overpasses. Thus, tank diameters exceeding 13 feet, as in large fermenters and storage vessels, are commonly field erected, at added onsite fabrication costs.

In most instances, baffle supports and braces are an integral part of the tank, whether attached by bolting or welding. Baffle MOC is usually similar to that of the tank. Various common baffle types, including their application-specific geometry and MOC, are described in Chapter 4, "Stirred Tank Reactors."

TABLE 20.1
Typical Cost of Mixing Vessels with 2:1 SEL Bottom in 2024 USD (full vacuum to 100psig rated, with shell jacket to 90psig rated)

Materials of Construction (MOC)	Cylindrical Volume [gal]	Tanks (w/standard internal surface polish) nominal diam. [in]	TL-TL height [in]	Initial Purchase Cost [USD]
Carbon Steel	500	48	72	56,000
	1000	60	72	65,000
	5000	113	113	135,000
Stainless Steel	500	48	72	90,000
	1000	60	72	110,000
	5000	96	140	350,000
Hastelloy	500	48	72	270,000
	1000	60	72	350,000
	5000	96	140	1,400,000
Glass-Lined CS	500	48	58	80,000
	1000	62	71	110,000
	5000	108	109	250,000

20.3.2 Agitator Shaft

In agitated vessels, a rotating shaft transmits power from the mixer drive to the impeller(s). Large mixing vessels may have a short drive shaft rigidly coupled to a longer main shaft, since this design provides flexibility in disconnecting the longer shaft for removal through a top nozzle. The shafts designed by mixer manufacturers are typically solid bars or machined pipes (hollow) that are mounted in thrust bearings. As mentioned earlier in Chapter 19, the shaft MOC must be chemically compatible with the working media. For example, shafts used in acidic environments typically have titanium cladding and glass lining. Hollow shafts are especially preferred in tall vessels to minimize downward weight exerted on the bearing(s) and to help reduce critical speed and resonance issues. Required shaft strength is of greatest importance, however, and must not be sacrificed. Accommodating the impeller hub locations must also be considered when designing a hollow shaft's wall thickness. Impeller hubs are generally welded to the shaft, in fixed positions. The typical cost, per linear foot, of a solid shaft varies from \$250 to \$5000 depending on its diameter (2″–6″) and MOC (carbon steel—316 stainless steel). Hastelloy shafts are approximately five times costlier than identical 316 stainless steel shafts. Hollow shafts are normally more economical (less costly) and appropriate for tall vessels without steady bearings.

Figure 20.1 shows impellers with different blade-to-hub, and hub-to-shaft mounting methods. Such methods depend on the torque and overhung thrust experienced by the shaft due to unsteady hydrodynamic forces on the blades and structural strength of shaft MOC at the working temperature. Thus, mixer shaft design is indirectly a function of power delivered to the impellers. The order of increasing cost-effectiveness for a few common impeller blade-to-hub and hub-to-shaft assembly methods is:

(a) Blades welded directly to the shaft, with or without a hub (typical of smaller agitated vessels)
(b) Blades welded to a simple sleeve-type hub and hub-to-shaft attachment with a key slot and set screws or pins
(c) Blades bolted to ears on a hub, which is mounted on the shaft using a key slot and a wedge-shaped block. Hardware used in fastening blades to the ear may be a simple bolt and nut connection, torqued to the specified locking torque value, based on hardware size. The effectiveness of double nuts, lock-washers, and safety wire are dependent on specific installation procedures. If there is potential for a fastener or impeller blade to dislodge from the assembly and damage tank internals, the nuts should be safety wired to the hub, as shown in Figure 20.2. Figures 20.2(a) and (b) feature lower-cost permanent attachments of (a) blades to the shaft and (b) blades to an earless-hub. They are ideal for tanks from which the impeller assembly can be easily removed through an open-top tank. Figure 20.2(c), which easily enables blade-hub attachment and removal, is common in closed-top tanks.

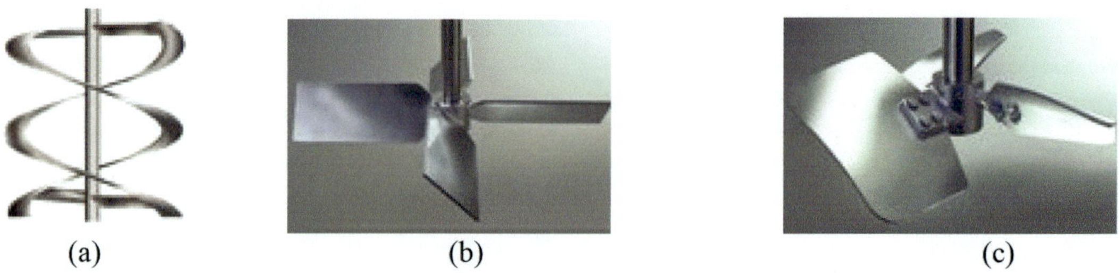

FIGURE 20.1 Blade-to-hub and hub-to-shaft attachments are: (a) weld-weld, (b) weld-key, (c) bolts-key.[1,2]

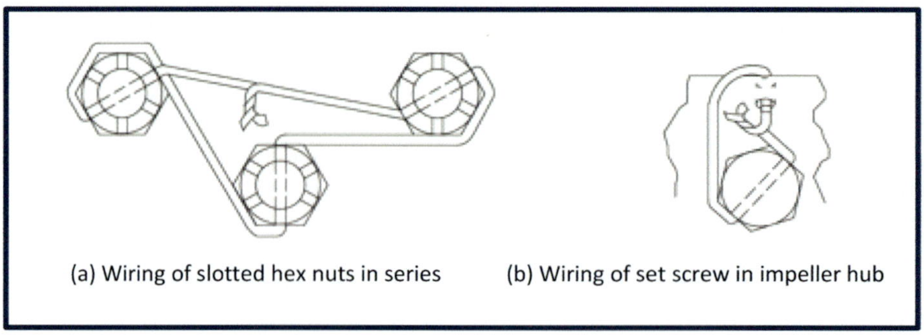

FIGURE 20.2 Safety wire installations to prevent bolt/blade fall-off from the impeller hub.[3]

In the case of Pfaudler's Cryolock® family of glass-lined agitators, impellers are directly mounted onto the shaft using a cryogenic fastening technique. Removal of a mounted impeller is illustrated in Figure 20.3.[4]

1. Removal begins by feeding liquid nitrogen into a hollow agitator shaft.
2. Cryogenic cooling shrinks the shaft and loosens the glass-to-glass joint.
3. The turbine slides off the shaft and onto a padded bottom head.
4. The separated impeller easily passes through the manway.

FIGURE 20.3 Photos demonstrating the method of removing Cryolock® impeller blades from the shaft.[4]

Cryogenic fluid such as liquid nitrogen is introduced into the hollow glass-steel shaft. Thermal contraction shrinks the shaft diameter as it chills. The glass-lined impeller assembly is then slid down the narrowed shaft. After cryo-fluid removal and shaft warming, thermal expansion restores the shaft to its original diameter. Shaft expansion is what locks the impeller assembly tightly onto the shaft. Reversing the order of the Figure 20.3 images (4-3-2-1) shows the technique of mounting an impeller on the shaft. This technique allows the impeller position to be varied on the shaft, although the impeller blade pitch is fixed.

In the DeDietrich GlasLock® system, hubs are integral to the glass-lined shaft. A tapered shaft on each individual impeller blade is inserted into a tapered hole in the hub (see Figure 20.4).[5] Warming, via a wrap-around heat blanket, expands the hub, so careful control of differential temperature and thermal expansion can form a predetermined blade-to-hub compression fit. This easy mounting and removal technique adds flexibility to change blade type or blade pitch angle prior to stem re-expansion. Hub locations on the shaft are predetermined and immobile, however.

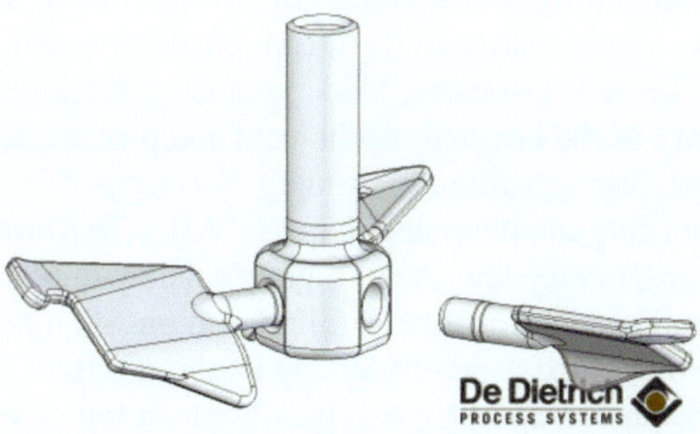

FIGURE 20.4 Schematic showing insertion of the GlasLock® blades into the hub holes.[5]

20.3.3 Bearings

The loads imposed on the mixer shaft are taken up by bearings within the gearbox, or by an independent bearing housing below the reducer. The bearings and gearing, which make up the key components of the drivetrain, must withstand axial thrust (upward from down-pumpers, or vice versa) and radial (bending) loads. These loads come from the process fluid and are imposed across the surface of impellers mounted on the cantilevered rotating shaft. Furthermore, as noted in Chapters 4 and 9, as long as the process torque requirement is met, the size of the gearbox or independent bearing housing may be reduced by using a bottom steady bearing and/or an intermediate bearing to reduce loads and allow a smaller diameter shaft. Design and MOC of such bearings are critical to the OpEx, which should be minimized as possible. Easy bearing installation and removal/repair must be factored into the design.

The use of submerged shaft-stabilizing bearings is strongly discouraged in media such as suspended crystal slurries, where the solids can attrit and then foul the bearings. Limit-rings (see Chapter 4) are an option to control shaft deflections caused by radial hydrodynamic loads, but they introduce transient transverse loading on the upper thrust bearing, which negatively impacts its serviceable life. Table 20.2 lists approximate costs for a small selection of steady bearings. Costs include the steady bearing pedestal, bushing, shaft wear sleeve, and hardware.

TABLE 20.2
Cost of Bottom-Mounted Shaft-Stabilizing Steady Bearings in 2024 USD

	Cost for Bushing Material [USD]		
Nominal Shaft Diameter [in]	Filled PTFE/ Carbon Steel	Filled PTFE/ Stainless Steel	Alternate Bushing Material Upgrade, e.g., graphite
2	4000	6000	+1500
4	6000	20,000	+2000
6	15,000	38,000	+3000
8	25,000	45,000	+4500

20.3.4 Impeller

Selecting the proper impeller(s) for a given process is vital to mixing and/or dispersion operations. Chapter 5 describes features of various commercially available generic and proprietary impellers and their typical applications. If the required mixing performance of a generic impeller, such as a pitch-blade turbine (PBT) or a flat-blade turbine (FBT), is not significantly different from that of custom-made impellers, manufacturing often prefers the former due to lower cost and greater availability with less inventory. As an alternative to the OEM, onsite or nearby machine shops sometimes fabricate needed flat blades. In such instances, blade geometry, thickness, and MOC must be truly replicated to maintain integrity of the original mechanical and process design. Replacing an existing proprietary impeller with a generic substitute is strongly discouraged, however, unless pre-approved through consultation with a mixing subject matter expert (SME). To illustrate, changing an operating crystallizer impeller from a proprietary low-shear, high-efficiency impeller to a standard PBT would likely degrade the process and product quality. Similarly, in gas–liquid dispersion, such as aerobic fermentation, an FBT must not replace a gas-dispersing Rushton turbine for the sake of cost savings. Because the cost of impellers (including hubs) is typically much less than the total cost of the agitation system, MET requirements override minor potential cost savings. Practicing engineers should see references 6 and 7 for mechanical design and cost estimation of mechanical agitators, respectively.

20.3.5 Prime Mover (Motor)

Motors, whether electric, hydraulic, or pneumatic, are included in agitation systems as the essential power source driving the impellers that rotate within the working media. Confined explosive environments may mandate hydraulic or pneumatic motors, depending on the required power level. Totally enclosed, fan-controlled (TEFC), and explosion-proof electric motors are widely used in production plants with power ratings ranging from 5 to 1500 hp. Typically, large-scale aerobic fermentations, ultra-high viscosity blending, and solids suspension applications require high-powered motors for their requisite power-per-mass input. The nominal cost of electric motors is based on their horsepower, as presented in Table 20.4 (shown later in this chapter).

20.3.6 Drive (Gearbox)

The gearbox is a critical component of any large-scale agitation system. Its internal gears reduce the motor speed to the required agitator shaft speed, while significantly increasing the shaft torque ($P = 2\pi NT \Rightarrow T \alpha N^{-1}$) to that needed to overcome hydrodynamic forces on the impeller(s). Typically, helical/spiral bevel gears with a high transmission efficiency are used. Frictional power loss is significant in the gearbox and bearings that support the shaft. Shaft power, P_s, is determined by subtracting frictional power loss, P_f, from the supplied motor power, P_m.

$$P_s = P_m - P_f \tag{20.1}$$

For the most part, mixer gearbox lubricants are engineered to provide a protective layer on the gear surfaces. This layer reduces metal-to-metal contact and conveys the generated frictional heat to the housing, where it can be dissipated to the atmosphere. The lubricant is held captive in the gearbox via either a dry well design or lip seals around the gearbox output shaft. Lip seals are considered a wear item, and they will allow leakage over time. Therefore, the gear oil must be compatible with the process fluid, and proper lubricant level should be monitored closely over time.

Mixer gearbox power ratings range from less than 1 hp to more than 1500hp. The typically acknowledged nominal AGMA speeds (rpm) are: 11, 13, 16, 20, 25, 30, 37, 45, 56, 68, 84, 100, 125, 155, 190, 230, 280. It is noted that AGMA allows manufacturers to deviate from stated nominal speed ratios by up to ± 3% for a single-reduction gear drive, and ± 4% for a double-reduction drive.

The gearbox is generally the highest capital cost item in an industrial-scale agitation system. Table 20.4 (shown later in this chapter) illustrates the approximate costs of some carefully selected industrial-scale gearboxes.

20.3.7 Variable Frequency Drive

Mixer shaft speed can be changed through the use of a variable frequency drive (VFD). The importance of the finer speed control provided by a VFD is evident in crystallizations or aerobic fermentation processes that often require a very specific speed to achieve required particle sizing, or to accommodate the required gassed and un-gassed power levels. To review discussions of these example applications, see Chapters 7 and 9. Although VFDs add cost, the resulting flexibility to change mixer shaft speed, rather than living with coarse AGMA speed increments, far outweighs their costs and explains their popularity. Table 20.3 lists some VFD prices for reference purposes.

TABLE 20.3
Approximate Costs of Industrial-Scale Variable Frequency Drives

VFD Power [hp] (460V, NEMA 1 Enclosure w/o line contactor)	2024 Cost [USD]
5	2000
10	2500
15	3500
25	5000
50	10000
75	15,000
100	20,000

20.4 AGITATION SYSTEM COST ESTIMATION

Table 20.4 presents a ballpark estimation in 2024 USD of agitation systems consisting of a prime mover (electric motor), gearbox, shaft, double mechanical seal, and a hydrofoil impeller. Prices for steady bearings and variable frequency drives, as listed previously in Tables 20.2 and 20.3, can be added to the costs from Table 20.4 for an estimated total cost of agitation systems incorporating all of these elements.

TABLE 20.4
Approximate Cost Estimation of Agitation Systems in USD

	Cost of Drive System (Gearbox)						
Torque [ft-lb] = $\left[\dfrac{Motor\ hp \times 5250}{Shaft\ rpm}\right]$	250	350	750	1500	2000	3000	4500
Cost [USD]	18,000	25,000	32,000	41,000	62,000	74,000	110,000
Open Tank/Beam Mount Mixer Drive							
	Add the following to drive system for the Double Mechanical Seal						
Mounting flange size (inch) at 150 psi	8	8	10	12	14	16	18
Cost of double mechanical seals [USD]	27,000	29,000	33,000	41,000	62,000	74,000	110,000

	Add the following for shaft cost per linear inch		
Shaft MOC	Carbon Steel	316 Stainless Steel	Hastelloy
Shaft Diameter (inch)		Cost {USD/inch]	
2	20	35	
3	30	80	
4	50	150	~5x the price of 316 SS.
5	60	200	
6	85	475	

	Add the following for motor cost (price includes motor coupling) [USD]	
Motor Type	TEFC/1800 rpm/230–460 volt, 3 phase, Inverter duty	Explosion Proof C1, Div1, Group D, Cl2 Group F&G
Motor power [hp]	Cost [USD]	
5	2,500	3,000
10	3,500	4,000
25	5,500	6,000
50	10,000	12,000
75	15,000	17,000
100	20,000	23,000

	Add the following for impeller cost (price includes motor coupling)		
Type of Impeller	Hydrofoil		
MOC of impeller	Carbon Steel	316 Stainless Steel	Hastelloy
Impeller diameter [inch]		Cost [USD]	
24–28	2,400	4,000	~five times the price of 316 SS.
29–33	2,900	4,700	
34–38	3,500	6,000	
39–43	4,500	7000	
44–48	6,000	9000	
49–53	7,800	11,000	
54–58	9,500	13,000	
59–63	11,000	16,000	
64–68	13,000	19,000	

20.5 AGITATION SYSTEM SEALS

Mechanical seals are elements used between the rotating shaft and a stationary surface to prevent or minimize the leakage of working media from inside the vessel. Single or double mechanical seals, with the latter being predominantly in the form of a cartridge, are used to lubricate interfaces and remove frictional heat. An external cooling system, which is commonly used in industrial-scale agitation systems, can be coupled with the cartridge (seal assembly) to regulate the temperature and pressure of the interfacial lubricant film and the barrier fluid. Smaller pilot/production-scale systems normally use a simple pressurized seal-pot for seal lubrication purposes. Primary rotor-stator seal materials must be compatible with the media and abrasion resistant to contaminants. Carbide/graphite (hard/soft) combinations are often the primary material of choice, while elastomeric and polymeric materials such as EPDM, silicon rubber, and Viton™ are often preferred for backup/secondary seal faces. The primary seal's material of construction must withstand the expected temperature of the process fluid. Figure 20.5 shows the cutaway section of an EagleBurgmann® double mechanical seal.[8] Multiple global agitator seal manufacturers, such as John Crane, Flowserve, SKF, and EagleBurgmann® Industries, support the chemical processing industries.

FIGURE 20.5 Cutaway image showing the interior of a gas-lubricated AZSZ double mechanical seal cartridge unit manufactured by EagleBurgmann® Industries.[8]

For double mechanical seal cartridges, easy replacement and maintenance are very important to users. The typical life span of a seal ranges from 3 to 7 years, depending on its application and maintenance protocol.

With shaft seals, inappropriate installation, incorrect alignment, insufficient lubrication, and excessive contamination may shorten their life span and cause premature failure. For further understanding of agitator shaft seals and lubricants/coolants, readers are referred to Chapter 19. Table 20.4 presents typical costs of double mechanical seals.

20.6 CONTROL VALVES

Control valves peripheral to a reactor maintain process variables such as feed flowrate, steam addition and re-circulation rates, and headspace pressure within required operating ranges to ensure appropriate mixing and product quality. Level detectors, load cells, temperature sensors, and pressure sensors provide feedback to the functioning control valve (CV). Various categories of CVs—namely, disk butterfly, globe (Figure 20.6), ball, and diaphragms with piston or rotary actuators—are externally used for different services associated with the reactor. Process engineers are advised to review reference 9 for control valve design and application feasibility, such as operating range, response time, materials of construction, durability, installation, maintenance, and cost.[9]

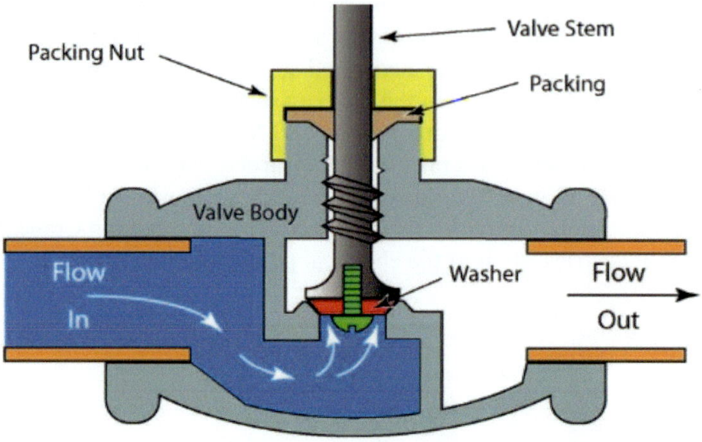

FIGURE 20.6 Cutaway section of a globe valve.[10]

[The actuator lifts the disc, with replaceable washer, from the seat to precisely control allowed flow volume.]

In many instances, control valves associated with mixing processes have disrupted continuous operations and forced production plants into unplanned shutdowns. These shutdowns ultimately increased product variability, productivity losses, and unnecessary maintenance costs. Oversized valves have a higher capital cost and tend to make operating mixing systems unstable, whereas undersized valves simply won't pass the required fluid flow through the line. Thus, correctly selected and sized control valves with actuators provide a significant cost-effective advantage to routine operations. The average price of industrial-scale control valves varies from $1,000 to $20,000, depending on their type, associated actuator/controller, accuracy, size, and material of construction.

Typically, the bi-directional globe valves shown in Figure 20.6 are widely used for the following benefits:

- Precise control of flowrate for optimizing process performance
- Design position and actuator ensure accurate modulation and pressure drop control
- Ease of operation, low leakage, operational reliability, and cost-effectiveness

20.7 STATIC AND ROTATING SPRAY BALLS AND NOZZLES

In most mechanically agitated tanks dealing with process fluids, tank internals are prone to fouling due to the adherence of solids and sticky substances. Thus, routine tank cleanings using high-velocity jets of hot water/steam/chemicals at the end of one or more batches is essential, especially in headspace regions that are more likely to experience deposits and crust formations. Automated devices such as static nozzles or rotating sprayers (see ball and jet examples in Figure 20.7) installed inside the vessel are widely used for such cleaning operations. The latter is typical of the pharmaceutical, healthcare, and food industries for clean-in-place (CIP) operations where ultra-cleanliness is vital to product quality and qualification.

(a) (b)

FIGURE 20.7 Photographs of Sani-Matic's in-tank rotating (a) spray ball and (b) spray jet.[11]

CIP spray balls are self-driven by the cleaning fluids. They are hollow spheres with internal vanes/slits and surface perforations which, in the presence of pressurized fluid, assist in rotating the ball and spraying the liquid. Normally, in production-scale vessels, there are three spray balls in the headspace, mounted 120 degrees apart. Their rotating sprays sweep the tank internals. Many chemical plants use manual high-pressure cleaning with spray nozzles or automated cleaning with built-in static jets aimed towards the dome region and other internal tank features. In the case of slurry reactors, retractable jet nozzles are used to alleviate potential solids plugging issues associated with permanent in-tank nozzles.

Globally, there are multiple spray/nozzle manufacturers such as Sani-Matic, Bete, Alfa Laval, Spraying Systems, etc. During the design phase of a tank, process engineers are advised to collaborate with the manufacturers in considering appropriate spray systems, including their mountings, fluid inlet pipe size, manifold, overhead nozzles, etc.

20.8 PROCESS FLUID LEVEL DETECTOR

The agitation system of a production-scale reactor is normally designed to accommodate three different process fluid levels, i.e., low, normal, and maximum. Such flexibility assists in reactor startup and in varying product demand. As a rule-of-thumb, the maximum allowed working media level in an STR is ~80% of the tank's straight-side height. This is preferred to avoid splashing of the process fluid onto the dome, to leave room for foams generated with low surface tension and hygroscopic fluids, and to avoid gas holdup in aerobic fermenters. In many instances, the aforementioned phenomena lead to plugging of overhead inlet(s) and outlet(s). Furthermore, maintenance of the appropriate process fluid level is essential to avoid overfilling and also vital to product quality.

Normally, reactors use in-tank level sensors to monitor and measure liquid levels. These devices typically utilize technologies such as electromagnetic waves (radar), sound waves (ultrasonic), hydrostatic head (fluid submergence), etc. Since radar sensors are extremely accurate and insensitive to product properties (temperature, pressure, density, conductivity, etc.), they are widely used in STRs.[12] Basically, radar sensors work by sending out electromagnetic waves and measuring how long it takes for them to bounce back from the working media surface. The information collected from the reflected waves is then used to identify distance to the surface. The measurement principle of Rosemont's fast-sweep frequency-modulated continuous wave (FMCW) radar technology[13] is described in Chapter 18 (Experimental Techniques in Mixing). Gems' ultrasonic level transmitters are mounted at the top of a tank and continuously transmit pulses of high-frequency sound waves away from the sensor. Once the sound waves hit the liquid surface, they return to the sensor.[14] These contactless devices are flange mounted and must be routinely calibrated. Practicing engineers should contact a manufacturer (e.g., Emerson, Gems Sensors, Flowline, etc.) to learn the pros and cons associated with different sensor types, materials of construction, preferred process fluid applications, relative tank sizes, etc.

REFERENCES

1. https://www.spxflow.com/en/lightnin/pc-impellers
2. https://www.chemineer.com/products/chemineer.html#impellers
3. Safety Wire Installation, SPXFLOW-Lightnin Standards, *INST.* NO. IT–3813, Rev. A, 1–18–02.
4. Pfaudler Cryolock® Mixing Systems. https://www.flexachem.com/wpcontent/uploads/2013/10/Mixing%20Systems.pdf
5. The GlasLock® Systems. https://www.dedietrich.com/en/solutions-and-products/reaction/mixing/glaslockr-system.
6. Fasano, J. B., Miller, J. L., Pasley, S. A., Consider mechanical design of agitators, *Chemical Engineering Progress*, vol. 91, pp. 60–71, 1995.
7. Meyer, W. S., Kime, D. L., Cost estimation for turbine agitators, *Chemical Engineering*, pp. 109–112, September 1976.
8. https://www.eagleburgmann.com/en/products/mechanical-seals/agitator-seals
9. Control Valve Handbook, *Emerson process management*, 4th. Edition, Fisher Controls International LLC, 2005.
10. https://www.fctvalve.com/libr/globe-valve
11. https://sanimatic.com/spray-devices-supply-tubes/
12. https://www.emerson.com/en-us/automation/measurement-instrumentation/common-applications/level-measurement-solutions-for-reactors-mixing-tanks
13. https://www.emerson.com/documents/automation/product-data-sheet-rosemount-5408-5408-sis-level-transmitters-en-202468.pdf. Rosemount 5408 Series, April 2022.
14. https://www.gemssensors.com/product/liquid-level-sensors/continuous-level-transmitters

Appendix A1
Book User-Interface for Working Examples

TABLE A1.1
Book Spreadsheet Links

Chapter	Page	ID	Spreadsheet Title	mixit-io spreadsheets*
1	5	SS1.1	Capillary Viscometry Data Analyses	**mixit-io-SS1.1**
1	7	SS1.2	Particle True and Apparent Densities	**mixit-io-SS1.2**
1	9	SS1.3	Mixture Thermal Conductivity	**mixit-io-SS1.3**
4	26	SS4.1	Vessel Sizing	**mixit-io-SS4.1**
4	39	SS4.2	Forces on Overhung Dip/Feed Pipe	**mixit-io-SS4.2**
4	39	SS4.3	Vortex—Outlet Liquid Submergence Depth	**mixit-io-SS4.3**
5	48	SS5.1	Impeller Characteristics	**mixit-io-SS5.1**
6	60	SS6.1	Agitation System Design-Blending	**mixit-io-SS6.1**
7	78	SS7.1	Just Suspension Speed	**mixit-io-SS7.1**
7	80	SS7.2	Sphericity of Rod-Shaped Particles	**mixit-io-SS7.2**
8	120	SS8.1	Reactor Design: L-L Dispersion	**mixit-io-SS8.1**
8	120	SS8.2	Sauter Mean Diameter Correlation Constant	**mixit-io-SS8.2**
11	180	SS11.1	Jet Mixer Design	**mixit-io-SS11.1**
12	203	SS12.1	Static Mixer Design	**mixit-io-SS12.1**
17	350	SS17.1	Heat Transfer System Jacket/Coil Design	**mixit-io-SS17.1**
19	403	SS19.1	Shaft Critical Speed	**mixit-io-SS19.1**

* URL: https://s3-eu-west-1.amazonaws.com/s3-euw1-ap-pe-ws4-cws-documents.ri-prod/9781032998619/mixit-io.zip
OR
https://www.routledge.com/9781032998619
Go to Support Materials on the page.

Note: (a) for SS7.1 and SS7.2, see SS7.1&2 (b) for SS8.1 and SS8.2, see SS8.1&2

INSTRUCTIONS FOR READERS

1. Download the "mixit-io zip" folder using one of the above URLs.
2. Unblock its contents (the spreadsheet files)
   Right Click on zip folder > Properties > Unblock > OK
3. Extract the files into a personal folder to enable their use
4. If the spreadsheets are still not accessible for use, enable Excel® / VBA Macro in the "Trust Center Settings". See relevant instructions for your PC and its Excel® Version.
5. Note: Transfer of folder/files to others is a violation of the copyright.

Appendix A2

Unit Conversions and Equivalents

Mass, length, and time (MLT) are globally recognized as the three fundamental dimensions characterizing the primary and derived units used by practicing engineers. Unlike time, mass and length units differ among professionals depending on their geographic location—imperial (British) "foot-pound-second" (FPS) and metric "meter-kilogram-second" (MKS) being the two most widely used systems in the chemical industries. Traditionally, US chemical industries have adapted the imperial FPS system along with a few derived units based on the MKS system. For example, foot and pound for length and mass, but Poise/centi-Poise for dynamic viscosity from the CGS (centimeter, gram, and second) system that is an off-shoot of the metric (MKS) system. Because the book authors have conscientiously used traditional/customary units practiced in the US chemical and allied industries, unit conversion tables are indispensable and imperative for accurate engineering calculations. Therefore, conversion tables are included in this appendix.

The weight of an object is determined by multiplying acceleration due to gravity, g, by its mass, m.

Gravitational acceleration is a function of elevation. In FPS and MKS systems, the mean sea-level values are 32.2 ft-s^{-2} and 9.81m-s^{-2}, respectively. The derived units of weight-force, i.e., pound-force (lb_f) and Newton (N), are simply the products of mass in units of pound (lb_m or lb) and kilogram (kg) with respect to the acceleration due to gravity as follows:

$$\text{Pound-force, } lb_f = lb_m \times 32.2\text{ft-s}^{-2} \text{ and Newton, } N = kg \times 9.81\text{m-s}^{-2}$$

Derived units of thermal properties normally encounter temperature, θ, as an essential parameter. In the English and metric systems, the units are respectively degrees Fahrenheit (°F) and degrees Celsius (°C). Corresponding absolute temperatures are Rankine, R (= 460 + °F) and Kelvin, K (= 273 + °C). The degree symbol is not typically used with Kelvin, and to some, should not be used with Rankine. Note that:

(a) Temperature difference, Δθ, in the absolute scales of R and K is equal to temperature difference in the non-absolute temperature values of °F and °C, respectively.
(b) Parameters per R and K are equivalent to the same parameters per °F and °C, respectively.
(c) Temperature conversions are: °C = 5/9 × (°F—32) and °F = (9/5 × °C) + 32. The freezing temperature of pure water (H_2O) at atmospheric pressure is 32 °F and 0 °C.

The online "Unit Converter" publication of equalsto.com[1] is an excellent resource for unit conversions. A comprehensive glossary of SI units and their conversions to other units of measurement is presented in reference 2.

TABLE A2.1
Unit Conversions and Equivalents

Units	MLT Notations	From English (FPS) System	To Metric (MKS) System	Multiply by
Primary Units				
mass	M	pound (lb_m also lb)	kilogram (kg)	0.454
length	L	foot (ft)	meter (m)	0.305
time	T	second (s)	second (s)	1.0
temperature	θ	Fahrenheit (°F)	Celsius (°C)	(5/9) (°F-32)
Derived Units				
area	L^2	square foot (ft^2)	square meter (m^2)	0.093
volume	L^3	cubic foot (ft^3)	cubic meter (m^3)	0.028
volumetric flowrate	L^3T^{-1}	ft^3/s	m^3/s	0.028
mass flowrate	MT^{-1}	lb_m/s or lb/s	kg/s	0.454
velocity	LT^{-1}	ft/s	m/s	0.305
acceleration	LT^{-2}	ft/s^2	m/s^2	0.305
weight	MLT^{-2}	pound-force (lb_f)	kilogram force (kgf)	0.454
force	MLT^{-2}	lb_f	Newton (N)	4.448
momentum	MLT-1	lb_m-ft/s or lb-ft/s	kg-m/s	0.138
density	ML^{-3}	lb_m/ft^3 or lb/ft^3	kg/m^3	16.018
specific weight	$ML^{-2}T^{-2}$	lb_f/ft^3	N/m^3	157.09
pressure or stress	ML^{-2}	lb_f/ft^2	N/m^2	47.88
torque	ML^2T^{-2}	lb_f-ft	N-m	1.356
torsion	ML^{-2}	lb_f/ft^2	N/m^2	47.88
energy or work	ML^2T^{-2}	ft-lb_f	N-m or Joule (J)	1.356
power	ML^2T^{-3}	ft-lb_f/s	N-m/s or Watt (W)	1.356
heat	ML^2T^{-2}	British thermal unit (Btu)	Joule (J)	1055
thermal power	ML^2T^{-3}	Btu/s	J/s or Watt	1055
thermal conductivity	$MLT^{-3}\theta^{-1}$	Btu/hr-ft.-°F	W/m-K or W/m-°C	1.731
specific heat	$L^2T^{-2}\ \theta^{-1}$	Btu/lb-R or Btu/lb-°F	J/kg-K or J/kg-°C	4.187
thermal diffusivity	L^2T^{-1}	ft^2/sec	m^2/s	0.093
heat transfer coefficient	$MT^{-3}\ \theta^{-1}$	Btu/hr.-ft^2-°F	W/m^2-K	5.674
enthalpy	L^2T^{-2}	Btu/lb_m or Btu/lb	J/kg	2326
entropy	$L^2T^{-2}\ \theta^{-1}$	Btu/lb_m-°F or Btu/lb-°F	J/kg-K	4187
heat flux	MT^{-3}	Btu/hr-ft^2	W/m^2	3.155
dynamic viscosity	$ML^{-1}T^{-1}$	lb/ft-s	Pa-s	1.488
kinematic viscosity	L^2T^{-1}	ft^2/s	m^2/s	0.093
mass transfer coefficient	$ML^{-2}T^{-1}$	lb_m/hr-ft^2 or lb/hr-ft^2	kg/sec-m^2	0.0013
diffusion coefficient	L^2T^{-1}	ft^2/s	m^2/s	0.093
surface tension	MT^{-2}	lb_f/ft	N/m	14.59

TABLE A2.2
US Customary Unit Conversion

Parameters	From	To	Multiply by
length/distance	inch (in)	centimeter (cm)	2.54
	mile	kilometer	1.61
	centimeter (cm)	meter	10^{-2}
	millimeter (mm)	meter	10^{-3}
	micron (μm)	meter	10^{-6}
	nanometer (nm)	meter	10^{-9}
	Angstrom (A°)	meter	10^{-10}
mass	pound	ounce (oz)	16
	kilogram	gram (gm)	10^{3}
	metric Ton (Tonne)	pound	2205
	imperial Ton (Ton)	pound	2240
	US Ton (short Ton)	pound	2000
time	hour	minute	60
	hour	second	3600
volume	cubic meter	liter	10^{3}
	US gallon	cubic meter	3.78×10^{-3}
	US gallon	liter	3.78
	Imperial gallon	liter	4.55
	barrel	US gallon	42
force	dyne	Newton	10^{-5}
surface tension	dyne/cm	Newton/m	10^{-3}
pressure	psi (lb_f/in^2)	Pascal (N/m^2) or Pa	6895
heat	calorie	Joule	4.18
power	horsepower (hp)	Watt	746
power per volume	hp/1000gal.	Watt/m^3	197
power per mass	hp/lb_m or hp/lb	Watt/kg	1643
dynamic viscosity	Poise	Pa-s	10^{-1}
	centi-Poise	Pa-s	10^{-3}
kinematic viscosity	Stoke	m^2/s	10^{-4}
	centi-Stokes	m^2/s	10^{-6}
torque	inch-ounce (in-oz)	ft-lb_f	5.21×10^{-3}

REFERENCES

1. https://www.equalsto.com/convert/
2. Definitions of SI Units and Other Units of Measurement, *Surface Engineering Forum*. https://www.gordonengland.co.uk/conversion/sidef.htm#btu

Appendix A3
Image and Table Reproduction Permissions

The authors acknowledge and sincerely thank the many equipment manufacturers who granted permission for us to reproduce images from their websites and literature. The following table names each of those companies and reports the figure number containing their image(s).

Figure No.	Credited to Image Source
3.3	Koflo Corporation
3.4	The Spray Nozzle People Ltd.
3.5	Statiflo International Ltd.
3.6	Ika Works, Inc.
3.7	Silverson Machines Inc.
3.8	SPX Flow, Inc.
3.9	Walker Process Equipment
4.3	SPX Flow, Inc.
4.4	NOV Inc.
4.5	EKATO Ruehr- und Mischtechnik GmbH
4.6	De Dietrich Process Systems, Inc.
4.7	GMM Pfaudler US Inc.
4.12	GMM Pfaudler US Inc.
5.2–3	EKATO Ruehr- und Mischtechnik GmbH
7.5	EKATO Ruehr- und Mischtechnik GmbH
7.6	SPX Flow, Inc., US
7.12a	Fusion Fluid Equipment LLC
7.12b	Conn & Company LLC
7.15	Silverson Machines Inc.
7.16a	Charles Ross & Son Company—"ROSS"
7.16b	OLIVER Y BATLLE S.A.U.
9.3	SPX Flow, Inc., US
11.8	Mixing Systems, Inc.
12.1a	NOV Inc., US
12.1b	Sulzer USA Inc.
12.3	NOV Inc., US
12.4	NOV Inc., US
12.5	StaMixCo LLC
12.7	NOV Inc., US
12.8	Inyo Process
12.12	Inyo Process
13.1	Silverson Machines Inc.
13.2	Ika Works, Inc.
13.4	NOV Inc., US
13.6	Silverson Machines Inc.
13.7	Silverson Machines Inc.

(Continued)

(Continued)

Figure No.	Credited to Image Source
13.8	EKATO Ruehr- und Mischtechnik GmbH
13.24a	Silverson Machines Inc.
13.25	SPX Flow, Inc., US
14.2a-1	NOV Inc., US
14.2a-2	SPX Flow, Inc., US
14.2c	EKATO Ruehr- und Mischtechnik GmbH
14.3	NOV Inc., US
14.4a	Fusion Fluid Equipment LLC
14.4b	SPX Flow, Inc., US
14.4c	EKATO Ruehr- und Mischtechnik GmbH
14.6	Charles Ross & Son Company—"ROSS"
14.7a	Charles Ross & Son Company—"ROSS"
14.7b	OLIVER Y BATLLE S.A.U.
14.8	Jaygo Inc.
14.9	Charles Ross & Son Company—"ROSS"
14.10	B&P Littleford
14.12	Hockmeyer Equipment Corporation
14.13	Charles Ross & Son Company
14.14	B&P Littleford
14.15	EKATO Ruehr- und Mischtechnik GmbH
14.16	EKATO Ruehr- und Mischtechnik GmbH
14.17	EKATO Ruehr- und Mischtechnik GmbH
15.5	Xylem Inc.*
15.6	Xylem Inc.*
15.7	Xylem-Sanitaire*
19.9	SPX Flow, Inc., US
20.1a	NOV Inc., US
20.1b	NOV Inc., US
20.1c	SPX Flow, Inc., US
20.2	SPX Flow, Inc., US
20.3	GMM Pfaudler US Inc.
20.4	De Dietrich Process Systems, Inc.
20.5	EagleBurgmann® Industries LP
20.6	FCT Valves
20.7	Armaturenwerk Hötensleben GmbH

* *permission pending*

Table Reproduction Permissions

The authors acknowledge and sincerely thank the many publishers who granted permission for us to reproduce tables from their books. The following table names each of those publishers, and reports the table number containing their table(s).

Table No.	Credited to Table Source*
1.1	Von Nostrand Reinhold / Society of Petroleum Engineers
8.9	John Wiley & Sons, Inc.
8.11	Cambridge University Press
8.12	Cambridge University Press
12.9	John Wiley & Sons, Inc.

* *Permissions received through Copyright Clearance Center, Inc. (CCC)*

Index

Note: Page numbers in *italics* indicate a figure and page numbers in **bold** indicate a table on the corresponding page.

D

E

H

I

For Product Safety Concerns and Information please contact our EU representative GPSR@taylorandfrancis.com Taylor & Francis Verlag GmbH, Kaufingerstraße 24, 80331 München, Germany